Progress in Nonlinear Differential Equations and Their Applications

Volume 4

Variational Methods

Proceedings of a Conference
Paris, June 1988

Edited by
Henri Berestycki
Jean-Michel Coron
Ivar Ekeland

With 13 Illustrations

1990

Birkhäuser
Boston · Basel · Berlin

Henri Berestycki
Mathématiques
Université Pierre et Marie Curie
75252 Paris Cedex 05
France

Ivar Ekeland
CEREMADE
Université de Paris IX-Dauphine
75775 Paris Cedex 16
France

Jean-Michel Coron
Département de Mathématiques,
Bâtiment 425
Université de Paris-Sud
Centre d'Orsay
91405 Orsay Cedex
France

Printed on acid-free paper.

Softcover reprint of the hardcover 1st edition 1990

ISBN 978-1-4757-1082-3 ISBN 978-1-4757-1080-9 (eBook)
DOI 10.1007/978-1-4757-1080-9

Camera-ready copy prepared by the authors.

9 8 7 6 5 4 3 2 1

PREFACE

In the framework of the "Année non linéaire" (the special nonlinear year) sponsored by the C.N.R.S. (the French National Center for Scientific Research), a meeting was held in Paris in June 1988. It took place in the Conference Hall of the Ministère de la Recherche and had as an organizing theme the topic of "Variational Problems."

Nonlinear analysis has been one of the leading themes in mathematical research for the past decade. The use of direct variational methods has been particularly successful in understanding problems arising from physics and geometry.

The growth of nonlinear analysis is largely due to the wealth of applications from various domains of sciences and industrial applications. Most of the papers gathered in this volume have their origin in applications: from mechanics, the study of Hamiltonian systems, from physics, from the recent mathematical theory of liquid crystals, from geometry, relativity, etc.

Clearly, no single volume could pretend to cover the whole scope of nonlinear variational problems. We have chosen to concentrate on three main aspects of these problems, organizing them roughly around the following topics:

1. Variational methods in partial differential equations in mathematical physics
2. Variational problems in geometry
3. Hamiltonian systems and related topics.

The conference on "Variational Problems" was sponsored by the following institutions:

—CNRS (Centre National de la Recherche Scientifique)
—DRET (Direction de la Recherche et des Etudes Techniques)
—CEA (Commissariat a l'Energie Atomique)
—CNES (Centre National d'Etudes Spatiales)
—SMAI (Société de Mathématiques Appliquées et Industrielles)

It has also benefitted from the support of Universities Paris IX (Dauphine), Paris XI (Orsay), and Paris XIII (Villetaneuse).

We thank the Steering Committee of the "Année non linéaire," Professor C. Bardos, O. Pironneau, and J.-P. Puel, for their moral and material help. We are very thankful to Michel Broué and the Département de Mathématiques et Informatique at the Ecole Normale Supérieure for its warm support throughout the organization. We wish to express our appreciation to Mrs. Ann Kostant for her assistance in the preparation of the manuscripts for the volume and to the Birkhäuser staff for their constant support and patience. And lastly, we wish to express gratitude to Mrs. A. Iapteff and Mrs. C. Pastadjian for their kind and competent secretarial assistance.

Paris, November 1989

Henri Berestycki
Jean-Michel Coron
Ivar Ekeland

CONTENTS

PART I

Partial Differential Equations and Mathematical Physics

THE (NON)CONTINUITY OF SYMMETRIC DECREASING REARRANGEMENT

Frederick J. Almgren, Jr. and Elliott H. Lieb
Department of Mathematics
Princeton University
Princeton, NJ 08544

ABSTRACT

The operation $\mathcal{R}$ of symmetric decreasing rearrangement maps $\mathbf{W}^{1,p}(\mathbf{R}^n)$ to $\mathbf{W}^{1,p}(\mathbf{R}^n)$. Even though it is norm decreasing we show that $\mathcal{R}$ is not continuous for $n \geq 2$. The functions at which $\mathcal{R}$ is continuous are precisely characterized by a new property called **co-area regularity**. Every sufficiently differentiable function is co-area regular, and both the regular and the irregular functions are dense in $\mathbf{W}^{1,p}(\mathbf{R}^n)$.

1. INTRODUCTION

Suppose $f(x^1, x^2) \geq 0$ is a continuously differentiable function supported in the unit disk in the plane. Its rearrangment is the rotationally invariant function $f^*(x^1, x^2)$ whose level sets are circles enclosing the same area as the level sets of f, i.e.

$$\mathcal{L}^2\{(x^1, x^2): f(x^1, x^2) > y\} = \mathcal{L}^2\{(x^1, x^2): f^*(x^1, x^2) > y\}$$

for each positive height y ($\mathcal{L}^n$ denotes Lebesgue measure over $\mathbf{R}^n$). Such rearrangment preserves $\mathbf{L}^p$ norms, i.e.

$$\int |f^*|^p \, d\mathcal{L}^2 = \int |f|^p \, d\mathcal{L}^2 \qquad (1 \leq p < \infty)$$

but decreases convex gradient integrals, e.g.

$$\int |\nabla f^*|^p \, d\mathcal{L}^2 \leq \int |\nabla f|^p \, d\mathcal{L}^2.$$

Now suppose that $f_j(x^1, x^2) \geq 0$ ($j = 1, 2, 3, \ldots$) is a sequence of continuously differentiable functions also supported in the unit disk which converge uniformly together with first derivatives to f, i.e.

$$f_j(x^1, x^2) \to f(x^1, x^2) \qquad \text{and} \qquad \nabla f_j(x^1, x^2) \to \nabla f(x^1, x^2)$$

uniformly in (x^1, x^2) as $j \to \infty$. It is not difficult to check that the symmetrized functions also converge uniformly. The real question is about convergence of the *derivatives* of the symmetrized functions. It is certainly plausible that they should converge strongly (we believed it for some time). Our principal new result is that *the derivatives of the symmetrized functions need not converge strongly*, e.g. for special f's and f_j's satisfying our conditions above it can happen that for every p

$$\liminf_{j\to\infty} \int |\nabla f_j^* - \nabla f^*|^p \, d\mathcal{L}^2 > 0.$$

Furthermore, we are able to characterize exactly those f's for which convergence is assured and for which it can fail.

The general notion of the symmetric decreasing rearrangement f^* of a function $f : \mathbf{R}^n \to \mathbf{R}^+$ is important in various parts of analysis. For example, various rotationally invariant variational integrals (like the gradient norms mentioned above) are not increased by symmetrization of competing functions. One is then free to search for a minimum among rotationally invariant decreasing functions (which are much easier to analyze since they are essentially functions of a single independent variable). A particular application of this technique has been in the computatation of optimal constants for Sobolev inequalities.

Some years ago W. Ni and L. Nirenberg raised the question whether the rearrangement map $\mathcal{R} : f \to f^*$ is strongly continuous in the $\mathbf{W}^{1,p}(\mathbf{R}^n)$ topology for all $1 \leq p < \infty$ (this would facilitate application of the "mountain pass lemma", for example). J-M. Coron [CJ] showed such strong continuity (and more) to be true in case $n = 1$, and we, at least, were led to the "obvious" conjecture that continuity holds for all n. We have settled this question [AL]—**rearrangement is not continuous in dimensions larger than one**. As indicated above, we can also identify precisely those f's at which the map $\mathcal{R}$ is continuous and those at which it is not. Our analysis has led us to isolate a property of functions which we call **co-area regularity** which deals with the behavior of functions on their critical sets. For $\mathbf{W}^{1,p}$ functions our main result is

THEOREM 1 [AL]. *For each $1 \leq p < \infty$ the rearrangement map $\mathcal{R}$ is $\mathbf{W}^{1,p}(\mathbf{R}^n)$ continuous at a function f if and only if f is co-area regular.*

Each $\mathbf{W}^{1,p}$ function on the line turns out to be necessarily co-area regular so that our theorem is consistent with Coron's result. For higher dimensional domains, however, there are always functions which are not co-area regular. In particular, in $\mathbf{R}^n$ $(n \geq 2)$

there are irregular functions in $\mathbf{C}^{n-1,\lambda}$ for each $0 < \lambda < 1$ (i.e. f's which are $n-1$ times continuously differentiable with $(n-1)^{th}$ derivatives which are Hölder continuous with exponent λ). In fact these irregular functions are dense in $\mathbf{W}^{1,p}(\mathbf{R}^n)$. However, each f with Lipschitz $(n-1)^{th}$ derivatives (i.e. $\lambda = 1$) is co-area regular.

In this note we shall briefly review symmetric rearrangement, introduce co-area regularity, sketch the construction of a co-area irregular function, give the reason that co-area irregularity implies lack of continuity of $\mathcal{R}$ in $\mathbf{W}^{1,p}$, and finally sketch the reason that co-area regularity implies continuity of $\mathcal{R}$. Our proof of continuity uses the theory of rectifiable currents in an essential way.

REMARK. One sometimes defines the symmetric decreasing rearrangment of vector valued functions $f : \mathbf{R}^n \to \mathbf{R}^m$ (as well as functions $\mathbf{R}^n \to \mathbf{R}^+$) by setting $f^* = |f|^*$. Sometimes it is also of interest to replace $\mathbf{W}^{1,p}$ norms by gradient energies associated with integrals of other convex integrands $\psi : \mathbf{R}^+ \to \mathbf{R}^+$, i.e. $\|\nabla f\|_p^p = \int |\nabla f|^p \, d\mathcal{L}^n$ is replaced by $\int \psi(|\nabla f|) \, d\mathcal{L}^n$. These two generalizations are carried out in [AL] but are omitted here for simplicity. *The conclusions about continuity remain the same.*

It is worth pointing out that although the map $\mathcal{R}$ is not continuous for $\mathbf{W}^{1,p}$ norms we show [AL]

THEOREM 2. *For each $0 < \alpha < 1$, each $1 \le p < \infty$, and each $n \ge 1$, the rearrangement map $\mathcal{R}$ is continuous on the fractional Sobolev space $\mathbf{W}^{\alpha,p}(\mathbf{R}^n)$.*

For $0 < \alpha < 1$ the norm $\|f\|_{\mathbf{W}^{\alpha,p}}$ is given by

$$\|f\|_{\mathbf{W}^{\alpha,p}}^p = \|f\|_p^p + \int\int |f(x) - f(y)|^p |x-y|^{-n-p\alpha} \, d\mathcal{L}^n x \, d\mathcal{L}^n y.$$

We have the curious conclusion that co-area regularity plays a role for $\mathbf{W}^{\alpha,p}$ only when $\alpha = 1$. Fractional derivatives, of course, are not a local construct.

2. REARRANGEMENTS AND CO-AREA REGULARITY

Rearrangements. We review the definition and basic properties of the symmetric decreasing rearrangment $f^* = \mathcal{R}f$ of a function $f : \mathbf{R}^n \to \mathbf{R}^+$. It is convenient to use the notation $\chi_{\{A\}} : \mathbf{R}^n \to \{0,1\}$ symbolically to denote a function which takes value 1 when the test A is passed and takes value 0 otherwise; e.g. $\chi_{\{f>y\}}(x)$ equals 1 in case

$f(x) > y$ and equals 0 otherwise. Also we associate to a fixed function f a radius function $R: \mathbf{R}^+ \to \mathbf{R}^+$ defined by requiring

$$\alpha(n)R(y)^n = \int \chi_{\{f>y\}} \, d\mathcal{L}^n \tag{2.1}$$

for each y; here $\alpha(n)$ is the volume of the unit ball in $\mathbf{R}^n$. We further denote by $\chi_R: \mathbf{R}^n \to \{0, 1\}$ the characteristic function of the open ball centered at the origin and of radius R. Finally, our **rearranged function**

$$f^*: \mathbf{R}^n \to \mathbf{R}^+$$

is defined by setting

$$f^*(x) = \int_{y>0} \chi_{R(y)}(x) \, d\mathcal{L}^1 y \tag{2.2}$$

for each x. It is immediate to check that f^* *is symmetric and decreasing*, i.e. $f(x) = f(z)$ if $|z| = |x|$ and $0 \le f(x) \le f(z)$ if $|x| \ge |z|$. It is also clear that f^* *is equimeasurable with* f, i.e.

$$\mathcal{L}^n(\{x : f(x) > y\}) = \mathcal{L}^n(\{x : f^*(x) > y\}) \tag{2.3}$$

for each $y > 0$.

Equation (2.3) implies immediately that *rearrangement preserves* $\mathbf{L}^p$ *norms*, i.e.

$$\|f\|_p = \|f^*\|_p. \tag{2.4}$$

Moreover [CG], *rearrangement is a contraction on* $\mathbf{L}^p$, i.e.

$$\|f - g\|_p \ge \|f^* - g^*\|_p \qquad \text{whenever } f, g \in \mathbf{L}^p. \tag{2.5}$$

In particular, $\mathcal{R}$ *is a continuous map from* $\mathbf{L}^p$ *into* $\mathbf{L}^p$.

The function space $\mathbf{W}^{1,p}(\mathbf{R}^n)$ consists of those functions f which belong to $\mathbf{L}^p(\mathbf{R}^n)$ and whose distribution gradients ∇f are functions belonging to $\mathbf{L}^p(\mathbf{R}^n, \mathbf{R}^n)$. It has long been known [B][BZ][H][K][L][PS][S1][S2][T] that $\mathcal{R}$ *is* $\mathbf{W}^{1,p}$ *norm non-increasing*, i.e.

$$\|\nabla f\|_p \ge \|\nabla f^*\|_p. \tag{2.6}$$

This implies that $\mathcal{R}f$ *also belongs to* $\mathbf{W}^{1,p}$. (Actually, when $p = 1$ it is not obvious that f^* is in $\mathbf{W}^{1,1}$ and not merely in $\mathbf{BV}$; this was proved by Hilden [H].) However, $\mathcal{R}$

is not a contraction mapping. Indeed, $\|\nabla f - \nabla g\|_p$ can be arbitrarily large compared to $\|\nabla f^* - \nabla g^*\|_p$. To see why this can happen, suppose that $f, g: \mathbf{R} \to \mathbf{R}^+$ are smooth functions with $f(x) = g(x)$ for $x \leq 0$ and $f(x) > g(x)$ for $x > 0$. Suppose also, for $x \leq 0$, that both ∇f (and hence ∇g) are very large in $\mathbf{L}^p$ norm while, for $x > 0$, both ∇f and ∇g are of order 1 in $\mathbf{L}^p$ norm. Then $\|\nabla f - \nabla g\|_p$ is of order one because of the cancellation for $x \leq 0$. On the other hand it is easy to arrange things so that the rearrangement destroys this cancellation so that $\|\nabla f^* - \nabla g^*\|_p$ will be large.

These facts suggest some of the subtlety of questions about the continuity of $\mathcal{R}$ on $\mathbf{W}^{1,p}$. We can phrase our question in the following way.

Given $f, f_1, f_2, \ldots$ in $\mathbf{W}^{1,p}$ with $f_j \to f$ in $\mathbf{W}^{1,p}$, is it true that $A_j = \|\nabla f_j^* - \nabla f^*\|_p$ ultimately converges to 0 as $j \to \infty$ even though A_j may be large for very many j's?

Co-area Regularity. Instead of the integral in (2.1) representing the full crossectional area at height y of the subgraph of our function f, consider the integral

$$\mathcal{G}_f(y) = \int \chi_{\{f>y\}} \chi_{\{\nabla f = 0\}} \, d\mathcal{L}^n \tag{2.7}$$

which, for each y, represents that part of the crossection of the subgraph associated with critical points of f. Since our function $\mathcal{G}_f : \mathbf{R}^+ \to \mathbf{R}^+$ is nonincreasing its distribution first derivative $\mathcal{G}'_f$ is a (negative) measure. Since a smooth function must be constant on any connected open set on which its gradient vanishes, there are many functions f for which the contributions to the integral in (2.7) come only from flat parts of the graph corresponding to those positive numbers y for which the set $\{x : f(x) = y\}$ has positive measure. Since there can be at most countably many such y's, the measure $\mathcal{G}'_f$ would then be singular with respect to $\mathcal{L}^1$ on $\mathbf{R}^+$. This situation is not the most general one, however, and there are "irregular" smooth functions f for which the measure $\mathcal{G}'_f$ has an absolutely continuous piece as well. Indeed, we have the following theorem.

THEOREM 3 [AL]. *For each $n \geq 2$ and each $0 < \lambda < 1$, there is (by construction) a positive constant C and a function $f : \mathbf{R}^n \to [0,1]$ with the following properties.*

(1) The function f belongs to $\mathbf{C}^{n-1,\lambda}(\mathbf{R}^n)$ and has support equal to the cube

$$\mathbf{Q} = \left\{ x : |x^i| \leq 1 \text{ for each } i = 1, \ldots, n \right\}$$

of side length 2.

(2) For each $0 < y \leq 1$,

$$\mathcal{G}_f(y) = C(1-y).$$

In particular, the measure $\mathcal{G}'_f$ is absolutely continuous with respect to $\mathcal{L}^1$. Thus f is co-area irregular.

It can be difficult to picture such a function. Somehow its gradient vanishes on a set of positive $\mathcal{L}^n$ measure containing no open subsets or flat spots, i.e. $\mathcal{L}^n(\{x: f(x) = y\}) = 0$ for every y. Furthermore, the image of the critical set is distributed uniformly over all y values in the range [0, 1].

Theorem 3 also tells us that the following definition is not an empty one

DEFINITION. *A function f in $\mathbf{W}^{1,p}$ is called* **co-area regular** *if and only if the measure $\mathcal{G}'_f$ (see (2.7)) is purely singular with respect to $\mathcal{L}^1$. Otherwise f is called* **co-area irregular.**

The term co-area in these definitions was suggested by H. Federer's "co-area formula" which gives an integral representation of the absolutely continuous function

$$y \mapsto \int \chi_{\{f>y\}} \chi_{\{\nabla f \neq 0\}} \, d\mathcal{L}^n.$$

A mild generalization [AL] of the Morse-Sard-Federer theorem shows that each f belonging to $\mathbf{C}^{n-1,1}$ is automatically co-area regular. An easy argument then shows

THEOREM 4 [AL]. *For each $n \geq 2$ and each $p \geq 1$, the co-area regular and the co-area irregular functions are each dense in $\mathbf{W}^{1,p}(\mathbf{R}^n)$.*

Questions of the behavior of functions on their critical sets have a substantial mathematical heritage both in theory and in examples. We here sketch the construction of a function f as in Theorem 3 when $n = 2$. First set $f(x) = 0$ for $x \notin \mathbf{Q}$. For $x \in \mathbf{Q}$ we will use 4-adic notation to express the values of our f, i.e. we will write

$$f(x) = \sum_{\ell=1}^{\infty} 4^{-\ell} a_\ell(x) \qquad \text{with } a_\ell(x) \in \{0,1,2,3\}.$$

First divide Q in the obvious way into four squares each of side length 1 and label these squares $S_0^{(1)}, S_1^{(1)}, S_2^{(1)}, S_3^{(1)}$ in clockwise order. Set $a_1(x) = j$ if $x \in S_j^{(1)}$ (don't worry about the boundaries of the $S_j^{(1)}$'s). Next, divide each $S_j^{(1)}$ into four squares each of side length $\frac{1}{2}$ and label these $S_{jk}^{(2)}$ (with $k = 0, 1, 2, 3,$) in the same clockwise order. Set $a_2(x) = k$ if $x \in S_{jk}^{(2)}$. The construction continues in the obvious way ultimately to define an f. For each $0 < a < b < 1$ we have $\mathcal{L}^2\left(f^{-1}(a, b)\right) = 4(b - a)$. At present our f is not even continuous much less smooth. We fix this up by modifying this construction. We replace each a_ℓ by a carefully constructed smooth function b_ℓ in our sum above. The support of each b_ℓ is contained within the $4^{\ell-1}$ squares on which $b_{\ell-1}$ assumes constant values, and b_ℓ assumes constant values on 4^ℓ squares nested within the $b_{\ell-1}$ constant value squares. The subgraph then resembles a union of step pyramids (like Zhoser not Cheops) with those at the ℓ-th level having bases on the tops of those at the $\ell - 1$-th level. With some effort one can construct the b_ℓ's so that $f \in \mathbf{C}^{1,\lambda}$ and $\{x : \nabla f = 0\}$ has positive measure. As expected the measure of the set $\{x : \nabla f = 0\}$ goes to zero as λ approaches 1.

3. REARRANGEMENT IS DISCONTINUOUS AT CO-AREA IRREGULAR FUNCTIONS

THEOREM 5 [AL]. *Suppose $n \geq 2$ and f is a co-area irregular function belonging to $\mathbf{W}^{1,p}(\mathbf{R}^n)$ Then there is a sequence $f_1, f_2, f_3, \ldots$ of functions in $\mathbf{W}^{1,p}(\mathbf{R}^n)$ such that $f_j \to f$ in $\mathbf{W}^{1,p}(\mathbf{R}^n)$ as $j \to \infty$ but $f_j^* \not\to f^*$. Moreover, for each $\epsilon > 0$, the f_j's can be chosen with the following properties.*

(1) The sequence of differences $f_j - f$ converges to zero in $\mathbf{L}^\infty(\mathbf{R}^n)$.

(2) There is a positive number Y such that

$$f_j(x) = f(x) \quad \text{whenever } f(x) < Y \text{ or } f(x) > Y + \epsilon$$

and

$$Y \leq f_j(x) \leq Y + \epsilon \quad \text{whenever} \quad Y \leq f(x) \leq Y + \epsilon.$$

(3) For $\mathcal{L}^n$ almost every x,

$$|\nabla f_j(x)| \leq \frac{3}{2}|\nabla f(x)| + \epsilon$$

(4) The measure of the set

$$\left\{x : \nabla f(x) \neq 0 \text{ and } \nabla f_j(x) \neq \nabla f(x)\right\}$$

converges to zero as $j \to \infty$.

If we do not require properties (2), (3), (4) then the difference $f_j - f$ can be chosen to belong to $\mathbf{C}^\infty$. If we drop all four properties then each f_j can be chosen to belong to $\mathbf{C}^\infty$. The basic idea behind the proof of Theorem 5 (omitting refinements (1), (2), (3), (4)) is the following. Let W be the characteristic function of the critical set of f, i.e. the set for which $\nabla f = 0$, and set

$$f_j(x) = f(x) + \frac{1}{2j} W(x) \sin(j f(x)) \tag{3.1}$$

for each x. Then clearly $f_j \to f$ in $\mathbf{L}^p$ as $j \to \infty$. For the gradients we compute formally

$$\nabla f_j(x) - \nabla f(x) = \frac{1}{2} W(x) \nabla f(x) \cos(j f(x)) + \frac{1}{2j} \nabla W(x) \sin(j f(x)). \tag{3.2}$$

The first term on the right side is zero since W vanishes when ∇f does not vanish. The second term on the right side in (3.2) is a bit problematic since ∇W is not p-th power summable. This defect, however, can be remedied (with some effort) by mollifying W in a j-dependent way so that $\|\nabla W\|_\infty < j^{1/2}$. The modified second term in (3.2) then goes to zero in $\mathbf{L}^p$ norm like $j^{-1/2}$. This establishes the $\mathbf{L}^p$ convergence of ∇f_j to ∇f.

Now define sets

$$K_j(y) = \left\{ x : f_j(x) > y \right\} \quad (j = 1, 2, 3, \ldots) \qquad \text{and} \qquad K(y) = \left\{ x : f(x) > y \right\}$$

for each y. Since the function $t \mapsto t + \frac{1}{2j}\sin(tj)$ is increasing (check the derivative) we infer that $K_j(y) = K(y)$ whenever m is an integer and $y = 2m\pi/j$. For these special y values we infer that the radius functions are equal, i.e. $R_j(y) = R(y)$ (recall (2.1)). On the other hand, if $0 < \sigma < 1$ and $y = (2m + \sigma)(\pi/j)$, then $R_j(y) \geq R(y)$ and, in general, $R_j(y) > R(y)$.

Think of the graphs of f_j^* and f^* parametrized by the height y instead of the radius $|x|$. When $y = 2m\pi/j$ the graphs intersect. When $y = (2m + \sigma)(\pi/j)$ and $0 < \sigma < 1$, the graph of f_j^* lies to the right of the graph of f^*. For our purposes it suffices to show that the numbers $B_j \equiv \|\nabla f_j^* - \nabla f^*\|_1$ are bounded away from zero. We then try to estimate the B_j's in terms of the distribution $\mathcal{G}_f$ from (2.7). Using the Schwarz inequality several times and a simple Sobolev inequality we are able to estimate

$$B_j \geq (\text{constant}) \int |h|^{1/2} \, d\mathcal{L}^1; \tag{3.3}$$

here $\mathcal{L}^1 \wedge h$ denotes the absolutely continuous part of our $\mathcal{G}_f'$.

It is reassuring that that the bound (3.3) above involves $|h|^{1/2}$ instead of $|h|$. This is so because "the square root of a *singular* measure is zero"; by this we mean that if the singular part of $\mathcal{G}'_f$ (which cannot contribute to the lack of convergence, as we assert in the next section) is approximated by absolutely continuous measures $\mathcal{L}^1 \wedge \tilde{h}^{(k)}$ $(k = 1, 2, 3, \ldots)$, then $\int |\tilde{h}^{(k)}|^{1/2}\, d\mathcal{L}^1$ converges to zero as $k \to \infty$.

4. REARRANGEMENT IS CONTINUOUS AT CO-AREA REGULAR FUNCTIONS

The proof [AL] that the co-area regularity of f implies $\mathbf{W}^{1,p}$ continuity of $\mathcal{R}$ at f is quite technical. We will attempt to outline some of the main ideas.

(1) Reduction to $\mathbf{W}^{1,1}$. Our first step is to establish the fact that continuity of $\mathcal{R}$ in $\mathbf{W}^{1,p}$ is implied by continuity of $\mathcal{R}$ in $\mathbf{W}^{1,1}$. This may seem surprising since ordinarily nothing can be inferred about $\|\nabla f_j^* - \nabla f^*\|_p$ from information about $\|\nabla f_j^* - \nabla f^*\|_1$. In the present case, however, our rearrangement operator $\mathcal{R}$ acts independently on slabs $\left\{x : Y_1 < f(x) < Y_2\right\}$. We can then surgically remove small, well chosen slabs from the f_j and f on which $|\nabla f_j^*|$ or $|\nabla f^*|$ is large. On these slabs we can control $\|\nabla f_j^* - \nabla f^*\|_p$ in terms of $\|\nabla f_j^*\|_p$ and $\|\nabla f^*\|_p$ and these quantities can, in turn, be controlled by $\|\nabla f_j\|_p$ and $\|\nabla f\|_p$ with use of the basic inequality (2.6). After these small slabs are removed, the f_j and f effectively have bounded gradients and then $\mathbf{W}^{1,1}$ convergence implies $\mathbf{W}^{1,p}$ convergence.

(2) The co-area formula and co-area regularity. The basic tool in our second step is H. Federer's co-area formula as extended by J. Brothers and W. Ziemer [BZ]. Suppose $f \in \mathbf{W}^{1,1}(\mathbf{R}^n)$ and g is a nonnegative Borel function. Then the slice integral

$$A(y) \equiv \int_{f^{-1}\{y\}} g\, d\mathcal{H}^{n-1} \tag{4.1}$$

exists for $\mathcal{L}^1$ almost every positive number y and we have the *co-area formula*

$$\int_{y>0} A\, d\mathcal{L}^1 = \int g|\nabla f|\, d\mathcal{L}^n; \tag{4.2}$$

here $\mathcal{H}^{n-1}$ denotes Hausdorff's (n-1)-dimensional measure over $\mathbf{R}^n$.

In one application of (4.2) we replace $f(x)$ by $F_t(x) = \max\{f(x), t\}$ (with $t > 0$), then replace $g(x)$ by $(|\nabla f(x)| + \delta)^{-1}$, then let $\delta \to 0+$, and finally use Lebesgue's monotone convergence theorem applied to each side of (4.2) to infer

$$\int_{y>t} \omega_f(y)\, d\mathcal{L}^1 y = \int \chi_{\{f>t\}} \chi_{\{\nabla f \neq 0\}}\, d\mathcal{L}^n \equiv \gamma_f(t) \tag{4.3}$$

where we have written

$$\omega_f(y) = \int_{f^{-1}\{y\}} |\nabla f|^{-1}\, d\mathcal{H}^{n-1}. \tag{4.4}$$

for each y. In other words, the basic distribution integral on the right side of (2.1) (call it $\sigma_f(y)$) breaks up naturally into two pieces

$$\sigma_f(y) = \gamma_f(y) + \mathcal{G}_f(y) \tag{4.5}$$

and (4.3) states that γ_f is absolutely continuous with derivative $-\omega_f$. The KEY POINT is: *the only absolutely continuous part of the measure* $-\sigma_f'$ *is* ω_f *if and only if* f *is co-area regular.*

(3) Currents and the lower semicontinuity of slice integrals. Suppose that we have a sequence f_j converging to f in $\mathbf{W}^{1,1}$ and that f is co-area regular. Henceforth we will omit the subscript f (e.g. σ_f will be denoted σ) when referring to f, and will use a subscript j when referring to f_j (e.g. σ_{f_j} will be denoted σ_j). We assert that

$$\liminf_{j\to\infty} \omega_j(y) \geq \omega(y) \qquad \text{for } \mathcal{L}^1 \text{ almost every } y. \tag{4.6}$$

To show this it suffices to prove that

$$\lim_{j\to\infty} \int_{f_j^{-1}\{y\}} g\, d\mathcal{H}^{n-1} = \int_{f^{-1}\{y\}} g\, d\mathcal{H}^{n-1} \qquad \text{for } \mathcal{L}^1 \text{ almost every } y. \tag{4.7}$$

whenever $g \in \mathbf{L}^\infty$. An approximation argument shows it is sufficient to prove (4.7) for $g \in \mathbf{C}_0^\infty$. It is here that we need to utilize the inherent current structure of the graph and subgraph of f and the f_j's and the inherent convergence as currents. To do this we form the $n+1$ dimensional current

$$Q = \mathbf{E}^{n+1} \llcorner \{(x, y) : x \in \mathbf{R}^n,\ y < f(x)\}$$

whose boundary $T = \partial Q$ is the current associated with the graph of f. The current T can then be sliced by the coordinate function y to obtain an $n-1$ dimensional slice current $T(y)$ corresponding to the level set $f^{-1}\{y\}$ for $\mathcal{L}^1$ almost every y. Likewise, we define $Q_j, T_j, T_j(y)$ for the various j's and further set $S_j = Q - Q_j$ with associated slice currents $S_j(y)$. Since "slicing commutes with boundaries" in the current setting we infer $\partial S_j(y) = T(y) - T_j(y)$ for almost every y.

Since the mass $\mathbf{M}$ of a current corresponds to its volume, we readily check that

$$\mathbf{M}(S_j) = \mathbf{M}(Q - Q_j) = \|f - f_j\|_1 \to 0 \quad \text{as } j \to \infty. \tag{4.8}$$

Since $\mathbf{M}(S_j) = \int \mathbf{M}(S_j(y))\, d\mathcal{L}^1 y$ for each j, there will be a subsequence (still denoted by j's) such that

$$\lim_{j\to\infty} \mathbf{M}(S_j(y)) = 0 \quad \text{for } \mathcal{L}^1 \text{ almost every } y. \tag{4.9}$$

Since $\partial S_j(y) = T(y) - T_j(y)$ we conclude the convergence of the $T_j(y)$'s to $T(y)$ for almost every y. The lower semicontinuity of mass under such convergence then implies

$$\liminf_{j\to\infty} \mathbf{M}(T_j(y)) \geq \mathbf{M}(T(y)) \quad \text{for } \mathcal{L}^1 \text{ almost every } y. \tag{4.10}$$

Using, for example, J. Michael's [M] Lipschitz approximation theorem we readily infer

$$\mathbf{M}(T_{(j)}(y)) = \mathcal{H}^{n-1}(f_{(j)}^{-1}(y)) \qquad \text{for } \mathcal{L}^1 \text{ almost every } y; \tag{4.11}$$

here (j) denotes either j or no j. We use the co-area formula again to infer

$$\int \mathbf{M}(T_{(j)}(y))\, d\mathcal{L}^1 y = \int |\nabla f_{(j)}|\, d\mathcal{L}^n. \tag{4.12}$$

However, $\int |\nabla f_j|\, d\mathcal{L}^n \to \int |\nabla f|\, d\mathcal{L}^n$ by the assumed $\mathbf{L}^1$ convergence of ∇f_j to ∇f.

The following is a general lemma. Suppose μ is a measure and $h, h_1, h_2, h_3, \ldots$ are nonnegative functions such that $\liminf_j h_j(x) \geq h(x)$ for μ almost every x. In case $\int h_j d\mu \to \int h d\mu$ as $j \to \infty$ then there is a subsequence $j(k)$ of the j's such that $h_{j(k)}(x) \to h(x)$ as $k \to \infty$ for μ almost every x.

We apply this lemma to the case at hand to infer that, for a further subsequence,

$$\lim_{j\to\infty} \mathbf{M}(T_j(y)) = \mathbf{M}(T(y)) \qquad \text{for } \mathcal{L}^1 \text{ almost every } y. \tag{4.13}$$

Equation (4.13), with a little more work, then leads to (4.7).

As an application of (4.7) we return to (4.4) and prove that

$$\liminf_{j\to\infty} \omega_j(y) \geq \omega(y) \qquad \text{for } \mathcal{L}^1 \text{ almost every } y. \tag{4.14}$$

This result is crucial for us. To prove it, we use (4.7) with $g_{(j)}(y) = (|\nabla f_{(j)}| + \delta)^{-1}$ (as in the proof of (4.4)) and then let $\delta \to 0$.

(4) Graph arc length as an invariant measure. The last main step in our proof is to combine (4.14), the co-area regularity of f, and the $\mathbf{W}^{1,1}$ convergence of the f_j's to f to show that the ∇f_j^*'s converge to ∇f^* in $\mathbf{L}^1$. Since $f_{(j)}^*(x)$ is really only a function of $r = |x|$, our considerations are essentially one-dimensional. (It is true that the real measure is $r^{n-1}dr$ and not dr, but this is merely a nuisance which one can handle.) Let us suppose then that $n = 1$ and we will denote d/dr by a prime.

Think of the graph of f^* (or f_j^*) which is a curve in $\mathbf{R}^2$. The geometrically invariant notion is not $f^{*\prime}$ (which is the quantity in which we are really interested) but rather the arc length derivative $(1 + (f^{*\prime})^2)^{1/2}$. The arc length can be computed in two different ways. The first way is to use the height y as parameter. Then the arc length of the graph of $f_{(j)}^*$ equals

$$\int (1 + (\rho_{(j)}(y))^2)^{\frac{1}{2}}\, d\mathcal{L}^1 y + \int d\nu_{(j)};$$

here $\nu_{(j)}$ is the singular part of the measure $-(d\sigma_{(j)}/dy)$ while $\mathcal{L}^1 \wedge \rho_{(j)}$ is the absolutely continuous part of $-(d\sigma_{(j)}/dy)$ *The crucial point is the following: The co-area regularity of f implies that $\rho(y) = \omega(y)$.* For f_j^*, all we can say is that $\rho_j(y) \geq \omega_j(y)$; but this is of no concern since, from (4.14), we have

$$\liminf_{j\to\infty} \rho_j(y) \geq \rho(y) \qquad \text{for } \mathcal{L}^1 \text{ almost every } y. \tag{4.15}$$

Concerning the singular components ν_j and ν one knows nothing. However, by the $\mathbf{L}^1$ convergence of f_j^* to f^* (see (2.5)) we can infer that the arcs converge pointwise, i.e. for any $0 < a < b$

$$\int_a^b \rho_j\, d\mathcal{L}^1 + \int_a^b d\nu_j \to \int_a^b \rho\, d\mathcal{L}^1 + \int_a^b d\nu. \tag{4.16}$$

It is then a simple exercise to show that (4.15), (4.16) alone imply *arc length convergence*, i.e.

$$\int (1 + \rho_j^2)^{1/2}\, d\mathcal{L}^1 + \int d\nu_j \to \int (1 + \rho^2)^{1/2}\, d\mathcal{L}^1 + \int d\nu. \tag{4.17}$$

Now think about this arc length convergence (4.17) in terms of the radius parameterization, i.e.

$$\int (1 + (f_j^{*\prime}(r))^2)^{1/2}\, d\mathcal{L}^1 r \to \int (1 + (f^{*\prime}(r))^2)^{1/2}\, d\mathcal{L}^1 r. \tag{4.18}$$

There is no singular part of the measure (since $f_{(j)}^{*\prime}$ is a function). Intuitively, it is clear (by drawing a few graphical examples) that arc length convergence implies $\mathbf{L}^1$ convergence of $f_j^{*\prime}$ to $f^{*\prime}$ because the function $t \mapsto (1+t^2)^{1/2}$ is strictly convex. This is indeed correct as the following general theorem [AL] states.

THEOREM 6. *Suppose $\psi : \mathbf{R}^n \to \mathbf{R}^+$ is a convex function. Suppose also that $f, f_1, f_2, f_3, \ldots$ are functions in $\mathbf{L}^1_{\text{loc}}(\mathbf{R}^n, \mathbf{R})$ having distributional gradients which are functions in $\mathbf{L}^1_{\text{loc}}(\mathbf{R}^n, , \mathbf{R}^n)$. Suppose that $\psi(\nabla f), \psi(\nabla f_1), \psi(\nabla f_2), \psi(\nabla f_3), \ldots$ also are functions in $\mathbf{L}^1(\mathbf{R}^n)$ and that $f_j - f \to 0$ in $\mathbf{L}^1(\mathbf{R}^n)$ as $j \to \infty$. Then (as has been known for some time [SJ])*

(1) $$\liminf_{j\to\infty} \int \psi(\nabla f_j)\, d\mathcal{L}^n \geq \int \psi(\nabla f)\, d\mathcal{L}^n.$$

(2) Suppose further that equality holds in (1) and that ψ is strictly convex (i.e. $\psi(x) + \psi(y) > 2\psi\left(\frac{x+y}{2}\right)$ whenever $x \neq y$). Uniform convexity is not assumed. Then $\psi(\nabla f_j) \to \psi(\nabla f)$ in $\mathbf{L}^1(\mathbf{R}^n)$ as $j \to \infty$. Furthermore, there is a subsequence $j(1), j(2), j(3), \ldots$ of $1, 2, 3, \ldots$ such $\nabla f_{j(k)}(x) \to \nabla f(x)$ for $\mathcal{L}^n$ almost every x as $k \to \infty$.

(3) Finally, suppose $\psi(\xi) \to \infty$ as $|\xi| \to \infty$ (e.g. our function $\xi \mapsto (1 + |\xi|^2)^{1/2}$). Then, for every measurable subset Ω of $\mathbf{R}^n$ of finite measure, $\nabla f_j \to \nabla f$ in $\mathbf{L}^1(\Omega, \mathbf{R}^n)$.

REFERENCES

[AL] F. Almgren and E. Lieb, *Symmetric decreasing rearrangement is sometimes continuous*, in preparation.

[B] C. Bandle, *Isoperimetric inequalities and applications*, Pitman (Boston, London, Melbourne) 1980.

[BZ] J. Brothers and W. Ziemer, *Minimal rearrangements of Sobolev functions*, Jour. Reine Angew. Math. **384**, 153-179 (1988).

[CG] G. Chiti, *Rearrangements of functions and convergence in Orlicz spaces*, Appl. Anal. **9**, 23-27 (1979).

[CJ] J-M. Coron, *The continuity of the rearrangement in* $\mathbf{W}^{1,p}(\mathbf{R})$, Ann. Scuol. Norm. Sup. Pisa, Ser 4, **11**, 57-85 (1984).

[H] K. Hilden, *Symmetrization of functions in Sobolev spaces and the isoperimetric inequality*, Manuscr. Math. **18**, 215-235 (1976).

[K] B. Kawohl, *Rearrangements and convexity of level sets in partial differential equations*, Lect. Notes in Math. **1150**, Springer (Berlin, Heidelberg, New York) 1985.

[L] E. Lieb, *Existence and uniqueness of the minimizing solution of Choquard's nonlinear equation*, Stud. Appl. Math. **57**, 93-105 (1977). See appendix.

[M] J. Michael, *Lipschitz approximations to summable functions*, Acta Math. **111**, 73-94 (1964).

[PS] G. Polya and G. Szegö, *Isoperimetric inequalities in mathematical physics*, Ann. Math. Stud. **27**, Princeton University Press (Princeton) (1951).

[SJ] J. Serrin, *On the definition and properties of certain variational integrals*, Trans. Amer. Math. Soc. **101**, 139-167, (1961).

[S1] E. Sperner, *Zur symmetrisierung von Funktionen auf Sphären*, Math. Z. **134**, 317-327 (1973).

[S2] E. Sperner, *Symmetrisierung für Funktionen mehrerer reeller Variablen*, Manuscr. Math. **11**, 159-170 (1974).

[T] G. Talenti, *Best constant in Sobolev inequality*, Ann. Pura Appl. **110**, 353-372 (1976).

COUNTING SINGULARITIES IN LIQUID CRYSTALS

Frederick J. Almgren, Jr. and Elliott H. Lieb
Department of Mathematics
Princeton University
Princeton, NJ 08544

ABSTRACT

Energy minimizing harmonic maps from the ball to the sphere arise in the study of liquid crystal geometries and in the classical nonlinear sigma model. We linearly dominate the number of points of discontinuity of such a map by the energy of its boundary value function. Our bound is optimal (modulo the best constant) and is the first bound of its kind. We also show that the locations and numbers of singular points of minimizing maps is often counterintuitive; in particular, boundary symmetries need not be respected.

1. INTRODUCTION

This note is an introduction to and summary of discoveries we have made about the singular behavior of

- A mathematical model of some liquid crystal geometries
- Dirichlet energy minimizing harmonic maps from regions in $\mathbf{R}^3$ to $\mathbf{S}^2$
- Energy minimizing configurations of a classical nonlinear sigma model ($\mathbf{R}^3 \to \mathbf{S}^2$).

These phenomena are different facets of a common mathematical analysis set forth in detail in our paper [AL]. There we study vector fields φ of unit length defined in a reasonable region Ω in $\mathbf{R}^3$. In coordinates we can thus write for each $x = (x^1, x^2, x^3)$ in Ω,

$$\varphi(x) = (\varphi^1(x), \varphi^2(x), \varphi^3(x)) \qquad \text{with} \qquad \sum_{i=1}^{3} \varphi^i(x)^2 = 1. \tag{1}$$

Since our target $\mathbf{S}^2$ is 2-dimensional we could, in principle, describe φ using two functions instead of our three constrained functions. It is easier, however, to work with three functions and a constraint.

The φ's important for us have distribution first derivatives which are square summable. (Caution: the space of such φ's satisfying (1) is not the completion of any space of smooth mappings $\Omega \to \mathbf{S}^2$.) The gradients of such φ's are defined for almost every x with norms represented by the formula

$$|\nabla\varphi(x)|^2 = \sum_{i=1}^{3} \sum_{\alpha=1}^{3} \left(\frac{\partial \varphi^i(x)}{\partial x^\alpha} \right)^2$$

which gives the value of **Dirichlet's integrand** at x. The integral of this integrand is is **Dirichlet's energy integral** of φ,

$$\mathcal{E}(\varphi) = \int_\Omega |\nabla\varphi|^2 dV,$$

with $dV = dx^1 dx^2 dx^3$. Critical points of this energy integral $\mathcal{E}$ are by definition **harmonic functions** and satisfy the associated **Euler-Lagrange partial differential equations**

$$-\Delta\varphi^i(x) = \varphi^i(x)|\nabla\varphi(x)|^2 \qquad (i = 1, 2, 3).$$

These equations state that a critical φ has vanishing Laplacian in directions in which it is unconstrained. Such an energy functional and associated partial differential equations appear in the physics literature under the rubric of **the nonlinear sigma model.**

Somewhat more generally, reasonable maps $\varphi: \mathcal{M} \to \mathcal{N}$ between Riemannian manifolds $\mathcal{M}$ and $\mathcal{N}$ (often submanifolds of Euclidean vector spaces) have a Dirichlet's energy integral

$$\mathcal{E}_{\mathcal{M}\mathcal{N}}(\varphi) = \int_\mathcal{M} |D\varphi|^2 \, dV_\mathcal{M}$$

of which ours is a special case. Alternatively, one can write

$$\mathcal{E}_{\mathcal{M}\mathcal{N}}(\varphi) = \int_\mathcal{M} g_{ij}(\varphi(x))\, G^{\alpha\beta}(x) \left(\frac{\partial \varphi^i}{\partial x^\alpha}(x) \right) \left(\frac{\partial \varphi^j}{\partial x^\beta}(x) \right) \, dV_\mathcal{M} x$$

where g is the metric on N, G is the metric on M and $dV_\mathcal{M} x = (\det G(x))^{1/2} dx$. Extremal mappings for such energies are also called **harmonic mappings** . Such mappings often are not continuous and there is an extensive mathematical theory about them.

The φ's mapping Ω to $\mathbf{S}^2$ which are important for us also have well defined **boundary functions** $\psi : \partial\Omega \to \mathbf{S}^2$ having boundary energy

$$\partial\mathcal{E}(\psi) = \int_{\partial\Omega} |\nabla_T \psi|^2 dA$$

which is finite; here $\nabla_T\psi$ is the tangential gradient of ψ and dA is surface area measure. Associated with such a ψ is the number

$$E(\psi) = \inf \{\mathcal{E}(\varphi) : \varphi \text{ has boundary value function } \psi\}.$$

We call φ an **energy minimizing map** for boundary value function ψ if and only if

$$\mathcal{E}(\varphi) = E(\psi).$$

If Ω is any reasonable bounded domain and ψ is any boundary value function of finite energy then there will always be at least one minimizer φ having ψ as boundary values (a compactness argument). Sometimes, however, there can be more than one minimizer. This is one of the fascinations of this simple nonlinear problem; if the target $\mathbf{S}^2$ were replaced by $\mathbf{R}^3$ (i.e. our constraint were removed) then the Euler-Lagrange partial differential equations are (the unconstrained) linear partial differential equations of Laplace, $\Delta\varphi^i = 0$ $(i = 1, 2, 3)$, for which uniqueness is well known.

If our domain Ω is all of $\mathbf{R}^3$ there is no boundary value function ψ, of course. We then say that $\varphi : \mathbf{R}^3 \to \mathbf{S}^2$ is a minimizer provided φ cannot be modified on a compact set K to decrease energy in a larger bounded open set containing K.

Liquid crystals. The connection of our energy minimizing φ's with liquid crystals requires explanation. We imagine that Ω is a container containing a liquid crystal. At points x in Ω the liquid determines a **directrix** $\mathbf{n}(x)$ lying in real projective space $\mathbf{RP}^2$. Since $\mathbf{RP}^2$ is obtained from $\mathbf{S}^2$ by identifying antipodal points, this means intuitively that $\mathbf{n}(x)$ is a unit vector like our $\varphi(x)$ except that its head is indistinguishable from its tail. For the liquid crystals with which we are concerned, the energy of $\mathbf{n}$ is defined analogously to our $\mathcal{E}$, e.g. zero energy corresponds to parallel alignment. Like our minimizing φ's (as we shall see), any minimizing $\mathbf{n}$ will be continuous except at isolated points. This means, in particular, that any minimizing $\mathbf{n}$ can locally be lifted to become a minimizing φ having the same energy; this lifting is global in case Ω is simply connected. (see [BCL], p. 686 for details). Thus, for simply connected Ω's, our original problem is equivalent to the liquid crystal problem. In any case, whether or not Ω is simply connected, our estimates on the number of singular points hold for these liquid crystal minimizers. Line singularities do not occur in our model because they would have infinite Dirichlet energy. They do occur in nature, but to model them one, effectively, has to fatten the line and treat it separately (much as in the liquid helium problem). A further complication for liquid crystals is that there are other, more appropriate, integrands which are quadratic in $\nabla\varphi$ and respect rotational symmetry. The general nematic liquid crystal integrand, for example, is of the form

$$K_1(\text{div } \varphi)^2 + K_2(\varphi \cdot \text{curl } \varphi)^2 + K_3(\varphi \wedge \text{curl } \varphi)^2.$$

Our Dirichlet's energy integrand corresponds (except for a fixed boundary term) to setting $K_1 = K_2 = K_3 = 1$ (see [BCL], p. 653). Our methods give information about such liquid crystal geometries (by a compactness argument) only when K_1, K_2 and K_3 are nearly equal.

2. BASIC FACTS ABUT MINIMIZERS

(A) Existence and regularity of minimizers. As we mentioned above, whenever we have a reasonable domain Ω and boundary function ψ of finite energy, there will always exist a minimizer φ having boundary values ψ. Such a result is included among the general analysis of Dirichlet's integral minimizing mappings between manifolds by R. Schoen and K. Uhlenbeck in their basic papers [SU1] [SU2]. They further showed that a minimizing φ in our context is a real analytic mapping except at isolated points of discontinuity (which are our singularities). Finally, they concluded that a minimizing φ assumes its boundary values smoothly when both $\partial\Omega$ and ψ are comparably smooth.

(B) Monotonicity of energy and tangential approximations. One of the basic technical properties of energy minimizing mappings is usually called **monotonicity**. Whenever φ is a minimizer in Ω, $y \in \Omega$, and $0 < r < s < R$ so that the ball $\mathbf{B}_R(y)$ also lies within Ω, then

$$\frac{1}{r}\int_{\mathbf{B}_r(y)} |\nabla\varphi|^2\, dV \le \frac{1}{s}\int_{\mathbf{B}_s(y)} |\nabla\varphi|^2\, dV.$$

For a proof, see [SU1]. (The absence of a corresponding monotonicity estimate is the main reason our analysis of liquid crystals is restricted to the $K_1 = K_2 = K_3$ case.) The monotonicity estimate leads fairly directly to the existence of certain tangential approximations to φ at each interior y. A major and deep development occurred in a paper of L. Simon [S] which for our problem guarantees the existence of a **unique tangential approximating mapping.** At regular points this approximating mapping is constant. For a singular point y of φ in Ω, Simon's result gives a **unique harmonic mapping** $f : \mathbf{S}^2 \to \mathbf{S}^2$ such that

$$\varphi(y + t\omega) \to f(\omega) \qquad \text{as} \qquad t \to 0+$$

uniformly for all ω's in $\mathbf{S}^2$ (see [AL]), i.e.

$$\varphi(x) \approx f\left(\frac{x-y}{|x-y|}\right)$$

for x's near y. The correspondence here is in several strong senses (see [AL]). In general, if $f: \mathbf{S}^2 \to \mathbf{S}^2$ and $F: \mathbf{R}^3 \to \mathbf{S}^2$ is defined by setting

$$F(x) = f\left(\frac{x}{|x|}\right)$$

for each $x \neq 0$ then f is harmonic if and only if F is.

Example: $f\left(\frac{x}{|x|}\right) = \frac{x}{|x|}$, i.e. f is the identity; see Figure 1.

(C) **Harmonic mappings between spheres and mapping degrees.** Any continuous mapping $\mathbf{S}^2 \to \mathbf{S}^2$ has a well defined topological degree measuring the number of times the first sphere covers the second, taking into account the orientations. Since the boundary functions ψ under consideration map $\mathbf{S}^2$ to $\mathbf{S}^2$ and have finite energy, they also have a well defined degree given by the Jacobian integral

$$\deg(\psi) = \frac{1}{4\pi} \int_{\partial\Omega} J(\psi)\, dA;$$

here $J(\psi)$ is the Jacobian (determinant) function of ψ whose sign is positive or negative at a point depending on whether $D\psi$ preserves or reverses orientations at that point. For continuous ψ's of finite energy these two notions of degree coincide.

All possible harmonic mappings from $\mathbf{S}^2$ to $\mathbf{S}^2$ have been classified for some time. In complex coordinates (resulting from stereographic projection of the $\mathbf{S}^2$'s onto $\mathbf{C}$) they are all of the form

$$f(z) = \frac{P(z)}{Q(z)} \quad \text{or} \quad f(z) = \frac{P(\overline{z})}{Q(\overline{z})}$$

corresponding to various complex polynomial functions P and Q which are relatively prime. The degree of these f's can been checked to be

$$\deg(f) = \begin{cases} \max(\deg(P), \deg(Q)) & \text{first case;} \\ -\max(\deg(P), \deg(Q)) & \text{second case.} \end{cases}$$

For these harmonic maps $f: \mathbf{S}^2 \to \mathbf{S}^2$ we also set $F(x) = f\left(\frac{x}{|x|}\right)$ as above and compute for each $0 < R < \infty$ that

$$\int_{|x|<R} |\nabla F|^2 \, dV = 8\pi R |\deg(f)|,$$

i.e. the energy does not depend on P and Q except via the degree.

(D). **Tangential approximations to minimizers.** Suppose $y \in \Omega$ is a singular point of a minimizer φ and the tangential approximation is of the form $F(x) = f\left(\frac{x}{|x|}\right)$ corresponding to one of the harmonic f's given in (C) above. By the **degree of the singular point** y we mean the mapping degree of the associated f. **Which of the possible f's actually occur?** This question was answered by H. Brezis, J-M. Coron, and E. Lieb in their paper [BCL]. The *only* f's that occur are rotations $\mathcal{R}$ and reflections of the f in the above example, i.e.

$$f(\omega) = \pm\mathcal{R}(\omega), \quad (\omega \in \mathbf{S}^2) \qquad \text{with } \deg(f) = \pm 1; \tag{2}$$

see Figure 1. This class does not even include all harmonic maps of degree ± 1. The proof proceeds by a construction of comparison functions. If $|\deg(f)| > 1$ then the energy of F can be decreased by splitting the singularity at the origin into two nearby singularities of lower degree. If $|\deg(f)| = 1$ and $f \neq \pm\mathcal{R}$ then the energy of F can be decreased by moving the singular point slightly.

The paper [BCL] also answered a question that in some sense is complementary to the minimization question we have been studying here. Suppose $y_1, \ldots y_n$ are *fixed* points in Ω and $d_1, \ldots, d_n$ are *fixed* degrees associated to these points (not necessarily ± 1).

What is the infimum of energies $\mathcal{E}(\varphi)$ among all φ's which are continuous except at the y_i's and map small spheres around each y_i with degree d_i?

The boundary function ψ is not fixed. This infimum is *not* achieved in general. The answer is shown in the Figure 2. Think of each singularity as a source or sink of flux and draw lines to carry the flux between singularities, or between a singularity and the boundary. Then

$$\inf \mathcal{E}(\varphi) = 8\pi \min\left\{\sum \text{lengths of lines}\right\}$$

where the minimum is over all possible ways of constructing the lines. A different proof of this result was later given by F. Almgren, W. Browder, and E. Lieb [ABL] using H. Federer's co-area formula in the context of currents. This is like quark confinement: a plus and minus quark have an energy proportional to their separation.

Figure 1. Here are shown representations of unit vector fields

$$F(x) = \left(\frac{x}{|x|}\right) \quad \text{and} \quad G(x) = \mathcal{R}\left(\frac{x}{|x|}\right)$$

in which $\mathcal{R}$ is a counterclockwise rotation through 45^0. Such arrays minimize Dirichlet's integral energy and are also observed as stable liquid crystals geometries [K].

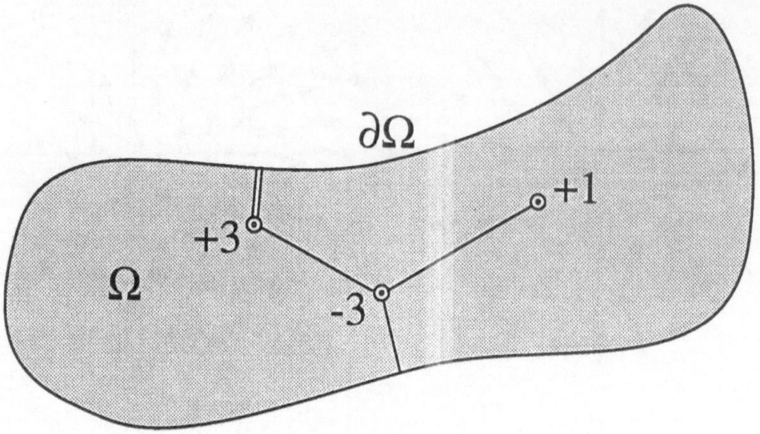

Figure 2. A region Ω is pictured here containing three prescribed singular points whose degrees (+3, -3, +1) are also prescribed. The least energy of unit vector fields having this singular behavior is the least total mass of oriented line segments connecting these singular points (as currents) either to each other or to the boundary. Such a least length array is illustrated.

From this result with *specified* singularities one is tempted to surmise that, in our original minimization problem, *potential* singularities would tend to annihilate each other (if of opposite degrees) or move to $\partial\Omega$. The number of singularities that will occur will be only that required by topology, i.e.

$$\sum_{\text{singularities}} \deg(\text{singularity}) = \deg(\psi) = \frac{1}{4\pi}\int_{\partial\Omega} J(\psi)\,dA.$$

This surmise is very wrong, as we shall see later in Example 3, and misled us for a long time. Arbitrarily many singularities (of mixed signs) can occur, even if the Jacobian $J(\psi)$ vanishes identically.

(E). **Boundary regularity and hot spots.** Our main estimates require an extension of the boundary regularity results indicated above in (A). These theorems take several pages merely to state precisely, but the essence of the matter is the following. Assume that $\partial\Omega$ is smooth and take a small patch $P \subset \partial\Omega$ which is roughly a 2-dimensional disk of radius R. One consequence of the boundary regularity theory mentioned in (A) is the following. There is a fixed $\varepsilon > 0$, independent of R, with the property that whenever the boundary function ψ satisfies

$$\int_P |\nabla_T\psi|^2\,dA < \varepsilon$$

then every minimizer φ is free of singularities in the region

$$K = \left\{x : x \in \Omega,\ \operatorname{dist}(x, P) > \frac{1}{2}R\varepsilon,\ \operatorname{dist}\left(x, P_{\frac{1}{2}}\right) < 2R\varepsilon\right\},$$

here $P_{\frac{1}{2}}$ is the concentric disk of radius $\frac{1}{2}R$. Note that ε is dimensionless. Our **hot spot boundary regularity** theorem (proved in [AL]) asserts the existence of a fixed number $0 < \delta << \varepsilon$ such that whenever $P' \subset P$ is a smaller subpatch of radius δR and

$$\int_{P\sim P'} |\nabla_T\psi|^2\,dA < \varepsilon$$

then φ is also free of singularities in the region K above. In other words arbitrarily large boundary energy in a very small disk P' cannot by itself induce singularities far away.

3. COUNTING SINGULARITIES

The principal question motivating our work in [AL] is this:

How many singular points $N(\psi)$ is it possible for a minimizing φ to have?

The following possibilities seem plausible at the outset:

$$N(\psi) \leq CE(\psi) \qquad \text{FALSE;}$$

$$N(\psi) \leq C \int_{\partial\Omega} |J(\psi)|\, dA \qquad \text{FALSE;}$$

$$N(\psi) \leq C\partial\mathcal{E}(\psi) \qquad \text{TRUE} \qquad \textbf{“The Linear Law”}.$$

here C is a constant, possibly depending on Ω.

The first possibility is false by counterexample—see below. The second possibility was suggested by the work in [BCL] and misled us for some time (had it been true it would have led to a beautiful geometric theory). In fact it is quite false as Example 1 below shows; in particular, $N(\psi)$ can be large while $J(\psi)$ vanishes identically.

Our main result, **The Linear Law**, is optimal (modulo the value of $C = C_\Omega$, of which we have no knowledge since our proof is by contradiction based on compactness arguments). It is, to our knowledge, the first bound of its kind.

The following example given by R. Hardt and F. H. Lin in [HL1] shows that $N(\psi)$ can indeed be proportional to $\partial\mathcal{E}(\psi)$. Choose N well separated small disks in $\partial\Omega$. Our ψ is constructed to wrap each disk D around the target sphere once (essentially by the inverse function to stereographic projection while preserving or reversing orientation as one chooses); each ∂D is mapped to the north pole. The complement of these disks in $\partial\Omega$ is mapped by ψ also to the north pole. Then $\partial\mathcal{E}(\psi) \approx CN$; the constant C is independent of the size of the disks since surface energy is scale invariant. Clearly the orientations of ψ on the disks can be arranged so that the total mapping degree of ψ is either zero or one. It is not hard to prove directly that any minimizing φ having ψ as boundary value function must have at least one singularity close to each tiny disk—otherwise $\mathcal{E}(\varphi)$ would be too large. Thus

$$N(\psi) \geq N \approx C^{-1}\partial\mathcal{E}(\psi).$$

Our first main new result (proved independently by Hardt and Lin in [HL2]) is that singularities cannot be very close if they are well inside Ω.

Theorem 1. *There is a universal constant C (independent of Ω and ψ) such that whenever y and z in Ω are singular points of a minimizer φ then* $\mathrm{dist}(y, z) \geq C\mathrm{dist}(y, \partial\Omega)$.

The idea of the proof is the following. Fix y and suppose the contrary. Then there will be a sequence of minimizing $\varphi^{(j)}$ with singular points at $z^{(j)}$ and at y such that $z^{(j)} \to y$ as $j \to \infty$. A compactness argument (contradicting the negation) and monotonicity (A) shows that the energy of φ in small balls of radius R about y is *uniformly* greater than $8\pi R$. The limit of a subsequence of the minimizers $\varphi^{(j)}$ is a minimizer φ which thus can have at worst a singularity of degree ± 1 at y (by equation (2) above). The tangential approximation theorem implies that the energy of the limit φ must be very close to $8\pi R$ for all small R's. This leads to a contradiction because of the continuity of Dirichlet's integral when minimizers converge.

A consequence of Theorem 1 together with equation (2) above is the following.

Theorem 2. (Complete classification of energy minimizing maps from $\mathbf{R}^3$ to $\mathbf{S}^2$.) *Suppose* $\varphi: \mathbf{R}^3 \to \mathbf{S}^2$ *is a minimizer. Then, either φ is a constant mapping or* $\varphi = \pm\mathcal{R}\left(\frac{x-y}{|x-y|}\right)$ *for some y and $\mathcal{R}$.*

Theorem 1 says that if there are many singularities they have to pile up near $\partial\Omega$. This leads to a difficult geometric-combinatorial problem on different scales proportional to δ^k, where δ is given in (E) above and $k = 1, 2, 3, \ldots$. We attempt to illustrate this in Figure 3. Referring to the ε and δ of (E) consider the points 1, 2, and 3 in Ω at distances $R\varepsilon, R\varepsilon\delta$, and $R\varepsilon\delta$ above a boundary patch P of radius R and two boundary patches P' and P'' of radii $R\delta$ inside P. The hot spot boundary regularity theorem gives us the following lower bounds for the energy of ψ in P if we consider the various possibilities of having singularities at positions 1, 2, or 3:

Positions occupied	Local boundary energy
(1 alone) or (2 alone) or (3 alone)	ε
(1 and 2) or (1 and 3) or (2 and 3)	2ε
(1 and 2 and 3)	2ε

The source of all our difficulties is that we cannot infer an energy 3ε if there are singularities at all three points.

If $S^{(k)}$ denotes the strip $\{x : x \in \Omega, \mathrm{dist}(x, \partial\Omega) \leq \varepsilon\delta^k\}$, we can effectively decompose each $S^{(k)}$ into cones of height $\varepsilon\delta^k$ and base radius δ^k. We then have a Cayley tree whose

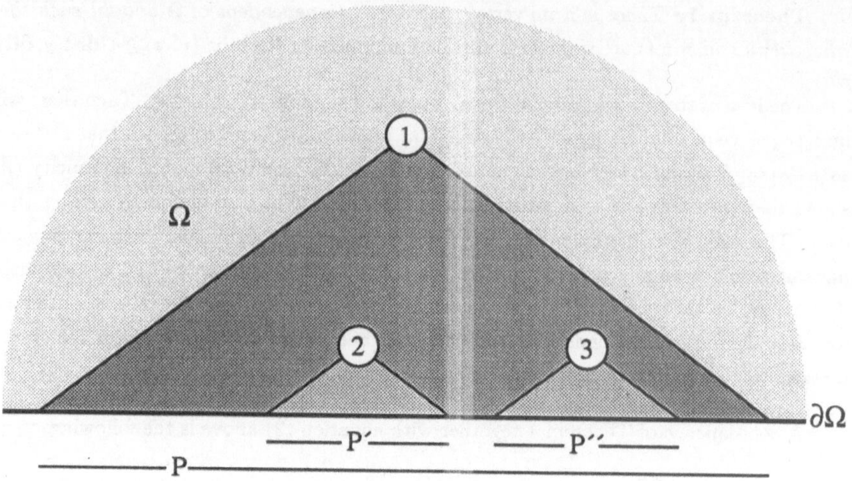

Figure 3. Pictured here are the "cones of influence" in Ω of three singular points. The presence of singular points 1, 2, 3 implies the presence of boundary energy in disks P, P′, P″ in ∂Ω. The problem is that these disks are not disjoint so that the total boundary energy is not a simple sum. Nesting of such cones induces a Cayley tree graph in which a combinatorial analysis overcomes this difficulty.

vertices represent these cones (i.e. a vertex of order $k+1$ is connected to a vertex of order k in the tree if the smaller cone is inside the larger one). A vertex is occupied if its cone has a singularity near the apex; otherwise it is unoccupied. Each occupied vertex gets an energy ε if and only if no more than one higher order vertex to which it is pathwise connected is occupied.

The actual details of decomposing each $S^{(k)}$ into cones so that due account is taken of overlaps (and all the other problems that will occur to the reader) involves a complicated covering and counting lemma. The final result is The Linear Law for $N(\psi)$ in terms of $\partial\mathcal{E}(\psi)$, as stated at the beginning of this section.

4. THREE EXAMPLES OF COUNTERINTUITIVE BEHAVIOR

Example 1. Zero Mapping Area. It is easy to prove for *any* Ω that if ψ takes values only in some closed hemisphere of $\mathbf{S}^2$ then φ has no singularities. We, however, are able to construct a single curve Γ in $\mathbf{S}^2$ which is a slight perturbation of the equator and, for each N, a smooth boundary value functions $\psi^N\colon \partial\Omega \to \mathbf{S}^2$ **having its image equal to** Γ such than any minimizer φ^N having boundary values ψ^N must have at least N singular points. In the example of [AL], Ω is taken to be a ball, but the details of Ω are not important. The Jacobian $J(\psi^N)$ of each ψ^N vanishes identically since its image is one dimensional.

The idea behind the construction appears in the following preliminary problem. Consider reasonable mappings $\varphi : \mathbf{D}^2 \to \mathbf{S}^2$ from the unit disk $\mathbf{D}^2$ in the plane having two dimensional Dirichlet's integral denoted by $\mathcal{E}_2(\varphi)$. Suppose $\Gamma \subset \mathbf{S}^2$ is a smooth embedding of a circle parametrized by a map $P\colon \partial\mathbf{D}^2 \to \Gamma$. The functions φ from $\mathbf{D}^2$ to $\mathbf{S}^2$ having boundary values P can be separated into two homological classes: the + class, in which, heuristically, φ "covers the top of $\mathbf{S}^2$ one more time than it covers the bottom" and, the − class in which φ "covers the bottom one more time than it covers the top"; see Figure 4.

Consider the two numbers

$$E^{\pm}(P) = \inf \{\mathcal{E}_2(\varphi) : \varphi = P \text{ on } \partial\mathbf{D}^2 \text{ and } \varphi \in \pm \text{ class}\}.$$

In general $E^+(P)$ will not be the same as $E^-(P)$.

We construct a single Γ having two different (homotopic) parametrizations P^+ and P^- such that

$$E^+(P^+) < E^-(P^+) + \varepsilon \quad \text{and} \quad E^-(P^-) < E^+(P^-) + \varepsilon$$

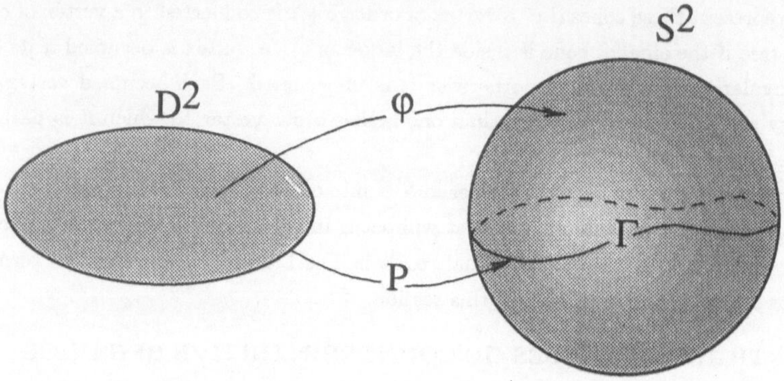

Figure 4. Illustrated here is one of two homologically distinct classes of mappings $\varphi: \mathbf{D}^2 \to \mathbf{S}^2$ corresponding to a given boundary parametrization $P: \partial\mathbf{D}^2 \to \Gamma$ (the curve Γ is a perturbation of the equator). A "+ function" is one which "covers the northern hemisphere". For some Γ's, the homology type preferred by a least energy mapping can change if the parametrization P is changed. This phenomenon leads ultimately to construction of least energy mappings from the ball to the sphere having many interior singularities but for which the boundary mapping of the sphere to the sphere has zero mapping area (its entire image lies within the curve Γ).

for some $\varepsilon > 0$. In other words if the parametrization of Γ changes from P^+ to P^- any *absolute* minimizer φ changes from lying in the $+$ class to lying in the $-$ class.

The next step is to let Ω be a very long solid tube T of radius 1 and length $N(L+1)$. (Actually, T is bent into a torus so that we can ignore the two ends.) As boundary function ψ we alternately paste P^- and P^+ on sections of length L (i.e. each cross-sectional disk has P^- or P^+ on its boundary). In the transitional regions of length 1 we smoothly interpolate between P^- and P^+ (which can be done since they are homotopic). In the transition regions ψ continues to take values only in Γ. See Figure 5.

If L is large enough (depending only on ϵ), it is believable (and we prove it) that φ must be mostly a $-$ function on the P^- disks and it must be mostly a $+$ function on the P^+ disks, for otherwise $\mathcal{E}(\varphi)$ would be unnecessarily large. But when φ switches from being a $-$ function to being a $+$ function φ must have a singularity for topological reasons. Thus, φ will have at least N singularities altogether.

The drawback to this tube example is that the domain T depends on N. To achieve the same result for a fixed domain $\Omega =$ unit ball, we first cut the surface ∂T longitudinally (i.e. perpendicular to the disks) and flatten it (key estimates here come from the conformal equivalence of the disk and the upper half plane and the fact that Dirichlet's integral in two dimensions is invariant under conformal reparametrizations of domains). This yields a strip of width 2π and length $N(L+1)$. We also rotate P^+ if necessary so that P^+ and P^- have the same value $\gamma \in \Gamma$ along the cut. Next we shrink the strip to width $(2\pi)^2/N(L+1)$ and length 2π. Finally we paste this strip (which is very narrow since N is large) along the equator of Ω and let $\psi : \partial\Omega \to \mathbf{S}^2$ be the old ψ in the strip and let $\psi(x) = \gamma$ for $x \in \partial\Omega$ but $x \notin$ the strip. A somewhat nerve wracking argument shows, as expected, that any minimizer $\varphi : \Omega \to \mathbf{S}^2$ must have at least N singularities close to the equator of Ω.

Example 2. Symmetry Breaking. When φ takes values in $\mathbf{R}^3$ instead of $\mathbf{S}^2$, any geometric symmetry of Ω and ψ is inherited by the minimizing φ. The reason is simply that minimizers are unique in the linear case ($\Delta\varphi = 0$). When, as in our case, φ takes values in $\mathbf{S}^2$, the symmetry of Ω and ψ can be broken by φ; obviously there must then be several minimizers.

Let Ω be the unit ball in $\mathbf{R}^3$ and let $\psi : \partial\Omega \to \mathbf{S}^2$ be the distortion of the identity map illustrated in Figure 6. In small caps N (resp. S) on $\partial\Omega$, ψ covers the northern (resp. southern) hemisphere of $\mathbf{S}^2$. The two maps are mirror images of each other. On the

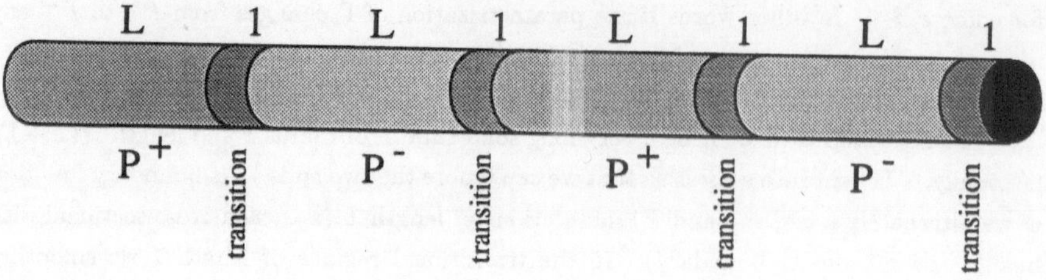

Figure 5. Illustrated here is a boundary value function $\psi: \partial\Omega \to \mathbf{S}^2$ for a long tube domain Ω. The image of ψ is a smooth curve Γ in $\mathbf{S}^2$. On crossectional circles of $\partial\Omega$ the boundary values alternate between intervals of P^+ mappings and intervals of P^- separated by transition intervals. Least energy maps $\varphi: \Omega \to \mathbf{S}^2$ with such boundary values map most crossections in P^+ regions to cover the northern hemisphere and map most crossections in P^- regions to cover the southern hemisphere. The minimizer φ therefore has at least one singular point near each transition region.

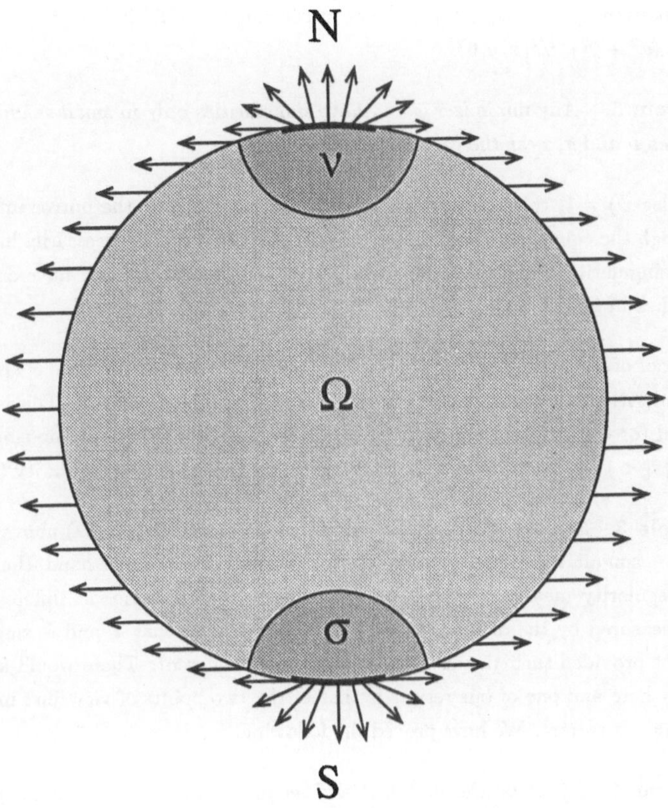

Figure 6. Here our domain Ω is the unit ball so that $\partial\Omega$ is the unit sphere. Pictured schematically is a special boundary value function $\psi\colon \partial\Omega \to \mathbf{S}^2$ having a mirror image symmetry through the equatorial plane. A small cap N around the north pole maps to cover the entire northern hemisphere of $\mathbf{S}^2$ while a small cap S around the south pole covers the entire southern hemisphere. The sphere less these two caps maps entirely to the equator. Longitude is preserved in each of these regions. No minimizing $\varphi\colon \Omega \to \mathbf{S}^2$ having boundary values ψ can possess such a symmetry since the (necessarily odd) number of singular points must be contained within one of the regions ν and σ near the poles.

rest of $\partial\Omega$ between N and S, ψ takes values in the equator of $\mathbf{S}^2$ in the obvious way, i.e. $\psi(x,y,z) = (x^2+y^2)^{-1/2}(x,y,0)$.

Theorem 3. *Any minimizer φ can have singularities only in small shaded regions in Ω, labelled ν and σ, near the caps N and S.*

Since $\deg(\psi) = 1$, this result implies that φ does not inherit the mirror image symmetry through the equatorial disk possessed by ψ. (Our function φ necessarily has an odd number of singularities, and, if φ were symmetric, it would necessarily have one on the equatorial disk in Ω.)

The proof of Theorem 3 has two parts. First we show that when N and S are small φ has no singularities in a concentric ball Ω' of radius $1-\varepsilon$ for some small ε. This is done by a variational (or comparison) argument. Second, we show that there are no singularities in $\{x : 1 \leq |x| < 1-\varepsilon \text{ and } \operatorname{dist}(x, \sigma\cup\nu) > \varepsilon\}$ by using the boundary regularity (F).

Example 3. Boiling Water. The [BCL] result mentioned in (D) above suggests that + and − singularities tend to annihilate each other. On the other hand, the hot spot boundary regularity mentioned in (E) above suggests that behavior at different length scales (as measured by the distance to $\partial\Omega$) is independent so that + and − singularities *could* coexist provided their distances to $\partial\Omega$ were very different. There would appear to be a conflict here and one of our results is that of the two points of view just mentioned the second one is correct. We have proved the following.

Theorem 4. *Let Ω be the unit ball and let $p_1, \ldots, p_M$ be any distinct points in $\partial\Omega$. Also let $N_1, \ldots, N_M$ be any positive integers and for each $i = 1, \ldots, M$ let A_i be any sequence of length N_i consisting of $+1$'s and -1's. Finally, let $\varepsilon > 0$. Then there is a smooth $\psi : \partial\Omega \to \mathbf{S}^2$ such that*

(i) $\partial\mathcal{E}(\psi) \leq \varepsilon + 8\pi\sum_{i=1}^{M} N_i$.

(ii) The minimizer φ is unique.

(iii) For each $i = 1, \ldots, M$ there are at least N_i singularities stacked nearly vertically above p_i (like bubbles in a pan of water that is about to boil), and these have the specified sequence of degrees given by A_i.

REFERENCES

[ABL] F. Almgren, W. Browder and E. Lieb, *Co-area, liquid crystals and minimal surfaces*, in *Partial Differential Equations*, ed. S.S. Chern, Springer Lecture Notes in Math. **1306**, 1-12 (1988).

[AL] F. Almgren and E. Lieb, *Singularities of energy minimizing maps from the ball to the sphere: examples counterexamples and bounds*, Ann. of Math., Nov. 1988. See also *Singularities of energy minimizing maps from the ball to the sphere*, Bull. Amer. Math. Soc. **17**, 304-306 (1987).

[BCL] H. Brezis, J-M. Coron and E. Lieb, *Harmonic maps with defects*, Commun. Math. Phys. **107**, 649-705 (1986).

[HL1] R. Hardt and F. H. Lin, *A remark on H^1 mappings*, Manuscripta Math. **56**, 1-10 (1986).

[HL2] R. Hardt and F. H. Lin, *Stability of singularities of minimizing harmonic maps*, J. Diff. Geom., to appear.

[K] M. Kléman, *Points, lignes, parois dans les fluides anisotropes et les solides cristalline*, Les Éditiones de Physique (Orsay), I, 36-37.

[S] L. Simon, *Asymptotics for a class of nonlinear evolution equations with applications to geometric problems*, Ann. of Math. **118**, 525-571 (1983).

[SU1] R. Schoen and K. Uhlenbeck, *A regularity theory for harmonic maps*, J. Diff. Geom. **17**, 307-335 (1982).

[SU2] R. Schoen and K. Uhlenbeck, *Boundary regularity and the Dirichlet problem of harmonic maps*, J. Diff. Geom. **18**, 253-268 (1983).

Relaxed Energies for Harmonic Maps

F. BETHUEL [1], H. BREZIS [2,3] and J.M. CORON[4]

INTRODUCTION

Let $\Omega \subset \mathbb{R}^3$ be an open bounded set such that $\partial\Omega$ is smooth. Set

$$H^1(\Omega; S^2) = \{u \in H^1(\Omega; \mathbb{R}^3); |u(x)| = 1 \text{ a.e.}\}$$

and

$$H^1_\varphi(\Omega; S^2) = \{u \in H^1(\Omega; S^2); u = \varphi \text{ on } \partial\Omega\},$$

where $\varphi: \partial\Omega \to S^2$ is a given boundary data.

If $u \in H^1(\Omega; S^2)$ is smooth on $\overline{\Omega}$ except at a finite number of point singularities in Ω, and if moreover, $\deg \varphi = \deg(u|_{\partial\Omega}) = 0$, then the length of a minimal connection connecting the singularities has been introduced by Brezis, Coron and Lieb in [BCL] and is given by

$$L(u) = \min_{\sigma \in \sigma_k} \sum_{i=1}^{k} d(p_i, n_{\sigma(i)})$$

where $(p_1, p_2, \ldots, p_k)$ are the singularities of positive degree (counted according to their multiplicity), $(n_1, \ldots, n_k)$ are the singularities of negative degree, d is the geodesic distance in Ω and the minimum is taken over all possible permutations σ of the integer $\{1, 2, \ldots, k\}$. (Since $\deg(u|_{\partial\Omega}) = 0$ the number of positive singularities is the same as the number of negative singularities.)

For any $u \in H^1(\Omega; S^2)$ the vector field $D(u)$, defined as follows,

$$D(u) = (u \cdot u_y \wedge u_z, u \cdot u_z \wedge u_x, u \cdot u_x \wedge u_y)$$

[1] CERMA, ENPC, La Courtine 93167 Noisy le Grand.

[2] Département de Mathématiques, Université P. et M. Curie, 4, pl. Jussieu, 75252 Paris Cedex 05.

[3] Department of Mathematics, Rutgers University, Hill Center, Busch Campus, New Brunswick, NJ 08903.

[4] Département de Mathématiques, Université Paris-Sud, 91405 Orsay Cedex.

plays an important role in [BCL]. If u is smooth except at a finite number of singularities (p_i, n_i) we recall (see [BCL], Appendix B) that

$$\operatorname{div} D(u) = 4\pi(\Sigma\delta_{p_i} - \Sigma\delta_{n_i}) \quad \text{in} \quad \mathcal{D}'(\Omega).$$

If in addition $\deg(u|_{\partial\Omega}) = 0$, we also know that (see [BCL], Section IV)

$$L(u) = \sup_{\substack{\xi:\Omega\to\mathbb{R} \\ \|\nabla\xi\|_\infty \le 1}} \left\{ \sum_{i=1}^{k} \xi(p_i) - \sum_{i=1}^{k} \xi(n_i) \right\}.$$

Note that

$$\sum_{i=1}^{k} \xi(p_i) - \sum_{i=1}^{k} \xi(n_i) = \int_\Omega \left(\sum_{i=1}^{k} \delta_{p_i} - \sum_{i=1}^{k} \delta_{n_i} \right) \xi = \frac{1}{4\pi} \int_\Omega \operatorname{div} D(u)\xi =$$
$$= -\frac{1}{4\pi} \int_\Omega D(u) \cdot \nabla\xi + \frac{1}{4\pi} \int_{\partial\Omega} (D(u) \cdot n)\xi \, d\sigma$$

where n denotes the outward normal to $\partial\Omega$. We also recall that $D(u) \cdot n$ depends only on $\varphi = u|_{\partial\Omega}$ and more precisely $D(u) \cdot n = \varphi \cdot \varphi_x \wedge \varphi_y$, where x, y are orthonormal coordinates on $\partial\Omega$ such that $(\frac{\partial}{\partial x}, \frac{\partial}{\partial y}, n)$ is a direct basis. It follows that if u is smooth except at a finite number of point singularities in Ω and $\deg(u|_{\partial\Omega}) = 0$ then

$$(1) \qquad L(u) = \frac{1}{4\pi} \sup_{\substack{\xi:\Omega\to\mathbb{R} \\ \|\nabla\xi\|_\infty \le 1}} \left\{ \int_\Omega D(u) \cdot \nabla\xi - \int_{\partial\Omega} D(u) \cdot n\xi \, d\sigma \right\}.$$

We shall use formula (1) as a *definition for L*. Clearly, it makes sense for any $u \in H^1(\Omega; S^2)$ with $\varphi = u|_{\partial\Omega} \in H^1(\partial\Omega; S^2)$ provided

$$(2) \qquad \deg(u|_{\partial\Omega}) = \frac{1}{4\pi} \int_{\partial\Omega} D(u) \cdot n = 0.$$

(The first equality in (2) is a definition but it coincides with the usual degree when u is C^1 on $\partial\Omega$, and can be extended to maps in $H^1(\partial\Omega; S^2)$ because of the density result of Schoen and Uhlenbeck [SU2]). Throughout Sections 1 and 2 of this paper, we assume (at least) that $\varphi \in H^1(\partial\Omega; S^2) \cap C^0(\partial\Omega; S^2)$ and $\deg(\varphi; \partial\Omega) = 0$. Our first result asserts that L is continuous on H^1_φ.

Theorem 1. *We have the following inequality*

$$(3) \quad |L(u) - L(v)| \le C\|\nabla u - \nabla v\|_2(\|\nabla u\|_2 + \|\nabla v\|_2) \quad \forall u, v \in H^1_\varphi.$$

The quantity L is very useful when dealing with questions of approximation of maps in $H^1(\Omega; S^2)$ by smooth maps from Ω into S^2. We recall that smooth maps from Ω into S^2 are not dense in $H^1(\Omega; S^2)$, (see [SU2], [BeZ], [Be1], [Be2]). We have the following result:

Theorem 2. *Let u be in $H^1_\varphi(\Omega; S^2)$. Then there exists a sequence of maps $v_n \in H^1_\varphi(\Omega; S^2) \cap C^0(\overline{\Omega}; S^2)$ such that:*

$$v_n \rightharpoonup u \quad \text{weakly in} \quad H^1$$

and

$$\overline{\lim} \int_\Omega |\nabla v_n|^2 \leq \int_\Omega |\nabla u|^2 + 8\pi L(u). \tag{4}$$

If, in addtion, $\varphi \in C^\infty(\partial\Omega; S^2)$ then one may take $v_n \in C^\infty(\overline{\Omega}; S^2)$.

As direct consequences of Theorem 2 we have the following corollaries

Corollary 1. *With the same assumptions as in Theorem 2 we have:*

$$\inf_{v \in H^1_\varphi \cap C^0 (resp.\ C^\infty)} \int_\Omega |\nabla(u - v)|^2 \leq 8\pi L(u). \tag{5}$$

Corollary 2. *We have:*

$$\inf_{u \in H^1_\varphi} \left\{ \int_\Omega |\nabla u|^2 + 8\pi L(u) \right\} = \inf_{v \in H^1_\varphi \cap C^0 (resp.\ C^\infty)} \int_\Omega |\nabla v|^2. \tag{6}$$

Clearly we have

$$\mu_\varphi \equiv \min_{u \in H^1_\varphi} \int_\Omega |\nabla u|^2 \leq \inf_{v \in H^1_\varphi \cap C^0} \int |\nabla v|^2 \equiv \overline{\mu}_\varphi.$$

As pointed out by Hardt and Lin [HL] (see also [B1]) one can construct smooth maps

$$\mu_\varphi < \overline{\mu}_\varphi.$$

They also raised the very interesting question (which is still unanswered) whether the infimum defining $\overline{\mu}_\varphi$ is achieved. The main difficulty comes from the fact that if (v_n) is a minimizing sequence for $\overline{\mu}_\varphi$ and $v_n \rightharpoonup v$

weakly in H^1, then v need not be continuous on $\overline{\Omega}$ (and hence it is not clear whether v is a minimizer for $\overline{\mu}_\varphi$). In trying to attack this problem and also in view of Theorem 2 it seems natural to introduce the following "energy":

$$F(u) = E(u) + 8\pi L(u), \quad \text{for } u \text{ in } H^1_\varphi \tag{7}$$

where $E(u) = \int_\Omega |\nabla u|^2 dx$ is the usual energy. Obviously F coincides with E on smooth maps, and by Corollary 2 we have

$$\inf_{u \in H^1_\varphi} F(u) = \inf_{v \in H^1_\varphi \cap C^0 \,(resp.\, C^\infty)} E(v). \tag{8}$$

The main interest of F lies in the following property:

Theorem 3. *F is sequentially lower semi-continuous on H^1_φ for the weak H^1 topology.*

As a consequence, we obtain:

Corollary 3. *We have*

$$\inf_{u \in H^1_\varphi} F(u) \quad \textit{is achieved.} \tag{9}$$

Also, every minimizing sequence for $\overline{\mu}_\varphi$ converges (up to a subsequence) weakly in H^1 to some minimizer for F on H^1_φ.

In view of Theorems 2 and 3, the function F is the largest sequentially lower semi-continuous function on H^1_φ which is less than E on $H^1_\varphi \cap C^0$ (resp. C^∞). This means that F plays the role of a "*relaxed energy.*"

Assuming one is able to show that a minimizer for (9) is continuous, this would answer the Hardt–Lin problem. Unfortunately we have only partial regularity results for the minimizers of F. We shall return to the question of regularity in a forthcoming paper. In Section 2 we discuss some properties of the minimizers for (9), in particular we have:

Theorem 4. *Every minimizer u for F in H^1_φ satisfies*

$$-\Delta u = u|\nabla u|^2 \quad \text{in} \quad \mathcal{D}'(\Omega; \mathbf{R}^3) \tag{10}$$

that is, u is weakly harmonic.

Assume that φ is a smooth map. By a result of [SU2] one knows that minimizers for E in H^1_φ are smooth except at a finite number of points.

Moreover, by [BCL], each singularity has degree ± 1 and the behavior of u near each singularity is well understood. In interesting situations the singular set of every minimizer of E on H^1_φ is not empty (otherwise the Hardt–Lin problem is irrelevant). We denote this property by (P_φ) and we have:

Theorem 5. *Property (P_φ) holds if and only if $\mu_\varphi < \overline{\mu}_\varphi$. In this case the set of minimizers for E on H^1_φ and the sets of minimizers for F on H^1_φ are disjoint.*

Note that Theorem 5 provides, in interesting situations, the existence of non-minimizing weakly harmonic maps with given boundary data φ. In fact, we shall prove in Section 3 the following:

Theorem 6. *Assume* $\deg(\varphi; \partial Omega) = 0$ *and P_φ holds or* $\deg(\varphi; \partial\Omega) \neq 0$, *then there exist infinitely many weakly harmonic maps having φ as boundary value.*

Finally, let us mention that a weaker form of the Hardt–Lin problem is still open: it is not known whether every smooth map φ (of degree zero) admits a smooth harmonic extension in Ω.

Some of the results in this paper answer questions raised in [B2].

1. Basic Properties of L and F

We start with:

Proof of Theorem 1. It is useful to set, for u, v in H^1_φ,

$$L(u,v) = \sup_{\substack{\xi:\Omega\to\mathbb{R} \\ \|\nabla\xi\|_\infty \leq 1}} \int_\Omega (D(u) - D(v))\nabla\xi = L(v,u). \tag{11}$$

Clearly we have, for u, v in H^1_φ

$$\int_\Omega D(u)\cdot\nabla\xi - \int_{\partial\Omega} D(u)\cdot n\,\xi d\sigma = \int_\Omega D(v)\cdot\nabla\xi - \int_{\partial\Omega} D(v)\cdot n\,\xi d\sigma + \int_\Omega (D(u) - D(v))\cdot\nabla\xi$$

(since $D(u)\cdot n$ depends only on φ) and therefore

$$|L(u) - L(v)| \leq L(u,v). \tag{12}$$

In order to estimate $L(u,v)$ we proceed as follows:

$$(D(u)-D(v))\cdot\nabla\xi = \mathrm{I}+\mathrm{II}+\mathrm{III}$$

where

$$\begin{aligned}
\mathrm{I} &= (u-v)\cdot[(u_y\wedge u_z)\xi_x+(u_z\wedge u_x)\xi_y+(u_x\wedge u_y)\xi_z]\\
\mathrm{II} &= v\cdot[(u_y-v_y)\wedge u_z\xi_x+(u_z-v_z)\wedge u_x\xi_y+(u_x-v_x)\wedge u_y\xi_z]\\
\mathrm{III} &= v\cdot[v_y\wedge(u_z-v_z)\xi_x+v_z\wedge(u_x-v_x)\xi_y+v_x\wedge(u_y-v_y)\xi_z].
\end{aligned}$$

By the Cauchy–Schwarz inequality, we clearly have:

$$\Big|\int_\Omega \mathrm{II}\Big| \le C\|\nabla(u-v)\|_2\|\nabla u\|_2$$

$$\Big|\int_\Omega \mathrm{III}\Big| \le C\|\nabla(u-v)\|_2\|\nabla v\|_2.$$

We now turn to the first term which we estimate using the following identity:

$$(13)\qquad \left\{\begin{aligned}
\int_\Omega \mathrm{I} = \frac{1}{2}\int_\Omega & u\cdot[(u-v)_y\wedge u_z+u_y\wedge(u-v)_z]\xi_x\\
&+\frac{1}{2}\int_\Omega u\cdot[(u-v)_z\wedge u_x+u_z\wedge(u-v)_x]\xi_y\\
&+\frac{1}{2}\int_\Omega u\cdot[(u-v)_x\wedge u_y+u_x\wedge(u-v)_y]\xi_z.
\end{aligned}\right.$$

This identity is easily established first when u,v are in $C^\infty(\overline{\Omega};\mathbf{R}^3)$, $u-v\in C_0^\infty(\Omega;\mathbf{R}^3)$ and $\xi\in C^\infty(\overline{\Omega})$: write, $(u_y\wedge u_z)=\frac{1}{2}(u\wedge u_z)_y+\frac{1}{2}(u_y\wedge u)_z$ etc..., integrate by parts, and note that:

$$(u\wedge u_z)\xi_{xy}+(u_y\wedge u)\xi_{xz}+(u\wedge u_x)\xi_{yz}+(u_z\wedge u)\xi_{xy}+(u\wedge u_y)\xi_{xz}+(u_x\wedge u)\xi_{yz}=0.$$

The general case ($u,v\in H^1_\varphi$ and $\nabla\xi\in L^\infty$) follows by density. We deduce from (13) that

$$\Big|\int_\Omega \mathrm{I}\Big| \le C\|\nabla(u-v)\|_2\|\nabla u\|_2.$$

This completes the proof of Theorem 1.

Proof of Theorem 2. Let us first assume that $\varphi\in C^\infty$. Set

$$R_\varphi^\infty=\Big\{u\in H^1_\varphi(\Omega;S^2);\exists a_1,a_2,\dots,a_N\in\Omega, u\in C^\infty\Big(\overline{\Omega}\setminus\bigcup_{i=1}^N\{a_i\};S^2\Big)\Big\},$$

that is, R_φ^∞ is the subset of maps in H_φ^1 which are smooth except at most at a finite number of points. We recall (see [BeZ], Theorem 4bis) that R_φ^∞ is dense in H_φ^1.

Given $u \in H_\varphi^1$ and given n, we fix some $u_n \in R_\varphi^\infty$ such that

$$\|\nabla(u_n - u)\|_2 < 1/n.$$

It follows that $\text{meas}\{u \in \Omega;\ |u_n(x) - u(x)| > n^{-1/2}\} \leq C/n$. We now apply Theorem 2 of [Be1] (see also the remark at the end of Section II.3 in [Be1]) to u_n and we find some map $v_n \in C^\infty(\overline{\Omega}; S^2)$ with $v_n = \varphi$ on $\partial\Omega$ such that:

$$\text{(14)} \qquad \begin{cases} E(v_n) \leq E(u_n) + 8\pi L(u_n) + 1/n, \\ \text{meas}\{x \in \Omega;\ v_n(x) \neq u_n(x)\} < 1/n. \end{cases}$$

Since $u_n \to u$ in H_φ^1 it follows that $L(u_n) \to L(u)$ (by Theorem 1) and therefore v_n is bounded in H^1. On the other hand

$$\text{meas}\{x \in \Omega;\ |v_n(x) - u(x)| > n^{-1/2}\} \leq (C+1)/n$$

and thus $v_n \to u$ a.e. We deduce that $v_n \rightharpoonup u$ weakly in H^1. Passing to the limit in (14) we obtain

$$\overline{lim}_{n\to+\infty} E(v_n) \leq E(u) + 8\pi L(u).$$

In the case where $\varphi \in C^0$ the proof is exactly the same except that R_φ^∞ is replaced by:

$$R_\varphi^0 = \left\{ u \in H_\varphi^1(\Omega; S^2);\ \exists a_1, a_2, \ldots, a_N \in \Omega, u \in C^0\left(\overline{\Omega} \setminus \sum_{i=1}^N \{a_i\}; S^2\right)\right\}.$$

Remark 1. Inequality (5) in Corollary 1 is optimal in the sense that given any $\delta > 0$ there is some $\varphi \in C^\infty(\partial\Omega; S^2)$ and some $u \in H_\varphi^1(\Omega; S^2)$ such that:

$$\text{(15)} \qquad \inf_{v \in H_\varphi^1 \cap C^\infty} \int |\nabla(u - v)|^2 \geq 8\pi(1 - 2\delta) L(u)$$

with $L(u) = 2 - \delta^2$ (when Ω is the unit ball). Here is an example.

Let Ω be the unit ball with north pole N and south pole S. Along the NS axis we place two ε-dipoles with the same orientation: $[p_1, n_1]$ is centered at N, $[p_2, n_2]$ is centered at S (for more details, see the Example in Section 2 of [B1]). For the map u obtained by gluing these dipoles we have, on Ω, $L(u) = 2 - \varepsilon$, and $E(u) \le 16\pi\varepsilon + 2\varepsilon$.

On the other hand we claim that

$$(16) \qquad E(u) + E(v) \ge 8\pi L(u), \quad \forall u \in H^1_\varphi, \ \forall v \in H^1_\varphi \cap C^0.$$

Indeed, since $v \in H^1_\varphi \cap C^0$, div $D(v) = 0$ and thus

$$\int_{\partial\Omega} D(u) \cdot n\xi d\sigma = \int_{\partial\Omega} D(v) \cdot n\xi d\sigma = \int_\Omega D(v) \cdot \nabla\xi.$$

Inserting this in (1) we obtain

$$\begin{aligned} L(u) &= \frac{1}{4\pi} \sup_{\substack{\xi:\Omega\to\mathbb{R} \\ \|\nabla\xi\|_\infty \le 1}} \int_\Omega [D(u) \cdot \nabla\xi - D(v) \cdot \nabla\xi] \\ &\le \frac{1}{4\pi} \int_\Omega |D(u)| + |D(v)| dx \le \frac{1}{8\pi}(E(u) + E(v)). \end{aligned}$$

Next, note that

$$\begin{aligned} \int |\nabla(u-v)|^2 dx &\ge -2\|\nabla u\|_2 \|\nabla v\|_2 + E(v) \ge (1-\delta)E(v) - \frac{1}{\delta}E(u) \\ &\ge (1-\delta)(8\pi L(u) - E(u)) - \frac{1}{\delta}E(u). \end{aligned}$$

We derive (15) by choosing $\varepsilon = \delta^2/2$.

There is a related notion of minimal connection defined as follows. Given $u \in H^1(\Omega; S^2)$ set

$$\tilde{L}(u) = \sup_{\substack{\xi:\Omega\to\mathbb{R} \\ \|\nabla\xi\|_\infty \le 1, \ \xi=0 \text{ on } \partial\Omega}} \int D(u) \cdot \nabla\xi.$$

Note that $\tilde{L}(u)$ makes sense for any u in $H^1(\Omega; S^2)$—even if $\deg(u|_{\partial\Omega}) \ne 0$, or even if $\deg(u|_{\partial\Omega})$ is not well defined. In the case where u is smooth on $\overline{\Omega}$, except at a finite number of point singularities in Ω, then $\tilde{L}(u)$ coincides with the length of a minimal connection, "allowing connections to $\partial\Omega$." More precisely, one adds (an unspecified number of) artificial singularities on the boundary in such a way that the new configuration has the same number of positive and negative singularities. Then one looks for the minimal length connection in the usual Euclidean sense. $\tilde{L}(u)$

corresponds to the infimum of these quantities when varying the positions and the number of artificial boundary points (this is a result in [BCL]).

Warning: If $\deg(u|_{\partial\Omega}) = 0$, $\tilde{L}(u) \leq L(u)$ with strict inequality in general.

There are variants of the previous results for $\tilde{L}$.

Theorem $\tilde{1}$. *We have*

$$|\tilde{L}(u) - \tilde{L}(v)| \leq C\|\nabla u - \nabla v\|_2(\|\nabla u\|_2 + \|\nabla v\|_2), \quad \forall\, u, v \text{ in } H^1(\Omega; S^2).$$

Theorem $\tilde{2}$. *Given u in $H^1(\Omega; S^2)$ there is a sequence of maps v_n in $C^\infty(\overline{\Omega}; S^2)$ such that:*

$$v_n \rightharpoonup u \quad \text{weakly in } H^1$$

and

$$\overline{\lim}_{n\to+\infty} \int_\Omega |\nabla v_n|^2 \leq \int_\Omega |\nabla u|^2 + 8\pi \tilde{L}(u).$$

Corollary $\tilde{1}$. *Given u in H^1, we have*

$$\inf_{v \in C^\infty(\overline{\Omega}; S^2)} \int |\nabla(u - v)|^2 \leq 8\pi \tilde{L}(u).$$

The proof of Theorem $\tilde{1}$ is exactly the same as the one of Theorem 1 (in all integrations by parts boundary integrals vanish because $\xi = 0$ on $\partial\Omega$). The proof of Theorem $\tilde{2}$ follows the same idea as the one of Theorem 2 except that one should now use Theorem 2bis of [Be1] instead of Theorem 2 of [Be1]. Note that here the v_n's do not coincide with u on $\partial\Omega$.

Remark 2. We do not know whether the inequality of Corollary $\tilde{1}$ is optimal in the sense of Remark 1 (if one uses the same u as in Remark 1, $\tilde{L}(u) = \varepsilon$).

In the proof of Theorem 3 we shall use the following lemma.

Lemma 1. *Let u, p_1, p_2, p_3, be 4 vectors of $\mathbf{R}^3$. Set (in a given basis e_1, e_2, e_3)*

$$V = (u \cdot p_2 \wedge p_3, u \cdot p_3 \wedge p_1, u \cdot p_1 \wedge p_2).$$

Then, we have

$$|V| \leq \frac{1}{2}|u|(|p_1|^2 + |p_2|^2 + |p_3|^2). \tag{17}$$

Proof. We may always assume that $|u| = 1$. Let R be a rotation such that

$$R(u) = e_3 = (0, 0, 1).$$

Recall that

$$Ru \cdot Rp_i \wedge Rp_j = u \cdot p_i \wedge p_j, \quad \text{for } i, j \text{ in } \{1, 2, 3, \}.$$

Write in the basis e_1, e_2, e_3

$$Rp_1 = (a_1, b_1, c_1), Rp_2 = (a_2, b_2, c_2), Rp_3 = (a_3, b_3, c_3).$$

Thus

$$V = (a_2 b_3 - a_3 b_2, a_3 b_1 - a_1 b_3, a_1 b_2 - a_2 b_1) = a \wedge b$$

with

$$a = (a_1, a_2, a_3) \quad \text{and} \quad b = (b_1, b_2, b_3).$$

We have

$$|V| \leq |a|\,|b| \leq \frac{1}{2}(|a|^2 + |b|^2) \leq \frac{1}{2}(|Rp_1|^2 + |Rp_2|^2 + |Rp_3|^2)$$

which leads to (17).

Proof of Theorem 3. Since a supremum of sequentially lower semi-continuous functions is sequentially lower semi-continuous it suffices to check that for any fixed $\xi : \Omega \to \mathbf{R}$ with $\|\nabla \xi\|_\infty \leq 1$ the function

$$u \in H^1_\varphi \mapsto F_\xi(u) = \int_\Omega |\nabla u|^2 dx + 2 \int_\Omega D(u) \cdot \nabla \xi \, dx$$

is sequentially lower semi-continuous for the weak H^1 topology (recall that the boundary integral $\int_{\partial\Omega} D(u) \cdot n \xi d\sigma$ depends only on φ).

Suppose that $u^n \rightharpoonup u$ weakly in H^1_φ and set $v^n = u^n - u$. We have, when $n \to +\infty$,

$$\int_\Omega |\nabla u^n|^2 = \int_\Omega |\nabla u|^2 + \int_\Omega |\nabla v^n|^2 + o(1).$$

Write

$$2\int D(u^n)\cdot\nabla\xi = A^n + B^n + C^n$$

where

$$A^n = 2\int u^n\cdot(u_y\wedge u_z\xi_x + u_z\wedge u_x\xi_y + u_x\wedge u_y\xi_z)$$

$$B^n = 2\int u^n\cdot(v_y^n\wedge u_z + u_y\wedge v_z^n)\xi_x + 2\int u^n\cdot(v_z^n\wedge u_x + u_z\wedge v_x^n)\xi_y$$

$$+ 2\int u^n\cdot(v_x^n\wedge u_y + u_x\wedge v_y^n)\xi_z$$

$$C^n = 2\int u^n\cdot(v_y^n\wedge v_z^n\xi_x + v_z^n\wedge v_x^n\xi_y + v_x^n\wedge v_y^n\xi_z)$$

clearly $A^n\to 2\int D(u)\cdot\nabla\xi$ since $u^n\rightharpoonup u$ weak * in L^∞. On the other hand, $B^n\to 0$ since for example $v_y^n\rightharpoonup 0$ weakly in L^2 and $(u^n\wedge u_z)\xi_x$ converges strongly in L^2 to $(u\wedge u_z)\xi_x$ (by dominated convergence). Finally we claim that:

$$|C^n|\leq\int|\nabla v^n|^2.$$

Indeed the integrand may be written as $V^n\cdot\nabla\xi$ where

$$V^n = (u^n\cdot v_y^n\wedge v_z^n, u^n\cdot v_z^n\wedge v_x^n, u^n\cdot v_x^n\wedge v_y^n),$$

and by Lemma 1, we have $|V^n|\leq\frac{1}{2}|\nabla v^n|^2$. We conclude that

$$\underline{\lim}_{n\to\infty}\int|\nabla u^n|^2 + 2\int D(u^n)\cdot\nabla\xi\geq\int|\nabla u|^2 + 2\int D(u)\cdot\nabla\xi.$$

This completes the proof of Theorem 3.

2. Some Properties of Minimizers for the F Problem

We first prove Theorem 4, i.e., every minimizer for the F problem is weakly harmonic.

Let $u\in H_\varphi^1$ and let $\psi\in C_0^\infty(\Omega;\mathbb{R}^3)$. Set $u(\psi) = \frac{u+t\psi}{|u+t\psi|}$. Note that $|u+t\psi| = 1 + 0(t)$ so that for t small enough $u(t)\in H_\varphi^1$. The following lemma will be used in the proof of Theorem 4.

Lemma 2. *We have for $|t|$ small enough*

$$L(u(t)) = L(u).$$

Proof. The conclusion is obvious if u is in R_φ^0 since $u(t)$ and u have the same singularities. In the general case, as in the proof of Theorem 2, there is a sequence $u_n \in R_\varphi^0$ such that $u_n \to u$ in H_φ^1. It is easy to see that $u_n(t) \to u(t)$ in H_φ^1 for every $|t|$ small enough. The conclusion then follows from the continuity of L (Theorem 1).

Proof of Theorem 4. Let u be a minimizer for F in H_φ^1. We have

$$E(u) + 8\pi L(u) \leq E(u(t)) + 8\pi L(u(t)) = E(u(t)) + 8\pi L(u)$$

(by Lemma 2). It follows that $\frac{d}{dt}E(u(t))|_{t=0} = 0$. This implies, by a standard computation, that u satisfies

$$\int \nabla u \cdot \nabla \psi = \int (u \cdot \psi)|\nabla u|^2$$

which says that u is weakly harmonic.

Proof of Theorem 5. First if $\mu_\varphi < \overline{\mu}_\varphi$ it is clear that (P_φ) holds since every smooth map has an energy which is at least $\overline{\mu}_\varphi$.

Conversely, we assume that (P_φ) holds.

Claim: *Let u be any minimizer for μ_φ. Then u cannot be a minimizer for F.*

We postpone the proof of the claim and complete the proof of Theorem 5. Suppose, by contradiction, that $\mu_\varphi = \overline{\mu}_\varphi$. By definition of $\overline{\mu}_\varphi$ there is a sequence $u_n \in H_\varphi^1 \cap C^0$ such that $E(u_n) \to \overline{\mu}_\varphi$. We may also assume that $u_n \rightharpoonup u$ weakly in H^1 and thus u is a minimizer for F, by Corollary 2. Since $\overline{\mu}_\varphi = \mu_\varphi$, u is also a minimizer for E in H_φ^1. This contradicts the claim.

Proof of the claim. Let $\eta(t)|_{t\in[-1,+1]}$ be a smooth family of diffeomorphisms from Ω into itself, satisfying $\eta(0) = \mathrm{Id}_\Omega$ and $\eta(t) = \mathrm{Id}|_{\partial\Omega}$ on $\partial\Omega$ for all $t \in [-1,+1]$. Set $u(t) = u \circ \eta(t)$. Since u is a minimizer for E we have:

$$\frac{d}{dt}E(u(t))|_{t=0} = 0.$$

(Any weakly harmonic map in H_φ^1 satisfying this condition for every $\eta(t)$ is called a stationary map.)

Suppose by contradiction that u is also a minimizer for F. Then we have $\frac{d}{dt}F(u(t))|_{t=0} = 0$, and therefore

$$\frac{d}{dt}L(u(t))|_{t=0} = 0. \tag{18}$$

By assumption (P_φ), u has a finite nonempty set of singularities. Let x_0 be one of the singularities of u. In a minimal connection x_0 is connected to some other singularity x_1 (both have degree ± 1). Let e be the unit vector tangent to x_0 to a geodesic curve connecting x_0 to x_1, pointing towards x_1. Let $X \in C^\infty(\mathbf{R}^3, \mathbf{R}^3)$ be any vector-field with support in a small neighbourhood of x_0, containing no other singularity of u other than x_0, and such that $X(x_0) = e$. Let $\eta(t)$ be the corresponding flow; the map $u(t)$ has the same singularities as u except that x_0 is replaced by $\eta(t)(x_0)$. Therefore $L(u(t)) = L(u) - t + o(t)$ as $t \to 0$. This contradicts (18).

3. The Existence of Infinitely Many (Weakly) Harmonic Maps

The Section is devoted to the proof of Theorem 6. Here we assume that φ is any map in $H^1(\partial\Omega; S^2) \cap C^0(\partial\Omega; S^2)$. Note that the definition of $L(u, v)$ given by (11) makes sense for any $u, v \in H^1_\varphi$ and for any φ (even if $\deg(\varphi; \partial\Omega) \neq 0$).

The proof relies on the following Lemmas:

Lemma 3. *Let u, v in H^1_φ be smooth except at a finite number of singularities, counted according to their multiplicity. Let (p_i) denote the positive singularities of u together with the negative singularities of v. Let (n_i) denote the negative singularities of u together with the positive singularities of v. Then $L(u, v)$ is the minimal connection associated to (p_i, n_i).*

Lemma 4. *Given v in H^1_φ set*

$$G(u) = E(u) + 8\pi L(u, v).$$

Then the conclusions of Theorem 3 and 4 hold, i.e., G is sequentially lower semi-continuous on H^1_φ and every minimizer of G is weakly harmonic.

Proof of Lemma 3. Integrating by parts in (11), we find

$$\begin{aligned} L(u, v) &= \frac{1}{4\pi} \sup_{\substack{\xi:\Omega\to\mathbf{R} \\ \|\nabla\xi\|_\infty \le 1}} \int_\Omega (\operatorname{div} D(u) - \operatorname{div} D(v))\xi \\ &= \sup_{\substack{\xi:\Omega\to\mathbf{R} \\ \|\nabla\xi\|_\infty \le 1}} \Sigma\xi(p_i) - \Sigma\xi(n_i), \end{aligned}$$

which is the length of the desired minimal connection.

Proof of Lemma 4. In order to prove that G is sequentially lower semi-continuous one uses exactly the same argument as in the proof of

Theorem 3. The fact that every minimizer of G is weakly harmonic is proved using the same method as in Theorem 4; it suffices to observe that:

$$|L(u_1,v)-L(u_2,v)|\le L(u_1,u_2)\le C\|\nabla(u_1-u_2)\|_2(\|\nabla u\|_2+\|\nabla v\|_2).$$

Proof of Theorem 6 in the case where $\deg(\varphi;\partial\Omega)=0$ and (P_φ) holds. Set for $\lambda\in(0,1]$ and $u\in H^1_\varphi$

$$F_\lambda(u)=\int_\Omega|\nabla u|^2+8\pi\lambda L(u).$$

The conclusions of Theorem 3, 4 and 5 are valid for the functional F_λ (the proofs are unchanged). For each $\lambda\in(0,1]$ one may minimize F_λ on H^1_φ. It is reasonable to conjecture that for every $\lambda<\lambda_0$ sufficiently small these minimizers are distinct (thus providing a "branch" of harmonic maps near a minimizer for E) and that they are smooth except on a set of low dimension (possibly isolated points). In fact, one can prove that, for some sequence $\lambda_n\to 0$ the minimizers of F_{λ_n} are distinct. Indeed, set

$$A_1=\min\{E(u); u \text{ is a minimizer for } F_1 \text{ in } H^1_\varphi\}.$$

It is clear, by Theorem 3, that this minimum is achieved by some u_1 which is weakly harmonic. We now construct a second weakly harmonic map u_2. By Theorem 5, $A_1>\mu_\varphi$. Let v_1 be a minimizer for μ_φ in H^1_φ. We claim that

$$L(v_1)\le\frac{1}{8\pi}(\mu_\varphi+\overline{\mu}_\varphi).$$

Indeed $E(v_1)+E(w)\ge 8\pi L(v_1)\ \forall w\in H^1_\varphi\cap C^0$ (see (16)) and then the conclusion follows by taking the infimum over w.

Let $0<\lambda_2<1$ be small enough so that

$$\mu_\varphi+\lambda_2(\mu_\varphi+\overline{\mu}_\varphi)<A_1.$$

By construction

$$\begin{aligned}F_{\lambda_2}(v_1)&=\int|\nabla v_1|^2+8\pi\lambda_2L(v_1)=\mu_\varphi+8\pi\lambda_2L(v_1)\\&\le\mu_\varphi+\lambda_2(\mu_\varphi+\overline{\mu}_\varphi)<A_1.\end{aligned}$$

Let u_2 be some minimizer for F_{λ_2} on H^1_φ. Since $F_{\lambda_2}(u_2)\le F_{\lambda_2}(v_1)<A_1$ it follows that $E(u_2)<A_1$ and hence u_2 is not a minimizer for F_1. Also

u_2 is not a minimizer for E (since Theorem 5 holds for F_{λ_2}). Hence u_2 is weakly harmonic and $u_2 \neq u_1$.

We now construct a third weakly harmonic map u_3. Let

$$A_2 = \min\{E(u); u \text{ is a minimizer for } F_{\lambda_2} \text{ on } H^1_\varphi\}.$$

By Theorem 5 (applied to F_{λ_2} instead of F) we have $A_2 > \mu_\varphi$. Let $\lambda_3 < \lambda_2$ be small enough so that

$$\mu_\varphi + \lambda_3(\mu_\varphi + \overline{\mu}_\varphi) < A_2.$$

We have (with the same v_1 as above)

$$F_{\lambda_3}(v_1) < A_2.$$

Let u_3 be some minimizer for F_{λ_3} on H^1_φ. Note that $E(u_3) < A_2 < A_1$. Thus $u_3 \neq u_2$ and $u_3 \neq u_1$ is a third weakly harmonic map. Iterating this construction we find infinitely many weakly harmonic maps.

Proof of Theorem 6 in case $\deg(\varphi; \partial\Omega) \neq 0$.

Case 1: There are infinitely many minimizers for E on H^1_φ. This case is trivial.

Case 2: There are finitely many minimizers $w_1, w_2, \ldots, w_k$. Every map in H^1_φ must have a singularity. We follow the same argument as above except that we replace $L(u)$ by $L(u,v)$ where $v \in R^\infty_\varphi$ is such that $L(w_i; v) \neq 0$, $i = 1, 2, \ldots, k$.

Remark 3. The existence of infinitely many minimizing harmonic maps for some special boundary data φ (axially symmetric) has been recently established by Hardt–Kinderlehrer–Lin [HKL].

Acknowledgements: This research was done during the visits of F. Bethuel and J.M. Coron at Rutgers University. They thank the Mathematics Department for its support and hospitality. We also thank R. Hardt for pointing out an error in an earlier version of this paper.

REFERENCES

[Be1] F. Bethuel, *A characterization of maps in $H^1(B^3, S^2)$ which can be approximated by smooth maps*, to appear.

[Be2] F. Bethuel, *Approximation dans des espaces de Sobolev entre deux variétés et groupes d'homotopie*, C.R. Acad. Sc. Paris **307** (1988),

293–296, and *The approximation problem for Sobolev maps between two manifolds*, to appear.

[BeZ] F. Bethuel and X. Zheng, *Density of smooth functions between two manifolds in Sobolev spaces*, J. Funct. Anal. **80** (1988), 60–75.

[B1] H. Brezis, *Liquid crystals and energy estimates for S^2-valued maps*, in *Theory and Applications of Liquid Crystals*, IMA Vol. **5** (J.L. Ericksen and D. Kinderlehrer eds.), Springer (1987).

[B2] H. Brezis, *S^k-valued maps with singularities*, in *Topics in the Calculus of Variations*, (M. Giaquinta ed.). Lecture Notes in Math. **1365**, Springer (1989), 1–30.

[BC] H. Brezis and J.M. Coron, *Large solutions for harmonic maps in two dimensions*, Comm. Math. Phys. **92** (1983), 203–215.

[BCL] H. Brezis, J.M. Coron, and E. Lieb, *Harmonic maps with defects*, Comm. Math. Phys. **107** (1986), 649–705.

[HKL] R. Hardt, D. Kinderlehrer, and F.H. Lin, *The variety of configurations of static liquid crystals*, to appear.

[HL] R. Hardt and F.H. Lin, *A remark on H^1 mappings*, Manuscripta Math. **56** (1986), 1–10.

[KW] H. Karcher and J.C. Wood, *Non-existence results and growth properties for harmonic maps and forms*, J. Reine Angew. Math. **353** (1984), 165–180.

[SU1] R. Schoen and K. Uhlenbeck, *A regularity theory for harmonic maps*, J. Diff. Geom. **17** (1983), 307–335.

[SU2] R. Schoen and K. Uhlenbeck, *Boundary regularity and the Dirichlet problem for harmonic maps*, J. Diff. Geom. **18** (1983), 253–268.

[W] J.C. Wood, *Non-existence of solutions to certain Dirichlet problems for harmonic maps*, preprint Leeds Univ. (1981).

Topological results on Fredholm maps and application to a superlinear differential equation

VITTORIO CAFAGNA

Introduction

Consider the equation $F(u) = h$, where $F : X \to Y$ is a smooth map between Banach spaces. By the Inverse Image Theorem regular solutions are isolated, so that only singular solutions may accumulate. Call $S = \{u \in X : F'(u) \text{ is not surjective}\}$ the singular set of F; then the problem of analyzing the structure of the solution set consists in trying to give a description of $F^{-1}(h) \cap S$. The point of view we adopt in this paper is to consider real-analytic Fredholm maps of index 0. In this case $F^{-1}(h) \cap S$ is a real-analytic set A such that $dim_u A = dim Ker F'(u)$. This implies in particular that pure one-dimensional components of A are real-analytic curves, whose tangent line at every point u is exactly $Ker F'(u)$. If moreover F is proper then these components are circles. One of the first questions one can raise is whether there are general conditions under which even singular solutions are isolated. Or, in other words, are there detectable obstructions to the accumulation of solutions ? It turns out that there actually is a variety of situations in which the solution set cannot contain circles. In this paper we give an exemple of such a situation: the two-point boundary value problem

$$(P) \qquad \begin{cases} u'' + u^3 = h & \text{in} \quad (0,1) \\ u(0) = u(1) = 0. \end{cases}$$

We prove that (despite the fact that the singular set of the map $F(u) = u'' + u^3$ is a very complicated object) the solutions are isolated for every right-hand side h. The idea of the proof is the following: first of all one remarks that $dim Ker F'(u) = 1$ for every singularity u of F, so that nonisolated solutions come in real-analytic curves. Then one proves that these curves are actually circles. Observe that F is not a proper map, so that *a priori* one could not say that they are circles, but there is a way to overcome this difficulty. In fact problem (P) has a variational structure: solutions are in 1-1 correspondence with critical points of a real-analytic Fredholm functional $J : H \to \mathbb{R}$, which, as it is well-known, satisfies the Palais-Smale assumption. Therefore $F^{-1}(h)$ can be seen as the union of compact slices $J^{-1}(c) \cap F^{-1}(h)$, where c ranges over the critical values of J. At this point one can use a general Morse-like result for real-analytic Fredholm functionals (Theorem 2. See also [6]), which ensures that the set of critical values is discrete, to deduce that $F^{-1}(h) \cap S$ is a priori the union of countably many circles. Then one concludes by showing that solutions cannot come in circles. This is due to the fact that the line $KerF'(u)$, winding up around a circle of solutions, must become parallel to every hyperplane in the space. On the other hand one can prove that there are forbidden directions, which the line $KerF'(u)$ cannot take.

This procedure of showing first, that nonisolated solutions come in circles, second, that circles cannot exist in the set of solutions seems to have a wide range of applicability. We refer to the paper [3], where this point of view is applied to a number of different problems.

Acknowledgement. The author wishes to thank J.J.Risler for useful discussions about the structure of real-analytic sets.

1. Two structure theorems for real-analytic maps

Let X and Y be Banach spaces and $F : X \to Y$ be a C^ω (real-analytic) Fredholm map of index 0. We say that u is a singularity of F if $dimKerF'(u) \geq 1$. Otherwise u is called a regular point. The set of singularities is denoted by S. Points of $F(S)$ are called singular values of F, while $Y \backslash F(S)$ is the set of regular values of F. The local behaviour of the map at a singularity $u \in S$ is completely described by the following local form theorem (note that the trivial case $n = 0$ is just the Inverse Function Theorem, so local behaviour of F at regular points is also included):

C^ω-Local Representative Theorem. *Let X and Y be Banach spaces, $F : X \to Y$ be a C^ω Fredholm map of index 0, and $u \in S$ be such that $dimKerF'(u) = codimImF'(u) = n$. Then there exist a Banach space Z, neighborhoods U of u in X, V of $F(u)$ in Y and C^ω- diffeomorphisms $\alpha : U \times \mathbb{R}^n \to Z$, $\beta : V \times \mathbb{R}^n \to Z$ such that $\tilde{F} = (\beta \circ F \circ \alpha^{-1})(t_1, ..., t_n, z) = (\tilde{f}(t_1, ..., t_n, z), z)$, where the local representative of F, $\tilde{F} : \mathbb{R}^n \times Z \to \mathbb{R}^n \times Z$ is a C^ω map such that $\frac{\partial \tilde{f}}{\partial t_i}(0, ..., 0, 0) = 0$, $\forall i = 1, ..., n$.*

A proof of the above theorem can be easily obtained by combining the standard smooth version of the Local Representative Theorem (see e.g. [1]) with the C^ω version of the Inverse Function Theorem (see e.g. [5]). Let us now give a first application of the Local Representative Theorem to the analysis of the structure of the set of solutions to a real-analytic Fredholm map.

Theorem 1. *Let X and Y be Banach spaces, $F : X \to Y$ be a C^ω Fredholm map of index 0 and h a point of Y. Call $A = A(h)$ the set of accumulation points of $F^{-1}(h)$ and let C be a component of A such that $dim Ker F'(u) = 1, \quad \forall u \in C$. Then C is a real-analytic 1-manifold.*

Proof ([3]). Let C be as above and $u \in C$. By the Local Representative Theorem we know that F near u looks like a C^ω map $\tilde{F} : \mathbb{R} \times Z \to \mathbb{R} \times Z$ of the kind $\tilde{F}(t, z) = (\tilde{f}(t, z), z)$. Then $F^{-1}(h)$ is near u a C^ω-diffeomorphic image of the set $\tilde{F}^{-1}(0) = \{(t, 0) \in \mathbb{R} \times Z : \tilde{f}(t, 0) = 0\}$. Being u an accumulation point of $F^{-1}(h)$, 0 is an accumulation point of the set of zeroes of the C^ω function $\tilde{f}(\cdot\, , 0) : \mathbb{R} \to \mathbb{R}$. Therefore $\tilde{f}(\cdot\, , 0)$ is identically zero, $\tilde{F}^{-1}(0) = \mathbb{R} \times \{0\}$ and $F^{-1}(h)$ is locally a C^ω-diffeomorphic image of the real line.

Observe that the tangent line to C at u is nothing but the kernel of the Fréchet derivative of F at u:

$$T_u C = Ker F'(u).$$

Let now H be a Hilbert space and $J : H \to \mathbb{R}$ a C^ω Fredholm functional of index 0 (recall that $J : H \to \mathbb{R}$ is said to be a Fredholm functional of index n if $J' : H \to H'$ is a Fredholm map of index n). We say that $u \in H$ is a *critical point* of J if $J'(u) = 0$, otherwise u is said to be *regular*. The set of critical points is denoted by C. A value $c \in \mathbb{R}$ is said to be *critical* if $c \in J(C)$, otherwise it is called a *regular value*. We say that a critical value $c \in \mathbb{R}$ is *Palais-Smale* (shortly *PS*) if every sequence $\{u_n\} \subset H$ of critical points s.t. $J(u_n) \to c$ has a converging subsequence. We are now in a position to state a second structure theorem, namely a Morse-like result on the non-accumulation of critical values.

Theorem 2. *Let H be a Hilbert space and $J : H \to \mathbb{R}$ be a C^ω Fredholm functional of index 0. Then PS critical values are isolated in the set of all critical values.*

Proof. By contradiction. Suppose that the PS value c is an accumulation point of critical values of J. Then there exists a sequence $\{u_n\} \subset H$

such that

$$J(u_n) \to c, \quad J'(u_n) = 0.$$

We claim that J is eventually constant on $\{u_n\}$, what is in contrast with being c an accumulation point of critical values. In fact, thanks to the PS assumption, (a subsequence of) $\{u_n\}$ converges to a point u_0. By the Local Representative Theorem we can reduce the study of the critical set of J in a neighborhood of u_0 to that of the critical set of a C^ω functional $\tilde{J} : \mathbb{R}^n \to \mathbb{R}$ in a neighborhood of 0 (here $n = dim Ker J''(u_0)$). Therefore it suffices to prove that if $\{x_n\} \subset \mathbb{R}^n$ is a sequence of critical points of $\tilde{J}$ s.t. $x_n \to 0$, then $\tilde{J}$ is eventually constant on $\{x_n\}$. This is an easy consequence of the analysis of the local structure of real-analytic sets. In fact one can prove (see e.g. [N]) that there is an $n_0 \in \mathbb{N}$, a neighborhood U of 0 in $\mathbb{R}^n$ and a C^ω submanifold V of the critical set C of $\tilde{J}$ s.t. $\forall n \geq n_0$, $x_n \in V \cap U$. Because $\tilde{J}$ is obviously constant on V this proves the claim.

2. An application

We will describe a result on the global structure of the solutions to a simple superlinear differential equation, in which both theorems of Section 1 are applied.

Consider the two-point boundary value problem

$$(P) \qquad \begin{cases} u'' + u^3 = h & \text{in} \quad (0,1) \\ u(0) = u(1) = 0. \end{cases}$$

It is well-known ([4]) that problem (P) has infinitely many solutions for every right-hand side h. We will prove that the solutions are actually isolated. Let $F : C_0^2([0,1]) \to C^0([0,1])$ be the map defined by $F(u) = u'' + u^3$. It is easy to see that F is a C^ω Fredholm map of index 0 s.t. $dim Ker F'(u) = 1, \quad \forall u \in S$ (see e.g. ([3]). Consider also the functional $J_h : H_0^1(0,1) \to \mathbb{R}$ defined by

$$J_h(u) = \int_0^1 (-\frac{1}{2}u'^2 + \frac{1}{4}u^4 - hu)dx.$$

Then solutions to problem (P) can be seen either as points in the preimage $F^{-1}(h)$ or as critical points of the C^ω Fredholm functional J_h. Both points of view are combined to yield a proof of the following

Theorem 3. $\forall h \in C^0([0,1])$ *the solutions to problem* (P) *form an unbounded sequence in* $C_0^2([0,1])$.

Proof. Let us begin by proving that the set of solutions is discrete. When combined with the known infinity results ([4]), this proves that the solutions form an infinite sequence. Unboundedness is discussed later on. Observe first of all that J_h satisfies the standard PS condition (see e.g. [3], Appendix), which in particular implies that every critical value is PS. Therefore Theorem 2 can be applied to deduce that the critical values of J_h are isolated. It suffices now to show that the critical slices $C_c = \{u \in H_0^1(0,1) : J_h'(u) = 0 \quad \text{and} \quad J_h(u) = c\}$ are finite sets, to obtain that the full critical set C is discrete. Again by the PS assumption we know that C_c's are compact. Therefore every C_c is in 1-1 correspondence with the set $F^{-1}(h) \cap L$, where L is a compact subset of $C_0^2([0,1])$. Suppose by absurd that $F^{-1}(h) \cap L$ contains an accumulation point u_0. As we already observed, $dim Ker F'(u_0) = 1$ so that Theorem 1 can be applied to show that $F^{-1}(h) \cap L$ contains a real-analytic 1-manifold M through u_0. Moreover, M, being a closed subset of a compact set, is itself compact, i.e., by the classification of compact 1-manifolds ([7]), a circle. We claim that this is actually impossible: solutions to problem (P) cannot come in circles. To see this, define a linear functional $I : C_0^2([0,1]) \to \mathbb{R}$ by $I(u) = u'(0)$ and consider the family I_t of affine hyperplanes defined by $I_t = \{u \in C_0^2([0,1]) : I(u) = u'(0) = t\}$. Simple geometrical considerations show that any circle in the space must be tangent to some hyperplane of the family. If we apply this observation to the M above, we obtain that there is a point $u \in M$ and a $t \in \mathbb{R}$ such that

$$KerF'(u) = T_u M < T_u I_t = I_0$$

i.e. there is a nonzero function $v \in C_0^2([0,1])$ such that sv satisfies the following Cauchy problem $\forall s \in \mathbb{R}$:

$$\begin{cases} sv'' + 3u^2(sv) = 0 \\ sv'(0) = 0 \\ sv(0) = 0 \end{cases}$$

which is manifestly absurd. It only remains to show that the sequences of solutions are unbounded. Suppose, by absurd again, that $F^{-1}(h)$ were contained in a ball. Then, by standard machinery, it would be compact and then contain an accumulation point (remember that it is an infinite set) and therefore, by the argument above, a circle. But this, as we already know, is impossible.

Remarks

1. Needless to say, Theorem 3 holds for more general real-analytic nonlinearities (see [3]).

2. One could raise the question if such a result generalizes to higher dimensions. Consider for exemple the problem

$$(P_1) \qquad \begin{cases} \Delta u + u^3 = h \text{ in } \Omega \\ u = 0 \text{ on } \partial\Omega \end{cases}$$

where $\Omega \in \mathbb{R}^2$ is a bounded set with smooth boundary. It is still true ([2]) that for every h there is an infinity of solutions. Define, as above, a C^ω Fredholm map $F : C_0^{2,\alpha}([\Omega]) \to C^{0,\alpha}([\Omega])$ by $F(u) = \Delta u + u^3$ and a C^ω Fredholm functional $J_h : H_0^1([\Omega]) \to \mathbb{R}$ by

$$J_h(u) = \int_\Omega (-\frac{1}{2} \parallel \nabla u \parallel^2 + \frac{1}{4} u^4 - hu) dx.$$

J_h still verifies the PS assumption, so that Theorem 2 is still applicable to prove that critical values cannot accumulate. On the other hand, it is obviously no longer true that $dim Ker F'(u) = 1,\ \forall u \in S$. In fact one can expect kernels of arbitrary large dimensions, so that the argument in Theorem 3, which relies heavily on the one-dimensionality of the set of nonisolated solutions is not applicable: *a priori* solutions might accumulate. But still something can be said. In fact nonisolated solutions form real-analytic sets, on which J_h is constant. Therefore the set of solutions is made up of a countable collection of compact real-analytic sets, with no *a priori* bound on the dimensions. It is the author's belief that, even if there is nothing against the existence of such agglomerates, they should be very instable objects, that are destroyed by small perturbations of the significant parameters of the problem. One could for exemple conjecture that an analogous of Theorem 3 for problem $(P1)$ is true for *generic* (in some suitable sense) domains.

REFERENCES

[1] R.Abraham and J.Robbin, *Transversal Mappings and Flows*, New York-Amsterdam, 1967, Benjamin.
[2] A.Bahri and P.L.Lions, Morse index of some min-max critical points. I. Application to multiplicity results, *preprint*.
[3] V.Cafagna and F.Donati, On the finite solvability of some nonlinear differential problems, *preprint*.
[4] H.Ehrmann, Über die Existenz der Lösungen von Randwertaufgaben bei gewönlicher nichtlineare Differentialgleichungen zweiter Ordnung, *Math. Ann*., **134** (1957), 167-194.
[5] S.Fučik, J.Nečas, J.Souček and V.Souček, *Spectral Analysis of Nonlinear Operators*, Berlin-Heidelberg-New York, 1973, Springer.
[6] S.Fučik, J.Nečas, J.Souček and V.Souček, New infinite dimensional versions of Morse-Sard theorem, *Boll. U.M.I.*, (4) **6** (1972), 317-322.
[7] J.W.Milnor, *Topology from the Differentiable Viewpoint*, Charlottesville, 1965, The University Press of Virginia.
[8] R.Narasimhan, *Introduction to the Theory of Analytic Spaces*, Berlin-Heidelberg-New York, 1966, Springer.

Vittorio Cafagna

Département de Mathématique et Informatique
Université Paris Nord
93430 Villetaneuse FRANCE
and
Istituto Matematico G.Castelnuovo
Università di Roma "La Sapienza"
00100 Roma ITALIA

Existence Results for some Quasilinear Elliptic Equations

Henrik Egnell

0. INTRODUCTION. In this paper we shall establish some existence results for the boundary value problem:

$$(0.1)\qquad \begin{cases} \Delta_m u + f(x,u) = 0 & \text{in } \Omega, \\ u > 0 & \text{in } \Omega, \\ u \in \mathcal{D}_0^{1,m}(\Omega) \end{cases}$$

Here Ω is an open connected set in $\mathbf{R}^n$ with smooth boundary and Δ_m is the m-Laplace operator defined by $\Delta_m u = \nabla \cdot (|\nabla u|^{m-2}\nabla u)$, where $1 < m < \infty$. Since we are considering positive solutions we will always assume that $f(x,s) = 0$ if $s \leq 0$. Furthermore, we will only consider the "super linear" case, that is when $f(x,s)$ grows faster than s^{m-1} in the s variable.

Throughout this paper we will assume that
(i) $f(x,s) \geq 0$, *for all* $x \in \Omega$ *and* $s \geq 0$.
(ii) *The function* $s \to f(x,s)$ *is continuous for a.e.* $x \in \Omega$.

The space of functions $\mathcal{D}_0^{1,m}(\Omega)$ is defined as the completion of $C_0^\infty(\Omega)$ in the norm $\|\nabla \cdot\|_m$ (here $\|\cdot\|_m$ is the standard norm in $L^m(\Omega)$). The resulting space is a Banach space, unless $\Omega = \mathbf{R}^n$ and $n \leq m$. In the exceptional case the space is not Hausdorff since it contains the constant functions.

One of the reasons to include the general case when $m \neq 2$ is that equation (0.1) has different properties in the cases when $m < n$, $m > n$ and the limiting case when $m = n$. If $m = 2$, then one case is just the one dimensional equation. The fact that the one dimensional case is different is not very surprising.

Let us mention that the methods in this paper also works in more general situations, when the m-Laplace operator is replaced by a more general operator of divergence form.

We shall also consider radial solutions of (0.1). In this case we replace $\mathcal{D}_0^{1,m}(\Omega)$ by $\mathcal{R}_0^{1,m}(\Omega) = \{u \in \mathcal{D}_0^{1,m}(\Omega) : u \text{ is radial }\}$.

We say that u solves (0.1) if u is positive in Ω, $u \in \mathcal{D}_0^{1,m}(\Omega)$ and satisfies

$$\int |\nabla u|^{m-2}\nabla u \cdot \nabla\varphi \, dx - \int f(x,u)\varphi \, dx = 0 \, ,$$

for all $\varphi \in \mathcal{D}_0^{1,m}(\Omega)$.

Let us remark that if $m \neq 2$, then the solutions are not in C^2 in general. In fact if u is a bounded solution and f is smooth then u is $C^{1+\alpha}$ and smooth at the points where $|\nabla u| \neq 0$ (see [TO]).

The boundary condition is $u = 0$ on $\partial\Omega$. However, if Ω is unbounded, we do not assume that the solution tends to zero as $x \to \infty$. Instead, we have an integrability or finite energy condition. That is we assume that $u \in \mathcal{D}_0^{1,m}(\Omega)$. If $n > m$ then radial solutions of (0.1) do tend to zero as $x \to \infty$. In many cases this holds also for nonradial solutions. On the other hand if $n \leq m$ and $\partial\Omega \neq \emptyset$ the condition means that the solution does not grow to fast at infinity.

If $n \leq m$ we can argue as in [NIS], to see that radial solutions of (0.1) can not tend to zero, as r tends to infinity. Assume that $u'(r_0) < 0$, then for $r > r_0$ we have $(|u'|^{m-2}u' r^{n-1})' \leq 0$. Thus integrating the inequality we find that $u'(r) \leq -Cr^{-\frac{n-1}{m-1}}$. Hence $u(r) \to -\infty$ as $r \to \infty$. This also shows that if f is smooth, then we can have no radial solutions of (0.1) if $\Omega = \mathbf{R}^n$ and $n \leq m$.

If $n \leq m = 2$, f is smooth and $\Omega = \mathbf{R}^n$, then the same argument as above can be used to show that we have no C^2 solutions in the nonradial case. To see this let u be a solution of (0.1) and let $\overline{u}(r)$ be the average of u over spheres with radius r and centered at the origin. Taking the average of the equation we find that: $\Delta\overline{u} + \overline{f(x,u)} = 0$. Now the same argument as above shows that this is not possible if $n \leq 2$.

Problems of the same type as (0.1) has been studied in many papers. Let us mention some of them: [BC1&2], [BL], [BLP], [BM], [KN], [NI1&2], [NIS], [NOS], [ON], [ST], [TOL], [TH1&2]. In some papers it is assumed that $f(x,s) < 0$ for small positive s.

In this paper we shall study equation (0.1) when f is singular in the space variable or has a certain asymptotic behaviour at infinity in the space variable. The cases when Ω is bounded and unbounded will be treated separately in Sections 1 and 2. We emphasize the case when Ω is unbounded. Typical examples are when $f(x,s) = b(x)s^p$ and Ω is a bounded set, the complement of a bounded set or $\mathbf{R}^n$.

The main tool used to prove the existence of solutions is the mountain pass lemma [AR].

MOUNTAIN PASS LEMMA. *Let X be a Banach space and let $\mathcal{F}$ be a real valued function in $C^1(X,\mathbf{R})$ satisfying the Palais-Smale condition (PS). Assume that there exist an open neighbourhood U of 0 and a point $u_0 \notin \overline{U}$ such that*

$$\max(\mathcal{F}(0), \mathcal{F}(u_0)) < c_0 = \inf_{u \in \partial U} \mathcal{F}(u).$$

Then the number $c = \inf_{P \in \mathcal{P}} \max_{u \in P} \mathcal{F}(u) \geq c_0$, is a critical value. Here $\mathcal{P}$ represents the set of all continuous paths from 0 to u_0.

In this paper we will use the functional
$\mathcal{F}(u) = \frac{1}{m}\int_\Omega |\nabla u|^m\,dx - \int_\Omega F(x,u)\,dx$, where $F(x,s) = \int_0^s f(x,t)\,dt$.

The Palais-Smale condition (PS) is as follows:

(PS). *Let $\mathcal{F} \in C^1(X, \mathbf{R})$. Any sequence $\{u_j\} \subset X$ such that $|\mathcal{F}(u_j)|$ is uniformly bounded and $\mathcal{F}'(u_j) \to 0$ in X^*, as $j \to \infty$, contains a subsequence which converges in X. Here X^* denotes the dual of X with the strong topology (i.e. uniform convergence on bounded sets).*

The equations studied in this paper are all simple in the sense that the Palais – Smale condition follows from the compactness of imbeddings between the appropriate function spaces. Thus we will not consider the difficult borderline case where we only have continuous imbeddings.

In Section 2 we also give a simple device to get improved asymptotic estimates of the solutions at infinity.

In the appendix we give some nonexistence results that show that the assumptions in the existence results in Sections 1 and 2 are needed. The main tool here is a generalized Pohozaev identity. For other nonexistence results see for example [TOL], [KN] or [NIS].

When finishing the manuskript I have learned that the ideas and methods used in this paper has also been used by Li and Ni in their work [LN1&2]. They also obtained results on solutions with infinite energy. Since we focus on the variational approach we will not get any results of that type.

Finally I would like to mention that asymptotic results for quasilinear equations has recently been obtained in [BI] and [PS2] (personal communication with Professor Serrin).

1. BOUNDED DOMAINS. In this section we give some remarks on how variational methods can be used to prove existence result when we have singular coefficients.

THEOREM 1. *Assume that $n > m$, Ω is a bounded domain and that*

(i) $f(x,s) \leq b(x)(|s|^p + 1)$, *where* $b \in \cap_{r<q} L^r(\Omega)$ *and* $m < p+1 < \frac{mn(q-1)}{(n-m)q}$.

(ii) *There exists a* $g \in \cup_{r>\frac{n}{m}} L^r(\Omega)$ *such that* $f(x,s)/g(x) = o(|s|^{m-1})$ *uniformly in* x, *as* $s \to 0$.

(iii) *There exists a function* $\varphi \in \mathcal{D}_0^{1,m}(\Omega)$ *such that*

$$\lim_{t\to\infty} t^{1-m} \int f(x, t\varphi)\varphi \, dx = \infty .$$

(iv) *There exist constants* $\theta \in (0, 1/m)$ *and* $M < \infty$ *such that*

$$\int_0^t f(x,s)\, ds \leq \theta f(x,t)t , \quad \textit{holds for } M \leq t \textit{ and } x \in \Omega .$$

Then (0.1) *has a solution.*

Remark:(a) If we add the term $a(x)u^{m-1}$ to the left side in the equation in (0.1). Then the theorem holds provided $a \in L^s(\Omega)$ for some $s > n/m$ and there exists a positive δ such that

$$\int (|\nabla\varphi|^m - a(x)\varphi^m)\,dx \geq \delta \int \varphi^m\,dx\,, \quad \text{for all } \varphi \in \mathcal{D}_0^{1,m}(\Omega)\,.$$

(b) In (i) we can have $f(x,s) \leq \sum_{i=1}^N b_i(x)(|s|^{p_i}+1)$, where each couple b_i, p_i satisfies the hypothesis given in (i).

Theorem 1 is a direct consequence of the mountain pass lemma and the following elementary compactness result.

LEMMA 2. *Let Ω be bounded and $n > m$. If $b \in \bigcap_{r<q} L^r(\Omega)$ and $1 \leq p+1 < \frac{mn(q-1)}{(n-m)q}$, then $\mathcal{D}_0^{1,m}(\Omega) \hookrightarrow L^{p+1}(\Omega, b(x)\,dx)$ is compact.*

PROOF: This follows immediately since

$$\mathcal{D}_0^{1,m}(\Omega) \hookrightarrow L^{mn\delta/(n-m)}(\Omega) \hookrightarrow L^{p+1}(\Omega, b\,dx),$$

where the first imbedding is compact and the second is continuous if we choose $\delta < 1$ near one. □

Remark: If $b(x) = |x|^\nu$, $0 \in \Omega$ and $-m \leq \nu \leq 0$. Then $b \in \bigcap_{r<n/|\nu|} L^r(\Omega)$, but the imbedding $\mathcal{D}_0^{1,m}(\Omega) \hookrightarrow L^{p+1}(\Omega, |x|^\nu\,dx)$ is not compact (but continuous) if $p+1 = \frac{m(n+\nu)}{n-m}$. This follows by a simple dilation argument.

Since the proof of Theorem 1 is just a standard application of the mountain pass lemma, it is omitted.

The case when $n < m$ is simple since $\mathcal{D}_0^{1,m}(\Omega)$ is continuously imbedded in a space of Hölder continuous functions. It is easy to see that the theorem holds if (i) and (ii) are replaced by

(i') $\sup_{t<s} f(\cdot,t) \in L^1(\Omega)$, for each positive s.

(ii') There exists a function $g \in L^1(\Omega)$ such that $\sup_{t<s} \frac{f(x,t)}{g(x)} = o(s^{m-1})$ uniformly in x, as $s \to 0$.

In fact this follows as a special case from Theorem 7 in the present paper.

If $n = m$ we can use the Moser – Trudinger inequality and compactness results for Orlicz – Sobolev spaces (see [AD] Chapter 8), to show that Theorem 1 holds if (i) is replaced by

(i") There exists a function $b \in \bigcup_{r>1} L^r(\Omega)$ and constants $\alpha \in \mathbf{R}$ and $p < \frac{n}{n-1}$ such that $f(x,s) \leq b(x)\exp(\alpha|s|^p)$.

THEOREM 1'. *If $n < m$ and (i) and (ii) are replaced by (i') and (ii') or if $n = m$ and (i) is replaced by (i"), then Theorem 1 holds.*

Example 1. Let us consider a typical example where $f(x,s) = b(x)s^p$, $0 \in \Omega$ and $b(x)$ is bounded if x is bounded away from the origin and $b(x) = O(|x|^\nu)$, as $x \to 0$.

If $n > m$ and $\nu \leq 0$, then Theorem 1 shows that there exists a solution if $m < p+1 < \frac{m(n+\nu)}{n-m}$. On the other hand if $b(x) = |x|^\nu$, $\nu > -n$, $p+1 \geq \frac{m(n+\nu)}{n-m}$ and Ω is starshaped with respect to the origin, then there exist no solutions of (0.1) that are locally bounded in $\overline{\Omega} \setminus \{0\}$. This follows immediately from the Pohozaev identity given in the appendix. Note that if $p+1 \leq \frac{nm}{n-m}$ and b is locally bounded in $\overline{\Omega} \setminus \{0\}$, then all solutions of (0.1) are locally bounded in $\overline{\Omega} \setminus \{0\}$.

If $n \leq m$ and f is as above, then there exists a solution for all $p \in]m-1, \infty[$, provided $b \in L^1(\Omega)$ and $n < m$ or $b \in L^r(\Omega)$ for some $r > 1$ and $n = m$. Clearly if (0.1) has a solution then b must be locally integrable.

Let us remark that if f is as in Example 1 above and $p+1 \neq m$ and $1 < p+1 < \frac{m(n+\nu)}{n-m}$, then there exists a solution. A solution can be obtained by minimizing the functional (Rayleigh quotient) $\|\nabla u\|_m^m / \|u\|_{p+1,b}^m$ over $H_0^{1,m}(\Omega)$.

If $p+1 = m$ and $\nu > -m$, then there exists a unique λ such that there is a positive solution of

$$\begin{cases} \Delta_m u + \lambda b u^{m-1} = 0 & \text{in } \Omega \\ u = 0 & \text{on } \partial\Omega\,. \end{cases}$$

However, in this paper we are only interested in the "super linear" case, that is when $p+1 > m$. For more general results for "sub linear" problems see [TH2].

If we have a singularity at only one point in the space variable, say the origin, then Theorem 11 and the theory of Orlicz spaces (cf [AD] Chapter 8) yield a slightly stronger result.

THEOREM 3. *Let $n > m$ and $\Omega = B_1(0)$, and assume that $f(x,s)$ is radial in the space variables and that the assumptions (ii) – (iv) in Theorem 1 hold. Furthermore, assume that $f(x,s) \leq b(x)g(s)$, where b is locally bounded in $\overline{\Omega} \setminus \{0\}$,*

$$\limsup_{x \to 0} \frac{b(x)}{|x|^\nu} = C_1 < \infty\,, \qquad \limsup_{s \to \infty} \frac{g(s)}{s^p} = C_2 < \infty\,,$$

with $p+1 = \frac{m(n+\nu)}{n-m} > m$, and at least one of C_1 and C_2 zero.
Then (0.1) has a radial solution.
If $\nu \leq 0$, then the same result holds for the nonradial problem, that is when Ω is any bounded domain and f is not required to be radial in the space variables.

In Example 1 above we pointed out that if Ω is starshaped w.r.t. the origin and $f(x,s) = |x|^\nu s^p$, where $p+1 = \frac{m(n+\nu)}{n-m}$, then there exist no solutions of (0.1) that are locally bounded in $\overline{\Omega} \setminus \{0\}$. This shows that the growth condition in Theorem 3 is best possible.

A slightly weaker form of Theorem 3 can be found in [NI2] and [TH1], (see also [BM]).

2. UNBOUNDED DOMAINS. First we specialize to the radial case. We will use the following notation $\tilde{f}(x,s) = \sup_{t \leq s} f(x,t)$.

THEOREM 4. *Assume that $f(x,s)$ is radial in the x variable, $\Omega = \mathbf{R}^n \setminus \overline{B}_1(0)$ or $\Omega = B_1(0)$. If $n > m$ we can also take $\Omega = \mathbf{R}^n$. Put*

$$\omega(x) = \begin{cases} |x|^{(m-n)/m} & \text{if } m \neq n\,, \\ |\log|x|| & \text{if } m = n\,. \end{cases}$$

Furthermore, assume that

(i) *$\tilde{f}(x, c\omega(x))\omega(x) \in L^1(\Omega)$ for each positive c, and*

$$\int_\Omega \tilde{f}(x, \varepsilon\omega(x))\varepsilon\omega(x)\,dx = o(\varepsilon^m)\,, \quad \text{as } \varepsilon \to 0\,.$$

(ii) *There exists a radial function $\varphi \in \mathcal{D}_0^{1,m}(\Omega)$ such that*

$$t^{1-m} \int f(x, t\varphi)\varphi\,dx \to \infty\,, \quad \text{as } t \to \infty\,.$$

(iii) *There exist constants $\theta \in (0, 1/m)$ and $M < \infty$ such that*

$$\int_0^t f(x, s\omega(x))\,ds \leq \theta f(x, t\omega(x))t\,, \quad \text{holds for all } t \geq M \text{ and } x \in \Omega\,.$$

Then (0.1) has a radial solution such that

$$u(x) = \begin{cases} o(\omega(x)) & \text{if } n > m\,, \\ O(\omega(x)) & \text{if } n \leq m\,, \end{cases} \quad \text{as } |x| \to \infty\,.$$

Before we prove the theorem let us give some examples.

Example 2: Let $f(x,s) = b(x)s^p$, where $p > m-1$ and b is an integrable radial nonnegative function.

If we apply Theorem 4 we find that (0.1) has a solution if

$$\int_\Omega b(x)|x|^{(m-n)(p+1)/m}\,dx < \infty\,. \tag{2.1}$$

If we also assume that $b(x) = O(|x|^{\nu_1})$, as $x \to 0$ and $b(x) = O(|x|^{\nu_2})$, as $x \to \infty$, we immediately get the following results.

If $\Omega = B_1(0)$ and $n > m$ then there exists a radial solution if $p + 1 < \frac{m(n+\nu_1)}{n-m}$. When $\nu_1 \geq 0$ this gives us the results in [NI2] and [TH1].

Assume that $\Omega = \mathbf{R}^n$, $n > m$ and that $\frac{m(n+\nu_2)}{n-m} < p + 1 < \frac{m(n+\nu_1)}{n-m}$. Then there exists a radial solution.
If $m = 2$, $-2 < \nu_2 < 0$ and $\nu_1 = 0$ this is the result in [NOS]. For another related result see [KN].

Let $\Omega = \mathbf{R}^n \setminus \overline{B}_1(0)$ and assume that one of the following conditions holds
(I) $n > m$ and $\frac{m(n+\nu_2)}{n-m} < p + 1$,
(II) $n = m$ and $\nu_2 < -n$,
(III) $n < m$ and $p + 1 < -\frac{m(n+\nu_2)}{m-n}$.
Then there exists a radial solution.

If we take $b(x) = \min(|x|^{\nu_1}, |x|^{\nu_2})$, where $\nu_2 < \nu_1$, then the Pohozaev identity given in the appendix shows that the results above are best possible.

Let us point out that the integrability condition (2.1) above is more general than the ones given in [KN], [NI2], [NOS], [TH2]. This is easily seen by choosing b to be zero on large sets near infinity or near the origin and large otherwise.

For the proof of Theorem 4 we need the following simple apriori estimate, which has been used by several authors (see for example [BL], [NI2], [TH1] or [NOS]).

LEMMA 5. *Let $u \in \mathcal{D}_0^{1,m}(\Omega)$ be radial and ω be as defined in Theorem 4. If $n \leq m$ we also assume that $0 \notin \overline{\Omega}$. Then $|u(x)| \leq C\|\nabla u\|_m \omega(x)$ for all $x \in \Omega$.*

Remark: If $n > m$, then it follows from the proof that $u(x) = o(|x|^{(m-n)/m})$ as $|x| \to \infty$.

PROOF: If $n > m$ and $\varphi \in C_0^1(\Omega)$ is radial then

$$|\varphi(r)| \leq \int_r^\infty |\varphi'(s)|\, ds \leq$$

$$\leq \left(\int_r^\infty |\varphi'(s)|^m s^{n-1}\, ds\right)^{1/m} \left(\int_r^\infty s^{-(n-1)m'/m}\, ds\right)^{1/m'} \leq$$

$$\leq C\|\nabla\varphi\|_m\, r^{(m-n)/m}.$$

The calculations when $n \leq m$ is similar but we start with $\varphi(r) = \int_0^r \varphi'\, ds$.
□

PROOF OF THEOREM 4: Define the function $F(x,s) = \int_0^s f(x,t)\, dt$ and the functional $\mathcal{F}(u) = \int(\frac{1}{m}|\nabla u|^m - F(x,u))\, dx$. Recall that $\mathcal{R}_0^{1,m}(\Omega)$ is the space $\{u \in \mathcal{D}_0^{1,m}(\Omega) \;:\; u \text{ is radial}\}$, with norm $\|\nabla \cdot\|_m$.

We shall verify the assumptions in the mountain pass lemma. In view of Lemma 5 and assumption (i) it is easy to verify that $\mathcal{F}$ is C^1 on $\mathcal{R}_0^{1,m}(\Omega)$. Clearly $\mathcal{F}(0) = 0$ and by (ii) we can find a function $\varphi \in \mathcal{R}_0^{1,m}(\Omega)$ such that $\mathcal{F}(t\varphi) < 0$ for t large. If $u \in \mathcal{R}_0^{1,m}(\Omega)$, then Lemma 5 and (i) yield

$$\int F(x,u)\,dx \le \int \tilde{f}(x,u)u\,dx \le$$
$$\le \int \tilde{f}(x, C\omega(x)\|\nabla u\|_m)C\omega(x)\|\nabla u\|_m\,dx = o(\|\nabla u\|_m^m),$$

for $\|\nabla u\|_m$ small. Thus $\mathcal{F}(u) = \frac{1}{m}\|\nabla u\|_m^m - o(\|\nabla u\|_m^m) > 0$, if $\|\nabla u\|_m$ is small.

To finish the proof we need to establish the (PS) condition. Let $\{u_j\} \subset \mathcal{R}_0^{1,m}(\Omega)$ be a given sequence such that:

$$\int (\tfrac{1}{m}|\nabla u_j|^m - F(x,u_j))\,dx = C + o(1) \tag{2.2}$$

$$\int (|\nabla u_j|^{m-2}\nabla u_j \cdot \nabla\varphi - f(x,u_j)\varphi)\,dx = \langle \xi_j, \varphi\rangle, \tag{2.3}$$

for all $\varphi \in \mathcal{R}_0^{1,m}(\Omega)$, where $\xi_j \to 0$ in $\left(\mathcal{R}_0^{1,m}(\Omega)\right)^*$ as $j \to \infty$.

If we subtract $\frac{1}{m}$(2.3) with $\varphi = u_j$ from (2.2) we obtain

$$\int (\tfrac{1}{m}f(x,u_j)u_j - F(x,u_j))\,dx = C - \tfrac{1}{m}\langle \xi_j, u_j\rangle + o(1).$$

Thus combining this with (i) and (iii) yields

$$\int f(x,u_j)u_j\,dx \le$$
$$\le \int_{\{x\,:\,u_j(x) > M\omega(x)\}} f(x,u_j)u_j\,dx + \int_{\{x\,:\,u_j(x)\le M\omega(x)\}} \tilde{f}(x,u_j)u_j\,dx \le$$
$$\le C_M + o(1)\|\nabla u_j\|_m.$$

If we insert this estimate in (2.3) with $\varphi = u_j$ we get

$$\|\nabla u_j\|_m^m \le C + o(1)\|\nabla u_j\|_m.$$

Thus $\{u_j\}$ is bounded in $\mathcal{R}_0^{1,m}(\Omega)$. Hence we can assume that $u_j \rightharpoonup u$ in $\mathcal{R}_0^{1,m}(\Omega)$ and $u_j \to u$ uniformly on compact sets not containing the origin, as $j \to \infty$.

Finally if we subtract $(1.2)_j$ from $(1.2)_k$, take $\varphi = u_j - u_k$ and use Lemma 5, we get

$$\int (|\nabla u_j|^{m-2}\nabla u_j - |\nabla u_k|^{m-2}\nabla u_k)\cdot(\nabla u_j - \nabla u_k)\,dx =$$
$$= o(1) + \int (f(x,u_j) - f(x,u_k))(u_j - u_k)\,dx \to 0$$

as $j,k \to \infty$. Thus $u_j \to u$ in $\mathcal{R}_0^{1,m}$ by Lemma 1 in [TO] (see also [TH2]).

Since $\mathcal{F}(u) > 0$ we conclude that $u \not\equiv 0$. Furthermore u satisfies $\Delta_m u = -f(x,u) \le 0$, and the strong maximum principle shows that u is strictly positive in Ω (see for example the appendix in [EG1]).

The estimates of the solutions for x large follows immediately from Lemma 5 and its remark. □

Before we proceed, we will give a result that improves the asymptotic estimate at infinity, given in Theorem 4. To do this we have to add some extra conditions on the function f.

Proposition 6. *Let $n > m$, Ω be radial and unbounded and let f be radial in the space variable. Assume that we can find constants $p > m-1$ and ν satisfying $p+1 > \frac{m(n+\nu)}{n-m}$ such that $f(r,s) \le Cr^{\nu}s^{p}$ holds for large r. Then each radial solution u of (0.1) has a finite positive limit $\lim_{r\to\infty} r^{\frac{n-m}{m-1}}u(r) = C$. If $\Omega = \mathbf{R}^n$, then $C = \frac{m-1}{n-m}\left(\int_0^\infty f(s,u(s))s^{n-1}\,ds\right)^{\frac{1}{m-1}}$.*

Proof: We can assume that $\Omega = \mathbf{R}^n$ and that $f(r,u)$ is locally bounded. Indeed both u and f can be redefined in a large ball, without changing the result.

If u is a radial solution of (0.1), then since $u(r) > 0$ and $u'(r) < 0$ and they both tend to zero as $r \to \infty$ we can integrate the equation to obtain:

$$u(r) = \int_r^\infty \left(s^{1-n} \int_0^s t^{n-1} f(t,u(t))\,dt \right)^{\frac{1}{m-1}} ds\,. \tag{2.4}$$

If $u(r) \le Cr^{-\beta}$ for large r we get using (2.4) and the assumptions on f (from Lemma 5 we know that $\beta \ge \frac{n-m}{m}$).

$$\begin{aligned}
u(r) &\le \int_r^\infty \left(s^{1-n} \int_1^s t^{n-1} f(t,u(t))\,dt + Cs^{1-n} \right)^{\frac{1}{m-1}} ds \\
&\le C\int_r^\infty s^{\frac{1-n}{m-1}}\,ds + D\int_r^\infty s^{\frac{1+\nu-p\beta}{m-1}}\,ds \\
&\le Cr^{\frac{m-n}{m-1}} + Dr^{\frac{m+\nu-p\beta}{m-1}}\,.
\end{aligned}$$

This gives an improved estimate $u(r) \le Cr^{-\alpha}$, where we must have $\alpha \le \frac{n-m}{m-1}$ and in this case we get by the estimate above

$$\alpha - \beta \ge \frac{-m-\nu+p\beta}{m-1} - \beta = \frac{(p-(m-1))\beta}{m-1} - \frac{\nu+m}{m-1}\,,$$

and if $\beta = \frac{n-m}{m}$ we get

$$\alpha - \beta \ge \frac{n-m}{(m-1)m}\left(p+1-\frac{m(n+\nu)}{n-m}\right) > 0\,.$$

Thus a simple iteration shows that we get $\alpha = \frac{n-m}{m-1}$.

Now it follows that $f(t, u(t))t^{n-1} \in L^1(\mathbf{R}_+)$ and the result follows immediately from (2.4). □

Let us mention an example taken from [NOS]. The function $u(r) = (1+r^2)^{-(n-2)/4}$ satisfies the equation $\Delta u + \frac{n-2}{2}(n + \frac{n-2}{2}r^2)(1+r^2)^{(p-1)(n-2)/4-2}u^p = 0$ in $\mathbf{R}^n$. Here $f(r,s)$ satisfies the condition in Proposition 6 with p and ν such that $p+1 = \frac{2(n+\nu)}{n-2}$ and the solution has the asymptotic behaviour $u(r) \sim r^{-(n-2)/2}$ at infinity. This shows that the iteration argument in the proof above can not work in the limit case. Note that in this example u can not be in $\mathcal{D}_0^{1,m}(\Omega)$.

When $m = 2$, McLeod, Ni and Serrin has been classifying the possible asymptotic behaviour of radial solutions of equations of the same type as (0.1) (personal communication with Professor Serrin).

Let us consider equation (0.1) in more general unbounded domains and with $f(x,s)$ nonradial in the space variable. When $n < m$ it is easy to generalize the argument above. If $n < m$, $\Omega^c \neq \emptyset$ and $u \in \mathcal{D}_0^{1,m}(\Omega)$, then the Sobolev imbedding theorem yields $|u(x)| \leq C\|\nabla u\|_m \, \mathrm{d}(x, \Omega^c)^\alpha$, where $\alpha = (m-n)/m$ and $\mathrm{d}(x,\Omega^c)$ is the distance between the point x and the set Ω^c. Thus the same argument as in the proof of Theorem 4 yields

THEOREM 7. *If $n < m$ and conditions (i)-(iii) in Theorem 4 hold with $\omega(x) = d(x,\Omega^c)^{\frac{m-n}{m}}$, then there exists a solution of (0.1) which has the asymptotic estimate $u(x) = O(\omega(x))$.*

If $n = m$ we see that the transformation $x \to x/|x|^2$ gives

$$\int_\Omega |\nabla u|^m \, dx \to \int_{\tilde\Omega} |\nabla u|^m \, dx\,,$$

$$\int_\Omega F(x,u)\, dx \to \int_{\tilde\Omega} F(x/|x|^2, u)|x|^{-2n}\, dx\,.$$

From this we find that if the closure of Ω does not contain the origin, then $\tilde\Omega$ is a bounded domain and we can apply Theorem 1' in Section 1. In fact if $\overline{\Omega} \neq \mathbf{R}^n$ then we can transform the problem so that Theorem 1' applies.

THEOREM 8. *Let $n = m$ and assume that $0 \notin \overline{\Omega}$. Furthermore we assume that (ii) – (iv) in Theorem 1 hold and that*

(i) *There exist constants $\alpha \in \mathbf{R}$, $p < \frac{n}{n-1}$ and a function b such that $b(x)|x|^{2n(r-1)/r} \in L^r(\Omega)$ for some $r > 1$ so that the estimate $f(x,s) \leq b(x)\exp(\alpha s^p)$, holds for a.e. $x \in \Omega$ and positive s.*

Then (0.1) has a solution.

Let us return to the case when $m < n$. First we state an imbedding result.

Lemma 9. *Let $n > m$, then the imbedding*

$$\mathcal{D}_0^{1,m}(\mathbf{R}^n) \hookrightarrow L^{p+1}(\mathbf{R}^n, |x|^\nu\,dx),$$

is continuous if $-m \le \nu \le 0$ and $p+1 = \frac{m(n+\nu)}{n-m}$. If $\nu > 0$, then the result holds if we restrict to radial functions.

Proof: This follows immediately using spherical symmetrization (if $\nu \le 0$) and a general inequality in [MA] (Theorem 2 in Section 1.3.1). Another proof that gives the best constant and the extremal function when $\nu > -m$, can be found in [EG2] (Lemma 7). □

Lemma 10. *Let b be a nonnegative function which is locally bounded in $\mathbf{R}^n \setminus \{0\}$ and such that*

$$b(x) = \begin{cases} o(|x|^{\nu_1}) & \text{as } x \to 0\,, \\ o(|x|^{\nu_2}) & \text{as } x \to \infty\,. \end{cases}$$

Then the imbedding $\mathcal{D}_0^{1,m}(\mathbf{R}^n) \hookrightarrow L^{p+1}(\mathbf{R}^n, b(x)\,dx)$, is compact if $\frac{m(n+\nu_2)}{n-m} \le p+1 \le \frac{m(n+\nu_1)}{n-m}$ and $p+1 < \frac{mn}{n-m}$.
If we restrict to radial functions the same result holds with the condition $p+1 < \frac{mn}{n-m}$ removed.

Proof: We have

$$\text{(2.5)} \qquad \int_{\mathbf{R}^n} |v|^{p+1}\, b\,dx \le \sup_{|x|\le r} \frac{b(x)}{|x|^{\nu_1}} \int_{\{x\,:\,|x|<r\}} |v|^{p+1}\,|x|^{\nu_1}\,dx + \\ + \sup_{|x|\ge R} \frac{b(x)}{|x|^{\nu_2}} \int_{\{x\,:\,|x|>R\}} |v|^{p+1}\,|x|^{\nu_2}\,dx + \\ + C \int_{\{x\,:\,r<|x|<R\}} |v|^{p+1}\,dx\,.$$

If we choose a bounded set $\{u_j\} \subset \mathcal{D}_0^{1,m}(\mathbf{R}^n)$, then since $p+1 < \frac{nm}{n-m}$ or the functions are radial we can choose a subsequence such that

$$\int_{\{x\,:\,r<|x|<R\}} |u-u_j|^{p+1}\,dx \to 0\,,$$

for every $0 < r < R < \infty$. Hence combining this fact with (2.5) ($v = u - u_j$) and Lemma 9 we find that $u_j \to u$ in $L^{p+1}(\Omega, b\,dx)$ as a subsequence $j \to \infty$. □

Now we can apply the mountain pass lemma in a standard way to prove the following existence result.

Theorem 11. *Assume that $n > m$ and that*

(i) *The estimate $f(x,s) \le \sum_{i=1}^{N} b_i(x)s^{p_i}$, holds for all $x \in \Omega$. Here each couple p_i, b_i satisfies the following conditions:*
  a) *b_i is locally bounded in $\overline{\Omega} \setminus \{0\}$.*
  b) *$m < p_i + 1 < \frac{mn}{n-m}$.*
  c) *There exist numbers ν_1 and ν_2 such that $b_i(x) = o(|x|^{\nu_1})$, for x small, $b_i(x) = o(|x|^{\nu_2})$, for x large and $\frac{m(n+\nu_2)}{n-m} \le p_i + 1 \le \frac{m(n+\nu_1)}{n-m}$.*

(ii) *There exists a $\varphi \in \mathcal{D}_0^{1,m}(\Omega)$ such that*

$$t^{1-m} \int f(x, t\varphi)\varphi \, dx \to \infty, \quad \text{as } t \to \infty .$$

(iii) *There exists a constant $\theta \in (0, \frac{1}{m})$ such that*

$$\int_0^t f(x,s)\, ds \le \theta f(x,t)t, \quad \text{for all } x \in \Omega \text{ and positive } t .$$

Then (0.1) has a solution.
If Ω is radial and f is radial in the space variable, then the result holds if we remove the condition $p_i + 1 < \frac{mn}{n-m}$ in (i) b).
In both cases the solutions tend uniformly to zero as $x \to \infty$.

Proof: The proof of existence of a solution is a standard application of the mountain pass lemma and is omitted.

The fact that the solution $u \to 0$ as $x \to \infty$, follows immediately from Lemma 5 if u is radial. In the nonradial case we know that there exists a $q < \frac{mn}{n-m} - 1$ such that $f(x,s) < Cs^q$, holds uniformly for large x. Thus we can apply a standard estimate (e.g. see [EG1] Proposition A.2) to obtain

$$\sup_{B_1(y)} u \le C\|u\|_{L^s(B_2(y))}, \quad \text{for } s > 1 \text{ and large } y.$$

Since $u \in L^{mn/(n-m)}(\Omega)$, we conclude that $u(x) \to 0$ uniformly, as $x \to \infty$. □

Remark: If $0 \notin \overline{\Omega}$ or if Ω is bounded then the corresponding decay condition in (i) is superfluous.

Example 3. Let $\Omega = \mathbf{R}^n$, $n > m$ and $f(x,s) = b(x)s^p$, where $p + 1 > m$. If b is bounded and $b(x) = o(|x|^\nu)$ as $x \to \infty$, then there is a solution if $\frac{m(n+\nu)}{n-m} \le p + 1 < \frac{mn}{n-m}$.

On the other hand if $b(x) = \min(1, |x|^\nu)$, $\nu \le 0$ then the Pohozaev identity shows that there are no solutions if $p + 1 \le \frac{m(n+\nu)}{n-m}$.

If we take b to be radial, then there is a radial solution if $\frac{m(n+\nu)}{n-m} \le p+1$.

These examples shows that that the growth condition in (i) in Theorem 11 is best possible (see also the appendix in the present paper).

Let us remark that the results for radial solutions in Theorem 10 is not contained in Theorem 4 and vice versa.

In Theorem 11 we used power growth estimates of the u variable in f. However, it is possible to get a more precise result using Orlicz spaces (see e.g. Theorem 3 in the present paper).

APPENDIX. We start with a variant of Pohozaev's identity.

THEOREM A1. *Assume that u solves (0.1) and $V \subset \Omega$ has a smooth boundary. Then if $u \in C^{1+\alpha}(\overline{V})$, $f(x,u(x)) \in C^1(\Omega)$ and $f(x,0)=0$ then*

$$\begin{aligned}
&\tfrac{m-1}{m}\int_{\partial V}|\nabla u|^m(x\cdot\mathbf{n})\,d\sigma - \int_V (nF + \tfrac{m-n}{m}fu + x\cdot\widehat{\nabla}F)\,dx = \\
&= \int_{\partial V}(|\nabla u|^{m-2}\partial_i u((x\cdot\mathbf{n})\partial_i u - n_i(x\cdot\nabla u)) - F\,(x\cdot\mathbf{n}))\,d\sigma + \\
&+ \tfrac{m-n}{m}\int_V (|\nabla u|^m - fu)\,dx\,.
\end{aligned}$$

Here $F(x,u)=\int_0^u f(x,s)\,ds$, $\widehat{\nabla}$ regards u as a constant and $\mathbf{n}=(n_1,\dots,n_n)$ is the outer normal of ∂V.

A proof can be found in [EG1] (the result used here is taken from Theorem 5', however as pointed out to me by Professor J. Serrin the signs on the right side of the identity in Theorem 5' is wrong). For other proofs and related results see also [GV], [ON], [PS1] and [TH1&2].

COROLLARY A2. *Assume that Ω is smooth and that $f(x,s)=b(x)g(s)$, where b is C^1 in $\Omega\setminus\{0\}$ and g is C^1 in $\mathbf{R}_+$. Then if u is locally bounded in $\overline{\Omega}\setminus\{0\}$ and solves (0.1) then it also satisfies the identity:*

$$\tfrac{m-1}{m}\int_{\partial\Omega}|\nabla u|^m(x\cdot\mathbf{n})\,d\sigma = \int_\Omega((nG(u)+\tfrac{m-n}{m}g(u)u)b(x)+(x\cdot\nabla b(x))g(u))\,dx,$$

where $G(u)=\int_0^u g(s)\,ds$.

PROOF: The corollary follows from Theorem A1 if we take $V=V_k=\{\,x\in\Omega\,:\,r_k<|x|<R_k\,\}$, with a proper choice of $r_k\to 0$ and $R_k\to\infty$. See [EG1] (Example 5) or [BL] (Proposition 1). □

If we take $b(x)=\min(|x|^{\nu_1},|x|^{\nu_2})$, $\nu_2<\nu_1$, then if u solves (0.1) Corollary A.2 yields (b is not C^1 here but it is easy to see that the Pohozaev identity still holds).

$$\begin{aligned}
\tfrac{m-1}{m}\int_{\partial\Omega}|\nabla u|^m(x\cdot\mathbf{n})\,d\sigma &= \int_{\Omega\setminus B_1(0)}\left(\tfrac{n+\nu_2}{p+1}+\tfrac{m-n}{m}\right)|x|^{\nu_2}u^{p+1}\,dx + \\
&+ \int_{\Omega\cap B_1(0)}\left(\tfrac{n+\nu_1}{p+1}+\tfrac{m-n}{m}\right)|x|^{\nu_1}u^{p+1}\,dx\,.
\end{aligned}$$

From this identity we immediately get the following results.

If $n > m$ and $\Omega = B_1(0)$, then there are no solutions if $p+1 \geq \frac{m(n+\nu_1)}{n-m}$.

If $\Omega = \mathbf{R}^n \setminus \overline{B}_1(0)$, then there are no solutions if $n > m$ and $p+1 \leq \frac{m(n+\nu_2)}{n-m}$ or $n < m$ and $p+1 \geq \frac{m(n+\nu_2)}{n-m}$ or $n = m$ and $\nu_2 \geq -n$.

If $n > m$, $\Omega = \mathbf{R}^n$ and $p+1 \notin]\frac{m(n+\nu_2)}{n-m}, \frac{m(n+\nu_1)}{n-m}[$, then there are no solutions. Note that if $\nu_1 = \nu_2 > -m$ then there is a solution if $p+1 = \frac{m(n+\nu_2)}{n-m}$ (e.g. see [EG2]).

ACKNOWLEDGEMENTS. The author would like to thank Professor Catherine Bandle for interesting discussions and useful remarks and Professor James Serrin, for pointing out the mistake in Theorem 5' in [EG1] (used in Theorem A1 in the present paper) and for telling me about recent results on asymptotics of solutions of quasilinear elliptic equations.

REFERENCES

[AD] R.A. ADAMS, Sobolev Spaces. Academic Press 1975.

[AR] A. AMBROSETTI, P.H. RABINOWITZ, Dual Variational Methods in Critical Point Theory and Applications. J. Funct. Anal. 14, 349 - 381, 1973.

[BC1] V. BENCI, G. CERAMI, Positive Solutions of Some Nonlinear Elliptic Problems in Exterior Domains. Arch. Rational Mech. Anal. 99, 283 - 300, 1987.

[BC2] V. BENCI, G. CERAMI, Existence of Positive Solutions of the Equation $-\Delta u + a(x)u = u^{(N+2)/(N-2)}$ in $\mathbf{R}^N$. To appear.

[BI] M.-F. BIDAUT-VERON, Local and Global Behaviour of Solutions of Quasi-Linear Emden-Fowler equations. To appear in Arch. Rational Mech. Anal..

[BL] H. BERESTYCKI, P.-L. LIONS, Nonlinear Scalar Field Equations, Part 1&2. Arch. Rational Mech. Anal. 82, 313 - 345 & 347 - 375, 1983.

[BLP] H. BERESTYCKI, P.-L. LIONS, L.A. PELETIER, An ODE Approach to the Existence of Positive Solutions for Semilinear Problems in $\mathbf{R}^n$. Indiana Univ. Math. J. 30, 141 - 157, 1981.

[BM] C. BANDLE, M. MARCUS, On the Structure of the Positive Radial Solutions for a Class of Nonlinear Elliptic Equations. To appear in Crelle J.

[EG1] H. EGNELL, Existence and Nonexistence Results for m-Laplace Equations Involving Critical Sobolev Exponents. Arch. Rational Mech. Anal. 104, 57 – 77, 1988.

[EG2] H. EGNELL. Elliptic Boundary Value Problems with Singular Coefficients and Critical Nonlinearities. Indiana Univ. Math. J. 38, 235 – 251, 1989.

[GV] M. GUEDDA, L. VERON, Quasilinear Elliptic Equations Involving Critical Sobolev Exponents. Nonlinear Analysis T.M.A. 13, 879 – 902, 1989.

[KN] T. KUSANO, M. NAITO, Positive Entire Solutions of Superlinear Equations. Hiroshima Math. J. 16, 361 – 366, 1986.

[LN1] Y. LI, W.- M. NI, On Conformal Scalar Curvature in $\mathbf{R}^n$. Duke Math. J. 57, 895 – 924, 1988.

[LN2] Y. LI, W.-M. NI, On the Existence and Symmetry Properties of Finite Total Mass Solutions of Matukuma equation, Eddington equation and their generalizations. To appear in Arch. Rational Mech. Anal..

[MA] V.G. MAZ'JA, Sobolev spaces. Springer Verlag 1985.

[NI1] W.-M. NI, On the Elliptic Equation $\Delta u + K(x)u^{(n+2)/(n-2)} = 0$, its Generalizations, and Applications in Geometry. Indiana Univ. Math. Journal 31, 493 – 529, 1982.

[NI2] W.-M. NI, A Nonlinear Dirichlet Problem on the Unit Ball and Its Applications. Indiana Univ. Math. Journal 31, 801-807, 1982.

[NIS] W.-M. NI, J SERRIN, Existence and Non-existence Theorems for Ground States of Quasilinear Partial Differential Equations. The Anomalous Case. Academia Nazinale dei Lincei. Atti di convegni 77, 231 – 257, 1986.

[NOS] E.S. NOUSSAIR, C.A. SWANSON, Positive Decaying Entire Solutions of Superlinear Elliptic Equations. Indiana Univ. Math. J. 36, 651 – 657, 1987.

[ON] M. ÔNTANI, Existence and Nonexistence of Nontrivial Solutions of Some Nonlinear Degenerate Elliptic Equations. J. Funct. Anal. 76, 140 – 159, 1988.

[PS1] P. PUCCI, J. SERRIN, A General Variational Identity. Indiana Univ. Math. J. 35, 681 – 703, 1986.

[PS2] P. PUCCI, J. SERRIN, Continuation and Limit Properties for Solutions of Strongly Nonlinear Second Order Differential Equations. To appear.

[ST] W.A. STRAUSS, Existence of Solitary Waves in Higher Dimensions. Comm. Math. Phys. 55, 149 - 162, 1977.

[TH1] F. DE THELIN, Quelques Résultats d'Existence et de Non-existence pour une E.D.P. Elliptique Non Linéaire. C. R. Acad. Sc. Paris, t. 299, Série I, 18, 911-914.

[TH2] F. DE THELIN, Résultats d'Existence et de Non Existence pour la Solution Positive et Bornee d'une E.D.P. Elliptique Non Linéaire. Ann. Faculté Sc. Toulouse VIII, 375 - 389, 1986 - 1987.

[TOL] J.F. TOLAND, On positive solutions of $-\Delta u = F(x, u)$. Mathematische Zeitschrift 182, 351 - 357, 1983.

[TO] P. TOLKSDORF, Regularity for a More General Class of Quasilinear Elliptic Equations. J. Diff. Eq. 51, 126 - 150, 1984.

Henrik Egnell
School of Mathematics
University of Minnesota
Minneapolis, MN 55455

A New Setting For Skyrme's Problem

MARIA J. ESTEBAN

Introduction

In [3] we gave some existence results for the general Skyrme's problem, without any symmetry assumption. This problem arises as a natural model when searching to identify baryons as solitons in meson field theory. For more details about the physics motivation see [1,7,9,10]. It can be defined as follows. A class of functions $\phi : \mathbf{R}^3 \to S^3$, X, has to be defined so that the Skyrme's energy functional

$$\mathcal{E}(\Phi) = \int_{\mathbf{R}^3} |\nabla\phi|^2 + |A(\phi)|^2 \, dx, \tag{1}$$

(with $A(\phi) = \left(\frac{\partial\phi}{\partial x_i} \wedge \frac{\partial\phi}{\partial X_j}\right)$ $i, j = 1, 2, 3$) is finitely defined in X. Then $\mathcal{E}$ has to be minimized in the subclass of X formed by the functions ϕ having a "topological degree" $d(\phi)$ equal to $k \in \mathbf{Z}$, where

$$d(\phi) = \frac{1}{2\pi^2} \int_{\mathbf{R}^3} \det(\phi, \nabla\phi) \, dx. \tag{2}$$

Physically $d(\phi)$ represents the baryonic number and has then to be an integer. So one has to consider a class X in which d takes naturally only integer values. Actually when $\phi \in C^1(\mathbf{R}^3, S^3)$, $d(\phi)$ is the topological degree of $\phi \circ E$, E being a stereographic projection mapping S^3 into $\mathbf{R}^3$. Therefore it is natural to define X as a class of functions which can be approached by smooth functions in a convenient sense. In [3] such a class was chosen where the corresponding density property was quite strong. Then existence results for the Skyrme's problem in that setting were given. Actually those results are valid only if one proves the approximation of finite energy functions by smooth functions in a rather strong way, and this is an open problem.

In this paper we define a new class X which takes into account the absence of information about the density of smooth functions in the set

$\{\phi \mid \mathcal{E}(\phi) < +\infty\}$. Then, for this new class we prove the same existence results as in [3].

The basic ideas used to perform the proofs in this paper are the same as in [3]. The changes appearing here are of two kinds. First, here we perform proofs which use only analytical arguments. In [3] some geometrical devices were used which were possible because we could work directly with smooth functions. On the other hand, some of the technical results we used in [3] have been adapted to fit in the following new setting:

Let $L > 0$ and let X^L be defined by

$$X^L = \{\phi : \mathbf{R}^3 \to S^3 \mid \mathcal{E}(\phi) < +\infty, \exists \phi_n \in C^1(\mathbf{R}^3, S^3) \text{ such that } \mathcal{E}(\phi_n) \leqq L \quad \forall\, n \quad \text{and (P) holds}\},$$

where

(P)
$$\begin{cases} \nabla\phi_n \underset{n}{\rightharpoonup} \nabla\phi, \; A(\phi_n) \underset{n}{\rightharpoonup} A(\phi) \quad \text{weakly in} \quad L^2(\mathbf{R}^3, \mathbf{R}^4), \\ \det(\phi_n, \nabla\phi_n) \underset{n}{\rightharpoonup} \det(\phi, \nabla\phi) \quad \text{in the narrow topology of measures.} \end{cases}$$

Then we observe that $d(\cdot)$ is integer valued in X^L and we define the subclasses

$$X_k^L = \{\phi \in X^L \mid d(\phi) = k\}, \qquad k \in \mathbf{Z} \tag{3}$$

and state Skyrme's problem as follows

$$I_k^L = \inf\{\mathcal{E}(\phi) \mid \phi \in X_k^L\}. \tag{4}$$

Moreover we say that $I_k^L = +\infty$ when $X_k^L = \emptyset$. As we see in Section 1 below this can happen if L is small. That is, maps with degree k will be shown to have an energy bounded from below by some constant $L'_k > 0$. Moreover, $X_k^L \neq \emptyset$ as soon as we can find some smooth function ϕ with degree k and energy less than or equal to L. Such functions can be easily found and that will define constants $\overline{L}_k > 0$ such that $X_k^L \neq \emptyset$ and $I_k^L < +\infty$ for all $L > \overline{L}_k$.

The main results we prove are the following.

Theorem 1. *I_1^L is achieved for every $L > \overline{L}_1$.*

Theorem 2. *Let $k \in \mathbf{Z}\setminus\{0, \pm 1\}$. Then for every $L > \overline{L}_k$, I_k^L is achieved whenever the following strict inequalities hold*

$$I_k^L < I_\ell^{L_1+} + I_{k-\ell}^{L-L_1+} \tag{5}$$

for all $\ell \in \mathbf{Z}\backslash\{0,k\}$, *for all* $L_1 \in (0,L)$, *where* $I_m^{C+} = \lim_{\eta \to 0+} I_m^{C+\eta}$.

This paper is organized as follows. In Section 1 we prove the auxiliary results we need. Then, we prove Theorems 1 and 2 in Section 2. Actually some of the auxiliary results have an independent interest. Section 3 is devoted to remarks on the lack of density, other possible choices for the classes to work with, open problems and extensions. In the appendix we deal with technical complicated cut-off results.

Finally let us define some notation. For any matrix or vector valued function F, we will always write $F \in E$ (scalar space) to indicate that each component of F is in E.

Moreover, for every measurable set $B \subset \mathbf{R}^N$, we denote by $|B|$ its n-dimensional Lebesgue measure. Finally the notation $\partial_i u$ is for $\frac{\partial u}{\partial x_i}$ and C will denote various positive constants.

Acknowledgement. I am very grateful to J. M. Dolbeault for many interesting comments on this work.

1. Some Auxiliary Results

We start this section by showing that all functions in X^L, for all $L > 0$, have a limit at infinity, at least in a weak sense.

Proposition 3. *For all* $\phi : \mathbf{R}^3 \to S^3$ *such that* $\nabla\phi \in L^2(\mathbf{R}^3, \mathbf{R}^4)$, *there exists* $P \in S^3$ *satisfying* $\phi - P \in L^6(\mathbf{R}^3, \mathbf{R}^4)$.

Remark 4. We may assume that up to a rotation $P = (1,0,0,0)$ for all ϕ in X^L, $L > 0$. Since rotations leave invariant the functionals $\mathcal{E}$ and d, this assumption is not restrictive and we will do it throughout this paper.

Proof of Proposition 3. For all $R > 0$ let $\bar{\phi}_R = \frac{1}{|B_R|}\int_{B_R} \phi(x)\,dx$. By using the Poincaré-Wirtinger inequality we know the existence of a positive constant C_0 such that

$$(6) \qquad \left(\int_{B_1} |\phi - \bar{\phi}_1|^6\,dx\right)^{1/6} \leqq C_0 \left(\int_{B_1} |\nabla\phi|^2\,dx\right)^{1/2}.$$

Let us now define $\phi^R : \mathbf{R}^3 \to S^3$ by $\phi^R(\cdot) = \phi(R\cdot)$. We observe that $(\overline{\phi^R})_1 = \bar{\phi}^R$ and then, we apply (6) to ϕ^R and obtain

$$(7) \qquad \left(\int_{B_R} |\phi - \bar{\phi}_R|^6\,dx\right)^{1/6} \leqq C_0 \left(\int_{B_R} |\nabla\phi|^2\,dx\right)^{1/2}, \quad \forall\, R > 0.$$

Now, since $\phi(\mathbf{R}^3) \subset S^3$, $P_R \in B_1$ for all $R > 0$. We can then find $P \in B_1$ such that $P_{R_n} \to P$ for some sequence $R_n \underset{n\to+\infty}{\longrightarrow} +\infty$. Therefore we can use (7) and Fatou's lemma to infer that $\phi - P \in L^6(\mathbf{R}^3)$. Indeed, $\nabla\phi$ is in $L^2(\mathbf{R}^3, \mathbf{R}^4)$ by our assumptions. On the other hand, there can not be two different P's satisfying $\phi - P \in L^6(\mathbf{R}^3)$ and hence $P = \lim_{R\to+\infty} P_R$ and $P \in S^3$.

QED

Next we prove two propositions which provide us with estimates for the value of I_k^L.

Proposition 5. *For all* $k \in \mathbf{Z}$,

$$I_k^L \geqq 12\pi^2|k| \qquad \text{for all } L > 0 \tag{8}$$

$$I_k^L \leqq 12\sqrt{2}\pi^2|k| \quad \text{for all } L > 12\pi^2\sqrt{2}|k|. \tag{8 bis}$$

Proof. First, if there is no ϕ in X^L such that $d(\phi) = k$, then by definition $I_k^L = +\infty$ and (8) is proved. In the opposite case, let ϕ be any function from $\mathbf{R}^3$ to S^3 such that $\mathcal{E}(\phi) < +\infty$ and $d(\phi) = k$. Then,

$$\begin{aligned} \mathcal{E}(\phi) &\geqq 2\left(\int_{\mathbf{R}^3} |\nabla\phi|^2\,dx\right)^{1/2}\left(\int_{\mathbf{R}^3} |A(\phi)|^2\,dx\right)^{1/2} \\ &\geqq 6\int_{\mathbf{R}^3} |\partial_1\phi \wedge \partial_2\phi \wedge \partial_3\phi|\,dx, \end{aligned}$$

by Schwarz and by Hölder inequalities. Moreover, since $|\det(\phi, \nabla\phi)| \leqq |\partial_1\phi \wedge \partial_2\phi \wedge \partial_3\phi|$ a.e., we have $|d(\phi)| \leqq \frac{1}{12\pi^2}\mathcal{E}(\phi)$, which proves (8).

On the other hand, (8 bis) is proved by Proposition 6 below when we observe that the inverse of some stereographic projection from S^3 into $\mathbf{R}^3$ is smooth enough and has energy $\mathcal{E}$ equal to $12\sqrt{2}\pi^2$.

QED

Proposition 6. *Let k, ℓ be any two integers and $L_1, L_2 > 0$. Then*

$$I_{k+\ell}^{(L_1+L_2)+} \leqq I_k^{L_1} + I_\ell^{L_2}, \tag{9}$$

where by I_m^{L+} we denote $\lim_{L^1 \to L+} I_m^{L^1}$.

Proof of Proposition 6. Whenever $I_k^{L_1}$ or $I_\ell^{L_2}$ are equal to $+\infty$, (9) is satisfied. Let us then consider the case $I_k^{L_1}, I_\ell^{L_2} < +\infty$.

Let $\varepsilon, \delta > 0$ and ϕ_1 (resp. ϕ_2) an ε - quasiminimum of $I_k^{L_1}$ (resp. $I_\ell^{L_2}$), i.e.,

$$|\mathcal{E}(\phi_1) - I_k^{L_1}| < \varepsilon, \quad |\mathcal{E}(\phi_2) - I_\ell^{L_2}| < \varepsilon$$

and $\phi_i \in X^{L_i}$, $i = 1,2$, $d(\phi_1) = k$, $d(\phi_2) = \ell$. By Proposition A2 in the appendix, there exist functions $\widetilde{\phi}_1$ and $\widetilde{\phi}_2$ which are constant outside some ball of radius $R(\delta)$ and such that

$$|\mathcal{E}(\widetilde{\phi}_1) - I_k^{L_1}| < \varepsilon + \delta, \quad |\mathcal{E}(\widetilde{\phi}_2) - I_\ell^{L_2}| < \varepsilon + \delta,$$

$\widetilde{\phi}_1 \in X_k^{L_1+\delta}$ and $\widetilde{\phi}_2 \in X_\ell^{L_2+\delta}$. Then we consider $M \in \mathbf{R}^3$ with $|M| > 2R(\delta)$ and define $\widetilde{\phi}$ by

$$\widetilde{\phi}(x) = \begin{cases} \widetilde{\phi}_1(x). & x \in B_{R(\delta)}, \\ \widetilde{\phi}_2(x - M), & x \in B^c_{R(\delta)}. \end{cases}$$

Then

$$d(\widetilde{\phi}) = k + \ell, \qquad \widetilde{\phi} \in X^{L_1+L_2+2\delta}$$

and

$$\mathcal{E}(\widetilde{\phi}) \leqq I_K^{L_1} + I_\ell^{L_2} + 2(\varepsilon + \delta).$$

Finally, since ε and δ are arbitrary positive constants the above inequality proves the proposition. QED

Remark 7. By using the same arguments as above we can prove that $I_{k+\ell}^{L_1+L_2} \leqq I_k^{L_1-} + I_\ell^{L_2-}$ (obvious notation).

We end this section by a series of results which prove the equi-integrability of 3-minors of matrix functions when the 2-minors are square integrable. These results are essential in the proof of Theorems 1 and 2, but can also have an independent interest.

Proposition 8. *Let $u : \mathbf{R}^3 \to \mathbf{R}^3$ a measurable function such that $\nabla u \in L^1_{\text{loc}}(\mathbf{R}^3)$ and $A_{ij}(u) = \frac{\partial u}{\partial x_i} \wedge \frac{\partial u}{\partial x_j}$ is in $L^2(\mathbf{R}^3)$ for all $i,j = 1,2,3$. Then $\det(\nabla u)$ is in $L^1_{\text{loc}}(\mathbf{R}^3)$ and*

$$(10) \qquad \int_B |\det(\nabla u)|\, dx \leqq |B|^{1/4} \left(\sum_{i,j=1}^{3} \int_B (A_{ij}(u))^2\, dx \right)^{3/4},$$

for all bounded measurable sets $B \subset \mathbf{R}^3$.

Proof. This proposition follows in a straightforward way from the Hölder inequality and the lemma below.

Lemma 9. *Let a, b, c be three vectors in $\mathbf{R}^3 \backslash \{\bar{0}\}$. Then,*

$$(11) \qquad |a \cdot b \wedge c| \leqq |a \wedge b|^{1/2} |a \wedge c|^{1/2} |b \wedge c|^{1/2}.$$

Proof. We may assume that the right hand side of (11) is positive. Indeed, if this were equal to 0, $a \cdot b \wedge c$ would also be null, and (11) would be proved. Then, inequality (11) is obviously equivalent to

$$h(a,b;c)\ell(a,b) \leqq \ell(a,b)^{1/2}\ell(a,c)^{1/2}\ell(b,c)^{1/2}, \tag{12}$$

where $h(a,b;c)$ we denote the ratio $\frac{|a\wedge b\cdot c|}{|a\wedge b|\ |c|}$ and by $\ell(a,b)$, $\frac{|a\wedge b|}{|a|\ |b|}$. Then we observe that $\ell(a,b) \in [0,1]$, and therefore $\ell(a,b) \leqq \ell(a,b)^{1/2}$. Moreover, $h(a,b;c) = |\sin\varphi|$ and $\ell(a,c)$ (resp. $\ell(b,c)$) is equal to $|\sin\beta|$ (resp. $|\sin\varphi|$), where φ is the angle between c and the plane defined by a and b, and β (resp. φ) is the angle between a and c (resp. b and c). Then it is obvious that $h(a,b;c) \leqq \min\{\ell(a,c), \ell(b,c)\}$. Thus, (12) follows. QED

Corollary 10. *Let $\phi : \mathbf{R}^3 \to S^3$ be any measurable function such that $A(\phi) \in L^2(\mathbf{R}^3, \mathbf{R}^4)$ and $\nabla\phi \in L^1_{\text{loc}}(\mathbf{R}^3)$. Then the function $|\partial_1\phi \wedge \partial_2\phi \wedge \partial_3\phi| \in L^1_{\text{loc}}(\mathbf{R}^3)$ and*

$$\int_B |\partial_1\phi \wedge \partial_2\phi \wedge \partial_3\phi|\, dx \leqq |B|^{1/4}\|A(\phi)\|^{3/2}_{L^2(B)}, \tag{13}$$

for any bounded measurable set $B \subset \mathbf{R}^3$.

Proof. As we explain in Remark A1 in the appendix if (φ, ξ, ∂) are the spherical coordinates associated with ϕ, we have

$$|\partial_1\phi \wedge \partial_2\phi \wedge \partial_3\phi| = |a(x)\cdot b(x) \wedge c(x)|,$$

with $a(x) = \nabla\varphi(x)$, $b(x) = (\cos\varphi(x))\nabla\xi(x)$, $c(x) = (\cos\varphi(x)\cos\xi(x))\nabla\theta$. Then we apply Proposition 8 to the function $u = (a,b,c)$ and obtain (13). Indeed, we use the form $|A(\phi)|^2$ takes in function of (φ,ξ,θ) (see again Remark A1).

QED

Corollary 11. *Assume that a sequence $\phi_n : \mathbf{R}^3 \to S^3$ converges towards some $\phi : \mathbf{R}^3 \to S^3$ in the following sense : $\nabla\phi_n \to \nabla\phi$, $A(\phi_n) \to A(\phi)$ in $L^2(\mathbf{R}^3, \mathbf{R}^4)$-weak. Then, if the sequence $\{|\partial_1\phi_n \wedge \partial_2\phi_n \wedge \partial_3\phi_n|\}_n$ satisfies uniformly Prokhorov's property, then*

$$\partial_1\phi_{n'} \wedge \partial_2\phi_{n'}\partial_3\phi_{n'} \underset{n'}{\to} \partial_1\phi \wedge \partial_2\phi \wedge \partial_3\phi \quad \text{in } L^1(\mathbf{R}^3)\text{-weak}, \tag{14}$$

$$\det(\phi_{n'}, \nabla\phi_{n'}) \underset{n'}{\to} \det(\phi, \nabla\phi) \quad \text{in } L^1(\mathbf{R}^3)\text{-weak} \tag{15}$$

for some subsequence $\{\phi_{n'}\}$.

Remark. Remember that the satisfaction of the Prokhorov's property in a uniform way means that for all $\varepsilon > 0$ there exists a compact set $K \subset \mathbf{R}^3$ such that

$$\int_{K^c} |\partial_1\phi_n \wedge \partial_2\phi_n \wedge \partial_3\phi_n|\, dx \leqq \varepsilon, \qquad \text{for all } n.$$

Proof of Corollary 11. Corollary 10 and the assumptions above imply that we can use the Dunford-Pettis criterium to infer that both $\{\partial_1\phi_n \wedge \partial_2\phi_n \wedge \partial_3\phi_n\}$ and $\{\det(\phi_n, \nabla\phi_n)\}$ are relatively compact in $L^1(\mathbf{R}^3)$-weak, since $|\phi_n| = 1$ a.e.. The only remaining thing is then the identification of the limits of converging subsequences.

First assume that (14) is already proved. Then (15) follows. Indeed for all $\varphi \in \mathcal{D}(\mathbf{R}^3)$,

$$\int_{\mathbf{R}^3} \varphi \det(\phi_{n'}, \nabla\phi_{n'})\, dx = \int_{\mathbf{R}^3} \varphi\phi_{n'} \cdot (\partial_1\phi_{n'} \wedge \partial_2\phi_{n'} \wedge \partial_3\phi_{n'})\, dx$$

and this converges to $\int_{\mathbf{R}^3} \varphi \det(\phi, \nabla\phi)\, dx$, since (14) holds and $\{\varphi\phi_{n'}\}$ converges towards ϕ in $L^\infty(\mathbf{R}^3)$-weak* and a.e.. Hence, $\{\det(\phi_{n'}, \nabla\phi_{n'})\}$ converges to $\det(\phi, \nabla\phi)$ in $\mathcal{D}'(\mathbf{R}^3)$, and therefore the limit in $L^1(\mathbf{R}^3)$- weak is the same. Note that we have assumed that $\{\varphi\phi_{n'}\}$ converges to ϕ in $L^\infty(\mathbf{R}^3)$-weak* and a.e.. Indeed it can always be assumed by extracting if necessary a subsequence.

We finally prove (14). Of course, it will be sufficient to prove that the convergence takes place in $\mathcal{D}'(\mathbf{R}^3)$. From Lemma 12 below we have that for all $\varphi \in (\mathcal{D}(\mathbf{R}^3))^4$,

$$\begin{aligned} 3&\int_{\mathbf{R}^3} \varphi \cdot \partial_1\phi_{n'} \wedge \partial_2\phi_{n'} \wedge \partial_3\phi_{n'}\, dx \\ &= -\int_{\mathbf{R}^3} \partial_1\varphi \cdot \phi_{n'} \wedge \partial_2\phi_{n'} \wedge \partial_3\phi_{n'}\, dx - \int_{\mathbf{R}^3} \partial_2\varphi \cdot \partial_1\phi_{n'} \wedge \phi_{n'} \wedge \partial_3\phi_{n'}\, dx + \\ &- \int_{\mathbf{R}^3} \partial_3\varphi \cdot \partial_1\phi_{n'} \wedge \partial_2\phi_{n'} \wedge \phi_{n'}\, dx. \end{aligned}$$

Now since $A(\phi_{n'}) \to A(\phi)$ in L^2-weak and $\phi_{n'} \to \phi$ in $L^2_{\text{loc}}(\mathbf{R}^3)$ (extracting a sequence if necessary), we obtain that

$$3\int_{\mathbf{R}^3} \varphi \cdot \partial_1\phi_{n'} \wedge \partial_2\phi_{n'} \wedge \partial_3\phi_{n'}\, dx \underset{n\to+\infty}{\longrightarrow} 3\int_{\mathbf{R}^3} \varphi \cdot \partial_1\phi \wedge \partial_2\phi \wedge \partial_3\phi\, dx,$$

where we have used again Lemma 12 below. This ends the proof. QED

Lemma 12. *Assume that $\phi : \mathbf{R}^3 \to S^3$ has finite energy $\mathcal{E}(\phi)$. Then for all $\varphi \in \mathcal{D}(\mathbf{R}^3)$,*

$$
\begin{aligned}
3\int_{\mathbf{R}^3} & \varphi \cdot \partial_1\phi \wedge \partial_2\phi \wedge \partial_3\phi \, dx = \\
&= - \int_{\mathbf{R}^3} \partial_1\varphi \cdot \phi \wedge \partial_2\phi \wedge \partial_3\phi \, dx \\
&\quad - \int_{\mathbf{R}^3} \partial_2\varphi \cdot \partial_1\phi \wedge \phi \wedge \partial_3\phi \, dx + \\
&\quad - \int_{\mathbf{R}^3} \partial_3\varphi \cdot \partial_1\phi \wedge \partial_2\phi \wedge \phi \, dx.
\end{aligned} \tag{16}
$$

Proof. If we write all the integrals in (16) as duality products then

$$
\begin{aligned}
\langle \partial_i\phi, \varphi \wedge \partial_j\phi \wedge \partial_k\phi \rangle = &- \langle\langle \phi \wedge \partial_i\varphi, \partial_j\phi \wedge \partial_k\phi \rangle\rangle + \\
&- \langle\langle \phi \wedge \varphi, \partial_i(\partial_j\phi \wedge \partial_k\phi) \rangle\rangle,
\end{aligned}
$$

where the right hand side has to be understood in the sense of the duality between $\mathcal{D}^{1,2}$ and its dual and $\langle\langle a \wedge b, c \wedge d \rangle\rangle = \langle a, b \wedge c \wedge d \rangle = a \cdot b \wedge c \wedge d \quad \forall a, b, c, d \in \mathbf{R}^4$. Hence (16) holds if we show

$$
\begin{aligned}
&\langle\langle \varphi \wedge \phi, [\partial_2(\partial_1\phi \wedge \partial_3\phi) - \partial_1(\partial_2\phi \wedge \partial_3\phi) - \partial_3(\partial_1\phi \wedge \partial_2\phi)] \rangle\rangle = 0, \\
&\qquad\qquad\qquad\qquad\qquad\qquad \text{for all } \varphi \in \mathcal{D}(\mathbf{R}^3).
\end{aligned} \tag{17}
$$

Then, the lemma will be proved as soon as we show that the distribution in between the brackets is null. Indeed, this distribution is in the dual space of $\mathcal{D}^{1,2}(\mathbf{R}^3)$ and $\varphi \wedge \phi \in \mathcal{D}^{1,2}$. Let us call $P(\phi)$ the expression between the brackets in (17). If ϕ were of class C^2, $P(\phi)$ would be equal to 0 everywhere, and hence also in $\mathcal{D}'$. Let us then consider a regularizing sequence $\rho_\varepsilon \in \mathcal{D}(\mathbf{R}^3)$ such that $\phi_\varepsilon = \phi * \rho_\varepsilon \in C^\infty(\mathbf{R}^3)$ and $\nabla\phi_\varepsilon \underset{\varepsilon \to 0}{\longrightarrow} \nabla\phi$ in $L^2(\mathbf{R}^3)$. Then $A(\phi_\varepsilon) \underset{\varepsilon \to 0}{\longrightarrow} A(\phi)$ in $L^1(\mathbf{R}^3)$. Moreover, for all $\psi \in \mathcal{D}(\mathbf{R}^3)$,

$$
\begin{aligned}
\langle P(\phi_\varepsilon), \psi \rangle = &- \langle \partial_1\phi_\varepsilon \wedge \partial_3\phi_\varepsilon, \partial_2\psi \rangle + \langle \partial_2\phi_\varepsilon \wedge \partial_3\phi_\varepsilon, \partial_1\psi \rangle \\
&+ \langle \partial_1\phi_\varepsilon \wedge \partial_2\phi_\varepsilon, \partial_3\psi \rangle
\end{aligned}
$$

and the right hand side converges to

$$
-\langle \partial_1\phi \wedge \partial_3\phi, \partial_2\psi \rangle + \langle \partial_2\phi \wedge \partial_3\phi, \partial_1\psi \rangle + \langle \partial_1\phi \wedge \partial_2\phi, \psi \rangle
$$

since $\psi \in \mathcal{D}(\mathbf{R}^3)$ and $A(\phi_\varepsilon) \underset{\varepsilon}{\to} A(\phi)$ in $L^1(\mathbf{R}^3)$. But the sum of these three duality products is again equal to $\langle P(\phi), \psi \rangle$. Therefore, $P(\phi_\varepsilon) \underset{\varepsilon \to 0}{\longrightarrow} P(\phi)$

in $\mathcal{D}'(\mathbf{R}^3)$. But $P(\phi_\varepsilon) = 0 \ \forall\, \varepsilon$, hence $P(\phi) = 0$ in $\mathcal{D}'(\mathbf{R}^3)$, and thus also in $(\mathcal{D}^{1,2}(\mathbf{R}^3))'$. QED

Remark 13. The above proof can be reproduced in the case when $\varphi \in L^\infty(\mathbf{R}^3) \cap \mathcal{D}^{1,2}(\mathbf{R}^3)$ and hence (16) also holds in this case.

2. Proofs of Theorems 1 and 2

Proof of Theorem 1. Let $\{\phi_n\}$ be a minimizing sequence for I_1^L. Then we can find a subsequence, still denoted by $\{\phi_n\}$, such that

$$(18)\qquad \begin{aligned} &\nabla\phi_n \underset{n}{\rightharpoonup} \nabla\phi, \quad A(\phi_n) \rightharpoonup A(\phi) \quad \text{weakly in} \quad L^2(\mathbf{R}^3), \\ &\phi_n \underset{n}{\rightarrow} \phi \quad \text{a.e. and in} \quad L^\infty(\mathbf{R}^3)\text{-weak}^*, \\ &\phi_n - P \underset{n}{\rightarrow} \phi - P \quad \text{in} \quad L^6(\mathbf{R}^3)\text{-weak}. \end{aligned}$$

Indeed, $A(\phi_n)$ will be convergent to some $\psi \in L^2$ and as has been remarked by several people (see [2,5,6,8]) the particular structure of A (2-minors of the $\nabla\phi$ matrix) implies that $\psi = A(\phi)$.

Hence from (18) we infer that $\mathcal{E}(\phi) \leqq I_1^L$. In order to prove that ϕ is a minimizer for $\mathcal{E}$ in X_1^L, one has to show that ϕ is in X_1^L. And that is not necessarily true since the problem defining I_1^L is invariant by translation in $\mathbf{R}^3$. To overcome this difficulty, we apply the concentration-compactness method (see [3] and [4]) to the sequence $f_n = |A(\phi_n)|^2 + |\nabla\phi_n|^2$. Then one has the following three possibilities (and only those) : up to subsequences,

(i) "*tightness*" : there exists $y_n \in \mathbf{R}^3$ such that for all $\varepsilon > 0$, $R > 0$, $\int_{B(y_n,R)^c} f_n\, dx \leqq \varepsilon$.

(ii) "*vanishing*" : for all $R > 0$, $\lim\limits_{n\to+\infty} \sup\limits_{y\in\mathbf{R}^3} \int_{B(y,R)} f_n\, dx = 0$.

(iii) "*dichotomy*" : there exists $y_n \in \mathbf{R}^3$ and $a \in (0,1)$ such that for all $\varepsilon > 0$, $\exists\, R$, $R_n \underset{n\to+\infty}{\longrightarrow} +\infty$ with

$$\left|\int_{B(y_n,R)} g_n\, dx - a\lambda_n\right| \leqq \varepsilon, \qquad \left|\int_{B(y_n,R_n)} g_n\, dx - (1-a)\lambda_n\right| \leqq \varepsilon,$$

where $\lambda_n = \|f_n\|_{L^1(\mathbf{R}^3)}$.

The proof will be done in three steps. First we show that neither (ii) nor (iii) may occur. Then we prove that I_1^L is achieved if (i) holds.

First step. Let $a_n = \sup\limits_{y\in\mathbf{R}^3} \int_{B(y,1)} |\nabla\phi_n|^2 + |A(\phi_n)|^2\, dx$. If (ii) holds, $a_n \underset{n\to+\infty}{\longrightarrow} 0$. Moreover by using the Sobolev imbedding theorems and the Hölder inequality

$$\int_{B(y,1)} |\phi_n - P|^{6\alpha}\, dx \leqq C \left(\int_{B(y,1)} |\phi_n - P|^6 + 6|\phi_n - P|^5 |\nabla \phi_n|\, dx \right)^\alpha ,$$

for all $\alpha \in (1, \frac{3}{2})$, for all $y \in \mathbf{R}^3$ and

$$b_n = \sup_{y \in \mathbf{R}^3} \int_{B(y,1)} |\phi_n - P|^6 + 6|\phi_n - P|^5 |\nabla \phi_n|\, dx \underset{n \to +\infty}{\longrightarrow} 0.$$

On the other hand we can find $m > 0$ and $\{x_i\}_{i \in \mathbf{N}} \subset \mathbf{R}^3$ such that $\mathbf{R}^3 = \bigcup_{i \in \mathbf{N}} B(x_i, 1)$ and every point in $\mathbf{R}^3$ is at most in m balls of the form $B(x_i, 1)$. Therefore, for all $\alpha \in (1, \frac{3}{2})$

$$\int_{\mathbf{R}^3} |\phi_n - P|^{6\alpha}\, dx \leqq C\, m\, b_n^{\alpha - 1} \int_{\mathbf{R}^3} |\phi_n - P|^6 + 6|\phi_n - P|^5 |\nabla \phi_n|\, dx$$

and since $||\phi_n - P||_{L^\infty(\mathbf{R}^3)} \leqq 2$, this implies that for any α in $(1, \frac{3}{2})$, $||\phi_n - P||_{L^{6\alpha}(\mathbf{R}^3)} \underset{n \to +\infty}{\longrightarrow} 0$.

Then let us define $A_{n,\varepsilon} = \{x \in \mathbf{R}^3 \mid |\phi_n - P| \geqq \varepsilon\}$. Since $\phi_n - P$ converges to 0 in $L^7(\mathbf{R}^3)$, $\varepsilon^7 |A_{n,\varepsilon}| \underset{n \to +\infty}{\longrightarrow} 0$. Moreover,

(19)
$$\left| \int_{A_{n,\varepsilon}^c} \det(\phi_n - P, \nabla \phi_n)\, dx \right| \leqq \varepsilon \int_{\mathbf{R}^3} |\partial_n \phi_n \wedge \partial_2 \phi_n \wedge \partial_3 \phi_n|\, dx \leqq C\varepsilon$$

(20)
$$\left| \int_{A_{n,\varepsilon}} \det(\phi_n - P, \nabla \phi_n)\, dx \right| \leqq 2 \int_{A_{n,\varepsilon}} |\partial_1 \phi_n \wedge \partial_2 \phi_n \wedge \partial_3 \phi_n|\, dx \leqq$$
$$\leqq 2\varepsilon^{-7/4} |A_{n,\varepsilon}|^{1/4} C,$$

for $\mathcal{E}(\phi_n)$ is uniformly bounded and we have used Corollary 10. Obviously we may choose ε small and then n large enough in order to have

$$\left| \int_{\mathbf{R}^3} \det(\phi_n - P, \nabla \phi_n)\, dx \right| < \pi^2,$$

but this is a contradiction since $d(\phi_n) = 1$, and $\int_{\mathbf{R}^3} \det(P, \nabla \phi_n)\, dx = 0$ for all n (see Remark 13). Hence the sequence $\{\phi_n\}$ cannot vanish in the sense above.

Second step. If dichotomy occurs we may use Propositions A2 to A5 in the Appendix to cut ϕ_n into two pieces, ϕ_n^1 and ϕ_n^2 as follows.

$$\begin{aligned}
&\phi_n^1 \equiv \phi_n \text{ in } B(y_n, R'), \ \phi_n^2 \equiv \phi_n \text{ in } \mathbf{R}^3 \backslash B(y_n, 80R') \\
&\phi_n^1 \in X^{L_1 + \frac{M}{R'}}, \ \phi_n^2 \in X^{L - L_1 + \frac{M}{R'}}, \ d(\phi_n^1) + d(\phi_n^2) = 1 \text{ and} \\
&|\mathcal{E}(\phi_n) - \mathcal{E}(\phi_n^1) - \mathcal{E}(\phi_n^2)| \leqq C\varepsilon, \\
&\mathcal{E}(\phi_n^1), \ \mathcal{E}(\phi_n^2) \geqq \alpha > 0 \quad \text{for all} \quad n,
\end{aligned}$$

where M, C and α are positive constants independent of n and of $\varepsilon, L_1 \in (0, L)$ and R' is any number between R and R_n . Then since $\{\phi_n\}$ is a minimizing sequence for I_1^L we have

$$I_1^L \geqq I_k^{L_1+} + I_{1-k}^{L-L_1+} \tag{21}$$

where $k \in \mathbf{Z}$. Indeed we obtain this by taking ε small and then by choosing R' large enough.

For $k \neq 0, 1$, (21) is false for all $L_1 \in (0, L)$. Indeed, we only have to use Proposition 5 to see this. On the other hand, if k were equal to 0, (resp. 1) for R' large enough ϕ_n^2 (resp. ϕ_n^1) would be an element of X_1^L with $\mathcal{E}(\phi_n^2) < I_1^L$ (resp. $\mathcal{E}(\phi_n^1) < I_1^L$), which contradicts I_1^L's definition.

Third step. Assume that ϕ_n is tight up to a translation $x \to x + y_n$, $y_n \in \mathbf{R}^3$ and define $\widetilde{\phi}_n$ by $\widetilde{\phi}_n(\cdot) = \phi_n(\cdot + y_n)$. Corollary 11 can be applied to the sequence $\{\widetilde{\phi}_n\}$ to obtain that for some subsequence, still denoted by $\{\widetilde{\phi}_n\}$,

$$\det(\widetilde{\phi}_n, \nabla\widetilde{\phi}_n) \underset{n \to +\infty}{\longrightarrow} \det(\phi, \nabla\phi) \qquad \text{in } L^1(\mathbf{R}^3)\text{-weak} \tag{22}$$

where ϕ is a function in $\mathcal{D}^{1,2}(\mathbf{R}^3, S^3)$ such that $\nabla\widetilde{\phi}_n \underset{n}{\to} \nabla\phi$, $A(\widetilde{\phi}_n) \underset{n}{\to} A(\phi)$ in $L^2(\mathbf{R}^3)$-weak and $\widetilde{\phi}_n \underset{n}{\to} \phi$ a.e. and in $L^\infty(\mathbf{R}^3)$-weak*. Thus I_1^L will be achieved by ϕ if we show that $\phi \in X^L$. Indeed, (22) implies that $d(\phi) = 1$.

Since ϕ_n (and therefore $\widetilde{\phi}_n$ also) is in X^L for all n, there exist sequences of C^1 functions $\{\psi_n^\ell\}_{\ell \in \mathbf{N}}$ which approach $\widetilde{\phi}_n$ as ℓ goes to $+\infty$ in the sense of (P). Moreover, any bounded set of $L^2(\mathbf{R}^3)$ is metrizable when we consider L^2 endowed with the weak topology. Hence we can find a sequence $\{\psi_n^{\ell_n}\}_n$ such that

$$\nabla\psi_n^{\ell_n} \underset{n}{\to} \nabla\phi, \quad A(\phi_n^{\ell_n}) \underset{n}{\to} A(\phi) \quad \text{in} \quad L^2(\mathbf{R}^3)\text{-weak.}$$

Then if this sequence were tight uniformly in n, it would converge towards ϕ in the sense of (P) and ϕ would be in X^L.

By definition of $\{\tilde{\phi}_n\}$, there exists $\bar{R} > 0$ such that

$$\int_{B_{\bar{R}}} |\partial_1\tilde{\phi}_n \wedge \partial_2\tilde{\phi}_n \wedge \partial_3\tilde{\phi}_n|\, dx \geqslant 2\pi^2 - \varepsilon_o, \forall\, n, \tag{23}$$

where ε_o is some small fixed positive constant. Moreover without loss of generality we may assume that

$$\int_{B_{\bar{R}}} |\partial_1\psi_n^\ell \wedge \partial_2\psi_n^\ell \wedge \partial_3\psi_n^\ell|\, dx \geqslant 2\pi^2 - 2\varepsilon_o, \forall\, n, \forall\, \ell, \tag{24}$$

and in particular for $\ell = \ell_n$. Then the sequence $f_n = |\partial_1\psi_n^{\ell_n} \wedge \partial_2\psi_n^{\ell_n} \wedge \partial_3\psi_n^{\ell_n}|$ cannot vanish in the sense of concentration-compactness (see (ii) above). Furthermore if $\{f_n\}$ is tight, by using the same arguments as above, $\{\psi_n^{\ell_n}\}$ will converge to ϕ in the sense of (P), and $\phi \in X^L$. The only remaining case is that of dichotomy for f_n.

Assume first that dichotomy arises, the translation sequence $\{y_n\}$ being bounded (see (i)-(iii) in the proof of Theorem 1).

Then we define a new sequence $\varphi_n \equiv T_R(\psi_n^{\ell_n}) \in C^1(\mathbf{R}^3, S^3)$ and φ_n will satisfy all the necessary properties : $E(\varphi_n) \leqslant L$ for all n, $\varphi_n \underset{n}{\rightarrow} \phi$ in the sense of (P). If $\{y_n\}$ were unbounded, we could cut f_n (and $\psi_n^{\ell_n}$) into two pieces f_n^1 and f_n^2, f_n^1 being basically supported in a finite ball centered at y_n for all n. Moreover, from (24), by taking ε_o small enough,

$$\int_{\mathbf{R}^3} f_n^2\, dx \leqslant \lambda_n - \pi^2, \quad \lambda_n = \int_{\mathbf{R}^3} |\partial_1\psi_n^{\ell_n} \wedge \partial_2\psi_n^{\ell_n} \wedge \partial_3\psi_n^{\ell_n}|\, dx.$$

On the other hand, it is easy to see that f_n^2 can be constructed in such a way that the corresponding $(\psi_n^{\ell_n})'^2$ have still degree equal to 1. Now, if $\{f_n^2\}$ is not tight around the origin we can go on with this procedure and either end the proof or reach a contradiction in a finite number of steps (remember that $\mathcal{E}(\psi_n^{\ell_n}) \leqslant L$ and that $d(\phi) = 1$ implies $\int_{\mathbf{R}^3} |\partial_1\phi \wedge \partial_2\phi \wedge \partial_3\phi|\, dx \geqslant 2\pi^2$). QED

Proof of Theorem 2. This proof is done quite similarly as the previous one. The only difference lies on the fact that inequality (5) forbids inequalities as (21) which appear when considering the possibility of dichotomy. The rest of the proof follows quite in the same way, by using the same arguments, and we will skip it. QED

3. Remarks and Open Questions

Here we discuss the new setting we have defined to solve Skyrme's problem. X_k^L's definition looks a little bit complicated, as well as the technical results involved. For instance, the main difficulty in the proof of Theorem 1 is to prove that ϕ is in X_1^L. This difficulty is two fold. First, one has to prove that we have found ϕ with $d(\phi)$ equal to 1. And this is a real problem, since the invariance of (4) by any translation in $\mathbf{R}^3$ is a source of lack of compactness and makes the constraint $d(\phi) = 1$ not compact. But the second part of the problem, i.e., to prove that $\phi \in X^L$ is technical and more or less artificial, but necessary once we have chosen the minimizing class to be X_1^L.

Another strange feature of this setting is that I_1^L is nonincreasing in L. But is it not constant very often? Theorem 1 proves that there are as many minimizers of $\mathcal{E}$ with degree 1 as different finite values taken by I_1^L. We do not think that many minimizers with different energy exist and therefore, our conjecture is that I_1^L takes only a finite number of finite values.

Other choices for the class to work in would be of course possible. The one we have made, which takes (P) as an approximation property seems to us one of the simplest. For example, we could have chosen a new class to work in as follows. Let $\psi \in C^1(\mathbf{R}^3, S^3)$ be such that $\mathcal{E}(\psi) < +\infty$ and $d(\psi) = 1$. Then for any $L > \mathcal{E}(\psi)$, for any set K weakly compact in $L^1(\mathbf{R}^3)$ which contains $\det(\psi, \nabla\psi)$, we define

$$\begin{aligned} X_L^K = \{\phi \mid \mathcal{E}(\phi) < +\infty, \quad & \exists \psi^\ell \in C^1(\mathbf{R}^3, S^3) \text{ s.t.} \\ & \nabla\psi^\ell \underset{\ell}{\rightharpoonup} \nabla\phi, \quad A(\psi^\ell) \underset{\ell}{\rightharpoonup} A(\phi) \quad \text{weakly in} \quad L^2(\mathbf{R}^3), \\ & \mathcal{E}(\psi^\ell) \leqslant L, \quad \det(\psi^\ell, \nabla\psi^\ell) \in K \ \forall \, \ell \,\} \end{aligned}$$

This class looks a little bit more artificial than X_L since we have added a new constraint with respect to $\det(\psi^\ell, \nabla\psi^\ell)$. The good feature of X_L^K is that it is metrizable for the L^2-weak topology w.r.t. $A(\phi)$, $\nabla\phi$ and metrizable for the L^1-weak topology for $\det(\phi, \nabla\phi)$. This simplifies considerably the end of the proof of Theorem 1, since it is easy to construct diagonal sequences in metrizable sets.

Our conjecture is that all these choices are only technical and not relevant as far as the minima for I_k are concerned.

All the difficulties we have, to choose the right class to minimize $\mathcal{E}$, are due to the lack of density results of smooth functions in the set of finite energy ϕ 's. If a convenient density result was proved, then all the classes we may consider are the same and the technique involved is easier and simpler. Without such a density property, all the classes X^L are *a priori* different and thus this allows us to obtain minimizers with possible different energies. Lavrentiev phenomena would then appear.

On the other hand, another way to look at the problem is to consider extended Skyrme's problems, with an energy $\mathcal{E}$ including more terms, taking into account other interactions between different mesons. Following this way, other models have been proposed and partially studied (see [1] and [9]). From the mathematical point of view, some of them are easily tractable with the techniques used here. Nevertheless technical difficulties remain open which are produced by the specificities of the new energy functionals.

Appendix

Remark A1. Let ϕ be a function in any class X_1^L for L large enough. Then by using spherical coordinates in $\mathbf{R}^4$ we can write ϕ as

$$\text{(a1)} \qquad \begin{aligned} \phi^1 &= \cos\varphi\cos\xi\cos\theta \\ \phi^2 &= \cos\varphi\cos\xi\sin\theta \\ \phi^3 &= \cos\varphi\sin\xi \\ \phi^4 &= \sin\varphi, \end{aligned}$$

φ, ξ being functions from $\mathbf{R}^3$ into $[-\frac{\pi}{2}, \frac{\pi}{2}]$ and $\theta : \mathbf{R}^3 \to [0, 2\pi]$. On the other hand, having rotated ϕ in order to have $\phi - P \in L^6(\mathbf{R}^3)$, $P = (1,0,0,0)$, we see that in some vague sense to be defined below, $\varphi(+\infty) = \xi(+\infty) = \theta(+\infty) = 0$. With this notation all the functionals relevant in this paper can be rewritten as follows

$$\text{(a2)} \qquad \begin{aligned} &\det(\phi, \nabla\phi) = \cos^2\varphi\cos\xi(\nabla\theta \cdot \nabla\xi \wedge \nabla\varphi) \\ &|\partial_1\phi \wedge \partial_2\phi \wedge \partial_3\phi| = \cos^2\varphi\cos\xi|\nabla\theta \cdot \nabla\xi \wedge \nabla\varphi| \\ &|\nabla\phi|^2 = |\nabla\varphi|^2 + \cos^2\varphi|\nabla\xi|^2 + \cos^2\varphi\cos^2\xi|\nabla\theta|^2 \\ &|A(\phi)|^2 = \sum_{i,j=1}^{3} [\cos^2\varphi(A_{ij}^{\varphi,\xi})^2 + \cos^2\varphi\cos^2\xi(A_{ij}^{\varphi,\theta})^2 \\ &\qquad + \cos^4\varphi\cos^2\xi(A_{ij}^{\xi,\theta})^2], \end{aligned}$$

where $A_{ij}^{f,g} = \frac{\partial f}{\partial x_i} \wedge \frac{\partial g}{\partial x_j}$. QED

The next two results deal with cutting-off functions $\phi : \mathbf{R}^3 \to S^3$ in order to obtain functions ψ which are constant either outside a ball B_R or inside it and which still map $\mathbf{R}^3$ into S^3.

Proposition A2. *Let $\phi : \mathbf{R}^3 \to S^3$ be a function with finite energy $\mathcal{E}(\phi)$. Then for every $R > 0$ there exists a function $T_R(\phi) : \mathbf{R}^3 \to S^3$ satisfying :*

$$\begin{aligned}
&\text{(a3)} \qquad T_R(\phi) \equiv \phi \ \text{ in } \ B_R, \quad T_R(\phi) = P \ \text{ in } \ \mathbf{R}^3 \backslash B_{8R} \\
&\text{(a4)} \qquad |\mathcal{E}(T_R(\phi)) - \mathcal{E}(\phi; B_R)| \leqq C\, \mathcal{E}(\phi; B_{8R} \backslash B_R),
\end{aligned}$$

where $C > 0$ is independent of ϕ and R, and where by $\mathcal{E}(\phi; B)$ we denote $\int_B |\nabla\phi|^2 + |A(\phi)|^2\, dx$, B being any measurable set of $\mathbf{R}^3$.

Proof. Let us construct $T_R(\phi)$ explicitly. Consider $m \in \mathcal{D}(\mathbf{R}^3)$ such that $0 \leqq m \leqq 1$, $m \equiv 0$ in B_2^c, $m \equiv 1$ in B_1 and then for all $R > 0$ define m_R by $m_R(\cdot) = m(\frac{\cdot}{R})$. Then we define $T_R(\phi)$ in spherical coordinates $(\varphi^R, \xi^R, \theta^R)$ as follows

$$\varphi^R = m_{4R}, \quad \xi^R = m_{2R}\xi, \quad \theta^R = m_R\theta.$$

This defines $T_R(\theta)$ completely. (a3) is obviously satisfied and (a4) is proved exactly as in [3] and therefore we will not do it here. QED

Proposition A3. *Consider $\phi : \mathbf{R}^3 \to S^3$ with $\mathcal{E}(\phi) < +\infty$. Then there exists a function $S_R(\phi) : \mathbf{R}^3 \to S^3$ satisfying*

$$\begin{aligned}
&\text{(a5)} \qquad S_R(\phi) \equiv \phi \ \text{ in } \ \mathbf{R}^3 \backslash B_{8R}, \quad S_R(\phi) \equiv P \ \text{ in } \ B_R \\
&\text{(a6)} \qquad |\mathcal{E}(S_R(\phi)) - \mathcal{E}(\phi; B_R^c)| \leqq C\, \mathcal{E}(\phi; B_{8R} \backslash B_R)
\end{aligned}$$

with $C > 0$ independent of ϕ and of R.

Proof. We change the proof of Proposition A2 by considering a new function $n \in \mathcal{D}(\mathbf{R}^3) : 0 \leqq n \leqq 1$, $n \equiv 0$ in $B_{1/2}$, $n \equiv 1$ in B_1^c which will replace m. QED

Finally we study the stability of X^L and of the convergence in the sense of (P) by cutting-off functions as above.

Proposition A4. *Assume that $\psi_n, \phi : \mathbf{R}^3 \to S^3$, ψ_n being of class C^1 and such that*

$$\text{(P)} \qquad \begin{aligned}
&\nabla\psi_n \underset{n}{\rightharpoonup} \nabla\phi, \quad A(\psi_n) \underset{n}{\rightharpoonup} A(\phi) \quad \text{weakly in} \quad L^2(\mathbf{R}^3), \\
&\det(\psi_n, \nabla\psi_n) \to \det(\phi, \nabla\phi) \quad \text{in the narrow topology of measures,}
\end{aligned}$$

then the same convergence properties are statisfied by $T_R(\psi_n)$ and $T_R(\phi)$ (resp. $S_R(\psi_n)$ and $S_R(\phi)$).

Proof. Let us do the proof only for the cut-off operator T_R. The same arguments apply to S_R.

Remark that the only thing to prove is that $\nabla T_R(\psi_n) \underset{n}{\rightarrow} \nabla T_R(\phi)$ and $A(T_R(\psi_n)) \underset{n}{\rightarrow} A(T_R(\phi))$ weakly in $L^2(\mathbf{R}^3)$. Indeed the sequence $\{T_R(\psi_n)\}$ being constant outside B_{8R}, Corollary 11 applies.

On the other hand the definition of T_R clearly shows that whenever $\psi_n \underset{n}{\rightarrow} \phi$ a.e., $T_R(\psi_n) \underset{n}{\rightarrow} T_R(\phi)$ a.e.. Therefore the only possible weak limit of subsequences of $\nabla T_R(\psi_n)$ (resp. $A(T_R(\psi_n))$) in $L^2(\mathbf{R}^3)$ is $\nabla T_R(\phi)$ (resp. $A(T_R(\phi))$). Then, since for some constants $C > 0$, $\mathcal{E}(T_R(\psi_n)) \leqq C\ \mathcal{E}(\psi_n) \leqq CL$, (P) holds.
QED

Proposition A5. *Let ϕ be any function in X^L. Then for all $R > 0$, $T_R(\phi)$ and $S_R(\phi)$ are in X^{L^1}, with $L^1 = L + \frac{CL}{R} + C\ \mathcal{E}(\phi; B_{8R}\backslash B_R)$, for some positive constant C.*

Corollary A6. *Consider a sequence $\{\phi_n\}$ in X^L such that $\mathcal{E}(\phi_n; B_{R_n}\backslash B_R) \leqq \varepsilon$ with $R_n \underset{n}{\rightarrow} +\infty$. Then for all $\delta > 0$ there exists $R' > R$ such that $\phi_n^{R'} \in X^{L+C\varepsilon+\delta}$.*

Proof of Proposition A5. Let $\phi \in X^L$ and let $\{\psi_j\}_{j\in\mathbf{N}} \in C^1(\mathbf{R}^3, S^3)$ be a sequence such that $\mathcal{E}(\psi_j) \leqq L$ and

(a7) $$\nabla\psi_j \underset{j}{\rightarrow} \nabla\phi,\ A(\psi_j) \underset{j}{\rightarrow} A(\phi) \quad \text{weakly in} \quad L^2(\mathbf{R}^3).$$

Then we truncate ψ_j by considering $T_R(\psi_j)$. Then the energy $\mathcal{E}$ of this function can be decomposed as follows

(a8)
$$\mathcal{E}(T_R(\psi_j)) = \mathcal{E}(\psi_j; B_R) + \int_{B_{8R}\backslash B_R} |\nabla T_R(\psi_j)|^2\,dx + \int_{B_{8R}\backslash B_R} |A(T_R(\psi_j))|^2\,dx.$$

From (a2) and the definition of T_R, we can easily show that the last term in (a8) is less than or equal to

(a9) $$\int_{B_{8R}\backslash B_R} |A(\psi_j)|^2\,dx + \frac{1}{R}0\,(\mathcal{E}(\psi_j; B_{8R}\backslash B_R)).$$

On the other hand we use again (a2) and (a7) to show that

(a10) $$\lim_{j\rightarrow+\infty} \int_{B_{8R}\backslash B_R} (|\nabla T_R(\psi_j)|^2 - |\nabla\psi_j|^2)\,dx \leqq \mathcal{E}(T_R(\phi); B_{8R}\backslash B_R).$$

Then (a4),(a9) and (a10) prove the proposition. QED

REFERENCES

[1] Adkins, G. S., C. R. Nappi, E. Witten, *Static properties of nucleons in the Skyrme model*, Nucl. Phys. **B228** (1983), pp. 552-566.

[2] Ball, J., *Convexity conditions and existence theorems in nonlinear elasticity*, Arch. Ration. Mech. Anal. **63** (1977), pp. 337-403.

[3] Esteban, M. J., *A direct variational approach to Skyrme's model for meson fields*, Comm. Math. Phys. **105** (1986), pp. 571-591.

[4] Lions, P. L., *The concentration-compactness principle in the calculus of variations ; Part I*, Ann. IHP Anal. Non Lin. **1** (1984), pp. 109-145; *Part II*, Ann. IHP Anal. Non Lin. **1** (1984), 223-283.

[5] Murat, F., *Compacité par compensation : conditions nécessaires et suffisantes de continuité faible sous une hypothèse de rang constant*, Ann. Soc. Norm. Super. Pisa, IV **VIII, 1** (1981), 69-102.

[6] Reshetnyak, Y. G., *Stability theorems for mappings with bounded excursions*, Sib. Math. J. **9** (1968), 499-512.

[7] Skyrme, T. H. R., *A non-linear field theory*, Proc. Roy. Soc. **A260** (1961), 127-138.

[8] Tartar, L., *Compensated compactness and applications to partial differential equations*, Non-linear analysis and mechanics : Heriot-Watt Symposium **IV**. Knops, R. J. (ed.), New York: Pitman (1979).

[9] Vinh Mau, G. S., M. Lacombe, B. Loiseau, W. N. Cottingham, and P. Lisboa, *The static baryon-baryon potential in the Skyrme model*, preprint.

[10] Witten, E., *Baryons in the* $1/N$ *expansion*, Nucl. Phys. **B160** (1979), pp. 57-115.

Maria J. Esteban
Analyse Numérique
Université Paris VI
4, place Jussieu
75252 Paris CEDEX 05
France

Relative Category and The Calculus of Variations

G. FOURNIER and M. WILLEM

1. Introduction

Contrary to Morse theory, Lusternik–Schnirelman theory is not applicable to functions which are unbounded from below. In order to overcome this difficulty, a notion of *relative category* was introduced in [6]. Under some assumptions, the following estimate is true:

$$\#\{u \in \varphi^{-1}([a,b]) : \varphi'(u) = 0\} \geq \text{cat}_{\varphi^b,\varphi^a}(\varphi^b)$$

where $\varphi^c = \varphi^{-1}(]-\infty, c])$.

The first aim of this paper is to define a notion of *relative cuplength* such that

$$\text{cat}_{X,Y}(X) \geq \text{cuplength}(X,Y) + 1.$$

Moreover, the following formula is useful in the study of asymptotically linear variational problems:

$$\text{cuplength}(X \times V, Y \times V) \geq \text{cuplength}(V) + \text{cuplength}(X,Y).$$

The final section concerns the existence of multiple periodic solutions of the Hamiltonian system

$$J\dot{u} + \nabla H(t,u) = h(t) \tag{1}$$

where ∇H is asymptotically linear at infinity:

$$\nabla H(t,u) = A_\infty(t)u + o(u), |u| \to +\infty.$$

Let N be the space of solutions of

$$J\dot{u} + A_\infty(t)u = 0$$
$$u(0) = u(T).$$

We prove that if $H_u''(t,u)$ is bounded, H satisfy a periodicity condition on N and h is orthogonal to N, then (1) has at least $k+1$ T-periodic solutions where k is the dimension of N. This theorem generalizes recent results due to K.C. Chang [2] and A. Fonda and J. Mawhin [5].

2A. Relative Category

In this section, we recall the definition of relative category and some elementary properties.

Definition 1. Let Y and A be closed subsets of a topological space X. The category of A in X relative to Y is the least integer k such that

$$A = \bigcup_{j=0}^{k} A_j$$

where, for $0 \leq j \leq k$, A_j is closed and there exists $h_j \in C([0,1] \times A_j, X)$ such that

(a) $h_j(0,u) = u$ for $x \in A_j$, $0 \leq j \leq k$,

(b) $h_0(1,u) \in Y$ for $x \in A_0$ and $h_0(t,y) = y$ for $y \in A_0 \cap Y$ and $t \in [0,1]$,

(c) $h_j(1,u) = u_j$ for $u \in A_j$ and some $u_j \in X$, $1 \leq j \leq k$.

The category of A in X relative to Y is denoted by $\text{cat}_{X,Y}(A)$. The Lusternik–Schnirelman category of A in X is defined by $\text{cat}_X(A) = \text{cat}_{X,\phi}(A)$. See [4] for a related notion of relative category.

Proposition 1 [6]. *Let A, B, Y be closed subsets of X.*

(i) *If $A \subset B$, then* $\text{cat}_{X,Y}(A) \leq \text{cat}_{X,Y}(B)$,

(ii) $\text{cat}_{X,Y}(A \cup B) \leq \text{cat}_{X,Y}(A) + \text{cat}_X(B)$,

(iii) *if there exists $h \in C([0,1] \times A, X)$ such that $h(t,y) = y$ for $y \in A \cap Y$ and $t \in [0,1]$, then* $\text{cat}_{X,Y}(A) \leq \text{cat}_{X,Y}(B)$ *where $B = h(1, A)$.*

2B. Relative Cuplength

Let us recall that, if X is an ANR (or in particular a manifold with boundary)

$$\text{cat}_X(X) \geq \text{cuplength}(X).$$

Using a notion of relative cuplength, we shall obtain a similar estimate for the relative category. We use the singular cohomology over the real field and we denote by $\cup$ as the cup product.

Definition 2. Let Y be a closed subset of a topological space X. The cuplength of X relative to Y is the largest integer n such that

$$\alpha_0 \cup \alpha_1 \cup \ldots \cup \alpha_n \neq 0$$

where $\alpha_0 \in H^*(X,Y)$, $\alpha_m \in H^*(X)$ and $* \geq 1$ for $1 \leq m \leq n$. The cuplength of X relative to Y is denoted by cuplength (X,Y). It is $-\infty$ if no such α_0 exists.

Theorem 1. *If X is an ANR then*

$$\mathrm{cat}_{X,Y}(X) \geq 1 + \mathrm{cuplength}(X,Y).$$

Proof. (1) Assume that $k = \mathrm{cat}_{X,Y}(X) \geq 1$. Let $A_0, \ldots, A_k$, $h_0, \ldots, h_k$ be as in Definition 1. Since $h_0(t,y) = y$ for $y \in A_0 \cap Y$ and $t \in [0,1]$, we can assume, without loss of generality, that $Y \subset A_0$.

(2) Consider the exact sequence

$$\ldots \to H^*(X, A_0) \xrightarrow{l_0^*} H^*(X,Y) \xrightarrow{i_0^*} H^*(A_0, Y) \to \ldots$$

where $i_0\colon (A_0, Y) \to (X,Y)$ and $l_0\colon (X,Y) \to (X, A_0)$ are inclusions. By assumption, the inclusion i_0 is homotopic to the map $h_0(1,\cdot)$ and $h_1(l, A_0) = Y$. Thus $i_0^* = 0$ and, by exactness, l_0^* is subjective.

(3) Consider now the exact sequence

$$\ldots \to H^*(X, A_j) \xrightarrow{l_j^*} H^*(X) \xrightarrow{i_j^*} H^*(A_j) \to \ldots$$

where $i_j\colon A_j \to X$ and $l_j\colon (X, \phi) \to (X, A_j)$ are inclusions. By assumption, for $j \geq 1$, the map i_j is homotopic to a constant map. Thus, for $* \geq 1$, $i_j^* = 0$ and, by exactness, l_j^* is surjective.

(4) The following diagram is commutative

$$\begin{array}{ccccccccc} H^*(X,A_0) & \otimes & H^*(X,A_1) & \otimes\ldots\otimes & H^*(X,A_k) & \xrightarrow{\cup} & H^*(X,X) \equiv \{0\} \\ \downarrow l_0^* & & \downarrow l_1^* & & \downarrow l_k^* & & \downarrow \\ H^*(X,Y) & \otimes & H^*(X) & \otimes\ldots\otimes & H^*(X) & \xrightarrow{\cup} & H^*(X,Y), \end{array}$$

because being an ANR we may consider all those couples to be excisive. Consider $\alpha_0 \in H^*(X,Y)$, $\alpha_j \in H^*(X)$ and $* \geq 1$ for $1 \leq j \leq k$. Using the surjectivity of l_0^* for $* \geq 1$ and $1 \leq j \leq k$, we obtain $\alpha_0 \cup \alpha_1 \cup \ldots \cup \alpha_k = 0$. The definition of relative cuplength implies that

$$\mathrm{cuplength}(X,Y) \leq k-1.$$

QED

Theorem 2. *Let X and V be topological spaces and let Y be a closed subset of X. If $H^*(X,Y)$ or $H^*(V)$ is of finite type, then*

$$\mathrm{cuplength}(X \times V, Y \times V) \geq \mathrm{cuplength}(V) + \mathrm{cuplength}(X,Y).$$

Proof. Let $n = \text{cuplength}(X,Y)$ and $k = \text{cuplength}(V)$. By definition, we have

$$\alpha_0 \cup \alpha_1 \cup \ldots \cup \alpha_n \neq 0$$

where $\alpha_0 \in H^*(X,Y)$, $\alpha_m \in H^*(X)$ and $* \geq 1$ for $1 \leq m \leq n$ and

$$\beta_1 \cup \ldots \cup \beta_k \neq 0$$

where $\beta_m \in H^*(V)$ and $* \geq 1$ for $1 \leq m \leq k$. Using the Künneth formula [9, p. 249], we obtain

$$\begin{aligned} 0 &\neq K[(\alpha_0 \cup \ldots \cup \alpha_n) \otimes (\beta_1 \cup \ldots \cup \beta_k)] \\ &= p_3^*(\alpha_0 \cup \ldots \cup \alpha_n) \cup p_2^*(\beta_1 \cup \ldots \cup \beta_k) \\ &= p_3^*\alpha_0 \cup p_1^*\alpha_1 \cup \ldots \cup p_1^*\alpha_n \cup p_2^*\beta_1 \cup \ldots \cup p_2^*\beta_k \end{aligned}$$

where $p_1 : X \times V \to X$, $p_2 : X \times V \to V$ and $p_3 : (X,Y) \times V \to (X,Y)$ are projections. The definition of relative cuplength implies that

$$\text{cuplength}(X \times V, Y \times V) \geq n + k. \qquad \text{QED}$$

Remark. If $\{X \times W, Y \times V\}$ is an excisive couple in $X \times V$ and if $H^*(X,Y)$ or $H^*(V,W)$ is of finite type, we obtain the following result by a similar proof:

$$\text{cuplength}((X,Y) \times (V,W)) \geq \text{cuplength}(X,Y) + \text{cuplength}(V,W).$$

3. Applications to Critical Point Theory

Let M be a C^2 Finsler manifold, i.e., a C^2 Banach manifold with a Finsler structure on its tangent bundle and let $\varphi \in C^1(M, \mathbf{R})$.

The function φ satisfies the *Palais–Smale condition on a closed subset S of M* if every sequence $(u_j) \subset S$ such that $(\varphi(u_j))$ is bounded and $\varphi'(u_j) \to 0$ contains a convergent subsequence.

The following result contains the basic application of relative category to critical point theory.

Theorem 3 [6]. *If $a < b$ are regular values of φ and if φ satisfies the Palais–Smale condition on $\varphi^{-1}([a,b])$, then*

$$\#\{u \in \varphi^{-1}([a,b]) : \varphi'(u) = 0\} \geq \text{cat}_{\varphi^b, \varphi^a}(\varphi^b).$$

Remarks. 1. The minimax characterization of the critical values is given by

$$c_j = \inf_{A \in \mathcal{A}_j} \sup_A \varphi$$

where

$$\mathcal{A}_j = \{A \subset \varphi^b : A \text{ is closed, } \operatorname{cat}_{\varphi^b, \varphi^a}(A) \geq j\}.$$

The proof depends only on Proposition 1 and on a standard deformation lemma.

2. If M is compact, we obtain the classical Lusternik–Schnirelman theorem ([7]) by setting

$$a = \min_M \varphi - 1, \qquad b = \max_M \varphi + 1.$$

3. We shall now generalize a theorem due to Chang ([1]). This result implies in particular a conjecture of Arnold, first proved by Conley and Zehnder ([3]).

Theorem 4. *Let L be a symmetric invertible operator defined on a Hilbert space X and let $\psi \in C^1(X \times V, \mathbf{R})$ where V is a C^2-compact manifold. Assume that L is positive definite on a subspace of positive finite codimension and that $\nabla_u \psi$ is compact and $\nabla_u \psi(u,v) = o(|u|)$ as $|u| \to \infty$ uniformly in v. Then the function*

$$\varphi(u,v) = (1/2)(Lu,u) + \psi(u,v)$$

has at least cuplength$(V) + 1$ *critical points.*

Lemma 1. *Let $a < 0 < b$ and define $\chi(u) = (1/2)(Lu,u)$. Then*

$$\operatorname{cuplength}(\chi^b, \chi^a) = 0.$$

Proof. (1) By assumption X is the orthogonal sum of X^+ and X^- and L is positive definite (resp. negative definite) on X^+ (resp. on X^-). Moreover, $k = \dim X^-$ is finite. By using the deformation

$$[0,1] \times X \to X : (t,u) \mapsto u^- + (1-t)u^+,$$

we obtain

$$H^p(\chi^b, \chi^a) \cong H^p(B^k, S^{k-1}) \cong \delta_{p,k}\mathbf{R}.$$

Thus cuplength$(\chi^b, \chi^a) \geq 0$.

(2) Since 0 is the only critical point of χ, it follows from Theorems 1 and 3 that

$$1 \geq \mathrm{cat}_{\chi^b,\chi^a}(\chi^b) \geq 1 + \mathrm{cuplength}(\chi^b,\chi^a). \qquad \text{QED}$$

Lemma 2. *There exists $\rho > 0$ and $\hat{\sigma} \in C^\infty(X,\mathbf{R})$ satisfying the following conditions:*

(a) $\nabla_u \varphi(u,v) = 0$ *implies* $\|u\| \leq \rho$,

(b) $\hat{\sigma}(u) = 1$ *if* $0 \leq \|u\| \leq \rho$ *and* $\hat{\sigma}(u) = 0$ *if* $\|u\| \geq 2\rho$,

(c) *the functional*

$$\hat{\varphi}(u,v) = (1/2)(Lu,u) + \hat{\sigma}(u)\psi(u,v)$$

is such that $\|\nabla_u \hat{\varphi}(u,v)\| \geq 1$ *if* $\rho \leq \|u\| \leq 2\rho$.

Proof. Similar to the proof of [8, Proposition V.9].

Proof of Theorem 4. (1) By Lemma 2, $\varphi'(u,v) = 0$ if and only if $\hat{\varphi}'(u,v) = 0$. Thus it suffices to prove that $\hat{\varphi}$ has at least cuplength $(V)+1$ critical points. It is easy to verify that $\hat{\varphi}$ satisfies the Palais–Smale condition on $M = X \times V$. Let us define

$$a = \inf_{B[0,2\rho]\times V} \hat{\varphi} - 1, \qquad b = \sup_{B[0,2\rho]\times V} \hat{\varphi} + 1.$$

Lemma 2 implies that $\hat{\varphi}^{-1}(]a,b[)$ contains the critical points of $\hat{\varphi}$ and that $\hat{\varphi}^a = \chi^a \times V$, $\hat{\varphi}^b = \chi^b \times V$.

(2) By Theorem 3, we have

$$\#\{(u,v) \in X \times V : \hat{\varphi}'(u,v) = 0\} \geq \mathrm{cat}_{\hat{\varphi}^b,\hat{\varphi}^a}(\hat{\varphi}^b). \tag{2}$$

Theorem 1 implies that

$$\begin{aligned} \mathrm{cat}_{\hat{\varphi}^b,\hat{\varphi}^a}(\hat{\varphi}^b) &\geq 1 + \mathrm{cuplength}(\hat{\varphi}^b,\hat{\varphi}^a) \\ &= 1 + \mathrm{cuplength}(\chi^b \times V, \chi^a \times V). \end{aligned} \tag{3}$$

It follows from Theorem 2 and Lemma 1 that

$$\begin{aligned} \mathrm{cuplength}(\chi^b \times V, \chi^a \times V) &\geq \mathrm{cuplength}(V) + \mathrm{cuplength}(\chi^b,\chi^a) \\ &= \mathrm{cuplength}(V). \end{aligned} \tag{4}$$

We obtain, from (2), (3) and (4),

$$\#\{(u,v) \in X \times V : \hat{\varphi}'(u,v) = 0\} \geq 1 + \text{cuplength}(V). \qquad \text{QED}$$

4. Periodic Solutions of Hamiltonian Systems

This section is devoted to the problem

$$(5) \qquad \begin{aligned} J\dot{x} + \nabla H(t,x) &= h(t) \\ x(0) &= x(T) \end{aligned}$$

where H, ∇H and H''_x are continuous on $[0,T] \times \mathbf{R}^{2N}$ and h is continuous on $[0,T]$.

We assume that there exists a symmetric matrix $A_\infty(t)$ which is continuous in t and such that

$$(6) \qquad \nabla H(t,x) = A_\infty(t)x + o(|x|), \qquad |x| \to +\infty,$$

uniformly in t. We denote by N the space of solutions of

$$\begin{aligned} J\dot{x} + A_\infty(t)x &= 0 \\ x(0) &= x(T). \end{aligned}$$

We assume that:

(A_1) $N = \text{span}(v_1, \ldots, v_k) \subset \mathbf{R}^{2N}$,

(A_2) $H(t, x + v_j) = H(t,x)$, $\forall t \in [0,1]$, $\forall x \in \mathbf{R}^{2N}$, $\forall j \in \{1, \ldots, k\}$.

(A_3) $\int_0^T (h(t), v_j)dt = 0$, $\forall j \in \{1, \ldots, k\}$,

(A_4) H''_x is bounded on $[0,T] \times \mathbf{R}^{2N}$.

Let us denote by G the additive group

$$\left\{ \sum_{j=1}^{k} z_j v_j : z_j \in \mathrm{Z},\ 1 \leq j \leq k \right\}.$$

Two solutions x_1 and x_2 of (5) are *geometrically distinct* if $x_1 - x_2 \notin G$.

Theorem 5. *Under the above assumptions, problem* (5) *has at least* $k+1$ *geometrically distinct solutions.*

Proof. In order to apply Theorem 4, we need a finite dimensional reduction since problem (5) is strongly indefinite. Define on $L^2 = L^2(0, T; \mathbf{R}^{2N})$ the linear operator L by

$$D(L) = \{x \in L^2 : \dot{x} \in L^2 \quad \text{and} \quad x(0) = x(T)\}$$

$$Lx(t) = J\dot{x}(t) + A_\infty(t)x(t)$$

and the nonlinear operator B by

$$Bx(t) = \nabla H(t, x(t)) - A_\infty(t)x(t).$$

Let $P: L^2 \to L^2$ be an orthogonal projector such that

$$N = \text{Ker } L \subset R(P) \subset D(L)$$

and

$$LPx = PLx$$

for every $x \in D(L)$. Problem (5) is then equivalent to the system

$$(7) \qquad \begin{aligned} &Ly + PB(y+z) = Ph \\ &Lz + (I-P)B(y+z) = (I-P)h, \end{aligned}$$

where $y \in R(P)$, $z \in D(L) \cap R(I-P)$. By assumption (A_4), there exists a projector P with finite rank satisfying the above assumptions and such that for every $y \in R(P)$ there exists a unique $z \in R(I-P) \cap D(L)$ which solves (7). Moreover the corresponding map $\Phi: y \to z$ is C^1 and $x = y + \Phi(y)$ is a solution of (5) if and only if y is a critical point of the C^2 functional

$$\begin{aligned} f(y) = \frac{1}{2}(L(y+\Phi y), y+\Phi y)_{L^2} + \int_0^T H(t, y(t) + \Phi y(t))dt - \\ (A_\infty(t)(y+\Phi y), y+\Phi y)_{L^2} - (h, y)_{L^2} - (h, \Phi(y))_{L^2}. \end{aligned}$$

(A complete proof is contained in [1] and [5].) It follows from (A_2) and (A_3) that

$$f(y + v_j) = f(y)$$

for every $y \in R(P)$ and $j \in \{1, \ldots, k\}$. Let us denote by X the orthogonal of N in $R(P)$. It suffices then to find critical points of the functional φ defined on $R(P)/G \cong X \times \mathbf{T}^k$ by

$$\varphi(u, v) = f(u + v).$$

By assumption (6), the functional

$$\psi(u,v) = \varphi(u,v) - \frac{1}{2}(Lu,u)$$

is such that $\nabla_u \psi(u,v) = o(|u|)$ as $|u| \to +\infty$ uniformly in v. Since X is finite-dimensional, $\nabla_u \psi$ is compact. Finally, assumption (A_1) implies the invertibility of L. Thus, by Theorem 4, φ has at least

$$\text{cuplength}(\mathbf{T}^k) + 1 = k + 1$$

critical points. QED

Remarks. (1) Theorem 5 generalizes Theorem 5 of [2] and Theorem 4 of [5] where it is assumed that $\nabla H(t,x) - A_\infty(t)x$ is bounded.

(2) If the critical points of φ are nondegenerate, Morse theory implies the existence of 2^k geometrically distinct solutions of (5).

(3) Let us consider two special cases:

(a) (Arnold conjecture). If H is periodic in each variable, then $k = 2N$ and (5) has at least $2N+1$ geometrically distinct solutions if $\int_0^T h(t)dt = 0$. This result was first proved in [3].

(b) Assume that H is periodic in $x_1, \ldots, x_N$ and that $A_\infty(t) = \text{diag}[0, B_\infty(t)]$. If $\int_0^T B_\infty(t)dt$ is invertible and if $\int_0^T h_n(t)dt = 0$, $n = 1, \ldots, N$, then (5) has at least $N+1$ geometrically distinct solutions. This result contains in particular Theorem 3 of [3] and the forced pendulum equation in Hamiltonian form.

REFERENCES

[1] K.C. Chang, *Infinite dimensional Morse theory and its applications,* Séminaire de Mathématiques supérieures, Presses de l'Université de Montréal, Montréal, 1985.

[2] K.C. Chang, *On the periodic nonlinearity and the multiplicity of solutions,* Nonlinear Analysis, TMA **13** (1987), 527–537.

[3] C. Conley and E. Zehnder, *The Birkhoff–Lewis fixed point theorem and a conjecture of V. Arnold,* Invent. Math. **73**(1983), 33–45.

[4] E. Fadell, *Cohomological methods in non-free G-spaces with applications to general Borsuk–Ulam theorems and critical point theorems for invariant functionals,* in Nonlinear functional analysis and its applications, D. Reidel, 1986, 1–45.

[5] A. Fonda and J. Mawhin, *Multiple periodic solutions of conservative systems with periodic nonlinearity,* preprint.

[6] G. Fournier and M. Willem, *Multiple solutions of the forced double pendulum equation,* Analyse Non Linéaire, Gauthier-Villars, Paris (1989), 259–281.

[7] L. Lusternik and L. Schnirelman, *Méthodes topologiques dans les problèmes variationnels,* Hermann, Paris, 1934.

[8] J. Mawhin and M. Willem, *Critical point theory and Hamiltonian systems,* Springer-Verlag, New York, 1989.

[9] E. Spanier, *Algebraic Topology,* McGraw-Hill, New York, 1966.

G. Fournier
Université de Sherbrooke
Département de Math. Info.
Sherbrooke, Québec, Canada J1K 2R1

and

M. Willem
Dept. Math.
Université Cath. de Louvain
1348 Louvain-la-Neuve
Belgium

Point and Line Singularities in Liquid Crystals

ROBERT M. HARDT*

A liquid crystal is generally understood to be a mesomorphic state of matter which flows like a liquid and which exhibits some anisotropic behavior. See [E], [EK], [C], [DG]. The liquid crystal phase usually lies between a solid phase and an isotropic liquid phase with phase transition being induced by temperature change. A static model typically involves a kinematic variable $n(x)$, called the *director*, which is a unit vector defined for x in a spatial region Ω.

It will be useful to consider the following statistical interpretation of the vector field n. An individual liquid crystal molecule is usually long and thin. Suppose that its direction is determined by a unit vector $\mathcal{Q}$. In statistical physics, one has a probability distribution of such $\mathcal{Q}$. Taking thermal averages, we will find that $\langle\mathcal{Q}\rangle = 0$ as the occurrence of $\mathcal{Q}$ is as equally likely as the occurrence of $-\mathcal{Q}$. However, one may employ second order averages $\langle\mathcal{Q}\otimes\mathcal{Q}\rangle = \langle l_i l_j\rangle$. Only the traceless part

$$Q = \langle\mathcal{Q}\otimes\mathcal{Q}\rangle - \frac{1}{3}\mathbf{1},$$

called the *order parameter*, contributes to the energy. Materials for which Q has three distinct eigenvalues are not of interest to us here. Suppose now that Q has two equal eigenvalues. This is characteristic of optically uniaxial materials such as most nematic liquid crystals. For these, the distinguished direction is determined by the eigenvector corresponding to the third eigenvalue. Here

$$Q = s\left[(n\otimes n) - \frac{1}{3}\mathbf{1}\right]$$

where $|n| = 1$ and $s = \frac{3}{2}\langle\mathcal{Q}\cdot n\rangle^2 - \frac{1}{2}$. Thus we see that the probability distribution of orientation matrices at a point x is determined by a vector $n(x) \in \mathbf{S}^2$ and a number $s(x) \in [-1/2, 1]$.

*Research partially supported by the National Science Foundation.

To understand the meaning of the weighting factor s we may consider three special cases:

$s = 1$: Here there is complete ordering with Q being parallel to n with probability 1.

$s = 0$: Here there is complete randonmess. The average angle that Q makes with n is $\cos^{-1}(1/\sqrt{3})$.

$s = -\frac{1}{2}$: Here Q has one degree of freedom, and is perpendicular to n with probability 1.

Most liquid crystal materials have $s \geq 0$ and constant solutions typically have $.3 \leq s \leq .7$.

If one assumes that s is *constant* throughout the material it is reasonable to look for critical points of a function $\int_\Omega W(n, \nabla n)dx$ where $W(n, \nabla n)$ represents the internal energy density of a configuration n. Assuming that W is a quadratic function of n and ∇n, that W is frame indifferent, (i.e. $W(\theta n, \theta \nabla n \theta^t) = W(n, \nabla n)$ for all $\theta \in \mathcal{O}(3)$), and that $W(-n, -\nabla n) = W(n, \nabla n)$, F. Frank [FF] showed that W must have the special form

$$2W(n, \nabla n) = \kappa_1(\operatorname{div} n)^2 + \kappa_2|n \cdot \operatorname{curl} n|^2 + \kappa_3|n \times \operatorname{curl} n|^2 \\ + (\kappa_2 + \kappa_4)\big(\operatorname{tr}(\nabla n)^2 - (\operatorname{div} n)^2\big).$$

The material constants κ_i depend on temperature. Here we assume that κ_1, κ_2, and κ_3 are positive. Some additional inequalities are suggested by thermodynamics [E]. The first three terms of W are said to describe the splay, twist, and bend of the vector field n. In the special case where these are balanced with $\kappa_1 = \kappa_2 = \kappa_3$ and $\kappa_4 = 0$, W equals $\frac{1}{2}|\nabla n|^2$, the ordinary energy density, and the corresponding critical points are harmonic maps from Ω to $\mathbf{S}^2$.

There are various physical and chemical treatments of the walls of the container for the liquid crystal which suggest the possibility of prescribing values of n on $\partial\Omega$. This Dirichlet boundary condition, called *strong-anchoring* in the liquid crystal literature, leads to the problem

$$\min_{n \in \mathcal{A}(\varphi)} \int_\Omega W(n, \nabla n)dx$$

where

$$\mathcal{A}(\varphi) = \{n \in H^1(\Omega, \mathbb{R}^3) : |n| = 1, n \mid \partial\Omega = \varphi\}$$

for fixed smooth boundary data φ. The existence and partial regularity of minimizers of this problem has been considered by R. Hardt, D. Kinderlehrer, and F.H. Lin in [HKL$_1$]. The existence theory uses the fact that the last term of W integrates to a boundary term only involving φ so that κ_4 may be altered without changing the minimizer. By [HKL$_1$] and [HKL$_2$] the minimizer is analytic off a closed set Sing(n) of dimension < 1. R. Schoen and K. Uhlenbeck [SU$_1$], [SU$_2$] had previously proven, for the harmonic map case $W = \frac{1}{2}|\nabla n|^2$, that Sing($n$) was at most a finite subset of Ω. The set Sing(n) consists of discontinuities of n and is frequently nonempty. It may be required by either the topological condition of φ having nonzero degree or by energy considerations [HL$_1$].

Actually much is now known about singular behavior of a minimizing harmonic map n from Ω to $\mathbf{S}^2$. Near $a \in \mathrm{Sing}(n)$ one has the estimate

$$\left| u(x) - \theta\left(\frac{x-a}{|x-a|}\right)\right| \leq c|x-a|^\alpha$$

for some rotation $\theta \in \mathcal{O}(3)$ and positive α. This follows from the classification of minimizing tangent maps by H. Brezis, J.-M. Coron, and E. Lieb [BCL], general results on asymptotic behavior near an isolated singularity of L. Simon [S], and a Jacobi field calculation of R. Gulliver and B. White [GW]. Motivating support for this classification was provided by experiment, the energy density bounds of [HKL$_2$], and the numerical work of M. Luskin, S.Y. Lin, and R. Cohen [CHKLL]. More recently, bounds on the number and behavior of singularities are treated in the works of R. Hardt and F.H. Lin [HL$_2$] and F. Almgren and E. Lieb [AL].

The classification of [BCL] implies that, up to a rotation, the map $x/|x| = (x_1, x_2, x_3)/\sqrt{x_1^2 + x_2^2 + x_3^2}$ is the unique homogeneous-degree-0 map from $\mathbb{R}^3$ to $\mathbf{S}^2$ that minimizes ordinary energy. This may or may not continue to hold for different choices of the material constants. F.H. Lin [L$_1$] showed the minimality for $\kappa_1 < \min\{\kappa_2, \kappa_3\}$ while F. Helein [H] established the nonminimality for $\kappa_1 > \kappa_2 + (1/8)\kappa_3$. S.Y. Lin independently obtained the nonminimality for $\kappa_1 \gg \kappa_2 = \kappa_3$ and, with R. Cohen, did some numerical work that indicates a spiralling behavior in the actual minimizer for $\kappa_1 = 3$, $\kappa_2 = \kappa_3 = 1$. Following [HKL$_1$], [HKL$_2$], it remains an open problem whether, a minimizer has only isolated singularities for a general choice of κ_i's.

An important difficulty associated with the above $s =$ constant model is the observed presence [FF] of line singularities, called *disclinations* in the literature. A simple line singularity such as

$$u(x) = (x_1, x_2, 0)/\sqrt{x_1^2 + x_2^2}$$

has $\int_{\mathbf{B}_1} W(u, \nabla u)dx = \infty$. Based on physicists' observations of line singularities, J. Ericksen [E] had suggested the possibility of the existence of an isotropic liquid core surrounding the singularity. In 1969 C. Dafermos [D] formulated and analysed a variational problem exhibiting such phases. Here, besides the variable n, there is a variable domain D which is determined roughly by points where W tries to become too large. In this theory regularity and even existence seems difficult to handle. A few years ago Ericksen suggested that one could also accommodate line singularities by considering the full order parameter Q as a variable, that is, with s as well as n being variables. Based on invariance principles and observed behavior, he reasoned that one should consider the density

$$\tilde{W}(Q,\nabla Q) = \tilde{W}(n,\nabla n,s,\nabla s) = s^2 W(n,\nabla n)+\kappa_5|\nabla s|^2+\kappa_6|\nabla s\cdot n|^2+W_0(s)$$

where $\kappa_5+\kappa_6>\kappa_1$ and the graph of W_0 should have the shape

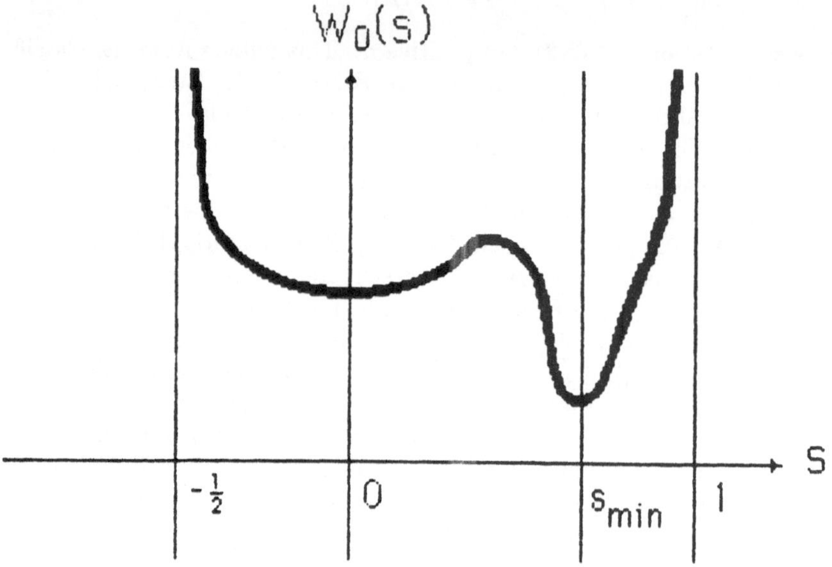

John Maddocks [M] used such a model to compute some planar solutions which exhibited line singularities and discussed various aspects of this new functional, including the inequality $\kappa_5+\kappa_6>\kappa_1$. The decreasing character of W_0 on $[-\frac{1}{2},0]$ make it reasonable to assume that $s\geq 0$ which we henceforth do.

F.H. Lin made the simple important observation that the vector field $u=sn$ has

$$\begin{aligned}|\nabla n|^2 &= \sum_{i,j}(sn^i)_j^2 = \sum_{i,j}(s_j n^i)^2 + 2ss_j n^i n^i_j + s^2(n^i_j)^2\\ &= |\nabla s|^2 + 0 + s^2|\nabla n|^2.\end{aligned}$$

So $\tilde{W}(Q, \nabla Q)$ is simply $|\nabla(s,u)|^2 + W_0(s)$ in case

$$(*) \qquad W(n) = \frac{1}{2}|\nabla n|^2, \quad \kappa_5 = 2, \quad \text{and } \kappa_6 = 0.$$

Note that a critical point of

$$\int_\Omega |\nabla(s,u)|^2 dx \quad \text{with the constraint } |u| = s \leq 1$$

is just a harmonic map into the cone

$$K = \{(t, v) \in [0,1] \times \mathbb{R}^3 : t = |v|\}.$$

Among the assumptions in $(*)$, the most restrictive is, as before, that $W(n) = \frac{1}{2}|\nabla n|^2$. For the analytic discussion to follow, a change of κ_5, or even κ_6, is easily accommodated. Note that trivially

$$\kappa_5|\nabla s|^2 \leq \kappa_5|\nabla s|^2 + \kappa_6|n \cdot \nabla s|^2 \leq (\kappa_5 + \kappa_6)|\nabla s|^2.$$

We will now describe various results of [L2] or [HL3] under the assumptions $(*)$. Thus we are interested in critical points (s, u) of

$$\int_\Omega (|\nabla(s,u)|^2 + W_0(s))dx \quad \text{for maps of } \Omega \text{ into } K$$

or equivalently critical points (s, n) of

$$\int_\Omega (2|\nabla s|^2 + s^2|\nabla n|^2 + W_0(s))dx \quad \text{for maps of } \Omega \text{ into } [0,1] \times \mathbf{S}^2.$$

Note that the variation $n \to (n + t\zeta)/|n + t\zeta|$ leads, as before, to the weak equation

$$\Delta n + \mathrm{S}^2|\nabla n|^2 n = 0$$

while the variation $(s, u) \to ((1 + t\eta)s, (1 + t\eta)u)$ leads to the equation

$$s\Delta s + u \cdot \Delta u - \frac{1}{2}sW_0'(s) = 0$$

or

$$\Delta s - \frac{1}{2}|\nabla n|^2 s - \frac{1}{2}W_0'(s) = 0 \quad \text{when } s \neq 0.$$

A reasonable boundary-value condition for the new problem which involves again prescribing the director field on $\partial\Omega$ is to have given:

$$n = \frac{u}{|u|} = \varphi \quad \text{and} \quad \frac{\partial s}{\partial \nu} = 0 \quad \text{on } \partial\Omega$$

where ν denotes the exterior unit normal of Ω. It is not difficult to show the existence of a $\tilde{W}$ energy minimizer satisfying this boundary condition. Unlike the situation with the $s =$ constant problem, we now have the

Theorem. *Any minimizer (s,u) is locally Hölder continuous on Ω and is smooth on $\Omega \sim s^{-1}\{0\}$.*

This may be established by using the above PDE to show that s^2 is essentially subharmonic, verifying that $H(r) = \int_{\partial \mathbf{B}_r(a)} s^2 dx$ satisfies

$$R^{-2}H(R) + r^{-2}H(r) = \int_r^R r^{-1}E(r)dx$$

where

$$E(r) = \int_{\mathbf{B}_r(a)} (|\nabla(s,u)|^2 + W_0'(s)s)dx,$$

and then seeing that n cannot have a discontinuity on $\{s > 0\}$. Alternately one may use arguments of S. Hildebrandt, H. Kaul, and K. Widman [HKW] by exploiting the strictly convex function t on the cone K. On the relatively open set $\Omega \sim s^{-1}\{0\}$, n, being continuous, is smooth by the higher regularity of harmonic maps [S]. The smoothness of s, and hence u, on $\Omega \sim s^{-1}\{0\}$ follows.

Thus $\mathrm{Sing}(n) \subset s^{-1}\{0\}$, and regularity questions now focus on the nature of this zero set. First we note that $s^{-1}\{0\}$ is in general nontrivial:

Remark. There exists a ball Ω and smooth boundary data φ on $\partial\Omega$ so that for the minimizer (s,u) of the above problem, the function s vanishes somewhere on Ω but is not identically zero.

The vanishing of s is guaranteed by simply having the degree of φ being nonzero. Fix a smooth $\psi\colon \partial\mathbf{B}_1 \to \mathbf{S}^2$ with degree $\psi \neq 0$. Let $\Omega = \mathbf{B}_R$ and $\varphi = \psi(R\cdot)$. For R sufficiently large,

$$\begin{aligned}\int_{\mathbf{B}_R} (|\nabla(0,u)|^2 + W_0(0))dx &\geq W_0(0)|\mathbf{B}_R| \\ &> (s_{\min})^2 R \int_{\partial\mathbf{B}_1} |\nabla_{\tan}\psi|^2 dA + W_0(s_{\min})|\mathbf{B}_R| \\ &= \int_{\mathbf{B}_R} (|\nabla(s_{\min}, \psi(x/|x|)|^2 + W_0(s_{\min}))dx \geq \text{infemum}.\end{aligned}$$

Thus, the minimizer (s,u) does not have $s \equiv 0$.

The next result is consistent with the observation of "line singularities."

Theorem. *For a minimizer* (s, u) *as above,* $\dim s^{-1}\{0\} \leq 1$.

The proof involves scaling. For $\lambda > 0$ one considers

$$n_\lambda(x) = n(\lambda, x), \quad s_\lambda(x) = \lambda^{-N} s(\lambda x), \quad u_\lambda(x) = \lambda^{-N} u(\lambda x).$$

The existence of a suitable N follows from the

Lemma (Monotonicity of frequency). *There is a positive constant* c *so that* $N(r)e^{cR}$ *is monotone increasing on* $[0, dist(a, \partial\Omega)]$ *where* $a \in \Omega$ *and*

$$N(r) = \int_{\mathbf{B}_r(a)} |(\nabla s, u)|^2 dx / H(r).$$

This type of monotonicity was used extensively by F. Almgren [A] in his study of multi-valued energy minimizing functions. N. Garofalo and F.H. Lin [GL] also employed it for unique continuation problems.

Monotonicity arguments show that a blow-up limit for a sequence $\lambda_i \to 0$ is a (nonconstant) homogeneous-degree-N harmonic map into the infinite cone

$$\tilde{K} = \{(t, v) \in [0, \infty) \times \mathbb{R}^3 : t = |v|\}.$$

As in the treatment of tangent maps for energy minimizing maps [SU_1] one proceeds essentially by induction on the number of independent variables. For a nonconstant solution of one variable $s^{-1}\{0\} = \emptyset$ as a minimizing geodesic on the cone $\tilde{K}$ does not pass through the vertex. For a solution depending on two variables it then follows, by an argument of DeGiorgi that $s^{-1}\{0\}$ is a discrete set. The Federer induction method then shows that $\dim s^{-1}\{0\} \leq 1$ in the general setting of three variables.

It remains an open problem to describe $s^{-1}\{0\}$ more precisely. Does $s^{-1}\{0\}$ have finite one dimensional measure?

To get a feeling for the variety of behaviors near singularities one may readily compute, by separating variables, the following harmonic maps into the cone $\tilde{K}$.

One may obtain point singularities by letting n correspond to the homogeneous-degree-0 extension of a rational function of degree d and $s_d(x) = |x|^{\sqrt{1+2d}-1}$.

One may obtain line singularities by letting $n(r, \theta, z) = j\theta$ and

$$s(r, \theta, z) = r^{-\frac{1}{2}+\frac{1}{2}\sqrt{1+4j}}$$

for a positive integer j, where (r, θ, z) denotes cylindrical coordinates. In constrast to the $s =$ constant model, the present model is consistent with the observed cancellation of line singularities [FG] and the existence [RWB], [K] of higher degree line singularities.

REFERENCES

[A] F.J. Almgren, *Q-valued functions minimizing Dirichlet's integral and the regularity of area minimizing rectifiable currents up to codimension 2*, preprint.

[AL] F.J. Almgren, Jr. and E. Lieb, *Singularities of energy minimizing maps from the ball to the sphere: Examples, counterexamples, and bounds*, Ann. Math., **128** (1988) 483–530.

[BCL] H. Brezis, J.-M. Coron, and E. Lieb, *Harmonic maps with defects*, Comm. Math. Physics **107** (1986), 649–705.

[C] S. Chandrasekhar, *Liquid Crystals*, Cambridge U. Press, 1977.

[CHKLL] R. Cohen, R. Hardt, D. Kinderlehrer, S.-Y. Lin, and M. Luskin, *Minimum energy configurations for liquid crystals: computational results in Theory and applications of liquid crystals*, IMA vol. 5, Springer, 1986, 99–122.

[D] C. Dafermos, *Disclinations in liquid crystals*, Quart. J. Mech. Appl. Math. **23.2** (1970), 49–64.

[DG] P.G. DeGennes, *The Physics of Liquid Crystals*, Clarendon Press, 1974.

[E] J. Ericksen, *Equilbrium theory of liquid crystals*, Adv. in liquid crystals 2, (G.H. Brown, ed.), Academic Press (1976), 233–298.

[EK] J. Ericksen and D. Kinderlehrer, ed., *Theory and Applications of Liquid Crystals*, IMA vol. 5, Springer, 1986.

[FF] F. Frank, *Liquid crystals*, Discuss. Faraday Soc. **25** (1958), 19–28.

[FG] G. Freidel, Annals Phys. 9e serie, **18** (1922), 273–474.

[GL] N. Garofalo and F.H. Lin, *Monotonicity properties of variational integrals, A_p-weights, and unique continuation*, Ind. U. Math. J. **35** (1986), 245–268.

[GW] R. Gulliver and B. White, *The rate of convergence of a harmonic map at a singular point*, Math. Annalen **283** (1989), 539–550.

[H] F. Hélein, *Minima de la fonctionnelle énergie libre des cristaux liquides*, to appear in Comp. Rend. A.S.

[HKL_1] R. Hardt, D. Kinderlehrer, and F.H. Lin, *Existence and partial regularity of static liquid crystal configurations*, Comm. Math. Physics **105** (1986), 547–570.

[HKL_2] R. Hardt, D. Kinderlehrer, and F.H. Lin, *Stable defects of minimizers of constrained variational principles,* Ann. Inst. H. Poincare, Anal. Nonlin. **5**, no. 4 (1988), 297–322.

[HKW] S. Hildebrandt, H. Kaul, and K. Widman, *An existence theory for harmonic mappings of Riemannian manifolds,* Acta. Math. **138** (1977), 1–16.

[HL_1] R. Hardt and F.H. Lin, *A remark on H^1 mappings,* Manuscripta Math. **56** (1986), 1–10.

[HL_2] R. Hardt and F.H. Lin, *Stability of singularities of minimizing harmonic maps,* to appear in J. Diff. Geom.

[HL_3] R. Hardt and F.H. Lin, *Line singularities in liquid crystals,* in preparation.

[L_1] F.H. Lin, *Une remarque sur l'application $x/|x|$,* Comp. Rend. A.S. **305-1** (1987), 529–531.

[L_2] F.H. Lin, *Nonlinear theory of defects in nematic liquid crystals—phase transition and flow phenomena,* preprint.

[M] J. Maddocks, *A model for disclinations in nematic liquid crystals,* IMA, vol. 5, Springer, 1986, 255–269.

[RWB] C. Robinson, J.C. Ward, and R.B. Beevers, Discuss. Faraday Soc. **25** (1958), 29–42.

[S] L. Simon, *Isolated singularities for extrema of geometric variational problems,* Springer Lecture Notes **1161**, 1985.

[SU_1] R. Schoen and K. Uhlenbeck, *A regularity theory for harmonic maps,* J. Diff. Geom. **17** (1982), 307–335.

[SU_2] R. Schoen and K. Uhlenbeck, *Boundary regularity and the Dirichlet problem of harmonic maps,* J. Diff. Geom. **18** (1983), 253–268.

Mathematics Department
Rice University
Houston, TX 77251

The Variety of Configurations of Static Liquid Crystals

ROBERT HARDT*, DAVID KINDERLEHRER*, and FANG HAU LIN*

Here we make several observations on a variety of special classes of harmonic maps from domains in $\mathbb{R}^3$ to $\mathbf{S}^2$. Such maps are relevant for the study of liquid crystals; see e.g. [HK$_1$], [HKL$_1$], [HKLu]. Part of our work is motivated by questions about the *nonuniqueness* and the *number* of harmonic maps having fixed boundary data. Here we show that the number may actually be infinite. For example, by Corollary 3.2 below *there exists a one parameter family of distinct energy-minimizers each having the same Dirichlet boundary data.*

In [HK$_2$, §3] we considered the subclass of *planar harmonic maps.* These are ones that have values in $\mathbf{S}^1 = \mathbf{S}^2 \cap \{z = 0\}$. With M. Luskin, we found an elementary example of a smooth planar harmonic map on a cube which is not energy-minimizing when considered as a map into $\mathbf{S}^2$. Thus, an energy-minimizer provided another stationary harmonic map with the same boundary values. In §1, we check that this energy minimizer $u = (u_1, u_2, u_3)$ is itself nonplanar. Hence, $(u_1, u_2, |u_3|)$ and $(u_1, u_2, -|u_3|)$ are both minimizers that differ from the original planar stationary harmonic map. Showing that $u_3 \not\equiv 0$ involves verifying the interesting fact that *on a connected region, there exists no nonconstant H^1 function whose trace assumes only a discrete set of values.* The above reflection symmetry was used in [SU$_3$] and occurs in Almgren and Lieb's [AL] example of nonuniqueness, which is based on the behavior of singularities. In another example, Hardt and Lin [HL$_3$, §5] also proved the existence of boundary data for which one minimizer has singularities while another minimizer does not.

Here our one parameter family of minimizers is constructed by breaking a putative *axial symmetry.* Here, for each positive integer n, a map u is called *n-axially symmetric* if, in cylindrical coordinates,

*Research partially supported by the National Science Foundation and the AFOSR through grant DMS–871881.

$$u(r, \theta, z) = (\cos n\theta \sin \varphi, \sin n\theta \sin \varphi, \cos \varphi)$$

for some real-valued function $\varphi(r, z)$. This implies that $u \circ R_\theta \equiv R_{n\theta} \circ u$ for any $\theta \in [0, 2\pi]$ where R_θ denotes rotation about the z-axis through an angle θ. We show that, *in spite of n-axially symmetric boundary data, an energy-minimizer u with singularities cannot be n-axially symmetric when $n \geq 2$.* In fact here the equation $u \circ R_\theta \equiv R_{n\theta} \circ u$ can hold for at most finitely many θ. So $\{R_{-n\theta} \circ u \circ R_\theta : -\varepsilon \leq \theta \leq \varepsilon\}$, for ε sufficiently small, is a family of distinct minimizers sharing the same boundary values. Energy minimizers must necessarily have singularities if the given boundary data has nonzero degree or is sufficiently oscillatory [HL_1]. In retrospect, the first indication of this symmetry breaking was in the numerical work of M. Luskin, S.Y. Lin, and R. Cohen [CHKLL]. Their results indicated that the specific 2-axially symmetric boundary data corresponding to the rational functional z^2 would give a minimizer having 2 off-axis singularities of degree 1. Concerning 1-axial symmetry, we also find in 5.2 certain 1-axially symmetric boundary data for which a minimizer cannot be 1-axially symmetric. For the general liquid crystal functional, F. Helein [H], and independently S.Y. Lin [Li], verified that the critical point $x/|x|$ is not a minimizing configuration for various ranges of constants. Using this, J. Ericksen c.f. [K] observed the necessity of rotational symmetry breaking and nonuniqueness for such liquid crystals. Here there is a two-parameter family of minimizers sharing the same boundary data.

The pure axially symmetric problem in which one considers critical points in the class of axially symmetric harmonic maps has been studied by D. Zhang [Z]. He obtained a smooth axially symmetric harmonic map corresponding to any given smooth axially symmetric boundary data that omits a neighborhood of the South Pole. §3 contains a slightly improved version of this result (perhaps optimal) in which one allows boundary data that can reach the South Pole but not wrap around. For a mapping that minimizers energy among axially symmetric maps, the singularities must be isolated (4.1). Also the singularities for such minimizers are classified, and a minimizer converges asymptotically to a unique tangent map on approach to a singular point (4.2). Examples with a prescribed number of singular points are given in §5. Some other constructions controlling the location and number of singularities in minimizing harmonic maps have been given in [HL_1] and [AL]. It is also observed in §6 that minimizers in the axially symmetric subclass, unlike absolute minimizers, do not satisfy the "Caccioppoli-type" estimate of [HL_2, 6.2] c.f. also [HKL_1], [HKL_2].

Harmonic maps between spheres with specified symmetries have been employed by B. Smith [S] to represent homotopy groups of spheres. As was observed recently by Ding [D], these may be constructed by minimizing energy among classes of maps with prescribed symmetries. W. Jager and H. Kaul [JK_2] and A. Baldes [B] have discussed rotational symmetry for harmonic maps to spheres and ellipsoids. See [CH] for references and more recent results on rotational symmetry. It is interesting to compare symmetry results of harmonic maps with those of minimal surfaces [M_2]. F. Morgan [M_1] has constructed in $\mathbb{R}^3$ a one parameter family of distinct minimal surfaces each having the same four circles as boundary. However, these minimal surfaces are necessarily unstable, and it is known that, for a boundary in the sphere invariant under a connected Lie group of symmetries, an area-minimizing hypersurface, even with singularities, inherits the symmetries of its boundary [La].

For recent developments in liquid crystal theory, see [LFH], [K], [Ha], and other paper in this volume.

1. Planar Harmonic Maps

Suppose Ω is a bounded domain in $\mathbb{R}^3$ and $v \in H^{1/2}(\partial\Omega, \mathbf{S}^1)$. Recall from [$HL_2$, 6.3] that there may exist no mapping $u \in H^1(\Omega, \mathbf{S}^1)$ whose trace on $\partial\Omega$ is v. However, if there is such a finite energy extension, then the resulting least-energy u is unique by the following remark:

For any $v \in H^{1/2}(\partial\Omega, \mathbf{S}^1)$, there exists at most one harmonic map $u \in H^1(\Omega, \mathbf{S}^1)$ with $u \mid \partial\Omega = v$.

To see this, recall [HK_2, §3] that a harmonic mapping $u \in H^1(\Omega, \mathbf{S}^1)$ has the form $(\cos\theta, \sin\theta, 0)$ where θ is a smooth harmonic function on Ω. Suppose u_1 and u_2 are two such harmonic mappings with $u_1 \mid \partial\Omega = u_2 \mid \partial\Omega$. Then, for the corresponding harmonic functions θ_1 and θ_2, the trace of $\theta_1 - \theta_2$ on $\partial\Omega$ has values in the discrete set $\{2\pi j : j \in \mathbf{Z}\}$. Thus the above remark follows from the:

1.1 Lemma. *On a Lipschitz domain $\Omega \subset \mathbb{R}^m$, there exists no nonconstant H^1 function v whose trace assumes only a discrete set of values.*

Proof. The lemma is local in character so we may suppose that a portion of Ω contains the half ball $G = \{x : x_m > 0,\ |x| < 1\}$ and that $\partial\Omega \cap \overline{G} = \Gamma = \{x_m = 0,\ |x'| < 1\}$. Replacing v by $\psi \circ v$ for a suitable smooth function ψ, we may assume that $v \mid \Gamma$ equals 0 or 1. Then in case $m = 2$, the comparison of the $H^{1/2}$ norm of $v \mid \Gamma$ with the Marcinkiewicz integral [St, p. 14] easily convinces the reader that $v \mid \Gamma$ is constant. Suppose inductively that the Lemma is true in dimensions $\leq m-1$. Since

$$\int_G |\nabla v|^2 dx < \infty,$$

there is, by Fubini's theorem, for almost every $a \in (-1,1)$, an $m-1$ dimensional domain

$$D(a) = G \cap \{x_1 = a\}$$

so that $\int_{D(a)} |\nabla v|^2 dx_2 \cdots dx_m < \infty$. Moreover, $v \mid \Gamma \cap \{x_1 = a\}$ equals 0 or 1. By induction $v \mid D(a)$ is constant. It is easy to check that the constant is independent of the choice of a. □

An old version of this fact, due to Wiener [C] states that a harmonic function in two variables with finite Dirichlet integral cannot suffer jump discontinuities on the boundary of its domain of definition.

In [HK_2, §3], it was shown that, *for sufficiently large q, the harmonic mapping*

$$n_q(x,y,z) = (\cos qz, \sin qz, 0)$$

on the unit cube Q in $\mathbb{R}^3$, *is not energy minimizing as a map into* $\mathbf{S}^2$. Choosing $u = (u_1, u_2, u_3)$ to minimize energy among maps in $H^1(Q, \mathbf{S}^2)$ with $u \mid \partial Q = n_q \mid \partial Q$, we conclude from the above remark that $u_3 \not\equiv 0$. Thus *There are at least three harmonic maps in* $H^1(Q, \mathbf{S}^2)$ *having trace* $n_q \mid \partial Q$; *these include* n_q *and the energy-minimizers* $(u_1, u_2, |u_3|)$ *and* $(u_1, u_2, -|u_3|)$.

2. A Continuum of Minimizers

Let $(r, \theta, z) \in \mathbb{R}^+ \times [0, 2\pi] \times \mathbb{R}$ be the standard cylindrical coordinates of $\mathbb{R}^3$, and let n be a positive integer. A map u from a subset of $\mathbb{R}^3$ into $\mathbf{S}^2$ is called *n-axially symmetric* if $u = u_\varphi$ where

$$u_\varphi(r, \theta, z) = (\cos n\theta \sin \varphi, \sin n\theta \sin \varphi, \cos \varphi) \tag{2.1}$$

for some real-valued function $\varphi(r, z)$. This implies that

$$u \circ R_\theta = R_{n\theta} \circ u$$

for any $\theta \in [0, 2\pi]$ where R_θ denotes rotation about the z-axis through an angle θ. Note than an n-axially symmetric u will be continuous at a point $(0,0,z)$ on the Z-axis only if $\varphi(0,0,z)$ equals an integer multiple of

π. Given φ, we will call u_φ the *n-axially symmetric extension* of φ. For example, if

$$\varphi = \text{arc cot}\left(r/[\sqrt{r^2+z^2}-z]\right)^n$$

on the half-disk $\{(r,z): 0 \le r^2 \le 1 - z^2\}$, then u_φ is the homogeneous-degree-0 extension of the conformal map of $\mathbf{S}^2$ given in sterographic projection by z^n.

2.1 Theorem. *Any energy minimizing map $u: \mathbb{B}^3 \to \mathbf{S}^2$ with singularities is* **not** *n-axially symmetric for $n \ge 2$. In fact, there are at most a finite number of angles $\theta \in [0, 2\pi)$ such that $u \circ R_\theta \equiv R_{n\theta} \circ u$.*

Proof. By [SU_1] the set of singularities of u is a discrete subset of $\mathbb{B} = \mathbb{B}^3$ characterized by density condition

$$\text{(2.2)} \qquad \operatorname{Sing} u = \left\{ a \in \mathbb{B}: \limsup_{r \to 0} r^{-1} \int_{\mathbb{B}_r(a)} |\nabla u|^2 dx > 0 \right\}.$$

First suppose that a is a singular point of u lying on the Z-axis and that $u_a(x) = \lim_{i\to\infty} u(\lambda_i(x-a))$ is a tangent map of u of a. Since this convergence is pointwise a.e., we would obtain the relation $u_a \circ R_\theta \equiv R_{n\theta} \circ u_a$ whenever $u \circ R_\theta \equiv R_{n\theta} \circ u$ was true. But the classification of [BCL] implies that such a tangent map must have the form $u_a(x) = g_a(x/|x|)$ for some rotation g_a of $\mathbb{R}^3$, from which we deduce that $u_a \circ R_\theta \not\equiv R_{n\theta} \circ u_a$ for *every* $\theta \in (0, 2\pi)$.

It remains to consider the case where there is a singular point a of u lying off the Z-axis. From (2.2) it is clear that each distinct angle $\theta \in [0, 2\pi)$ satisfying $u \circ R_\theta \equiv R_{n\theta} \circ u$, then gives rise to a distinct point $R_\theta(a)$ in the intersection of $\operatorname{Sing} u$ with the horizontal circle $\{x_1^2 + x_2^2 = a_1^2 + a_2^2,\ x_3 = a_3\}$. Thus the number of such θ is finite. □

2.2 Corollary. *For any nonsmooth energy minimizing map $u: \mathbb{B}^3 \to \mathbf{S}^2$ whose trace on $\partial\mathbb{B}^3$ is n-axially symmetric with $n \ge 2$,*

$$\{R_{-n\theta} \circ u \circ R_\theta : -\varepsilon \le \theta \le \varepsilon\}$$

forms, for ε sufficiently small, a one parameter family of distinct energy-minimizing maps each having the same Dirichlet boundary data.

3. Regular Axially Symmetric Harmonic Maps

For the remainder of the paper, we will consider critical points of the energy functional on axially symmetric maps. For simplicity, axial symmetry will mean 1-axial symmetry. All of the results, with a few changes of formulas, continue to hold for n-axial symmetry. See 6.3.

For a real-valued function φ on the half-disk $D = \{(r,z): 0 \le r^2 \le 1 - z^2\}$, one readily computes the energy of the axially symmetric extension of φ,

$$\mathsf{E}(\varphi) = \int_{\mathbf{B}} |\nabla u_\varphi|^2 dx = \int_D [r(\partial\varphi/\partial r)^2 + r(\partial\varphi/\partial z)^2 + r^{-1}\sin^2\varphi]\, dr\, dz.$$

Thus φ is a critical point for E if it satisfies the equation

$$\partial/\partial r(r\partial\varphi/\partial r) + \partial/\partial z(r\partial\varphi/\partial z) + \frac{1}{2}r^{-1}\sin 2\varphi = 0 \text{ in } D. \tag{3.1}$$

Dong Zhang has proven [Z] that, for any $\delta > 0$, equation (3.1) with the boundary condition

$$\varphi = g \quad \text{on} \quad \partial D \tag{3.2}$$

has a solution $\varphi \in C^0(\overline{D}) \cap C^\infty(D)$ for any Lipschitz function

$$g: \partial D \to [0, \pi - \delta] \quad \text{with} \quad g(0,z) \equiv 0.$$

Note that the axially symmetric extension u_φ of any $C^0(\overline{D}) \cap C^\infty(D)$ solution φ of (3.1) is a continuous, hence real analytic, harmonic map from $\mathbf{B}$ to $\mathbf{S}^2$. Thus we describe such a φ as a *regular solution.*

3.1 Theorem. *Problem* (3.1)–(3.2) *has a regular solution φ for any Lipschitz function*

$$g: \partial D \to [-\pi, \pi] \quad \textit{with} \quad g(0,z) \equiv 0.$$

We follow closely the argument of [Z] and use the additional:

3.2 Lemma. *Suppose $0 < \eta < \frac{1}{2}$, $0 < \delta < 1 - 2\eta$, and $\Omega_\delta = D \cap \{r > \delta\}$. For any Lipschitz function*

$$g: \partial\Omega_\delta \to [-\pi, \pi] \quad \textit{with} \quad |g| \le \pi - \eta \quad \textit{on} \quad \partial\Omega_\delta \cap \{\delta < r < \delta + 2\eta\},$$

an E-minimizer φ with $\varphi|_{\partial\Omega_\delta} = g$ has

$$|\varphi| \le \pi - c(\eta) \quad \textit{on} \quad \{r = \delta + \eta\}$$

for some positive $c(\eta)$ depending only on η.

Proof. If the Lemma were false for some positive $\eta < \frac{1}{2}$, then there would exist a sequence $\delta_i \in (0, 1-2\eta)$, Lipschitz functions

$$g_i\colon \partial\Omega_{\delta_i} \to [-\pi, \pi] \text{ with } |g_i| \le \pi - \eta \text{ on } \partial\Omega_{\delta_i} \cap \{\delta_i < r < \delta_i + 2\eta\},$$

and E-minimizers φ_i with $\varphi_i \mid \partial\Omega_{\delta_i} = g_i$ but

$$\|\varphi_i\|_{L^\infty\{r=\delta_i+\eta\}} > \pi - i^{-1}.$$

Note that $\|\varphi_i\|_{L^\infty} \le \pi$ by the minimality of φ_i. Thus, by elliptic regularity theory, we may assume, by passing to a subsequence, that

$$\begin{aligned}&\delta_i \to \delta_0 \in [0, 1-2\eta], \quad \varphi_i \to \varphi_0 \text{ in } C^2_{loc}(\Omega_{\delta_0}), \text{ and}\\ &\varphi_i \to \varphi_0 \text{ uniformly on } \{r = \delta_0 + \eta\}.\end{aligned}$$

Moreover, φ_0 will satisfy (3.1) on Ω_{δ_0},

$$\begin{aligned}&\|\varphi_0\|_{L^\infty} \le \pi, \ \|\varphi_0\|_{L^\infty\{r=\delta_0+\eta\}} = \pi, \text{ and}\\ &|\varphi_0| \le \pi - \eta \text{ on } \partial\Omega_{\delta_0} \cap \{\delta_0 < r < \delta_0 + 2\eta\}.\end{aligned}$$

These facts will contradict the Hopf maximum principle. In fact, the function $v = \varphi_0 - \pi$ is nonpositive in Ω_{δ_0} and (as we may assume) vanishes at some point $x_0 \in \Omega_{\delta_0}$. Since v satisfies an elliptic equation of the form

$$a_{ij}(x)(\partial^2/\partial x_i \partial x_j)v + b_i(x)(\partial/\partial x_i)v + c(x)v = 0$$

in a neighborhood of x_0, $v \equiv 0$ near x_0. (Note that the sign of $c(x)$ does not play a role here.) By analytic continuation, $\varphi_0 - \pi = v \equiv 0$ in Ω_{δ_0}, contradicting the boundary condition. □

Proof of Theorem 3.1. Since g is Lipschitz continuous on ∂D with $g(0, z) \equiv 0$, we may choose a positive $\eta < \frac{1}{2}$ such that $|g| \le \pi - 2\eta$ on $\partial D \cap \{r < 3\eta\}$. Modifying [Z], we choose, for a positive $\varepsilon \ll c(\eta)$, functions $\Phi^\pm_\varepsilon = \Phi^\pm_\varepsilon(r)$ satisfying

$$\begin{aligned}&d/dr(r d\Phi^\pm_\varepsilon/dr) - \tfrac{1}{2} r^{-1} \sin 2\Phi^\pm_\varepsilon = 0 \text{ on } \{0 < r < 2\eta\},\\ &\Phi^\pm_\varepsilon(2\eta) = \pm(\pi - \varepsilon),\\ &\Phi^\pm_\varepsilon \text{ are smooth with } |\Phi^\pm_\varepsilon| \ge \pi - \varepsilon \text{ on } \{r \ge 2\eta\}, \text{ and}\\ &\Phi^\pm_\varepsilon \equiv \pm\pi \text{ on } \{r \ge 2\eta + \varepsilon\}.\end{aligned}$$

As in [Z], $|\Phi^\pm_\varepsilon| \ge |g|$ on ∂D when ε is sufficiently small, and $\Phi^\pm_\varepsilon(r) \to \pm\pi$ as $\varepsilon \to 0$, for all $r > 0$.

Next we let φ be a minimizer of E subject to the constraints:

$$\varphi \mid \partial D = g \quad \text{and} \quad \Phi_\varepsilon^-(r) \le \varphi(r,z) \le \Phi_\varepsilon^+(r).$$

We claim that φ is then a regular solution of (3.1) on all of D. As in [Z], φ is a regular solution on $D \sim \{2\eta \le r \le 2\eta + \varepsilon\}$. Moreover, since $|\varphi| \le \pi$ and $\varphi \mid \partial D = g$, we may apply the maximum principle and the above Lemma (with g replaced by $\varphi \mid \partial D_\delta$, $0 < \delta < 2\eta$) to conclude that φ must satisfy in addition, the inequality $|\varphi| \le \pi - c(\eta) < \pi - \varepsilon$ on $\{\eta < r < 3\eta\}$. In particular, $\Phi_\varepsilon^- < \varphi < \Phi_\varepsilon^+$, and hence φ satisfies (3.1) also on the middle strip $\{\eta < r < 3\eta\}$. □

4. Minimizers in the Axially Symmetric Class

Suppose $v \in H^{1/2}(\mathbf{S}^2, \mathbf{S}^2)$ is axially symmetric and

$$\mathcal{U}(v) = \{w \in H^1(\mathbf{B}, \mathbf{S}^2) \colon w \text{ is axially symmetric, } w \mid \partial\mathbf{B} = v\}$$

is nonempty. Note in particular that $v(x/|x|) \in \mathcal{U}(v)$ in case $v \in H^1$. An energy minimizing sequence $\{u_i\}$ in $\mathcal{U}(v)$ contains a subsequence convergent weakly in H^1 to a map $u \in H^1(\mathbf{B}, \mathbf{S}^2)$ with $u \mid \partial\mathbf{B} = v$. Moreover, u is axially symmetric because the convergence of u_i to u is pointwise a.e. By lower semi-continuity, u minimizes energy in $\mathcal{U}(v)$. For the remainder of this section we discuss the regularity of such a *minimizer among axially symmetric maps.*

4.1 Lemma. *There is a positive constant ε_0 such that if u is energy minimizing among axially symmetric maps with $\int_{\mathbf{B}} |\nabla u|^2 dx \le \varepsilon_0$, then u is smooth on $\mathbf{B}_{1/2}$.*

Proof. As before, $u = u_\varphi$ for a suitable $\varphi \colon D \to \mathbb{R}$ and

$$\int_{\mathbf{B}} |\nabla u|^2 dx = \mathsf{E}(\varphi) = \int_D \left[r(\partial\varphi/\partial r)^2 + r(\partial\varphi/\partial z)^2 + r^{-1}\sin^2\varphi\right] dr\, dz.$$

Since

$$\begin{aligned} |\nabla \cos\varphi| &= |\sin\varphi| \cdot |\partial\varphi/\partial r| + |\sin\varphi| \cdot |\partial\varphi/\partial z| \\ &\le \frac{1}{2}\left[r\left(|\partial\varphi/\partial r|^2 + |\partial\varphi/\partial z|^2\right) + r^{-1}\sin^2\varphi\right], \end{aligned}$$

one has the estimate

$$\int_D |\nabla \cos\varphi| dr\, dz \le \frac{1}{2}\mathsf{E}(\varphi). \tag{4.1}$$

For $0 < \rho < 1$, let

$$D_\rho = \{(r, z): 0 \le r^2 \le \rho^2 - z^2\} \quad \text{and} \quad \partial^+ D_\rho = \{(r, z): 0 \le r^2 = \rho^2 - z^2\}.$$

The finiteness of $\mathbf{E}(\varphi)$ and Fubini's theorem imply that for almost all ρ, $\varphi \mid \partial^+ D_\rho$ is continuous with $\varphi(0, \pm\rho) = k_\pm \pi$ for some integer $k_\pm$. By (4.1) and Fubini's theorem, we may also insist that $\frac{1}{2} < \rho < 1$ and

$$\int_{\partial^+ D_\rho} |\nabla \cos \varphi| dr\, dz \le \mathbf{E}(\varphi).$$

In particular, $\operatorname{osc}_{\partial^+ D_\rho}(\cos \varphi) \le 2\mathbf{E}(\varphi)$, $k_+ = k_-$, and u maps $\partial \mathbf{B}$ continuously into a small neighborhood of the North or South Pole. As in [SU$_3$] one sees that, by minimality, $u \mid D_\rho$ must have image in the corresponding open hemisphere. Following [JK$_1$] and [HKW], $u \mid D_\rho$ is the unique harmonic map into this hemisphere, with respect to its boundary values, and $u \mid D_\rho$ is smooth. □

4.2 Theorem. *Suppose $u: \mathbf{B} \to \mathbf{S}^2$ is energy minimizing among axially symmetric maps. Then $u \mid \mathbf{B}$ is real analytic away from a set of isolated points on the Z-axis. If $u \mid \partial \mathbf{B}$ is Lipschitz, then u is also Hödler continuous in a neighborhood of $\partial \mathbf{B}$.*

Proof. We follow many of the arguments used in the partial regularity theory [SU$_1$] for absolute energy minimizers. By the ellipticity of (3.1) off the Z-axis and by Theorem 4.1,

$$S = \{a \in \mathbf{B}: \limsup_{r \to 0} r^{-1} \int_{\mathbf{B}_r(a)} |\nabla u|^2 dx > 0\}$$

is a relatively closed subset of $\{r = 0, -1 < z < 1\}$, and $u \mid \mathbf{B}$ is real analytic off of S. Moreover, S has one dimensional Hausdorff measure zero because $u \in H^1$ [SU$_1$, 2.7]. To see that S is actually only an isolated set, we first note that homogeneous extension from spheres centered on the Z-axis preserves axial symmetry. So minimality again gives a monotonicity formula as in [SU$_1$, 2.4] [HL$_2$, 4.1]. We then may find, for each $a \in S$, a sequence $\lambda_i \to 0$ so that the maps $u_i = u(\lambda_i(x - a))$ converge weakly in H^1 to an axially symmetric map u_0 that is homogeneous of degree 0, i.e., $u_0(x) = u_0(x/|x|)$. Moreover, since $u_0 \mid \mathbf{S}^2 \sim \{(0,0,-1), (0,0,1)\}$ is a harmonic map of finite energy, $u_0 \mid \mathbf{S}^2$ is a smooth harmonic map by [SaU].

At this stage in the argument of [SU$_1$, 4.6] the convergence of the u_i to u_0 is shown to be *strong* in H^1. This strong convergence then is used first to establish the energy minimality of u_0 and second, along with the small-energy-regularity lemma [SU$_1$, 3.1], to show that singularities are only isolated in dimension 3. With Lemma 4.1 above, these arguments will carry over once we establish this strong convergence.

For this purpose, first note that the energy bounds provided by the monotonicity formula and elliptic regularity theory imply that $u_i \to u_0$ in $C^2_{loc}(\overline{\mathbf{B}} \sim \{r = 0\})$. For a small positive δ, let $C_\delta^\pm$ denote the solid polar regions

$$C_\delta^\pm = \{x \colon 1 - \delta < |x| < 1,\ \ |(x_1, x_2)| < \delta|x|\}.$$

On the shell $\{1 - \delta \le |x| \le 1\}$, we use the linear interpolation between u_i and u_0,

$$Q_i(x) = \delta^{-1}(|x| + \delta - 1)u_i + \delta^{-1}(1 - |x|)u_0,$$

to define a comparison map

$$\begin{aligned}
&v_i(x) = u_0(x) \quad \text{for} \quad |x| \le 1 - \delta,\\
&v_i(x) = Q_i(x)/|Q_i(x)| \ \text{ for } 1 - \delta \le |x| \le 1 \ \text{ and } \ |(x_1, x_2)| \ge \delta|x|,\\
&v_i(x) = \text{the homogeneous-degree-0 extension of } v_i \mid \partial C_\delta^\pm\\
&\qquad \text{with center } \left(0, 0, \pm\left(1 - \frac{1}{2}\delta\right)\right) \qquad \text{for } x \in C_\delta^\pm.
\end{aligned}$$

Clearly v_i is axially symmetric with $v_i \mid \partial\mathbf{B} = u_i$. From the minimality of u_i and the C^2-convergence in $\overline{\mathbf{B}} \sim C_\delta^\pm$, we find, for i sufficiently large, that

$$\int_{\mathbf{B}} |\nabla u_i|^2 dx \le \int_{\mathbf{B}} |\nabla v_i|^2 dx \le \int_{\mathbf{B}_{1-\delta}} |\nabla u_0|^2 dx + c\Lambda\delta$$

where $\Lambda = \sup_{i, 1-\delta \le \rho \le 1} \int_{\partial \mathbf{B}_\rho} |\nabla u_i|^2 dx$. By the monotonicity formula, Λ is finite, and we may let $i \to \infty$ and $\delta \to 0$ to conclude that

$$\limsup_{i \to \infty} \int_{\mathbf{B}} |\nabla u_i|^2 dx \le \int_{\mathbf{B}} |\nabla u_0|^2 dx.$$

Combined with the weak H^1 convergence, this implies the desired strong convergence:

$$\limsup_{i \to \infty} \int_{\mathbf{B}} |\nabla u_i - \nabla u_0|^2 =$$

$$\limsup_{i \to \infty} \int_{\mathbf{B}} |\nabla u_i|^2 dx - \int_{\mathbf{B}} |\nabla u_0|^2 dx - 2\int_{\mathbf{B}} \langle \nabla u_i - \nabla u_0, \nabla u_0 \rangle dx \le 0 - 0.$$

Turning now to the question of boundary regularity for given Lipschitz Dirichlet data, we first note that the ellipticity of (3.1) off the Z-axis leads readily to the local Hölder continuity of u near points of $\partial\mathbb{B} \sim \{(0,0,1),(0,0,-1)\}$. To deduce Hölder continuity also near the two poles, we modify, as in [SU$_2$] or [HL$_2$, §5], the above interior regularity discussion. The continuity of $u \mid \partial\mathbb{B}$ implies that $u(0,0,\pm 1) \in \{(0,0,1),(0,0,-1)\}$.

In the boundary version of Lemma 4.1, one assumes, for some positive ρ_0, that both the oscillation of u on $\partial\mathbb{B} \cap \overline{\mathbb{B}}_{\rho_0}(0,0,\pm 1)$ and the normalized energy

$$\rho_0^{-1} \int_{\mathbb{B}\cap\mathbb{B}_{\rho_0}(0,0,\pm 1)} |\nabla u|^2 dx$$

are small. One finds as before that u then has small oscillation on $\overline{\mathbb{B}} \cap \partial\mathbb{B}_\rho(0,0,\pm 1)$ for a suitable $\rho \in \left(\frac{1}{2}\rho_0, \rho_0\right)$. Thus u maps $\partial(\mathbb{B} \cap \overline{\mathbb{B}}_\rho(0,0,\pm 1))$ close to the pole $u(0,0,\pm 1)$. Minimality again implies that the entire polar cap $\overline{\mathbb{B}} \cap \overline{\mathbb{B}}_\rho(0,0,\pm 1)$ is mapped by u into the corresponding open hemisphere and Hölder continuity on the cap follows. □

4.3 Lemma (tangent maps). *The only homogeneous-degree-*0 *axially symmetric maps from* $\mathbb{B}$ *to* $\mathbf{S}^2$ *that minimize energy among axially symmetric maps are* $+\Lambda$, $-\Lambda$, $+\Omega$, *and* $-\Omega$ *where*

$$\Lambda(x) = (x_1, x_2, x_3)/|x| \quad \textit{and} \quad \Omega(x) = (x_1, x_2, -x_3)|x|.$$

Proof. Recall that any finite energy harmonic map v from $\mathbf{S}^2$ to $\mathbf{S}^2$ determines, via stereographic projection, a function $f_v : \mathbb{C} \to \mathbb{C}$, which is a rational function of z or $\overline{z}$. Assuming also that v is axially symmetric, we infer from (2.1) with $n = 1$ that f_v maps $\{0,\infty\}$ to $\{0,\infty\}$ and the real axis to itself. Moreover, the restriction of f_v to a circle centered at the origin is simply multiplication by a real number. These facts imply that f_v equals either λz or $\lambda/\overline{z}$ with $0 \neq \lambda \in \mathbb{R}$. Finally, by computing the energy of the homogeneous extension from different points on the Z-axis as in [BCL] or [HL, 5.7], we infer from energy minimality that $|\lambda| = 1$. The four remaining functions $+z$, $-z$, $+1/\overline{z}$, and $-1/\overline{z}$ correspond to respectively $+\Lambda$, Ω, $-\Lambda$, and $+\Omega$. Being rotations of $x/|x|$, these four minimize energy among all maps, in particular among axially symmetric maps, having the same boundary data [BCL]. □

4.3 Theorem. *There is a positive* $\alpha < 1$ *so that if* $u : \mathbb{B} \to \mathbf{S}^2$ *is energy minimizing among axially symmetric maps and* a *is a singular point of* u *in* $\mathbb{B}$, *then* $|u(x) - u_0(x-a)| \leq C|x-a|^\alpha$ *where* u_0 *is either* $+\Lambda$, $-\Lambda$, $+\Omega$, *or* $-\Omega$.

Proof. As observed in the proofs of 4.1 and 2.1, a must lie on the Z-axis and any tangent map u_0 of u at a must be axially symmetric. The strong convergence implies as in [SU_1, 5.2] that u_0 must also minimize energy among axially symmetric maps. The uniqueness of this tangent map and the decay estimate follow by repeating arguments of L. Simon [S] and R. Gulliver and B. White [GW]. Here the Jacobi fields are axially symmetric and integrable via axially symmetric maps.

5. The Number of Singularities in Minimizers

By the singularity classification of [BCL], an energy-minimizing map from $\mathbf{B}$ to $\mathbf{S}^2$ whose boundary data has positive degree k will necessarily have at least k singular points. In [HL_1] we constructed degree-0 examples of boundary data whose energy minimizers also must have singularities. Here we give some examples of minimizers for which the number of singularities is precisely k.

First we will sketch one construction which involves a dumbell-shaped domain. This has been worked out independently by Libin Mou. Let $\Omega_\varepsilon = B_- \cup H_\varepsilon \cup B_+$ where

$$B_\pm = \mathbf{B}_1((0,0,\pm 2)), \quad H_\varepsilon = \{x \in \mathbb{R}^3 : x_1^2 + x_2^2 < \varepsilon^2, -1 < x_3 < 1\}.$$

Analogous with [HL_1], we choose uniformly Lipschitz boundary data $g_\varepsilon \in \mathcal{C}^\infty(\partial\Omega_\varepsilon, \mathbf{S}^2)$ so that

$$g_\varepsilon(x) = \pm(x - (0,0,\pm 2))/|x - (0,0,\pm 2)| \text{ on } \partial B_\pm \sim B_{2\varepsilon}((0,0,\pm 1)),$$
$$g_\varepsilon(x) = (0,0,-1) \text{ on } \partial H_\varepsilon \sim (B_- \cup B_+).$$

Let u_ε be an energy minimizing map with $u_\varepsilon \mid \partial\Omega_\varepsilon = g_\varepsilon$. Thanks to the uniqueness [BCL] of the energy minimizer $x/|x|$ and the strong convergence of minimizers, we see that $u_\varepsilon \mid B_\pm$ converges strongly in H^1 to $\pm(x-(0,0,\pm 2))/|x-(0,0,\pm 2)|$. Using [$HL_3$] or [AL], including the uniform boundary regularity lemma, we see that, for ε sufficiently small,

$$u_\varepsilon \mid (B_\pm \sim B_{1/8}(0,0,\pm 1))$$

has precisely one singularity. Concerning the remaining middle region of Ω_ε, we note that the strong convergence also implies that, for ε sufficiently small, u_ε maps the small spherical cap $B_\pm \cap \partial B_{1/4}((0,0,\pm 1))$ near the South Pole $(0,0,-1)$. For such ε, u_ε thus maps the entire boundary of the small dumbell

$$D_\varepsilon = B_{1/4}((0,0,-1)) \cup H_\varepsilon \cup B_{1/4}((0,0,1))$$

into the open Southern Hemisphere. By minimality $u_\varepsilon \mid D_\varepsilon$ has image in this hemisphere and so [HKW] is free of singularities. We conclude that u_ε has precisely 2 singularities.

Adding additional balls one readily obtains examples with an arbitrary number of singularities. Libin Mou has begun making such a bridge construction using nonminizing configurations involving higher degree singularities by modifying arguments of N. Smale [Sm] for minimal hypersurfaces.

Finally we give an axially symmetric construction which will illustrate some of the differences between minimizers among all maps and minimizers in the axially symmetric class. For convenience we will work with the fixed cylindrical domain

$$U = \{x \in \mathbb{R}^3 : x_1^2 + x_2^2 < 1, -1 < x_3 < 1\}$$

and the corresponding orbit space $E = \{(r,z) \in \mathbb{R}^2 : 0 \le r < 1,\ -1 < z < 1\}$.

5.1 Theorem. *Suppose $k \in \{1,2,\ldots\}$ and $v\colon \partial U \to \mathbf{S}^2$ is a Lipschitz axially symmetric mapping which corresponds (as in §3) to a function $g\colon (\partial E) \cap \{r > 0\} \to \mathbb{R}$ for which*

$$g(r,1) \equiv 0,\ g(r,-1) \equiv k\pi,\ \textit{and } g(1,t) \textit{ is monotonic on } \{-1 \le t \le 1\}.$$

Then the singular set of a mapping u that minimizes energy among axially symmetric maps with trace v has precisely k singular points.

Proof. Suppose $u = u_\varphi$ with $\varphi\colon E \to \mathbb{R}$, as in §3. By considering the one dimensional monotone rearrangement, in the Z direction, as discussed by B. Kawohl [Ka, Th. 2.13], we deduce that $\varphi(r,\cdot)$ is monotonic on $\{-1 \le t \le 1\}$ for every positive $r < 1$. Moreover, by the regularity of φ away from $\{r = 0\}$, the total variation of $\varphi(r,\cdot)$ on $\{-1 \le t \le 1\}$ remains $k\pi$ as $r \downarrow 0$. From 4.3 and 4.2 we conclude that u has precisely k singularities, whose degrees alternate between 1 and -1 as one descends the Z-axis. □

5.2 Corollary. *For k sufficiently large, a mapping w that minimizes energy among all maps having trace v, as in 5.1, is* **not** *axially symmetric. Moreover, $\{R_{-\theta} \circ w \circ R_\theta : -\varepsilon \le \theta \le \varepsilon\}$ forms, for ε sufficiently small, a one parameter family of distinct energy-minimizing maps each having trace v.*

Proof. By the uniform boundary regularity theorem [HL3, 2.1],

$$(\operatorname{Sing} w) \cap \mathbf{B}_\delta((0,0,\pm 1)) = \emptyset$$

for some $\delta > 0$, independent of k because v is constant on neighborhoods of $(0,0,1)$ and $(0,0,-1)$ that are independent of k. Finally by the interior bounds of [HKL$_2$], [AL] or [HL$_3$, §4],

$$\operatorname{card}\{a \in \operatorname{Sing} w : \operatorname{dist}(a, \partial U) > \delta\} < N_\delta$$

for some constant N_δ depending only on δ (actually $N_\delta = c/\delta$).

For $k \geq N_\delta$ we conclude that some singularity must lie off of the Z-axis and hence, as in 2.1, w is not axially symmetric, and the equation $w \circ R_\theta \equiv R_\theta \circ w$ holds for at most a finite number of angles $\theta \in [0, 2\pi)$.

6. Final Remarks

6.1 *Caccioppoli Inequality.* Unlike a minimizer in the full class of mappings, a minimizer in the axially symmetric subclass does not satisfy the "Caccioppoli-type" estimate of [HL$_2$, 6.2]. In fact, such an estimate would lead, as in [HKL$_2$], [AL] or [HL$_3$, §4], to energy bounds, uniform boundary regularity, and estimates on the total number of singularities, which would, as in 5.2, contradict the construction of 5.1.

6.2 *The dumbell.* In [HL$_1$], we obtained degree-0 boundary data for which the energy minimizer has an arbitrary positive even number of singularities. This involved using translations of the singular maps $+x/|x|$ and $-x/|x|$, and here the axially symmetric boundary data was monotone in the sense of 5.1. If we replace $-x/|x|$ by the other axially symmetric degree -1 singular tangent map $\Omega(x) = (x_1, x_2, -x_3)/|x|$, then we can achieve the same conclusion but now with axially symmetric boundary data for which the corresponding g, as in §3, satisfies $|g| \leq \pi$. We can also get $|g| < \pi - \delta$ for some positive at the expense of adding some small energy in the large middle region. With Zhang's theorem we thus have, as in [HL$_3$, §5], a specific example of boundary data for which there is one regular and one singular harmonic map.

6.3 *n-axially symmetric maps.* The results of §3–§5 easily carry over for n-axially symmetric maps. In the formula for $\mathsf{E}(\varphi)$ one should add the factor n^2 to the $r^{-1}\sin^2\varphi$ term. Then $\frac{1}{2}$ should be replaced by $\frac{1}{2}n^2$ in (3.1). This does not substantially change the proofs of 3.1, 3.2, 4.1, or 4.2 as the types of the P.D.E.'s and O.D.E.'s considered are unchanged. In 4.3, the classification of harmonic maps from $\mathbf{S}^2$ to $\mathbf{S}^2$ and the n-axially symmetric constraint now imply that the tangent map at a singularity must correspond to either z^n, $-z^n$, $\bar{z}^{-n}$, or $-\bar{z}^{-n}$. These new tangent maps should be used in the statement and proof of 4.3 and in the constructions of §5. In 5.1

note that the boundary map v was degree 1 if k was odd and degree 0 if k was even. Now it is either degree n or degree 0.

REFERENCES

[AL] F. J. Almgren, Jr. and E. Lieb, *Singularities of energy minimizing maps from the ball to the sphere: Examples, counterexamples, and bounds*, Ann. Math. **128** (1988), 483–530.

[B] A. Baldes, *Stability and uniqueness properties of the equator map from a ball into an ellipsoid*, Math. Z. **185** (1984), 505–516.

[BCL] H. Brezis, J.-M. Coron, and E. Lieb, *Harmonic maps with defects*, Comm. Math. Physics **107** (1986), 649–705.

[C] R. Courant, *Dirichlet's principle*, Interscience, 1950.

[CH] J.-M. Coron and F. Helein, *Harmonic diffeomorphisms, minimizing harmonic maps, and rotational symmetry*, preprint, Univ. Paris Sud, Orsay, 1988.

[CHKLL] R. Cohen, R. Hardt, D. Kinderlehrer, S.Y. Lin, and M. Luskin, *Minimum energy configurations for liquid crystals; computational results*, in *Theory and Applications of Liquid Crystals*, IMA, **5**, Springer, 1986.

[D] Ding, *Symmetric harmonic maps between spheres*, preprint.

[E] J. Ericksen, *Static theory of point defects in nematic liquid crystals*, preprint.

[GW] R. Gulliver and B. White, *The rate of convergence of a harmonic map at a singular point*, Math. Ann. **283** (1989), 539–550.

[Ha] R. Hardt, *Point and line singularities in liquid crystals*, this volume.

[He] F. Helein, *Minima de la fonctionelle éenergie libre des cristaux liquides*, Calcul des Variations, to appear in C.R.A.S.

[HK_1] R. Hardt and D. Kinderlehrer, *Mathematical questions of liquid crystal theory*, in *Theory and Applications of Liquid Crystals*, IMA, **5**, Springer, 1986.

[HK_2] R. Hardt and D. Kinderlehrer, *Mathematical questions of liquid crystal theory*, in College de France Seminar, vol. IV(3), 1987.

[HKL_1] R. Hardt, D. Kinderlehrer, and F.H. Lin, *Existence and partial regularity of static liquid crystal configurations*, Comm. Math. Physics **105** (1986), 547–570.

[HKL_2] R. Hardt, D. Kinderlehrer, and F.H. Lin, *Stable defects of minimizers of constrained variational problems*, Ann. Inst. Henri Poincare, Analyse non lineaire, **5**, no. 4 (1988), 297–322.

[HKLu] R. Hardt, D. Kinderlehrer, and M. Luskin, *Remarks about the mathematical theory of liquid crystals*, in *Calculus of Variations and*

Partial Differential Equations, S. Hildebrandt, D. Kinderlehrer, and M. Miranda, ed., Lecture Notes in Math. **1340**, 123–138.

[HKW] S. Hildebrandt, H. Kaul, and K. Widman, *An existence theory for harmonic maps of Riemannian manifolds,* Acta Math. **188** (1977), 1–16.

[HL_1] R. Hardt and F.H. Lin, *A remark on H^1 mappings,* Manuscripta Math. **56** (1986), 1–10.

[HL_2] R. Hardt and F.H. Lin, *Mappings minimizing the L^p norm of the gradient,* Comm. Pure & Appl. Math. **40** (1987), 555–588.

[HL_3] R. Hardt, and F.H. Lin, *Stability of singularities of minimizing harmonic maps,* to appear in J. Diff. Geom. **28** (1988).

[JK_1] W. Jager and H. Kaul, *Uniqueness and stability of harmonic maps and their Jacobi fields,* Manuscripta Math. **28**(4) (1979), 269–271.

[JK_2] W. Jager and H. Kaul, *Rotationally symmetric harmonic maps from a ball into a sphere and the regularity problem for weak solutions to an elliptic systems,* J. Reine Angew. Math. **343** (1983), 146–161.

[K] D. Kinderlehrer, *Recent developments in liquid crystal theory,* Proceedings of Colloq. Lions.

[Ka] B. Kawohl, *Rearrangements and convexity of level sets in PDE,* Lecture Notes in Math. **1150**, Springer-Verlag, Berlin-Heidelberg-New York, 1985.

[La] H.B. Lawson, *The equivariant Plateau problem,* Trans. A.M.S. **173** (1972), 231–250.

[LSY] Lin, San-Yih, *Numerical analysis of liquid crystal problems,* Thesis, U. of Minnesota, 1987.

[LFH] Lin, Fang-hua, *Nonlinear theory of defects in nematic liquid crystals —phase transition and flow phenomena,* to appear in Comm. Pures & Appl. Math.

[LM] J.L. Lions and E. Magenes, *Nonhomogeneous boundary value problems and applications I,* Springer-Verlag, Berlin-Heidelberg-New York, 1962.

[M_1] F. Morgan, *A smooth curve in $\mathbb{R}^3$ bounding a continuum of minimal surfaces,* Arch. Rat. Mech. Anal. **75** (1981), 193–197.

[M_2] F. Morgan, *Finiteness of the number of stable minimal hypersurfaces with a fixed boundary,* Indiana U. Math. J. **35**(4) (1986), 779–833.

[S] L. Simon, *Isolated singularities of extrema of geometric variational problems,* Lecture Notes in Math. **1186**, Springer-Verlag, Berlin-Heidelberg -New York, 1986.

[SaU] J. Sacks and K. Uhlenbeck, *The existence of minimal immersions of 2-spheres,* Ann. of Math. **113** (1981), 1–24.

[Sm] R.T. Smith, *Harmonic maps of spheres,* Amer. J. Math. **97**(2) (1975), 364–385.

[Sma] N. Smale, *Minimal hypersurfaces with many isolated singularities,* Ann. Math. **130** (1989), 603–642.

[St] E. Stein, *Singular integrals and differentiability properties of functions,* Princeton, 1970.

[SU_1] R. Schoen and K. Uhlenbeck, *A regularity theory for harmonic maps,* J. Diff. Geom. **17** (1982), 307–335.

[SU_2] R. Schoen and K. Uhlenbeck, *Boundary regularity and the Dirichlet problem of harmonic maps,* J. Diff. Geom. **18** (1983), 253–268.

[SU_3] R. Schoen and K. Uhlenbeck, *Regularity of minimizing harmonic maps into the sphere,* Inventiones Math. **78** (1984), 89–100.

[Z] D. Zhang, *Axially symmetric harmonic maps,* preprint, Univ. Cal. San Diego, March, 1987.

Robert Hardt
Mathematics Department
Rice University
Houston, TX 77251

David Kinderlehrer
School of Mathematics
University of Minnesota
Minneapolis, MN 55455

Fang Hau Lin
Courant Institute of Mathematical Sciences
New York University
New York, NY 10012

Existence of Multiple Solutions of Semilinear Elliptic Equations in R^N

YANYAN LI

Abstract

We study the existence of multiple solutions of semilinear elliptic equations in R^N with growth of nonlinearities below the critical Sobolev exponent.

Acknowledgement

This paper is part of the author's thesis. The author is greatly indebted to his thesis advisor Professor Nirenberg for his encouragement, useful suggestions and comments.

Introduction

In this paper we present some results on the existence of multiple solutions of the following equations:

$$-\Delta u + u - g(x,u) = 0, \quad x \in R^N \tag{1}$$

In Section 1 we discuss it only for a model case:

$$\begin{cases} -\Delta u + u - q(x)|u|^{\sigma}u = 0 & \text{in } R^N \\ 0 < \sigma < \frac{4}{N-2} & \text{if } N \geq 3 \\ 0 < \sigma < +\infty & \text{if } N = 1,2 \\ q \in C(R^N) & \\ q \in L^{\infty}(R^N) + L^{\beta}(R^N) & \\ q(x) > 0 & \text{for some } x \in R^N \end{cases} \tag{2}$$

Where

$$\beta = \frac{2N}{2N - (N-2)(\sigma+2)} \quad \text{if } N \geq 3$$
$$\beta > 1 \quad \text{if } N = 2$$
$$\beta = 1 \quad \text{if } N = 1$$

In Section 2 we use the same idea to deal with the more general case (1).

For similar problems on a bounded domain of R^N ,there have been many results. (see [18] and the references there.) We usually use variational method to deal with semilinear elliptic equations and apply Ljusternik-Schnirelmann type theory to explore the invariance of functionals under group action. Since we are working in infinite dimensional space we need some compactness hypothesis on the functional. One of such hypothesis is Palais-Smale condition. If we study semilinear elliptic equations with growth of nonlinearities below critical exponent on a bounded domain of R^N ,usually the associated functional satisfies the Palais-Smale condition. But this is not true any more for equations on R^N . The main difference here is that $H^1(R^N)$ does not imbed to $L^p(R^N)$ compactly for $2 \leq p < \frac{2N}{N-2}$, $N \geq 3$; $2 \leq p < +\infty, N = 1,2$. Therefore we need to analyze on which interval does the functional satisfies the Palais-Smale condition. This has been studied by P. L. Lions (see [15]

,[16] and the references there). Here we give a somewhat different proof for this. In order to obtain multiple solutions we make use of the properties of the nonlinearity to prove that certain minmax values we have constructed really lie in the interval where the associated functional satisfies Palais-Smale condition. The existence of one positive solution of (1) has been studied by P. L. Lions,M. Esteban, Weiyue Ding,Weiming Ni ,Dong Zhang and some others (see [9],[16] ,[15],[22] ,[10] and the references there). The existence of at least two solutions has been studied by Xiping Zhu (see [23]). For the existence of infinitely many solutions there have been work by Berestycki and Taubes (see [5]). Our reselt is different from theirs in the way that we make use of the properties of the nonlinearity in bounded region while their results concern hypotheses at infinity.

In a separate paper (see [14]) we apply this method to study the existence of multiple solutions of the standing waves of Schrödinger equations. In this case we need to analyze where the constrained functional satisfies Palais-Smale condition and construct apropriate minmax values in the interval where the constrained functional satisfies Palais-Smale condition.

1 Existence of Multiple Solutions of (2)

We study the existence of multiple solutions of (2) and therefore (2) is assumed throught this section. First we point out where the difficulty lies. Consider the corresponding problem on a bounded domain:

$$\begin{cases} -\Delta u + u - q(x)|u|^\sigma u = 0 & \text{in } \Omega \subset R^N \\ 0 < \sigma < \frac{4}{N-2} & \text{if } N \geq 3 \\ 0 < \sigma < +\infty & \text{if } N = 1,2 \\ q(x) \in C(\Omega) \cap L^\beta(\Omega) & \\ q(x) > 0 & \text{for some } x \in \Omega \end{cases} \tag{3}$$

We can apply Ljusternik-Schnirelmann type theory to prove that (3) has infinitely many solutions in $H_0^1(\Omega)$ (See [18]). The difference between (2) and (3) lies in the fact that the embedding from $H^1(R^N)$ to $L^{\sigma+2}(R^N)$ is no more compact. This is easily seen by taking any nontrivial function $u \in H^1(R^N)$ and letting $u_n(x) := u(x+n)$,clearly u_n converges weakly in $H^1(R^N)$ but not strongly in $L^{\sigma+2}(R^N)$ as n goes to $+\infty$.

If $q(x)$ in (2) satisfies some further hypotheses,the kind of loss of compactness as above will not occur,therefore the standard Ljusternik-Schnirelmann theory will apply. More precisely we have the following two propositions:

Propsition 1.1: In addition to (2) if we assume that $\lim_{|x|\to\infty} q(x) = 0$, then (2) has infinitly many solutions in $H^1(R^N)$.

The above result is essentially proved in [20]. There is only a slight difference, namely,we assume here that $q(x)$ is positive somewhere not that,as in [20], positive everywhere.But only a minor modification is needed.

Proposition 1.2: In addition to (2) if we assume that $q(x)$ is a radially symmetric function, namely, $q(x) = q(y)$ for $|x| = |y|$, then (2) has infinitely many solutions in $H^1(R^N)$.

The proof of Proposition 1.2 rests on the fact that the embedding from $H^1(R^N)$ to $L^p(R^N)$ $(2 < p < \frac{2N}{N-2}$ if $N \geq 3$, $2 < p < +\infty$ if $N = 1,2)$

is actually compact among radially symmetric functions (See [21]). For a proof see [9] and the references there. In fact stronger results are proved in [9] in the radially symmetric case.

If we do not assume further hypotheses on $q(x)$, (2) may not have any nontrivial solutions in $H^1(R^N)$. We give such an example due to M. J. Esteban and P. L. Lions (See [10]):

Example: Let $q(x)$ be a smooth bounded positive function defined on R^N with bounded derivatives and q_{x_1} be nonnegative but not identically equal to zero,where q_{x_1} denotes the partial derivative of q in x_1 direction. Then any solution of

$$-\triangle u + u - q(x)|u|^\sigma u = 0$$

which belongs to $L^p(R^N)$ for some $p > \sigma + 1$ has to be identically equal to zero. Where $0 < \sigma < \frac{4}{N-2}$ if $N \geq 3$, $0 < \sigma < +\infty$ if N=1,2.

Proof: Let $u \in L^p(R^N)$ $(p > \sigma + 1)$ be a solution. It follows from the standard bootstrap method that $u \in H^1(R^N)$. Multiply the equation by u_{x_1} and integrate by parts,we obtain :

$$\int q_{x_1} |u|^{\sigma+2} = 0$$

which implies that u has to vanish on an open set. By some well known unique continuation results, u is identically equal to zero.

The example above shows that the loss of compactness we encounter is genuine, not only because of the method we use.

The existence of one positive solution of (2) under some further hypotheses on $q(x)$ has been discussed by M. J. Esteban, P.L.Lions,Weiyue Ding andWeiming Ni and some others. (See [9],[10],[16] and the references there)

According to their results, either of the following additional hypotheses on $q(x)$ will give rise to at least one positive solution of (2):

$$\begin{cases} \lim_{|x|\to\infty} q(x) = q_\infty > 0 & \\ & \\ q(x) > q_\infty & x \in R^N \end{cases} \tag{4}$$

There exist positive constants $R, \alpha > 0$,such that,

$$\begin{cases} \lim_{|x|\to\infty} q(x) = q_\infty > 0 \\ q(x) > q_\infty + |x|^{-\alpha} \qquad |x| > R, \quad x \in R^N \end{cases} \tag{5}$$

There are other known hypotheses on $q(x)$ which will ensure the existence of at least one positive solution. Since they are basically of the same nature we are not going to list them all here. See the references above for more information. One can also find a survey on the subject in [17],[16].

The existence of at least two pairs of nontrivial solutions of (2) has been studied by Xiping Zhu [23] by using a technique in [6] and the concentration compactness principle introduced by P. L. Lions in [15].

In the following we are going to apply minmax principle to study the existence of multiple solutions of (2) under some further hypotheses on $q(x)$.

Theorem1.1: In addition to (2),let q_∞ be a positive constant, n be a positive integer. Suppose that:

$$\limsup_{|x|\to\infty} q(x) \le q_\infty \tag{6}$$

and there exist n functions $u_j \in H^1$ with disjoint supports, $j = 1, \ldots, n$,such that,

$$\int q(x)|u_j|^{\sigma+2} > 0 \qquad j = 1, \ldots, n \tag{7}$$

$$\sum_{j=1}^{n} \frac{(\int |\nabla u_j|^2 + u_j^2)^{\frac{\sigma+2}{\sigma}}}{(\int q(x)|u_j|^{\sigma+2})^{\frac{2}{\sigma}}} < S_\infty^{\frac{\sigma+2}{\sigma}} \tag{8}$$

where

$$S_\infty := \inf_{u\in H^1\setminus\{0\}} \frac{\int |\nabla u|^2 + u^2}{(\int q_\infty |u|^{\sigma+2})^{\frac{2}{\sigma+2}}}$$

Then (2) has at least n pairs of nontrivial solutions in $H^1(R^N)$ and one of them is positive.

Remark 1.1: Qualitatively Theorem1.1 implies that the larger $q(x)$ is in the bounded region comparing to $\lim_{|x|\to\infty} \sup q(x)$,the more solutions we are able to obtain.

Remark 1.2: The existence of one positive solution is classical. See [9],[15],[16] and [17].

Theorem 1.1 has the following corollary:

Corollary 1.1: In addition to (2) if $\lim_{|x|\to\infty} \sup q(x) \leq 0$,then (2) has infinitely many solutions.

Proof of the corollary 1.1:

Since $q(x)$ is continuous and positive somewhere,there exists an open set $O \subset R^N$,such that, $q(x) \geq c > 0 \, \forall x \in O$,where c is some constant.

For any integer n , take n disjoint balls $B_1, \ldots, B_n$, $\bigcup_{j=1}^n B_j \subset O$ and any functions $u_j \in H_0^1(B_j) \setminus \{0\}$ $(j = 1, \ldots, n)$.

$$C_j := \frac{(\int |\nabla u_j|^2 + u_j^2)^{\frac{\sigma+2}{\sigma}}}{(\int q(x)|u_j|^{\sigma+2})^{\frac{2}{\sigma}}} > 0 \qquad j = 1, \ldots, n$$

Choose $q_\infty > 0$ so small that

$$\sum_{j=1}^{n} C_j < S_\infty^{\frac{\sigma+2}{\sigma}}, \quad j = 1, \ldots, n.$$

According to Theorem 1.1 (2) has at least n pairs of nontrivial solutions. The conclusion of the corollary follows from the fact that we can choose n to be arbitrarily large.

Corollary 1.2: Suppose that $q(x) \in L^\infty(R^N) \cap C^\infty(R^N), \psi(x) \in C_0^\infty(R^N) \setminus \{0\}$ and $\psi(x) \geq 0$. Consider:

$$\begin{cases} -\Delta u + u - q_\lambda(x)|u|^\sigma u = 0 & \text{in } R^N \\ 0 < \sigma < \frac{4}{N-2} & \text{if } N \geq 3 \\ 0 < \sigma < +\infty & \text{if } N = 1,2 \end{cases}$$

Where $q_\lambda(x) = q(x) + \lambda\psi(x)$.

The number of solutions of the above equation tends to $+\infty$ as λ tends to $+\infty$.

In order to prove Theorem 1.1 we first introduce some notations:

$$\begin{aligned}
\|u\| &:= \int_{R^N} |\nabla u|^2 + u^2 \\
J(u) &:= \frac{1}{2}\int_{R^N} |\nabla u|^2 + u^2 - \frac{1}{\sigma+2}\int_{R^N} q(x)|u|^{\sigma+2} \\
J_\infty(u) &:= \frac{1}{2}\int_{R^N} |\nabla u|^2 + u^2 - \frac{1}{\sigma+2}\int_{R^N} q_\infty|u|^{\sigma+2} \\
V_\infty &:= \{u \in H^1(R^N) \setminus \{0\} : 0 =< J'_\infty(u), u >\} \\
m_\infty &:= \inf_{u \in V_\infty} J_\infty(u)
\end{aligned}$$

where $< \cdot, \cdot >$ denotes the paring between $H^1(R^N)$.

Remark 1.3: By simple calculation we have

$$< J'_\infty(u), u >= \int |\nabla u|^2 + u^2 - \int q_\infty |u|^{\sigma+2}$$

Remark 1.4: m_∞ and S_∞ are actually attained by some positive radially symmetric functions. Therefore m_∞ and S_∞ are positive numbers.

Lemma 1.1:

$$S_\infty = \{(\frac{1}{2} - \frac{1}{\sigma+2})m_\infty\}^{\frac{\sigma}{\sigma+2}}$$

Proof: $\forall \epsilon > 0$, there exists $u \in H^1(R^N) \setminus \{0\}$,such that,

$$\frac{\int |\nabla u|^2 + u^2}{(\int q_\infty |u|^{\sigma+2})^{\frac{2}{\sigma+2}}} \leq S_\infty + \epsilon \tag{9}$$

Let $\lambda > 0$ be defined by

$$\int |\nabla u|^2 + u^2 = \lambda^\sigma \int q_\infty |u|^{\sigma+2}$$

Then we have

$$\lambda^\sigma\left(\int q_\infty|u|^{\sigma+2}\right)^{\frac{\sigma}{\sigma+2}} \leq S_\infty + \epsilon \tag{10}$$

Let

$$w := \lambda u$$

clearly

$$w \in V_\infty$$

Therefore according to the definition of m_∞ and (10) we have

$$\begin{aligned} m_\infty &\leq J_\infty(w) \\ &= \left(\frac{1}{2} - \frac{1}{\sigma+2}\right)\int q_\infty|w|^{\sigma+2} \\ &= \left(\frac{1}{2} - \frac{1}{\sigma+2}\right)\left\{\lambda^\sigma\left(\int q_\infty|u|^{\sigma+2}\right)^{\frac{\sigma}{\sigma+2}}\right\}^{\frac{\sigma+2}{\sigma}} \\ &\leq \left(\frac{1}{2} - \frac{1}{\sigma+2}\right)(S_\infty + \epsilon)^{\frac{\sigma+2}{\sigma}} \end{aligned}$$

Let $\epsilon \to 0$,we have

$$S_\infty \geq \left\{\left(\frac{1}{2} - \frac{1}{\sigma+2}\right)^{-1} m_\infty\right\}^{\frac{\sigma}{\sigma+2}} \tag{11}$$

On the other hand, $\forall \epsilon > 0$, there exists some $w \in V_\infty$ with $J_\infty(w) \leq m_\infty + \epsilon$,namely,

$$\begin{aligned} \int |\nabla w|^2 + w^2 &= \int q_\infty|w|^{\sigma+2} \\ \epsilon + m_\infty &\geq \left(\frac{1}{2} - \frac{1}{\sigma+2}\right)\int |\nabla w|^2 + w^2 \end{aligned}$$

Therefore

$$\begin{aligned} S_\infty &\leq \frac{\int |\nabla w|^2 + w^2}{\left(\int q_\infty|w|^{\sigma+2}\right)^{\frac{2}{\sigma+2}}} \\ &= \left(\int |\nabla w|^2 + w^2\right)^{\frac{\sigma}{\sigma+2}} \\ &\leq \left\{\left(\frac{1}{2} - \frac{1}{\sigma+2}\right)^{-1} m_\infty\right\}^{\frac{\sigma}{\sigma+2}} \end{aligned}$$

Let $\epsilon \to 0$,we have

$$S_\infty \geq \left\{\left(\frac{1}{2} - \frac{1}{\sigma+2}\right)^{-1} m_\infty\right\}^{\frac{\sigma}{\sigma+2}} \tag{12}$$

The conclusion of Lemma 1.1 follows from (11) and (12).

To look for $H^1(R^N)$ solutions of (2) we use a variational approach, namely, to look for critical points of $J(u)$. It is well known that $J \in C^1(H^1(R^n\ ), R)$ under the assumption of Theorem 1.1. Due to the fact that the embedding from $H^1(R^N)$ to $L^p(R^N)$ $(2 \le p < \frac{2N}{N-2}$ if $N \ge 3$, $2 \le p$ if $N = 1,2)$ is not compact,J does not, in general,satisfies $(PS)_c$ for all real values c. But we can prove that J satisfies $(PS)_c$ for suitable values c.

Definition 1.1: We say that J satisfies $(PS)_c$ for some real value c,if for any sequence $\{u_n\} \subset H^1(R^N)$ satisfying

$$J(u_n) \to c$$

$$J'(u_n) \to 0 \quad \text{strongly in } H^{-1}(R^N)$$

there exists a subsequence of $\{u_n\}$ which converges strongly in $H^1(R^N)$.

Lemma 1.2 : Under the assumption of Theorem 1.1, J satisfies $(PS)_c$ for any value c lying in the open interval $(-\infty, m_\infty)$.

Remark 1.2: Lemma 1.2 is sharp in the sense that if $q(x) \to q_\infty$ as $|x| \to \infty$, J does not satify $(PS)_{m_\infty}$.

Lemma 1.2 has been proved by P. L. Lions (see [16]). For completeness we give a somewhat different proof.

Proof of Lemma 1.2:

Let $\{u_n\} \subset H^1(R^N)$ be a sequence satifying:

$$J(u_n) \to C' < m_\infty$$

$$J'(u_n) \to 0 \quad \text{strongly in } H^{-1}(R^N)$$

namely

$$\frac{1}{2}\int |\nabla u_n|^2 + u_n^2 - \frac{1}{\sigma+2}\int q(x)|u_n|^{\sigma+2} \to C' \tag{13}$$

$$|< J'(u_n), v >| \le o(1)\|v\| \qquad \forall v \in H^1(R^N) \tag{14}$$

Substitute u_n for v in (14) we have

$$|< J'(u_n), u_n >| \le o(1)\|u_n\|$$

namely

$$-o(1)\|u_n\| \le \int |\nabla u_n|^2 + u_n^2 - \int q(x)|u_n|^{\sigma+2} \le o(1)\|u_n\| \tag{15}$$

Multiply (15) by $-(\sigma+2)^{-1}$ and add it to (13) we have

$$(\frac{1}{2} - \frac{1}{\sigma+2})\|u_n\|^2 \le C + o(1)\|u_n\|$$

Clearly $\{\|u_n\|\}$ is bounded.

According to some well known facts in functional analysis,there exists $u_\infty \in H^1(R^N)$, such that,along a subsequence of $\{u_n\}$(still being denoted as $\{u_n\}$),

$$u_n \rightharpoonup u_\infty \qquad \text{weakly in} \qquad H^1(R^N)$$

$$u_n \to u_\infty \text{ a.e. in } R^N$$

We claim that u_∞ satisfies:

$$< J'(u_\infty), u_\infty >= 0 \tag{16}$$

In fact,we have

$$\begin{aligned} o(1) &= \int \nabla u_n \nabla u_\infty + u_n u_\infty - \int q(x)|u_n|^\sigma u_n u_\infty \\ &= o(1) + \|u_\infty\|^2 - \int q(x)|u_n|^\sigma u_n u_\infty \\ &= o(1) + \|u_\infty\|^2 - \int q(x)|u_\infty|^{\sigma+2} \end{aligned}$$

The last equality follows from some elementary real analysis argument. Next we want to prove the strong convergence of $\{u_n\}$ in $H^1(R^N)$. There are only two posibilities for $\{u_n\}$:

(1): $\forall\delta > 0$, there exists $\tilde{R} > 0$,such that, $\forall n > \tilde{R}$,

$$\int_{|x|\ge\tilde{R}} |\nabla u_n^2| + u_n^2 < \delta$$

(2): There exists $\delta_0 > 0$,such that, $\forall \tilde{R} > 0$, there exists $n = n(\tilde{R}) \geq \tilde{R}$ with

$$\int_{|x|\geq\tilde{R}} |\nabla u_n|^2 + u_n^2 \geq \delta_0$$

Case (1) corresponds to the case that there is no fixed amount of positive mass of $\{u_n\}$ slipping away to infinity,which is easier to handle. Let us deal with it first,more precisely,we prove that case (1) leads to the strong convergence of $\{u_n\}$ in $H^1(R^N)$.

$\forall \delta > 0$,there exists $\tilde{R} > 0$, such that , $\forall n > \tilde{R}$ we have

$$\int_{|x|\geq\tilde{R}} |\nabla u_n|^2 + u_n^2 < \delta \tag{17}$$

By the weak lower semicontinuity of the above integral and the fact that $u_n \rightharpoonup u_\infty$ weakly in $H^1(R^N)$,we have

$$\int_{|x|\geq\tilde{R}} |\nabla u_\infty|^2 + |u_\infty|^2 \leq \delta \tag{18}$$

Since $J'(u_n) \to 0$ strongly in $H^{-1}(R^N)$, we have:

$$|\int \nabla u_n \nabla v + u_n v - \int q(x)|u_n|^\sigma u_n v| \leq o(1)\|v\| \qquad \forall v \in H^1(R^N) \tag{19}$$

Use Hölder inequality,Sobolev inequality,(17),(18) and the property of $q(x)$ we obtain $\forall v \in H^1(R^N)$ that

$$\begin{cases} |\int_{|x|\geq\tilde{R}} q(x)|u_n|^\sigma u_n v| + |\int_{|x|\geq\tilde{R}} q(x)|u_\infty|^\sigma u_\infty v| & \leq \; c\delta^{\frac{\sigma+1}{2}}\|v\| \\ |\int_{|x|\leq\tilde{R}} q(x)|u_n|^\sigma u_n v - \int_{|x|\leq\tilde{R}} q(x)|u_\infty|^\sigma u_\infty v| & \leq \; o(1)\|v\| \end{cases} \tag{20}$$

Combine (19) with (20) we obtain

$$|\int \nabla u_n \nabla v + u_n v - \int q(x)|u_\infty|^\sigma u_\infty v| \leq C\delta^{\frac{\sigma+1}{2}}\|v\| + o(1)\|v\| \tag{21}$$

Where C is some constant independent of n and δ.

Since $\delta > 0$ can be chosen arbitrarily small,(21) implies the strong convergence of $\{u_n\}$ in $H^1(R^N)$.

In order to prove Lemma 1.2 we only need to rule out case (2). we argue by contradiction. If case (2) occurs, for that δ_0,take $\widetilde{R} = l = 1, 2, 3, \dots$,there exists $n_l \geq l$,such that,

$$\int_{|x| \geq l} |\nabla u_{n_l}|^2 + u_{n_l}^2 \geq \delta_0 \tag{22}$$

Take $\epsilon > 0$ so small that

$$C\epsilon + C' < m_\infty$$

Here and later,$C > 0$ denotes some constant which will be determined in the following calculation, C is independent of ϵ.

Since $u_\infty \in H^1(R^N)$ and $\limsup\limits_{|x| \to R_0} q(x) \leq q_\infty$, there exists $R_0 > 0$, such that,

$$\int_{|x| \geq R_0} |\nabla u_\infty|^2 + u_\infty^2 + |q(x)| |u_\infty|^{\sigma+2} < \epsilon \tag{23}$$

$$q(x) \leq q_\infty + \epsilon \qquad\qquad |x| \geq R_0 \tag{24}$$

Because of the boundedness of $\{\|u_{n_l}\|\}$, there exists some integer j_ϵ,such that,

$$j_\epsilon \cdot \epsilon > \|u_{n_l}\|^2 \qquad\qquad l = 1, 2, 3, \dots \tag{25}$$

Consider the annuli $\{I_k\}_1^{j_\epsilon}$,where

$$I_k = \{x \in R^N : R_0 + k - 1 \leq |x| \leq R_0 + k\} \quad k = 1, \dots, j_\epsilon$$

Because of (25), for any l , there exists some $I \subset \{I_1, \dots, I_{j_\epsilon}\}$, such that,

$$\int_I |\nabla u_{n_l}|^2 + u_{n_l}^2 < \epsilon \tag{26}$$

There are only finitely many annuli but infinitely many $\{u_{n_l}\}$, there must exists at least one annulus $I \subset \{I_1, \dots, I_{j_\epsilon}\}$ for which (26) holds for infinitely many l. Take such a subsequence and , for simplicity,still denote it as $\{u_{n_l}\}$. Let us denote the above annulus as

$$I = \{x \in R^N : R \leq |x| \leq R + 1\}, \qquad R \geq R_0$$

Construct a function $\rho_1 \in C^\infty(R^N)$ satisfying:

$$\rho_1(x) = \begin{cases} 1 & |x| \le R \\ 0 & |x| \ge R+1 \\ \text{between 0 and 1} & R \le |x| \le R+1 \end{cases}$$

$$|\nabla \rho_1(x)| \le 2 \qquad \forall x \in R^N$$

Let

$$\rho_2(x) = 1 - \rho_1(x) \qquad \forall x \in R^N$$

For $l = 1, 2, 3, \ldots,$ let

$$v_l = \rho_1(x) u_{n_l}$$
$$w_l = \rho_2(x) u_{n_l}$$

We have

$$u_{n_l} = v_l + w_l$$

$$\operatorname{supp} v_l \subset B_{R+1} \qquad \operatorname{supp} w_l \subset R^N \setminus B_R$$

With (26) and the properties of ρ_1, it is easy to see that

$$\int |q(x)||v_l|^{\sigma+1}|w_l| + \int |q(x)||v_l||w_l|^{\sigma+1} + \int |v_l||w_l| + \int |\nabla v_l||\nabla w_l| \le C\epsilon$$

Therefore we have

$$o(1) =< J'(u_{n_l}), v_l >=< J'(v_l), v_l > +O(\epsilon)$$

Here and later ,$O(\epsilon)$ denotes some quantity bounded by $C\epsilon$

Similarly we have

$$o(1) =< J'(w_l), w_l > +O(\epsilon)$$

namely

$$\int |\nabla v_l|^2 + v_l^2 = \int q(x)|v_l|^{\sigma+2} + O(\epsilon) \tag{27}$$

$$\int |\nabla w_l|^2 + w_l^2 = \int q(x)|w_l|^{\sigma+2} + O(\epsilon) \tag{28}$$

On the other hand

$$C' + o(1) = J(u_{n_l}) = J(v_l) + J(w_l) + O(\epsilon)$$

Use (27),we have

$$J(v_l) = (\frac{1}{2} - \frac{1}{\sigma+2})\|v_l\|^2 + O(\epsilon) \geq O(\epsilon)$$

Hence we have

$$C' \geq \frac{1}{2}\int |\nabla w_l|^2 + w_l^2 - \frac{1}{\sigma+2}\int q(x)|w_l|^{\sigma+2} + O(\epsilon) \tag{29}$$

We are going to use (28) and (29) to get a contradiction , which means that case (2) never occurs.

The following formulas hold only for large l ,at least $l > R+1$.It follows from (22),(28) and the choice of ϵ that

$$\int |\nabla w_l|^2 + w_l^2 \geq \delta_0$$

$$\int q_\infty |w_l|^{\sigma+2} \geq \frac{\delta_0}{2} \tag{30}$$

Let

$$\lambda := \lambda_{l,\epsilon} := \{\frac{\int |\nabla w_l|^2 + w_l^2}{\int q_\infty |w_l|^{\sigma+2}}\}^{\frac{1}{\sigma}}$$

Use (28) and (30),we have

$$\lambda \leq 1 + C\epsilon \tag{31}$$

Let $\xi_l = \lambda w_l$, then $\xi_l \in V_\infty$. Use (28),(31) we have that

$$m_\infty \leq (\frac{1}{2} - \frac{1}{\sigma+2})\int q(x)|w_l|^{\sigma+2} + C\epsilon \tag{32}$$

Combine (28) and (29) we have

$$C' \geq (\frac{1}{2} - \frac{1}{\sigma+2})\int q(x)|w_l^{\sigma+2} - C\epsilon \tag{33}$$

It follows from (32) and (33) that $m_\infty \leq C' + C\epsilon$,which contradicts to the choice of ϵ. This concludes the proof.

Proof of Theorem 1.1:
Let

$$X_n = \{t_1u_1 + \ldots + t_nu_n : (t_1, \ldots, t_n) \in R^N\}$$

where $u_1, \ldots, u_n$ are those functions in the hypotheses. Clearly $X_n \cap \{u \in H^1(R^N) : J(u) \geq 0\}$ is bounded. Because of Lemma 1.2 and Theorem A in the Appendix ,to conclude the proof we only need to get the following estimates:

$$\sup_{u \in X_n} J(u) < m_\infty \tag{34}$$

According to Lemma 1.1 and (5),there exists $\tilde{S}_\infty < S_\infty$,such that

$$\sum_{j=1}^{n} \frac{(\int |\nabla u_j|^2 + u_j^2)^{\frac{\sigma+2}{\sigma}}}{(\int q(x)|u_j|^{\sigma+2})^{\frac{2}{\sigma}}} < \tilde{S}_\infty^{\frac{\sigma+2}{\sigma}} \tag{35}$$

Take any $u \in X_n$, $u = t_1u_1 + \ldots + t_nu_n$, $(t_1, \ldots, t_n) \in R^N$.We have

$$\left\{ \begin{array}{lcl} \int q(x)|u|^{\sigma+2} & = & \sum_{j=1}^{n} |t_j|^{\sigma+2} B_j \\ \int |\nabla u|^2 + u^2 & = & \sum_{j=1}^{n} t_j^2 A_j \end{array} \right. \tag{36}$$

where

$$A_j = \int |\nabla u_j|^2 + u_j^2, \quad B_j = \int q(x)|u_j|^{\sigma+2}, \quad j = 1, \ldots, n$$

By Hölder's inequality we have

$$\begin{aligned} \sum t_j^2 A_j &\leq \{\sum (\frac{A_j}{B_j^{\frac{2}{\sigma+2}}})^{\frac{\sigma+2}{\sigma}}\}^{\frac{\sigma}{\sigma+2}} \{\sum |t_j|^{\sigma+2} B_j\}^{\frac{2}{\sigma+2}} \\ &< \tilde{S}_\infty \{\sum t_j^2 A_j\}^{\frac{2}{\sigma+2}} \end{aligned}$$

Hence

$$\sup_{u \in X_n} J(u) < m_\infty \tag{37}$$

Theorem 1.1 follows from Lemma 1.1 and Theorem A in the Appendix .

2 Existence of multiple solutions of more general equations

In this section we study the equation (1). Suppose that $g(x,u)$ satisfies the following hypotheses:

$$g(x,u) \quad \text{is continuous and odd in } u \tag{38}$$

$$|g(x,u)| \leq b_1(x)|u| + b_2(x)|u|^{\sigma+1} \tag{39}$$

where

$$\begin{array}{ll} (x,u) \in R^N \times R & \\ 0 < \sigma < \frac{4}{N-2} & \text{if } N \geq 3 \\ 0 < \sigma < +\infty & \text{if} N = 1,2 \\ b_1 \in L^\alpha(R^N) + L^\infty(R^N) & \\ b_2 \in L^\beta(R^N) + L^\infty(R^N) & \\ \alpha = \frac{N}{2} & \text{if } N \geq 3 \\ \alpha > 1 & \text{if } N = 2 \\ \alpha = 1 & \text{if } N = 1 \\ \beta = \frac{2N}{2N-(N-2)(\sigma+2)} & \text{if } N \geq 3 \\ \beta > 1 & \text{if } N = 2 \\ \beta = 1 & \text{if } N = 1 \end{array}$$

There exists $\theta \in (0, \frac{1}{2})$,such that,

$$G(x,u) \leq \theta u g(x,u) \qquad \forall (x,u) \in R^N \times R \tag{40}$$

where $G(x,u) := \int_0^u g(x,s)ds$

$$g(x,u)u^{-1} \text{ is nondecreasing in } u \qquad \forall u \geq 0 \quad x \in R^N \tag{41}$$

Remark 2.1: (40) and (41) follow from the following stronger assumption:

$$g \in C^1 \text{ and there exists } \tilde{\theta} \in (0,1), \text{such that,}$$
$$g(x,u) \leq \tilde{\theta} u g_u(x,u) \qquad \forall (x,u) \in R^N \times R^+$$

Suppose that $h : R^N \times R \to R$ satisfies the following properties: There exists $R_1 \gg 1$, such that

$$g(x,u)u \le h(x,u)u \qquad |x| > R_1, u \in R \tag{42}$$

$$\begin{cases} h \in C(R^N \times R, R), \quad u(\cdot) \mapsto h(\cdot, u(\cdot))u(\cdot) \\ \text{is a continuous map from } H^1(R^N) \text{ into } L^1(R^N). \end{cases} \tag{43}$$

There exists $\epsilon_1 > 0$, such that

$$h(x,tu)tu \ge t^{2+\epsilon_1} h(x,u)u \qquad t \ge 1, x \in R^N, u \in R. \tag{44}$$

For all $w \in H_0^1(R^N) \cap L^\infty(R^N)$, $\operatorname{supp} w \subset R^N \setminus B_{R_2}$:

$$\lim_{t\to 0^+} \sup \frac{1}{t^2} \int h(x,tw)tw = 0 \tag{45}$$

Where $R_2 \gg 1$ is some positive constant.

Let

$$\begin{aligned} H(x,u) &:= \int_0^u h(x,s)ds && (x,u) \in R^N \times R \\ J_g(u) &:= \frac{1}{2}\int |\nabla u|^2 + u^2 - \int G(x,u)dx && u \in H^1(R^N) \\ J_h(u) &:= \frac{1}{2}\int |\nabla u|^2 + u^2 - \int H(x,u)dx && u \in H^1(R^N) \end{aligned}$$

According to (42),for any $u \in H^1(R^N)$ with $\operatorname{supp} u \subset R^N \setminus B_{R_2}$,we have

$$\int G(x,u)dx \le \int H(x,u)dx \tag{46}$$

Let

$$m_h := \inf_{u \in V_h} J_h(u)$$

where

$$V_h := \{u \in H^1(R^N) \setminus \{0\} : \int |\nabla u|^2 + u^2 - \int h(x,u)u = 0\}$$

It is well known that $J_g, J_h \in C^1(H^1(R^N), R)$ under (38),(39),(42).

Theorem 2.1: In addition to (38) through (45), there exist n functions with disjoint supports $u_j \in H_0^1(R^N) \setminus \{0\}$, $j = 1, \ldots, n$, such that,

$$\sup_{u \in X_n} J_g(u) < m_h$$

where n is a positive integer and

$$X_n := \{t_1 u_1 + \ldots + t_n u_n : (t_1, \ldots, t_n) \in R^n\} \subset H^1(R^N)$$

is a n dimensional space.

Then (1) has at least n pairs of nontrivial solutions in $H^1(R^n)$ and one of them is positive.

In the following we denote J_g as J.

Remark 2.1 The existence of one positive solution under the hypotheses is classical,see [16] ,[9] and the references there.

Lemma 2.1: Under the hypotheses of Theorem 2.1, J satisfies $(PS)_c$ for $c \in (-\infty, m_h)$.

Proof: Let $\{u_n\} \subset H^1(R^N)$ be a sequence satifying:

$$J(u_n) \to C' < m_h$$

$$J'(u_n) \to 0 \quad \text{strongly in } H^{-1}(R^N)$$

By a similar argument as in the proof of Lemma 1.2 we obtain the boundedness of $\{\|u_n\|\}$ and the existence of $u_\infty \in H^1(R^N)$,such that,along a sequence of $\{u_n\}$ (still being denoted as $\{u_n\}$)

$$\begin{cases} u_n \rightharpoonup u_\infty & \text{weakly in } H^1(R^N) \\ u_n \to u_\infty & \text{a.e. in } R^N \end{cases} \tag{47}$$

$$< J'(u_\infty), u_\infty >= 0 \tag{48}$$

There are only two possibilities for $\{u_n\}$.

Case (1): $\forall 0 < \delta < 1$, there exists $\widetilde{R} > 0$, such that, $\forall n \in \widetilde{R}$,

$$\int_{|x| \geq \widetilde{R}} |\nabla u_n|^2 + u_n^2 < \delta$$

Case (2): There exists $\delta_0 > 0$, such that, $\forall \tilde{R} > 0$, there exists $n = n(\tilde{R}) \geq \tilde{R}$,such that

$$\int_{|x|\geq \tilde{R}} |\nabla u_n|^2 + u_n^2 < \delta_0$$

As in the proof of Lemma 1.2 we will prove that Case (1) leads to the strong convergence of $\{u_n\}$ in $H^1(R^N)$. In fact, $\forall \delta > 0$,there exists $\tilde{R} > 0$,such that, $\forall n > \tilde{R}$ we have

$$\int_{|x|\geq \tilde{R}} |\nabla u_n|^2 + u_n^2 < \delta \tag{49}$$

By the lower semicontinuity of the above integral and (47) we have

$$\int_{|x|\geq \tilde{R}} |\nabla u_\infty|^2 + u_\infty^2 \leq \delta \tag{50}$$

Since $J'(u_n) \to 0$ strongly in $H^1(R^N)$ we have

$$|\int \nabla u_n \nabla v + u_n v - \int g(x, u_n)v| \leq o(1)\|v\| \quad v \in H^1(R^n) \tag{51}$$

Use Hölder inequality,Sobolev inequality,Rellich lemma on bounded domains and (39)we have ,for any $v \in H^1(R^N)$

$$\begin{cases} |\int_{|x|\geq \tilde{R}} g(x, u_n)v| + |\int_{|x|\geq \tilde{R}} g(x, u_\infty)v| & \leq & C\delta^{\frac{1}{2}}\|v\| \\ |\int_{|x|\leq \tilde{R}} g(x, u_n)v - \int_{|x|\leq \tilde{R}} g(x, u_\infty)v| & \leq & o(1)\|v\| \end{cases} \tag{52}$$

Where C is some constant independent of δ and n. Combine (51) and (52) we have

$$|\int \nabla u_n \nabla v + u_n v - \int g(x, u_\infty)v| \leq C\delta^{\frac{1}{2}}\|v\| + o(1)\|v\| \quad v \in H^1(R^N) \tag{53}$$

Since $\delta > 0$ can be chosen arbitrarily small,(53) implies the strong convergence of $\{u_n\}$ in $H^1(R^N)$.

To conclude the proof we only need to rule out Case (2) . As before we argue by contradiction .If Case (2) occurs, for that δ_0,take $\tilde{R} = l = 1, 2, 3, \ldots$,there exists $n_l \geq l$,such that

$$\int_{|x|\geq l} |\nabla u_{n_l}|^2 + u_{n_l}^2 \geq \delta_0 \tag{54}$$

Take $\epsilon > 0$ so small that

$$C\epsilon + C' < m_h$$

Here and later,$C > 0$ denotes some constant being determined in the following calculation,which is independend of ϵ.

According to (43) and the fact that $u_\infty \in H^1(R^N)$, there exists $R_0 > \max\{R_1, R_2\}$,such that

$$\int_{|x|\geq R_0} |\nabla u_\infty|^2 + u_\infty^2 + |q(x)||u_\infty|^{\sigma+2} < \epsilon \tag{55}$$

$$g(x,u)u \leq h(x,u)u \qquad \forall |x| > R_2, u \in R \tag{56}$$

As in the proof of Theorem 1.1 we can find $R \geq R_0, I = \{x \in R^N : R \leq |x| \leq R+1\}$ a subsequence of $\{u_{n_l}\}$, which will still be denoted it as $\{u_{n_l}\}$,such that,

$$\int_I |\nabla u_{n_l}|^2 + u_{n_l}^2 < \epsilon \tag{57}$$

Let

$$v_l = \rho_1(x)u_{n_l}, \qquad w_l = \rho_2(x)u_{n_l} \;.\; l = 1,2,3;\ldots$$

we have

$$u_{n_l} = v_l + w_l, \quad \operatorname{supp} v_l \subset B_{R+1}, \quad \operatorname{supp} w_l \subset R^N \setminus B_R, \quad l = 1,2,3,\ldots$$

where ρ_1, ρ_2 are the same functions as in the proof of Theorem 1.1.

With (57) and the properties of ρ_1, ρ_2 and g it is not difficult to see ,by using Sobolev embedding theorem,that

$$\begin{aligned} < J'(v_l), v_l > &= o(1) + O(\epsilon) &(58)\\ < J'(w_l), w_l > &= o(1) + O(\epsilon) &(59)\\ J(v_l) + J(w_l) &= C' + o(1) + O(\epsilon) &(60) \end{aligned}$$

Notice that according to (40),(48),(58) and Sobolev embedding theorem we have

$$J(v_l) = \frac{1}{2}\int_{B_{R+1}} \{g(x,v_l)v_l - G(x,v_l)\} + o(1) + O(\epsilon)$$

$$\int_{B_{R+1}} g(x,u_\infty)u_\infty \geq o(1) + O(\epsilon)$$

$$J(v_l) \geq o(1) + O(\epsilon)$$

It then follows that

$$J(w_l) \leq C' + o(1) + O(\epsilon) \tag{61}$$

Here and later,the formulas hold only for large l .

Writing (59) and (61) more explicitly we have

$$\int |\nabla w_l|^2 + w_l^2 = \int g(x, w_l)w_l + O(\epsilon) \tag{62}$$

$$\frac{1}{2}\int |\nabla w_l|^2 + w_l^2 - \int G(x, w_l) \leq C' + O(\epsilon) \tag{63}$$

According to (54) ,for $l > R + 1$,

$$\int |\nabla w_l|^2 + w_l^2 \geq \delta_0.$$

For $\epsilon > 0$ small,use (62) we have

$$\int g(x, w_l) \geq \frac{1}{2}\delta_0. \tag{64}$$

Consider

$$\xi(t) = \frac{1}{t^2}\int h(x, tw_l)tw_l \quad t \in R.$$

For $t = 1 + C\epsilon > 1$,where C is large but independent of ϵ, we deduce from (43),(44) and (64) that

$$\xi(t) \geq t^{\epsilon_1}\int g(x, w_l)w_l \geq \int g(x, w_l)w_l + O(\epsilon) = \int |\nabla w_l|^2 + w_l^2.$$

On the other hand, $\lim_{t\to 0^+} \xi(t) = 0$ according to (45). By the continuity of ξ,there exists some $t = t(l, \epsilon) \leq 1 + C\epsilon$,such that,

$$\xi(t) = \int |\nabla w_l|^2 + w_l^2$$

Let $\eta_{l,\epsilon} := tw_l$,then $\eta_{l,\epsilon} \in V_h$, hence

$$\begin{aligned} m_h &\leq J_{h,\epsilon}(\eta_l) \\ &= \frac{1}{2}t^2\int |\nabla w_l|^2 + w_l^2 - \int H(x, tw_l) \\ &\leq \frac{t^2}{2}\int g(x, w_l)w_l + O(\epsilon) - \int G(x, tw_l) \end{aligned}$$

It follows immediately from (41) that

$$\frac{t^2}{2}\int g(x,w)w - \int G(x,tw) \text{ is nondecreasing for } 0 \le t \le 1.$$

Therefore if $t \le 1$,

$$m_h \le J(w_l) + O(\epsilon) \le C' + O(\epsilon)$$

If $t \ge 1$, remember that $t \le 1 + C\epsilon$,by the mean value theorem we have

$$m_h \le C' + O(\epsilon)$$

This is a contradiction to the choice of ϵ.

Proof of Theorem 2.1 follows immediatly from the application of Theorem A of the Appendix by letting $E = H^1(R^N), J = J_g$.

3 Appendix : A Minmax Lemma

Let E be a real Banach space and $\Sigma(E) \equiv \Sigma$ the collection of $A \subset E \setminus \{0\}$ with A closed in E and symmetric with respect to the origin,namely, $-x \in A$ whenever $x \in A$. The set $A \subset E$ is said to have genus n (denoted by $\gamma(A) = n$) if there exists an odd map $\varphi \in C(A, R^n \setminus \{0\})$ and n is the smallest integer having this property. If $A = \phi$, we write $\gamma(A) = 0$ and if there is no such n, we set $\gamma(A) = +\infty$. For the properties of the genus,see [18].

Suppose that E is an infinite dimensional Banach space, $J \in C^1(E, R)$, satisfying:
For some positive integer n, positive constant C',

$$J \text{ satisfies } (PS)_c \quad \forall c \in (0, C') \tag{65}$$

$$J \text{ is even,namely } J(-u) = J(u) \tag{66}$$

There exists $\rho, \alpha > 0$, such that

$$\begin{cases} J > 0 & \text{in } B_\rho \setminus \{0\} \\ J \geq \alpha & \text{on } \partial B_\rho \end{cases} \tag{67}$$

There exists a $n-$dimensional subspace X_n of E ,such that

$$\begin{cases} X_n \cap \widehat{A}_0 \text{ is bounded} \\ \sup_{u \in X_n} J(u) < C' \end{cases} \tag{68}$$

where

$$\widehat{A}_0 := \{u \in E : J(u) \geq 0\}$$

Let

$$\Gamma^* := \{h : \ h \text{ is a homeomorphism of } E \text{ onto } E, \\ h(0) = 0, h(B_1) \subset \widehat{A}_0, h \text{ is odd } \}$$

where B_1 is the unit ball of E.
Let

$$\Gamma_m := \{K \subset E : K \text{ is compact,symmetric with respect to } 0, \text{ and for all } h \in \Gamma^*, \gamma(K \cap h(S)) \geq m\}$$

where

$$S := \partial B_1 \equiv \{u \in E : \|u\|_E = 1\}$$

Observe that if $h \in \Gamma^*, h(S) \subset E \setminus \{0\}$ and is closed and symmetric. Therefore $K \cap h(S) \in \Sigma(E)$. For the properties of Γ_m, see [18].

Now we state a theorem without proof. For a proof ,see [18].

Theorem A: Let E be a Banach space, $J \in C^1(E, R)$ satisfying (65) through (68). let

$$b_m = \inf_{K \in \Gamma_m} \max_{u \in K} J(u) \quad m = 1, \ldots, n$$

Then we have
(1) $0 < \alpha \leq b_1 \leq \ldots \leq b_n \leq C' < +\infty$ and $b_1, \ldots, b_n$ are critical values of J.
(2) If $b_m = b_{m+1}$ for some $m \in \{1, \ldots, n-1\}$, then J has infinitely many critical points corresponding to b_m.

References

[1] S. Agmon,A. Douglis and L. Nirenberg, Estimates near the boundary for solutions of elliptic partial differential equations satifying general boundary conditions, Comm. Pure Appl. Math. 12 (1959),pp. 623-727.

[2] S. Agmon,A. Douglis and L. Nirenberg, Estimates near the boundary for solutions of elliptic partial differential equations satifying general boundary conditions, Comm. Pure Appl. Math. 17 (1964),pp. 35-92.

[3] H. Berestycki and P. L. Lions, Nonolinear scarlar field equations. Arch. Rat. Mech. Anal., I,82(1983) p. 313-346; II,82 (1983) p. 347-376.

[4] H. Brezis and L. Nirenberg, Positive solutions of nonlinear elliptic equations involving critical Sobolev exponents. Comm. Pure. Appl. Math. 36 (1983) pp 437-477

[5] H. Berestycki and C. Taubes ,In preparation.

[6] G. Cerami,S. Solimini and M. Struwe, Some existence results for superlinear elliptic boundary value problems involving critical exponents. J. Func. Anal. 69 (1986) pp 289-306

[7] C. V. Coffman, A nonlinear boundary value problem with many positive solutions. J. Diff. Eqn. 54 (1984) 429-437.

[8] Weiyue Ding, On a conformally invariant equation on R^N.

[9] Weiyue Ding and Weiming Ni, On the existence of positive entire solutions of a semilinear elliptic equation. Arch. Rat. Mech. Anal. 91 (1986),pp 288-308

[10] M. J. Esteban and P. L. Lions, Existence and nonexistence results for semilinear elliptic problems in unbounded domains. Proc. Roy. Soc. Edim., 93 (1981) pp. 1-14

[11] B. Gidas, W. Ni and L. Nirenberg, Symmetry of positive solutions of nonlinear elliptic equations in R^N. in "Mathematical Analysis and Applications" Part A(L. Nachbin ed.) pp 370-401 Academic Press Orlando Fl. 1981.

[12] D. Gilbarg and N. S. Trudinger, Elliptic partial differential equations,2nd edition , Grundlehoen der mathematischen Wissenschaften 224,Springer-Verlag.

[13] T. Kato, Perturbation theory for linear operators, 2nd edition , Grundlehoen der mathematischen Wissenschaften 132,Springer-Verlag.

[14] Y. Y. Li , Existence of multiple standing wave solutions of Schrödinger equations. Preprint

[15] P. L. Lions, The concentration-compactness principle in the calculus of variations. The locally compact case. Ann. Inst. Henri. Poincare. Vol 1 (1984) pp 102-145 and pp 223-283

[16] P. L. Lions, On positive solutions of semilinear elliptic equations in unbounded domsins. Preprint

[17] Weiming Ni, Some aspects of semilinear elliptic equations. Lecture notes, Published by the National Tsing Hua University,Taiwan.

[18] P. H. Rabinowitz, Variational methods for nonlinear eigenvalue problems, Edicioni Cremonese,Roma (1974),pp 141-195

[19] P. H. Rabinowitz, A global theorem for nonlinear eigenvalue problems and applications. Contrib. Nonlinear Fcl. Anal. Academic Press. 1971,pp 11-36

[20] C. A. Stuart,Bifurcation for Dirichlet problems without eigenvalues, Proc. London Math. Soc. (3) 45 (1982),pp 169-192

[21] W. A. Strauss, Existence of solitary waves in higher dimensions. Comm. Math. Phys. 55 (1977) 149-162

[22] Dong Zhang,

[23] Xiping Zhu, Multiple entire solutions of a semilinear elliptic equation. Preprint

Mathematics Department
Rutgers University
New Brunswick, NJ 08903

Lagrange Multipliers, Morses Indices and Compactness

P.L. LIONS*

SUMMARY

I. Introduction

The purpose of this paper is to emphasize two observations on variational problems. In some sense, they are independent even if they can be combined in some problems. The first one concerns the so-called strict subadditivity inequalities in the concentration-compactness method as introduced by the author [23], [24]. See also for some extensions, or applications of these arguments: M.J. Esteban and P.L. Lions [19], [20], M.J. Esteban [17], [18], M. Weinstein [31], P.L. Lions [25], H. Berestycki and P.L. Lions [9], A. Bahri and P.L. Lions [4], P.L. Lions [26], D. Gogny and P.L. Lions [21]. Roughly speaking, this method shows that, for various minimization problems in unbounded domains with constraints, all minimizing sequences are converging if and only if a certain strict subadditivity inequality holds. This inequality involves the infimum of the minimization problem as a function of the "level of the constraint" and corresponds to the possible losses of compactness which are basically due to the effect of

*Consultant at CEA — Service de Physique Générale.

"unbounded translations" or "concentrating-diluting dilations." In some sense, the strict subadditivity inequalities represent the energy balance preventing (and this is necessary and sufficient) losses of compactness. The subadditivity comes into the picture because losses of compactness, when they occur, split the minimizing sequences into various parts which are either "infinitely away from each other" (translations) or "live in different scales" (dilations). Of course, remains the question of checking these inequalities and this is by large an open question: in [23], [24], this was done on several examples by scaling arguments (dilations and or multiplications). We present here a few examples on which one checks these inequalities using in a fundamental way the Euler–Lagrange equations and the associated Lagrange multipliers. The fact that Lagrange multipliers should play a role is no big surprise since, after all, a Lagrange multiplier is formally the derivative of the infimum with respect to the constraint. One useful remark in this direction taken from [26] is that all the "pieces of minimizing sequences" when compactness is lost inherit of the *same* Lagrange multiplier. Then, one can try to check the strict inequalities by a careful *interaction computation,* very much in the spirit of A. Bahri and J.M. Coron [3], H. Berestycki and C. Taubes [11]. Then, in the examples we present below, the "interaction of various pieces sharing the same Lagrange multiplier" decreases the sum of energies of the individual pieces provided this *Lagrange multiplier is not degenerate* ($\neq 0$), leading in that case to a strict subadditivity condition. We hope this strategy will become clear when applied to the specific examples below. However, it might be worth emphasizing that the only general remark is the common value of Lagrange multipliers while the interaction computation and the verification of the non-degeneracy of the Lagrange multiplier require a specific analysis on each problem—even if some general recipes may be invoked...

The second observation we wish to make in this paper—and this is the subject of Part III—concerns the Palais–Smale condition and Morse indices of critical points. As it is well-known, the "min-max" approach to the existence of critical point(s) of functionals—as introduced and developed by P.H. Rabinowitz, A. Ambrosetti...—relies on a compactness condition called the Palais–Smale condition: this condition requests the compactness of "approximated critical points" near the studied min-max level. On the other hand, for functionals satisfying the Palais–Smale condition, there now exists several results showing that one can find at a given min-max level a critical point whose Morse index—up to nullity—is precisely the "topological index" of the sets involved in the min-max principle. We refer the reader to A. Bahri [2], A. Bahri and P.L. Lions [5], C. Viterbo [30], C.V. Coffman [12], A.C. Lazer and S. Solimini [22], S. Solimini [28] for such results.

The observation we want to emphasize in Part III is that this knowledge of Morse indices can be used to check a generalized Palais–Smale condition, get compactness and thus existence results. The idea is to approximate a functional—for which the Palais–Smale condition is not clearly satisfied—by functionals satisfying the Palais–Smale conditions. In this way, we create at a certain approximated min-max level a critical point with a given Morse index. Hence, we obtain a *particular Palais–Smale sequence*—i.e. a sequence of approximated critical points of the original functional—with a fixed Morse index. And we claim that this additional information may be used to gain compactness, in fact we will present an example where this additional information is even necessary for compactness—this example concerns the Hartree–Fock equations in Atomic and Molecular Physics and is taken from P.L. Lions [26]. To illustrate this viewpoint, we present in Part III three results. The first one, taken from A. Bahri and P.L. Lions [6], is not directly related to existence questions and shows that for superlinear elliptic problems the energy of critical points (or their L^∞ norm) remains bounded if and only if the Morse index also remains bounded. The second example concerns superlinear convex Hamiltonian systems and we obtain with the above strategy new existence results for periodic orbits with a prescribed minimal period—this result is due to I. Ekeland and P.L. Lions and is developed and extended to more general situations in V. Coti-Zelati, I. Ekeland and P.L. Lions [14], see also I. Ekeland [15]. The final example is the one mentioned above—it was in fact the first time, to our knowledge, that compactness was obtained in this way.

Acknowledgement. The author wishes to thank the Air France Company for the various flights during which the results of Part I were obtained (corollaries of the free champagne?).

II. Strict Sub-Additivity Inequalities and Lagrange Multipliers

II.1 Some Nonlinear Eigenvalue Problems. In this section, we consider the following problem

$$(1) \qquad I_\lambda = \operatorname{Inf}\left\{ \int_{\mathbb{R}^N} |\nabla u|^2 + u^2 dx / u \in H^1(\mathbb{R}^N), \int_{\mathbb{R}}^N F(x,u)dx = \lambda \right\}$$

where $\lambda > 0$, $N \geq 3$, F is nonnegative (for example), even, $F(x,t)$ is twice continuously differentiable with respect to t and $\partial^2 F/\partial t^2$ is locally Hölder continuous (for some exponent $\alpha > 0$) in t uniformly in x and uniformly continuous in x uniformly for t bounded. We will also always assume in this section—and we will not recall these assumptions—that

$$(2)\quad \begin{cases} F(x,0) = \frac{\partial}{\partial t}F(x,0) = \frac{\partial^2}{\partial t^2}F(x,0) = 0 \text{ on } \mathbf{R}^N \\ \frac{\partial F}{\partial t}(x,t)t|t|^{-\ell} \to 0 \text{ as } |t| \to \infty, \text{uniformly in } x, \text{ where } \ell = \frac{N+2}{N-2}, \end{cases}$$

$$(3)\qquad F(x,t) \to F(t) \text{ as } |x| \to \infty$$

$$(4)\qquad F(t) > 0 \text{ for some } t > 0, \frac{\partial F}{\partial t}(x,t) \geq 0 \text{ on } \mathbf{R}^N \times [0,\infty).$$

And we recall a typical application of the concentration-compactness argument. Denoting by

$$(5)\quad I_\lambda^\infty = \operatorname{Inf}\left\{\int_{\mathbf{R}^N} |\nabla u|^2 + u^2 dx / u \in H^1(\mathbf{R}^N), \int_{\mathbf{R}^N} F(u)dx = \lambda\right\},$$

then we have the

Theorem. [23]. 1) *Every minimizing sequence of* (1) (*resp.* (5)) *is relatively compact in* $H^1(\mathbf{R}^N)$ *up to a translation if and only if the following condition holds*

$$(6)\qquad I_\lambda < I_\alpha + I_{\lambda-\alpha}^\infty \quad \textit{for all } \alpha \in (0,\lambda)$$

(*resp.*

$$(7)\qquad I_\lambda^\infty < I_\alpha^\infty + I_{\lambda-\alpha}^\infty \quad \textit{for all } \alpha \in (0,\lambda)).$$

In particular, if (7) *holds, there exists a minimum of* (5).

2) *Every minimizing sequence of* (1) *is relatively compact in* $H^1(\mathbf{R}^N)$ *if and only if the following condition holds*

$$(8)\qquad I_\lambda < I_\alpha + I_{\lambda-\alpha}^\infty \quad \textit{forall } \alpha \in [0,\lambda).$$

In particular, if (8) *holds, there exists a minimum of* (1).

At this point, several remarks are in order: first of all, checking (6) or (7) is not obvious unless we assume (see [23]) for some $\theta < 1/2$

$$(9)\qquad \theta t \frac{\partial F}{\partial t}(x,t) \geq F(x,t) \quad \text{for } x \in \mathbf{R}^N,\ t \in \mathbf{R}.$$

Next, concerning the problem (5), one knows there exist *some* minimizing sequences which are relatively compact (and thus that a minimum exists) indeed, this was proven by W. Strauss [29]—see also C.V. Coffman [13], H. Berestycki and P.L. Lions [10] for related arguments—using Schwarz symmetrization. It is then a natural question to ask whether (7) holds. Finally, it is clear that the difference between (6) and (8) is only the condition

$$I_\lambda < I_\lambda^\infty. \tag{10}$$

Even if this condition may not be easy to check, one possible strategy is to insert a conveniently chosen minimum of (5)—recall that any translate of a minimum of (5) is still a minimum—in the problem (1) and compute the "energy balance."

Before presenting results and a method for the verification of (6)–(7), let us make a final general remark taken from [24].

Fact 1. If $I_{\bar\lambda} < I_{\bar\lambda}^\infty$ for some $\bar\lambda > 0$, there exists $\lambda \in (0, \bar\lambda]$ such that (6) holds. Indeed, if (6) does not hold for $\bar\lambda$, there exists some $\alpha \in (0, \bar\lambda)$ such that

$$I_{\bar\lambda} = I_\alpha + I_{\bar\lambda-\alpha}^\infty.$$

Then, if we denote by λ the smallest of these α, we still have

$$I_{\bar\lambda} = I_\lambda + I_{\bar\lambda-\lambda}^\infty$$

and we have

$$I_{\bar\lambda} < I_\alpha + I_{\bar\lambda-\alpha}^\infty \quad \text{for } \alpha \in (0, \lambda).$$

We now claim that (6) holds. Indeed we have on one hand

$$I_{\bar\lambda} = I_\lambda + I_{\bar\lambda-\lambda}^\infty < I_{\bar\lambda}^\infty \leq I_\lambda^\infty + I_{\bar\lambda-\lambda}^\infty$$

the first inequality being true by assumption, while the other is "always" true by the concentration-compactness argument [23].

On the other hand, we have by assumption for all $\alpha \in (0, \lambda)$

$$I_{\bar\lambda} = I_\lambda + I_{\bar\lambda-\lambda}^\infty < I_\alpha + I_{\bar\lambda-\alpha}^\infty \leq I_\alpha + I_{\bar\lambda-\lambda}^\infty + I_{\lambda-\alpha}^\infty$$

and the claim is proved.

Fact 2. If $\{\lambda > 0 / I_\lambda < I_\lambda^\infty\} \neq \emptyset$, we have just seen that $\{\lambda > 0/$ (6) holds$\}$ is also non-empty and we claim that it is an open set, provided we assume that there does not exist $u \in H^1(\mathbf{R}^N)$, $u \not\equiv 0$ such that $\frac{\partial F}{\partial t}(x, u) = 0$ on $\mathbf{R}^N$. We only sketch the proof which relies on the following fact that we do not recall here: $I_\lambda^\infty \lambda^{-1} \to +\infty$ as $\lambda \to 0+$. Indeed, if (6) holds and if there exists a sequence λ_n converging to λ for which (6) does not hold, we reach a contradiction as follows. We denote by α_n the smallest α for which $I_{\lambda_n} = I_\alpha + I_{\lambda_n - \alpha}^\infty$, by continuity $\alpha_n \to_n \lambda$ and by the above argument, there exists a minimum u_n of I_{α_n} and without loss of generality—taking a subsequence if necessary—we may assume that u_n converges in $H^1(\mathbf{R}^N)$ to a minimum u of I_λ. In view of (4), there exist Lagrange multipliers θ_n, θ ($\theta_n, \theta \neq 0$) such that

$$-\Delta u_n + u_n = \theta_n f(x, u_n) \text{ in } \mathbf{R}^N, -\Delta u + u = \theta f(x, u) \text{ in } \mathbf{R}^N \tag{11}$$

where we denote by $f(x,t) = \frac{\partial F}{\partial t}(x,t)$. Next, we choose $t_n > 0$ such that

$$\int_{\mathbf{R}^N} F(x, t_n u_n) dx = \lambda_n, \tag{12}$$

such a t_n exists for n large enough by the implicit functions theorem and we have

$$\left\{(t_n - 1) \int_{\mathbf{R}^N} f(x, u_n) u_n dx \right\} (\lambda_n - \alpha_n)^{-1} \to_n 1$$

i.e. in view of (11)

$$(t_n - 1) \frac{I_{\alpha_n}}{\theta_n} (\lambda_n - \alpha_n)^{-1} \to_n 1. \tag{13}$$

But then, we may reach a contradiction observing that we have

$$\frac{I_{\lambda_n - \alpha_n}^\infty}{\lambda_n - \alpha_n} = \frac{I_{\lambda_n} - I_{\alpha_n}}{\lambda_n - \alpha_n} \leq \frac{t_n^2 - 1}{\lambda_n - \alpha_n} \int_{\mathbf{R}^N} |\nabla u_n|^2 + u_n^2 dx \to_n 2\theta$$

while the left hand side goes to $+\infty$ since $\lambda_n - \alpha_n$ goes to 0.

We may now state two results concerning the verification of (6)–(7).

Theorem 1. *The strict subadditivity inequality* (7) *holds. Therefore, all minimizing sequences of* (5) *are relatively compact in* $H^1(\mathbf{R}^N)$ *up to a translation.*

Theorem 2. *If we assume that there exist a positive constant* C_K *(for all* $K \geq 1$*) such that*

$$F(x,t) \geq F(t) - C_K e^{-2|x|} |x|^{-\alpha} \quad \textit{for } |x| \geq R, \quad |t| \leq R, \qquad \textit{for some } R \in (0, \infty) \textit{ and some } \alpha > \frac{2N-1}{2}, \tag{14}$$

$$
(15)\qquad \begin{aligned} f(x,t) \geq f(t) - C_K e^{-\frac{1}{\delta}|x|}|x|^{-\beta} \ \textit{for}\ |x| \geq R, t \in [0,K], \\ \textit{for some}\ R \in (0,\infty), \end{aligned}
$$

and some $\beta > \frac{N+1}{2}$, *where* $\delta \in (0,1]$ *is an Hölder exponent for* $\partial f/\partial t$ *near* 0.

Then, the strict subadditivity inequality (6) *holds. Therefore, all minimizing sequences of* (1) *are relatively compact in* $H^1(\mathbf{R}^N)$ *up to a translation. If in addition* $I_\lambda < I_\lambda^\infty$, *all minimizing sequences of* (1) *are relatively compact in* $H^1(\mathbf{R}^N)$ *and there exists a minimum.*

Remarks. (1) Even if the conditions (14), (15) can be relaxed a bit—as it will be clear from the proof below, some conditions of that sort are needed in our approach.

(2) One can (try to) check the condition $I_\lambda < I_\lambda^\infty$ following the strategy mentioned above. In particular, it holds if $F(x,t) \geq F(t)$ on $\mathbf{R}^N \times \mathbf{R}$ and $F(x,t) \not\equiv F(t)$ or if

$$
F(x,t) \geq (1 + \lambda e^{-2|x|}|x|^{-\alpha})F(t)
$$

for $|x| \geq R$, $|t| \leq K$ and some $\lambda > 0$, $\alpha > 0$, $R, K > 0$.

Open Problem 1. Is the above result true without conditions like (14)–(15)?

Proof of Theorem 1. We follow the approach described in the Introduction. Assume by contradiction that (7) does not hold. Then, let $\tilde{u}_n$ be a minimizing sequence which is not relatively compact up to a translation. Applying Ekeland's variational principle as in [26], we deduce that there exists another minimizing sequence u_n such that $u_n - \tilde{u}_n$ goes to 0 in $H^1(\mathbf{R}^N)$—hence u_n is not relatively compact up to a translation—and u_n satisfies

$$
(16)\quad -\Delta u_n + u_n - \theta_n f(u_n) = \varepsilon_n \to_n 0 \ \text{ in } H^{-1}(\mathbf{R}^N), \text{ for some } \theta_n \geq 0.
$$

Reiterating the concentration-compactness argument or analyzing (16) (as in P.L. Lions [26], V. Benci and G. Cerami [7], A. Bahri and P.L. Lions [4]...) we deduce that (taking subsequences) θ_n converges to some $\theta > 0$ and there exist $m \geq 2$, $\alpha_1, \ldots, \alpha_m > 0$, sequences y_j^n for $1 \leq j \leq m$ and elements u_j of $H^1(\mathbf{R}^N)$ for $1 \leq j \leq m$ such that

$$
(17)\qquad \sum_{j=1}^m \alpha_j = \lambda, \quad |y_j^n - y_k^n| \to_n \infty \ \text{ for } 1 \leq j \neq k \leq m,
$$

$$(18) \qquad u_n - \sum_{j=1}^{m} u_j(\cdot + y_j^n) \to_n 0 \quad \text{in } H^1(\mathbb{R}^N)$$

$$(19) \qquad -\Delta u_j + u_j = \theta f(u_j) \text{ in } \mathbb{R}^N, \ u_j \not\equiv 0, \ \int_{\mathbb{R}^N} F(u_j)dx = \alpha_j.$$

Observe also that without loss of generality we may assume that $\tilde{u}_n$ and thus u_n, and u_j $(1 \leq j \leq m)$ are nonnegative on $\mathbb{R}^N$.

In fact, one technical point is to check that θ_n remains bounded: this is done by remarking that, if $\theta_n \to_n +\infty$, the concentration-compactness analysis would yield some $u_1 \in H^1(\mathbb{R}^N)$, $u_1 \not\equiv 0$ such that

$$f(u_1) = 0 \quad \text{in } \mathbb{R}^N.$$

Therefore, $\nabla F(u) = 0$ a.e. in $\mathbb{R}^N$ i.e. $F(u)$ is constant thus $F(u) \equiv 0$, a contradiction which shows the boundedness of θ_n.

Next, we observe that clearly

$$I_\lambda^\infty = \sum_{j=1}^{m} \int_{\mathbb{R}^N} |\nabla u_j|^2 + |u_j|^2 dx \geq \sum_{j=1}^{m} I_{\alpha_j}^\infty \geq I_\lambda^\infty .$$

Hence, we deduce that

$$(20) \qquad u_j \text{ is a minimum of } I_{\alpha_j}^\infty$$

$$(21) \qquad \sum_{j\in J} I_{\alpha_j}^\infty = I_{\alpha_J}^\infty \text{ where } \alpha_J = \sum_{j \in J} \alpha_j \\ \text{and } J \text{ is any subset of } \{1, \ldots, m\}.$$

Therefore, it is enough to show that $I_{\alpha_1}^\infty + I_{\alpha_2}^\infty > I_\alpha^\infty$ where $\alpha = \alpha_1 + \alpha_2$ in order to reach the desired contradiction. Observe at that point that we have thus found "subminima" with the *same* Lagrange multiplier.

In order to show the above strict inequality, we use u_1 and u_2 as follows: let e be any unit vector, and set $u_2^n = u_2(\cdot + ne)$, $\tilde{u}^n = u_1 + u_2^n$. We now wish to estimate the quantity $\int_{\mathbb{R}^N} F(\tilde{u}^n)dx$ (this is the main interaction computation or energy balance). Observe also that by elliptic estimates $u_1, u_2 \in L^\infty$ and thus $0 \leq u_1, u_2 \leq R_0$ in $\mathbb{R}^N$ for some $R_0 > 0$. We will need the following lemma.

Lemma 3. *If $F \in C^{2,\alpha}([0,\infty))$ for some $\alpha > 0$ and $F(0) = F'(0) = F''(0) = 0$, then there exists a positive constant C such that for all $a, b \in [0, R_0]$*

$$|F(a+b) - F(a) - F(b) - F'(a)b - F'(b)a| \le C(a \wedge b)^\alpha ab. \tag{22}$$

Proof. Denoting by $G(a,b)$ the left-hand side of (22), we see that

$$\frac{\partial^2}{\partial a \partial b} G(a,b) = F''(a+b) - F''(a) - F''(b).$$

Therefore, since $F''(0) = 0$ and $F \in C^{2,\alpha}$,

$$\left| \frac{\partial^2}{\partial b \partial a} G(a,b) \right| \le C_1 (a \wedge b)^\alpha \quad \text{for } a, b \in [0, R_0].$$

Integrating this bound in b and then in a we find since $F(0) = F'(0) = F''(0) = 0$

$$|G(a,b)| \le C_1 (a \wedge b)^\alpha ab.$$

Applying Lemma 3 we find

$$\begin{aligned} &\left| \int_{\mathbb{R}^N} F(u^n)dx - \alpha - \int_{\mathbb{R}^N} f(u_1)u_2^n dx - \int_{\mathbb{R}^N} f(u_2^n)u_1 dx \right| \\ &\qquad \le C \int_{\mathbb{R}^N} (u_1 \wedge u_2^n)^\alpha u_1 u_2^n dx. \end{aligned} \tag{23}$$

Next, we recall that u_1, u_2 behave like $e^{-|x|}|x|^{-(N-1)/2}$ at infinity, i.e.

$$u_i(x) \exp|x| \, |x|^{\frac{N-1}{2}} \to c_i > 0, \qquad \text{for } i = 1, 2. \tag{24}$$

This combined with (23) yields

$$\begin{aligned} &\int_{\mathbb{R}^N} F(\tilde{u}^n)dx - \alpha - \int_{\mathbb{R}^N} f(u_1)u_2^n dx - \int_{\mathbb{R}^N} f(u_2^n)u_1 dx \\ &\qquad = o\left(e^{-n} n^{-\frac{N-1}{2}}\right) \end{aligned} \tag{25}$$

while we have by the equations (19)

$$\Delta_n = \int_{\mathbb{R}^N} f(u_1)u_2^n dx = \int_{\mathbb{R}^N} f(u_2^n)u_1 dx \tag{26}$$

and

$$\Delta_n e^n n^{\frac{N-1}{2}} \to_n C_0 = c_1 \int_{\mathbb{R}^N} f(u_2) e^{x_1} dx > 0 \tag{27}$$

The asymptotic behaviours in (25) and (27) follow from simple considerations—see also [4] for similar computations.

Next, we wish to find t_n (close to 1) such that

$$\int_{\mathbb{R}^N} F(v^n) dx = \alpha \tag{28}$$

where $v^n = t_n \tilde{u}^n$. This follows from the implicit functions theorem and one finds

$$(t_n - 1) \int_{\mathbb{R}^N} f(\tilde{u}^n) \tilde{u}^n dx = -2\Delta_n + 0(\Delta_n^2)$$

and one checks

$$\int_{\mathbb{R}^N} f(\tilde{u}^n) \tilde{u}^n dx \to_n \int_{\mathbb{R}^N} f(u_1) u_1 dx + \int_{\mathbb{R}^N} f(u_2) u_2 dx = \frac{1}{\theta} I_\alpha^\infty .$$

Therefore, we have

$$(t_n - 1) = -\frac{2}{I_\alpha^\infty} \theta \Delta_n + o(\Delta_n). \tag{29}$$

In order to conclude, we just observe that because of (28)

$$\begin{aligned} I_\alpha^\infty \leq \int_{\mathbb{R}^N} |\nabla v_n|^2 + |v_n|^2 dx &= t_n^2 \{ I_\alpha^\infty + 2\theta \Delta_n \} \\ &= I_\alpha^\infty - 4\theta \Delta_n + 2\theta \Delta_n + o(\Delta_n) \end{aligned}$$

and the right-hand side is strictly less than I_α^∞ for large n. This contradiction proves Theorem 1.

Proof of Theorem 2. Following the proof of Theorem 1, we find $\alpha \in (0, \lambda)$, $u_1, u_2 \in H^1(\mathbb{R}^N) \not\equiv 0$ such that, for some $\theta > 0$,

$$-\Delta u_1 + u_1 = f(x, u_1) \text{ in } \mathbb{R}^N, \quad -\Delta u_2 + u_2 = f(u_2) \text{ in } \mathbb{R}^N, \tag{30}$$

$$\int_{\mathbb{R}^N} F(x, u_1) dx = \alpha, \quad \int_{\mathbb{R}^N} F(u_2) dx = \lambda - \alpha, \tag{31}$$

$$\int_{\mathbb{R}^N} |\nabla u_1|^2 + u_1^2 dx = I_\alpha, \quad \int_{\mathbb{R}^N} |\nabla u_2|^2 + u_2^2 dx = I^\infty_{\lambda-\alpha} \tag{32}$$

and

$$I_\lambda = I_\alpha + I^\infty_{\lambda-\alpha} \tag{33}$$

(the reason for the presence of "only one piece u_2" is the fact that (7) holds by Theorem 1!). We may then argue as we did in Theorem 1, the only new estimate being on the differences

$$\int_{\mathbb{R}^N} F(x, u_2^n) - F(u_2^n) dx, \quad \int_{\mathbb{R}^N} f(x, u_2^n) u_1 - f(u_2^n) u_1 dx.$$

We claim that we have

$$\int_{\mathbb{R}^N} F(x, u_2^n) - F(u_2^n) dx \geq -o\left(e^{-n} n^{-\frac{N-1}{2}}\right) = -o(\Delta_n) \tag{34}$$

$$\int_{\mathbb{R}^N} f(x, u_2^n) u_1 - f(u_2^n) u_1 dx \geq -o\left(e^{-n} n^{-\frac{N-1}{2}}\right) = -o(\Delta_n). \tag{35}$$

In order to prove (34), we use (14) as follows

$$\begin{aligned}
\int_{\mathbb{R}^N} F(x, u_2^n) - F(u_2^n) dx &\geq -C \int_{|x| \leq n/2} (u_2^n)^{2+\delta} dx \\
&\quad - C \int_{|x| \geq n/2} e^{-2|x|} |x|^{-\alpha} dx \\
&\geq -C \int_{|y| \geq n/2} |u_2|^{2+\delta} dy - C e^{-n} n^{-(\alpha - N)} \\
&\geq -C \int_{|y| \geq n/2} e^{-(2+\delta)|y|} |y|^{-(N-2)(2+\delta)} dy - C e^{-n} n^{-(\alpha-N)} \\
&\geq -C \exp\left\{-\left(1 + \frac{\delta}{4}\right) n\right\} - C e^{-n} n^{-(\alpha - N)} \\
&= -o\left(e^{-n} n^{-\frac{N-1}{2}}\right)
\end{aligned}$$

for some $\delta > 0$: recall that $F \in C^{2,\delta}$ for some $\delta > 0$ and that $\alpha - N > \frac{N-1}{2}$ by (14).

We now prove (35): in view of (15) we may write for some $\gamma \in (0,1)$ to be determined

$$\int_{\mathbb{R}^N} f(x,u_2^n)u_1 - f(u_2^n)u_1 dx \geq -C\int_{|x|\leq \gamma n} (u_2^n)^{1+\delta} u_1 dx + $$
$$- C\int_{|x|\geq \gamma n} e^{-\frac{1}{\delta}|x|}|x|^{-\beta} u_1 dx$$
$$\geq -C\exp\{-(1+\delta)(1-\gamma)n\}n^{-\frac{N-1}{2}(1+\delta)}$$
$$- C\exp\left\{-\left(\frac{1}{\delta}+1\right)\gamma n\right\}n^{-(\beta-2)}.$$

And we conclude choosing $\gamma = \frac{\delta}{1+\delta}$ so that $(1+\delta)(1-\gamma) = 1$, $(\frac{1}{\delta}+1)\gamma = 1$, recalling that $\beta - 2 > \frac{N-1}{2}$.

II.2 TDFW Problem. In this section, we present some results taken from [26] on the so-called Thomas–Fermi–Dirac–Von Weizäcker problem:
(36)

$$I_\lambda = \mathrm{Inf}\left\{\int_{\mathbb{R}^3} \frac{1}{2}|\nabla u|^2 - \sum_{j=1}^{M} \frac{1}{2}\frac{z_j}{|x-\bar{x}_j|}u^2 + F(u)dx \right.$$
$$\left. + \frac{1}{4}\int\int_{\mathbb{R}^3\times\mathbb{R}^3} u^2(x)u^2(y)|x-y|^{-1}dxdy / u \in H^1(\mathbb{R}^3),\ \int_{\mathbb{R}^3} u^2 dx = \lambda\right\}$$

where $M \geq 1$, $z_j > 0$, $\bar{x}_j \in \mathbb{R}^3$ (for $1 \leq j \leq M$), F is twice continuously differentiable, even, and $F(0) = F'(0) = F''(0) = 0$ and F satisfies

$$(37) \qquad F^-(t)|t|^{-10/3} \to 0, \quad F''(t)|t|^{-4} \to 0 \text{ as } |t| \to \infty.$$

We also introduce
(38)

$$I_\lambda^\infty = \mathrm{Inf}\left\{\int_{\mathbb{R}^3} \frac{1}{2}|\nabla u|^2 + F(u)dx + \frac{1}{4}\int\int_{\mathbb{R}^3\times\mathbb{R}^3} u^2(x)u^2(y)|x-y|^{-1}dxdy \right.$$
$$\left. / u \in H^1(\mathbb{R}^3),\ \int_{\mathbb{R}^3} u^2 dx = \lambda\right\}.$$

Again, as in Section II.1, conditions (6) and (8) are necessary and sufficient for the relative compactness (up to a translation in the case of (6)) of all minimizing sequences of (36). Checking (8) requires some additional work and was completed in [26] as follows.

Theorem [26]. i) *The strict subadditivity condition* (8) *holds if* λ *is small enough.*

ii) *If we assume in addition that* $Z = \sum_{j=1}^{M} z_j \geq \lambda$ *and that* F *satisfies*

$$F'(t)^+ t^{-2/3} \text{ is bounded for } t \in [0, t_0], \text{ for some } t_0 > 0, \tag{39}$$

then the condition (8) *holds.*

Open Problem 2. The example of main physical interest is $F(t) = c_1|t|^{10/3} - c_2|t|^{8/3}$ with $c_1, c_2 > 0$. In this case, is it possible to sharpen the condition on Z and λ? How does it depend upon the geometry of $\{\bar{x}_j \mid 1 \leq j \leq M\}$? Let us also point out that a more detailed analysis is possible when $c_2 = 0$: see R. Benguria, H. Brézis and E.H. Lieb [8].

We will now reprove here the above result. Let us only mention that the proof of part i) is very much in the spirit of the proofs of Facts 1 and 2 in Section II.1, while the proof of part ii) uses the strategy mentioned in the Introduction and followed in Section II.1 in the proofs of Theorems II.1 and II.2. More precisely, one has first to prove the "nondegeneracy of Lagrange multipliers": this is done in [26] using a tricky spectral argument combined with the condition $Z \geq \lambda$. Then, if we argue by contradiction as in the proof of Theorem 1, one obtains the existence of $\theta > 0$, $m \geq 2$, $u_1, \ldots, u_m \in H^1(\mathbb{R}^3)$, $\alpha_1 \geq 0$, $\alpha_2, \ldots, \alpha_m > 0$ such that

$$-\Delta u_1 + V u_1 + f(u_1) + \left(u_1^2 * \frac{1}{|x|}\right) u_1 + \theta u + 1 = 0 \text{ in } \mathbb{R}^3, u_1 \geq 0 \text{ in } \mathbb{R}^3, \tag{40}$$

$$\begin{aligned} -\Delta u_j + f(u_j) + \left(u_j^2 * \frac{1}{|x|}\right) u_j + \theta u_j = 0 \text{ in } \mathbb{R}^3, \\ u_j \geq 0 \text{ in } \mathbb{R}^3 \text{ for } 1 \leq j \leq m \end{aligned} \tag{41}$$

$$\begin{aligned} &\mathcal{E}(u_1) = I_{\alpha_1}, \quad \mathcal{E}^\infty(u_j) = I_{\alpha_j}^\infty \quad (2 \leq j \leq m), \\ &\int_{\mathbb{R}^3} u_j^2 dx = \alpha_j \quad (1 \leq j \leq m), \end{aligned} \tag{42}$$

where $V = -\sum_{i=1}^{M} \frac{z_i}{|x - \bar{x}_i|}$, $\mathcal{E}$ and $\mathcal{E}^\infty$ denote respectively the functionals to be minimized in (36) and (38). Finally, we have

$$I_\lambda = I_{\alpha_1} + \sum_{j=2}^{m} I_{\alpha_j}^\infty, \quad \lambda = \sum_{j=1}^{m} \alpha_j. \tag{43}$$

The final step consists in considering (for example)

$$\tilde{u}^n = u_1 + u_2(\cdot + ne_2) + \sum_{j=3}^{m} u_j(\cdot + n^2 e_j), \quad v^n = \frac{\tilde{u}^n}{|\tilde{u}^n|_{L^2}}\sqrt{\lambda}.$$

Then, straightforward computations (in fact simpler than those used in the proof of Theorem 1) show that

$$(44) \qquad I_\lambda \leq \mathcal{E}(v^n) = I_{\alpha_1} + \sum_{j=2}^{m} I_{\alpha_j}^\infty + \{-Z\alpha_2 + \alpha_2(\lambda - \alpha_2)\}\frac{1}{n} + o\left(\frac{1}{n}\right)$$

and we easily reach a contradiction since we have $Z \geq \lambda > \lambda - \alpha_2$.

II.3 Other Examples. In this section, we will present two more examples. The first one is the rotating stars problems (see J.F.G. Auchmuty and R. Beals [1], P.L. Lions [27], [23]...)

$$(45) \qquad \begin{aligned} I_\lambda = \operatorname{Inf}\Big\{ &\int_{\mathbb{R}^3} \frac{1}{q} p^q + Vp\,dx - \frac{1}{2}\int\int_{\mathbb{R}^3\times\mathbb{R}^3} p(x)p(y)|x-y|^{-1}dx\,dy \\ &/ p \in L^q(\mathbb{R}^3) \cap L^1(\mathbb{R}^3),\ p \geq 0 \text{ a.e.}, \int_{\mathbb{R}^3} p\,dx = \lambda \Big\} \end{aligned}$$

where $q > 4/3$, $V \in L^{q'}(\mathbb{R}^3) \cap C_0(\mathbb{R}^3)$ with $q' = \frac{q}{q-1}$ (for example). The corresponding problem at infinity is

$$(46) \qquad \begin{aligned} I_\lambda^\infty = \Big\{ &\int_{\mathbb{R}^3} \frac{1}{q} p^q dx - \frac{1}{2}\int\int_{\mathbb{R}^3\times\mathbb{R}^3} p(x)p(y)|x-y|^{-1}dx\,dy \\ &/ p \in L^q(\mathbb{R}^3) \cap L^1(\mathbb{R}^3),\ p \geq 0 \text{ a.e.}, \int_{\mathbb{R}^3} p\,dx = \lambda \Big\}. \end{aligned}$$

This problem was studied in particular in [23] and (7) was shown to hold for (46) by a scaling argument. Again, (6) is the necessary and sufficient condition for the relative compactness up to a translation in $L^1 \cap L^q$ all minimizing sequences of (45). And we have the

Theorem 4. *Assume that V satisfies the following condition: for each $R \in (0, \infty)$, there exists a sequence of points y_n such that $|y_n| \to_n \infty$ and*

$$(47) \qquad \int_{y_n + B_R} V^+ dz = o(|y_n|^{-1}).$$

Then, the strict subadditivity inequality (6) *holds and all minimizing sequences of* (45) *are relatively compact up to a translation in* $L^1 \cap L^q$. *Therefore, all minimizing sequences of* (45) *are relatively compact in* $L^1 \cap L^q$ *if and only if* $I_\lambda < I_\lambda^\infty$.

Remark. The last condition $I_\lambda < I_\lambda^\infty$ holds as soon as $\operatorname{diam}\{V < 0\} = \infty$.

Open Problem 3. Is a condition like (47) necessary for the above result to hold?

Sketch of Proof. Since (7) holds (see [23]), arguing as in the proofs of Theorems 1,2, we obtain the existence of $\alpha \in (0, \lambda)$, $p_1, p_2 \in L^1_+ \cap L^q$, $\theta \in \mathbb{R}$ such that

$$\int_{\mathbb{R}^3} p_1 dx = \alpha, \quad \int_{\mathbb{R}^3} p_2 dx = \lambda - \alpha, \quad \mathcal{E}(p_1) = I_\alpha, \quad \mathcal{E}^\infty(p_2) = I_{\lambda-\alpha}^\infty$$

(denoting by $\mathcal{E}$, $\mathcal{E}^\infty$ the functionals to be minimized in (45), (46) respectively),

$$p_1^{q-1} + V - p_1 * \frac{1}{|x|} + \theta \geq 0 \text{ a.e. in } \mathbb{R}^3, \; = 0 \text{ a.e. on } \{p_1 > 0\}, \tag{48}$$

$$p_2^{q-1} - p_2 * \frac{1}{|x|} + \theta \geq 0 \text{ a.e. in } \mathbb{R}^3, \; = 0 \text{ a.e. on } \{p_2 > 0\}. \tag{49}$$

If we introduce—as it is usual—the potentials U_1, U_2 given by

$$U_i = p_i * \frac{1}{|x|}, \quad \text{for } i = 1, 2, \tag{50}$$

we see that (48), (49) may be written as

$$-\Delta U_1 = 4\pi(U_1 - \theta - V)^{+\gamma} \text{ in } \mathbb{R}^3, \quad -\Delta U_2 = 4\pi(U_2 - \theta)^{+\gamma} \text{ in } \mathbb{R}^3 \tag{51}$$

where $\gamma = (q-1)^{-1} < 3$ since $q > 4/3$. From (51) (or the fact that p_1, p_2 are minima) one deduces easily that necessarily $\theta \geq 0$. If $\theta = 0$, we see from (51) that

$$U_2(x) \geq \frac{\nu}{|x|} \quad \text{for } |x| \geq 1, \text{ for some } \nu > 0.$$

Hence, we have, using again (51), for some $\nu' > 0$

$$-\Delta U_2 \geq \nu' \frac{1}{|x|^\gamma} \text{ for } |x| \geq 1,\ U_2 \geq 0,\ U_2 \in L^p(\mathbf{R}^3) \text{ for all } p > 3.$$

And we easily reach a contradiction since $\gamma < 3$: the contradiction shows that $\theta > 0$. Next, using elliptic estimates, one shows that $U_1, U_2 \in C_0$, $p_1, p_2 \in L^\infty$, therefore p_1 and p_2 have compact support say Supp $p_i \subset B_R$ $(i = 1, 2)$ for some $R \in (0, \infty)$.

We then consider: $p_n = p_1 + p_2(\cdot - y_n)$ where y_n is selected from the condition (47). For n large enough, we have easily

$$\int_{\mathbf{R}^3} p_n\, dx = \lambda,$$

$$\mathcal{E}(p_n) = I_\lambda + \int_{\mathbf{R}^3} V\, p_2(\cdot - y_n) dx - \frac{1}{|y_n|}\alpha(\lambda - \alpha) + o\left(\frac{1}{|y_n|}\right)$$

and since $p_2 \in L^\infty$, Supp $p_2 \subset B_R$

$$\int_{\mathbf{R}^3} V\, p_2(\cdot - y_n) dx \leq C \int_{y_n + B_R} V^+ = o\left(\frac{1}{|y_n|}\right), \text{ in view of (47).}$$

And this concludes the proof of Theorem 4.

We now turn to our final example taken from some models in Nuclear Physics (see [21]):

(52)

$$I_\lambda^\infty = \text{Inf}\left\{ \int_{\mathbf{R}^3} \frac{1}{2}|\nabla u|^2 + F(u) dx + \frac{1}{4} \int\int_{\mathbf{R}^3 \times \mathbf{R}^3} u^2(x) u^2(y) V(x - y) dx dy \right.$$
$$\left. /u \in H^1(\mathbf{R}^3),\ \int_{\mathbf{R}^3} u^2 dx = \lambda \right\}$$

where $F \in C^{2,\delta}$ (for some $\delta > 0$), F is even, $F(0) = F'(0) = F''(0) = 0$, satisfies (3.7) and

$$F'(t) \leq C t^3 \quad \text{for } t \in (0, t_0), \text{ for some } C > 0, t_0 > 0. \tag{53}$$

We assume that the potential V satisfies (for example)

$$\begin{cases} V \in L^1 \cap L^\infty,\ V^+ \text{ has compact support, } V \text{ is even,} \\ \int_{|z| \geq R} V^- dz = 0(e^{-\delta R}) \text{ as } R \to +\infty, \text{ for all } \delta > 0. \end{cases} \tag{54}$$

Then, we have the

Theorem 5. *All minimizing sequences of I_λ^∞ are relatively compact in $H^1(\mathbf{R}^3)$ up to a translation if and only if $I_\lambda^\infty < 0$.*

Open Problem 4. Generalize this result to systems, in particular the Hartree–Fock systems—see [21]—, with conditions on λ, V, F and the number of components of the system.

Sketch of Proof. The first step is to show that if $u \geq 0$, $u \in H^1(\mathbf{R}^3)$ solves

$$-\Delta u + F'(u) + (u^2 * V)u = 0 \quad \text{in } \mathbf{R}^3 \tag{55}$$

then $u \equiv 0$. Indeed, one first observes that, by elliptic estimates, $u \in C_0^2(\mathbf{R}^3)$. Then, remarking that (55) yields

$$-\Delta u + au = 0 \quad \text{in } \mathbf{R}^3, u \geq 0, a \in L^\infty$$

one sees that either $u \equiv 0$, or $u > 0$ in $\mathbf{R}^3$ and the Harnack inequality holds

$$\sup_{z+B_R} u \leq C_R \inf_{z+B_R} u, \quad \text{for all } z \in \mathbf{R}^3 \tag{56}$$

for some constant $C_R > 0$ depending only on R, and R is arbitrary in $(0, \infty)$. Since F satisfies (53) and V^+ has compact support, we deduce from (56) that either $u \equiv 0$ or $u > 0$ in $\mathbf{R}^3$ and

$$-\Delta u \geq -C_0 u^3 \quad \text{on } \mathbf{R}^3, \text{ for some } C_0 \geq 0. \tag{57}$$

Therefore, if $u \not\equiv 0$, we may choose $\nu > 0$ small enough such that for γ fixed in $(1, 3/2)$, $w = \nu/|x|^\gamma$ satisfies

$$w \leq u \quad \text{on } \{|x| = 1\}, \tag{58}$$

$$-\Delta w + C_0 w^3 = C_0 \nu^3 \frac{1}{|x|^{3\gamma}} - \nu\gamma(\gamma - 1)\frac{1}{|x|^{\gamma+2}} \leq 0 \quad \text{for } |x| \geq 1. \tag{59}$$

Hence, by an easy application of the maximum principle, one concludes

$$u \geq w \quad \text{for } |x| \geq 1,$$

a contradiction with the fact that $u \in L^2(\mathbf{R}^3)$ since $\gamma < 3/2$. Therefore, $u \equiv 0$.

We may then follow the arguments used in the proof of Theorem 1 and we obtain $\alpha_1, \alpha_2 > 0$, $\theta > 0$, $u_1, u_2 \in C_0^2(\mathbf{R}^3)$, $u_1, u_2 > 0$ in $\mathbf{R}^3$ such that

$$(60) \quad I_\alpha^\infty = I_{\alpha_1}^\infty + I_{\alpha_2}^\infty < 0 \text{ where } \alpha = \alpha_1 + \alpha_2, \int_{\mathbf{R}^3} |u_i|^2 dx = \alpha_i \ (i = 1, 2)$$

$$(61) \qquad \mathcal{E}(u_i) = I_{\alpha_i}^\infty, \ -\Delta u_i + f'(u_i) + \theta u_i + (u_i^2 * V) u_i = 0 \ \text{ in } \mathbf{R}^3$$

where $\mathcal{E}$ is the functional to be minimized in (52). Then, proving Theorem 5 follows from obtaining a contradiction through the "interaction computation." To this end, we need to obtain a precise asymptotic bahaviour of u_i $(i = 1, 2)$ at infinity. This is done in two steps remarking first that since $\frac{F'(u_i)}{u_i}$ and $u_i^2 * V$ go to 0 as x goes to $+\infty$, we obtain easily

$$(62) \qquad u_i(x) \le C_\varepsilon \frac{e^{-(\omega - \varepsilon)|x|}}{|x|} \quad \text{for large } |x|, \text{ for all } \varepsilon > 0 (i = 1, 2)$$

where $\omega = \theta^{1/2}$. Therefore, by a straightforward computation, we will deduce

$$(63) \qquad u_i(x) \exp\{\omega |x|\} |x| \to c_i \ \text{ as } |x| \to \infty \ (i = 1, 2)$$

where $c_i > 0$ $(i = 1, 2)$, provided we show that

$$(64) \quad F'(u_i) \exp\{\omega|x|\}|x|, \ ((u_i^2 * V) u_i) \exp\{\omega|x|\}|x| \in L^1(\mathbf{R}^3) \ (i = 1, 2).$$

Since $F'(0) = F''(0) = 0$ and $F \in C^{2,\delta}$ (for some $\delta > 0$), the first part of (64) is immediately deduced from (62) while the second part is also deduced from (62) if we show

$$(65) \qquad |u_i^2 * V| \le C e^{-\delta|x|} \quad \text{for some } \delta > 0, C > 0 \ (i = 1, 2).$$

This is shown as follows

$$\int u_i^2(y) |V(x-y)| dy \le C \int_{|y| \ge |x|/2} u_i^2(y) dy + C \int_{|z| \ge |x|/2} |V(z)| dz$$

and we conclude using (62) and (54).

We now copy the proof of Theorem 1 introducing $\tilde{u}^n$ as before and $v^n = \frac{\tilde{u}^n}{|\tilde{u}^n|_{L^2}} \sqrt{\alpha}$. Clearly, we have (as in the proof of Theorem 1)

$$(66) \qquad \int_{\mathbf{R}^3} |u^n|^2 dx = \alpha + 2 \int_{\mathbf{R}^3} u_1 u_2^n dx = \alpha + 2\delta_n$$

(67) $$\mathcal{E}(v^n) = \mathcal{E}(\tilde{u}^n) - \theta\delta_n + 0(\delta_n^2)$$

(68)
$$\begin{aligned}\mathcal{E}(\tilde{u}^n) = I_\alpha^\infty + \theta\delta_n + \frac{1}{2}\Big\{\int_{\mathbb{R}^3} & F'(u_1)u_2^n + F'(u_2^n)u_1 \\ & + (u_1^2 * V)u_1u_2^n + ((u_2^n)^2 * V)u_2^n u_1\, dx\Big\} \\ & + \frac{1}{2}\int_{\mathbb{R}^3}(u_1^2 * V)(u_2^n)^2 + 2((u_1u_2^n) * V)u_1u_2^n dx + o\Big(\frac{e^{-\omega n}}{n}\Big).\end{aligned}$$

Therefore, we deduce from all these estimates and from (61) and (63)

(69)
$$\begin{aligned}\mathcal{E}(v^n) = I_\alpha^\infty + 0(\delta_n^2) + o\Big(\frac{e^{-\omega n}}{n}\Big) + C_0\frac{e^{-\omega n}}{n} + \\ + \frac{1}{2}\Big\{\int_{\mathbb{R}^3}(u_1^2 * V)(u_2^n)^2 + 2((u_1u_2^n) * V)u_1u_2^n dx\Big\}\end{aligned}$$

where

$$C_0 = \frac{1}{2}\int_{\mathbb{R}^3}(-\Delta u_1 + 2u_1)exp(-\omega x\cdot e) + (-\Delta u_2 + 2u_2)exp(\omega x\cdot e)dx > 0.$$

Again, because of (63), one checks that $\delta_n^2 = o(e^{-\omega n}/n)$ and

$$\int_{\mathbb{R}^3}((u_1u_2^n) * V)\, u_1u_2^n dx \le \max_{\mathbb{R}^3}(u_1u_2^n)\delta_n\|V\|_{L^\infty} = o\Big(\frac{e^{-\omega n}}{n}\Big).$$

We will have reached the desired contradiction if we show that

(70) $$(u_1^2 * V)(x) = o(e^{-\omega|x|}|x|^{-1}) \quad \text{as } |x|\to\infty.$$

But we have for some $\mu\in(0,1)$ to be determined later on

$$|(u_1^2 * V)(x)| \le C\int_{|y|\ge\mu|x|}u_1^2(y)dy + C\int_{|z|\ge(1-\mu)|x|}V(z)dz$$

and we conclude using (63), (54), choosing μ fixed in $\left(\frac{1}{2},1\right)$.

III. Morse Indices and Compactness

III.1 Superlinear Elliptic Equations. We present here a result taken from A. Bahri and P.L. Lions [6] which shows that, for solutions of superlinear elliptic equations, bounds on solutions are equivalent to bounds on the Morse indices. More precisely, let Ω be a bounded smooth domain in $\mathbb{R}^N$ ($N \geq 2$) and let u_n be a sequence of, say, $C^2(\bar{\Omega}) \cap C_0(\bar{\Omega})$ solutions of

(71) $$-\Delta u_n = f_n(x, u_n) \quad \text{in } \Omega, \ u_n = 0 \text{ on } \partial\Omega$$

where f_n is continuous in (x,t), continuously differentiable with respect to t, $\frac{\partial f_n}{\partial t}$ is continuous in (x,t) uniformly in n (for example) and f_n satisfies

(72) $$f_n(x,t)|t|^{-(p-1)}t \to \lambda > 0 \ \text{ as } |t| \to +\infty,$$

uniformly in n, $x \in \bar{\Omega}$, where $1 < p < \frac{N+2}{N-2}$ ($p < \infty$ if $N = 2$). We denote by $i(u_n)$ the Morse index of u_n i.e. the number of negative eigenvalues of $(-\Delta - \frac{\partial f_n}{\partial t}(x, u_n))$ (with Dirichlet boundary conditions) and by

$$I_n = \int_\Omega \frac{1}{2}|\nabla u_n|^2 - F_n(x, u_n)dx$$

where $\frac{\partial F_n}{\partial t}(x,t) = f_n(x,t)$ on $\bar{\Omega} \times \mathbb{R}$, $F_n(x,0) = 0$ on $\bar{\Omega}$.

Then, we have the

Theorem 6. *The following assertions are equivalent:*

$$\text{i) } I_n \to_n \infty, \quad \text{ii) } \|u_n\|_{L^\infty(\Omega)} \to_n +\infty, \quad \text{iii) } i(u_n) \to_n +\infty.$$

The equivalence between i) and ii) follows easily from (72). And iii) implies ii) by trivial considerations on the spectrum of elliptic operators. The only real difficulty is to prove that ii) implies iii) and we refer to [6] for the proof (which is long and difficult!).

Open Problem 5. Prove or disprove the same result assuming only

(73) $$f_n(x,t)t^{-1} \to +\infty \text{ as } |t| \to +\infty, \text{ uniformly in } n, x \in \bar{\Omega}$$

(74) $$f_n(x,t)|t|^{-\frac{4}{N-2}}t^{-1} \to 0 \text{ as } |t| \to \infty, \text{uniformly in } n, x \in \bar{\Omega}.$$

III.2 Convex Superquadratic Hamiltonian Systems. We now consider Hamiltonian systems of the form

(75) $$\dot{z} = JH'(z)$$

where $H \in C^2(\mathbf{R}^N \times \mathbf{R}^N)$, $z = (p,q) \in \mathbf{R}^N \times \mathbf{R}^N$, $J = \begin{pmatrix} 0 & I_N \\ -I_N & 0 \end{pmatrix}$, $N \geq 1$.

We assume that H is convex and satisfies

$$H(z) = o(|z|^2) \quad \text{as } |z| \to 0 \tag{76}$$

$$\bar{\lambda}(H''(z)) \to +\infty \quad \text{as } |z| \to +\infty \tag{77}$$

where $\bar{\lambda}$ denotes the maximal eigenvalue of the Hessian $H''(z)$.

The following result is due to I. Ekeland and the author: extensions of this result together with variants including the case of second-order equations are given in V. Coti-Zelati, I. Ekeland and P.L. Lions [14].

Theorem 7. *For each $T \in (0,\infty)$, there exists a periodic solution of* (75) *with minimal period given by T.*

Open Problem 6. Is the same result true if we replace (77) by

$$|H'(z)|\,|z|^{-1} \to +\infty \quad \text{as } |z| \to +\infty. \tag{78}$$

Remark. The above result was first shown by I. Ekeland and H. Hofer [16] in the case when H satisfies for some $\theta \in (0, \frac{1}{2})$

$$\theta z \cdot H'(z) \geq H(z) \quad \text{for } |z| \text{ large.} \tag{79}$$

Notice that (77) and (79) cannot be compared: there are examples of Hamiltonians satisfying one of the two conditions but not the other.

We will not reprove this result here: let us only sketch the proof. By approximating H (at infinity), one obtains, using the result of I. Ekeland and H. Hofer [16], periodic orbits which are critical points of "mountain pass type" for the dual variational formulation. Then, by [16], their Morse index is either 0 or 1. Next, one observes that critical points with Morse indices bounded from above, for Hamiltonians satisfying (uniformly) (77), are uniformly bounded: this is easily shown by some spectral arguments. At this point, one sees that the orbits created by this approximation are already periodic orbits of the original problem and one concludes.

III.3 HF Equations in Atomic and Molecular Physics. The results of this section are taken from [26]. One considers the problem of finding solutions of Hartree–Fock equations for Coulomb systems. Mathematically, one looks for critical points of

$$\mathcal{E}_K(\varphi_1, \ldots, \varphi_N) \tag{80}$$

where

$$K = \left\{ (\varphi_1, \ldots, \varphi_N) \in H^1(\mathbf{R}^3)^N \mid \int_{\mathbf{R}^3} \varphi_i \varphi_j^* dx = \delta_{ij} \right\},$$

z^* denotes the complex conjugate of z, and

$$\begin{aligned} \mathcal{E}(\varphi_1, \ldots, \varphi_N) = \int_{\mathbf{R}^3} \tau + Vp\, dx + \frac{1}{2} \int\int_{\mathbf{R}^3 \times \mathbf{R}^3} \{p(x)p(y) \\ - |p(x,y)|^2\} |x-y|^{-1} dx dy \end{aligned} \tag{81}$$

where

$$\tau = \sum_{i=1}^{N} |\nabla \varphi_i|^2, \; p = \sum_{i=1}^{N} |\varphi_i|^2, \; p(x,y) = \sum_{i=1}^{N} \varphi_i(x) \varphi_i^*(y),$$

$$V = -\sum_{j=1}^{M} z_j |x - \bar{x}_j|^{-1}, \; z_j > 0, \; \bar{x}_j \in \mathbf{R}^3 \; (1 \le j \le M).$$

The main result is the following

Theorem [26]. *If $Z = \sum_{j=1}^{M} z_j \ge N$, the functional $\mathcal{E}$ restricted to K admits infinitely many distinct critical points.*

Open Problem 7. Find sharper conditions on Z, N, $\bar{x}_j$, z_j $(1 \le j \le M)$ insuring the existence of such solutions.

Remark. Observing that if $(\varphi_1, \ldots, \varphi_N)$ is a critical point of $\mathcal{E}|_K$, then $U(\varphi_1, \ldots, \varphi_N)$ is also a critical point of $\mathcal{E}|_K$ for any unitary matrix U on $\mathbb{C}^N$, we need to mention that, in the above theorem, distinct means that the solutions we build are not unitary transforms of others.

Again, we will not reprove this result: let us only mention that this is precisely a situation when the usual Palais–Smale condition does not hold. However, when $Z \ge N$, one recovers the compactness if we have the spectral information contained in the Morse index. Physically, losing the compactness means that one or several electrons vanish at infinity ("fall into the continuum", mathematically this means that in the concentration-compactness terminology [23], some part of p vanishes). However, if this happens, this or these electrons "see" a Coulomb potential which at infinity looks like $-\frac{Z}{|x|} + p' * \frac{1}{|x|} \cong -\frac{Z-N'}{|x|}$ (where p' denotes the density of the remaining electrons and N' denotes the number of those). Hence, if $Z \ge N$, $Z > N'$ and such Coulomb potentials have infinitely many bound states. So, if we are interested in the k-th excited state, such a loss of compactness should not occur (for any fixed $k \geq 1$). Mathematically the Morse index reflects precisely that we are looking at the k-th excited state and allows to justify the above heuristic physical argument. In order to make a complete proof, one follows precisely the strategy mentioned in the Introduction and the compactness of the approximated critical points is proven along the above line of arguments.

REFERENCES

[1] J.F.G. Auchmuty and R. Beals, *Variational solutions of some nonlinear free boundary problems,* Arch. Rat. Mech. Anal. **43** (1971), 255–271.

[2] A. Bahri, *Critical points at infinity in some variational problems,* to appear in Longman-Pitman Research Notes.

[3] A. Bahri and J.M. Coron, *On a nonlinear elliptic equation involving the Sobolev exponent,* preprint.

[4] A. Bahri and P.L. Lions, *On the existence of a positive solution of semilinear elliptic equations in unbounded domains,* preprint.

[5] A. Bahri and P.L. Lionns, *Morse indices of some min-max critical points, I, Applications to multiplicity results,* Comm. Pure Appl. Math., **61** (1988), 1027–1037.

[6] A. Bahri and P.L. Lions, in preparation.

[7] V. Benci and G. Cerami, *Positive solutions of semilinear elliptic problems in exterior domains,* preprint.

[8] R. Benguria, H. Brézis, and E.H. Lieb, *The Thomas–Fermi–von Weizäcker theory of atoms and molecules,* Comm. Math. Phys. **79** (1981), 167–180.

[9] H. Berestycki and P.L. Lions, *Continua of vortex rings,* in preparation.

[10] H. Berestycki and P.L. Lions, *Nonlinear scalar field equations,* I and II, Arch. Rat. Mech. Anal. **82** (1983), 313–375.

[11] H. Berestycki and C.H. Taubes, in preparation.

[12] C.V. Coffman, *Lyusternik–Schnirelman theory: complementary principles and the Morse index,* Nonlinear Anal. T.M.A. **12** (1988), 507–529.

[13] C.V. Coffman, *Uniqueness of the ground state solution for $\Delta u - u + u^3 = 0$ and a variational characterization of other solutions,* Arch. Rat. Mech. Anal. **46** (1972), 81–95.

[14] V. Coti-Zelati, I. Ekeland, and P.L. Lions, *Index estimates and critical points of functionals not satisfying Palais-Smale*, preprint.

[15] I. Ekeland, *Convexity methods in Hamiltonian systems,* in preparation.

[16] I. Ekeland and H. Hofer, *Periodic solutions with prescribed minimal period for convex autonomous Hamiltonian systems,* Invent. Math. **81**

(1985), 155–188.

[17] M.J. Esteban, *A direct variational approach to Skyrme's model for meson fields,* Comm. Math. Phys. **105** (1986), 187–195.

[18] M.J. Esteban, this volume.

[19] M.J. Esteban and P.L. Lions, *Stationary solutions of nonlinear Schrödinger equations with an external magnetic field,* in *Partial Differential Equations and the Calculus of Variations,* Birkhäuser, Basel, 1989.

[20] M.J. Esteban and P.L. Lions, Γ *convergence and the concentration-compactness method for some variational problems with lack of compactness,* preprint.

[21] D. Gogny and P.L. Lions, *Hartree–Fock theory in Nuclear Physics,* RAIRO Model. Math. et Anal. Num. **20** (1986), 571–637.

[22] A.C. Lazer and S. Solimini, *Nontrivial solutions of operator equations and Morse indices of critical points of min-max type,* preprint.

[23] P.L. Lions, *The concentration-compactness principle in the calculus of variations. The locally compact case,* I and II. Ann. I.H.P. Anal. Non Lin. **1** (1984), 109–145 and **1** (1984), 223–282. See also C.R. Acad. Sci. Paris, **294** (1982), 261–264, and in *Contributions to Nonlinear Partial Differential Equations,* Pitman, London, 1983.

[24] P.L. Lions, *The concentration-compactness principle in the calculus of variations. The limit case,* Rit. Mat. Iberoamer. **1** (1985), 145–201 and **1** (1985), 45–121. See also C.R. Acad. Sci. Paris, **296** (1983), 645–648 and in *Séminaire Goulaouic–Meyer–Schwartz,* **82–83**, Ecole Polytechnique, Palaiseau.

[25] P.L. Lions, *On positive solutions of semilinear elliptic equations in unbounded domains.* In *Nonlinear Diffusion Equations and Their Equilibrium States,* Springer, New York, 1988.

[26] P.L. Lions, *Solutions of Hartree–Fock equations for Coulomb systems,* Comm. Math. Phys. **109** (1987), 33–97.

[27] P.L. Lions, *Minimization problems in* $L^1(\mathbf{R}^N)$, J. Funct. Anal. **41** (1981), 236–275.

[28] S. Solimini, *Morse index estimates in min-max theories,* preprint.

[29] W. Strauss, *Existence of solitary waves in higher dimensions,* Comm. Math. Phys. **55** (1977), 149–162.

[30] C. Viterbo, *Indice de Morse des points critiques obtenus par minimax,* Ann. I.H.P. Anal. Non Lin. **5** (1988), 221–226.

Ceremade
Université Paris-Dauphine
Place de Lattre de Tassigny
75775 Paris Cedex 16, France

Elliptic equations with critical growth and Moser's inequality

J.B. McLeod & L.A. Peletier

1. Moser's inequality

Let Ω be a bounded domain in $\mathbf{R}^2$ and let

$$M = \{u \in H_0^1(\Omega) : \int_\Omega |\nabla u|^2 \leq 1\}. \tag{1.1}$$

Then

$$S \stackrel{\text{def}}{=} \sup_M \int_\Omega e^{\alpha u^2} \leq c|\Omega|, \tag{1.2}$$

where c does not depend on Ω, whenever

$$\alpha \leq 4\pi. \tag{1.3}$$

This is *Moser's inequality* for $\mathbf{R}^2$. It is optimal, in the sense that if $\alpha > 4\pi$, $S = \infty$. The supremum (1.2) was actually first proved in 1967 by Trudinger for sufficiently small values of α [T], but the optimal bound for α was established in 1971 by Moser [M]. Quite recently, in 1985, there was a further development: somewhat surprisingly it was shown by Carleson and Chang, that not only for $\alpha < 4\pi$, but also for $\alpha = 4\pi$, the supremum is achieved [CC].

The proofs of these results were all quite elementary. It was observed that without loss of generality it could be assumed that $u \geq 0$, and by a symmetrization argument, that it was sufficient to consider domains and functions with radial symmetry. Thus, the problem was reduced to a variational problem involving functions $v : \overline{\mathbf{R}}^+ \to \overline{\mathbf{R}}^+$.

If $u : \Omega \to \mathbf{R}$ is a maximizer of (1.1), (1.2), it is a solution of the problem

$$(\mathrm{I}) \begin{cases} -\Delta u = \lambda u e^{\alpha u^2}, \quad u > 0 & \text{in} \quad \Omega \\ u = 0 & \text{on} \quad \partial\Omega \\ \int_\Omega |\nabla u|^2 = 1, & \end{cases}$$

where $\lambda \in \mathbf{R}$ is a Lagrangian multiplier. By the maximum principle, $\lambda > 0$. From the results of Trudinger, Moser and Carleson and Chang, we can now conclude that Problem (I) has a solution u_α if $\alpha \leq 4\pi$, and that this solution corresponds to a maximum of $\int_\Omega e^{\alpha u^2}, u \in M$. It should be emphasized, however, that there may be other solutions, not corresponding to a global maximum.

Suppose $\alpha < 4\pi$, and u_α is a maximizer of (1.1), (1.2) and thus a solution of Problem (I). What happens as $\alpha \uparrow 4\pi$, and why is it that $u_{4\pi}$ still exists? This is surprising as an analogous problem may show.

Consider the variational problem in $\mathbf{R}^3$:

$$\Sigma = \sup_M \int_\Omega u^{p+1}, \tag{1.4}$$

where Ω is a bounded domain in $\mathbf{R}^3$ and M is as before. It is well known that

$$\Sigma < \infty \quad \text{if} \quad p \leq 5.$$

For $p < 5$ the supremum is achieved, but for $p = 5$ it is not.

To see what happens as $p \uparrow 5$, set

$$p = 5 - \epsilon,$$

and $\Omega = B$, B being the unit ball in $\mathbf{R}^3$. Then if $\epsilon > 0$, (1.4) has a unique maximizer u_ϵ, which, modulo an appropriate multiplicative factor, satisfies

$$\begin{cases} -\Delta u = 3u^{5-\epsilon}, u > 0 & \text{in} \quad B \\ u = 0 & \text{on} \quad \partial B. \end{cases}$$

Letting $\epsilon \to 0$ one finds that [AP1,BP]

$$\epsilon u_\epsilon^2(0) \to \frac{32}{\pi}$$

and

$$\epsilon^{-1/2} u_\epsilon(x) \to \pi \sqrt{\frac{\pi}{2}} G_0(x) \quad \text{if} \quad x \neq 0,$$

where G_0 is the Green's function with Dirac mass at the origin:

$$\begin{cases} -\Delta G_0 = \delta_0 & \text{in} \quad B \\ G_0 = 0 & \text{on} \quad \partial B. \end{cases}$$

Thus, as $\epsilon \to 0$, the solution "concentrates" at the origin, and finally ceases to exist at $\epsilon = 0$.

2. The ODE problem

After symmetrization, the admissible functions $u(r), r = |x|$, have the properties

(a) $u \in C^1([0,1])$;

(b) $u(r) > 0,\ u'(r) \leq 0$ for $0 \leq r < 1$ and $u(1) = 0$;

(c) $2\pi \int_0^1 \{u'(r)\}^2 r dr \leq 1$.

If we set

$$t = -2\log r, \quad w(t) = \sqrt{4\pi} u(r)$$

then

$$2\pi \int_0^1 u'^2 r dr = \int_0^\infty w'^2 dt,$$

$$\int_B e^{\alpha u^2} = \pi \int_0^\infty e^{\beta w^2 - t} dt,$$

where

$$\beta = \frac{\alpha}{4\pi}.$$

Thus the set of admissible functions becomes

$$X = \{w \in C^1([0,\infty)) : w(0) = 0, w' \geq 0 \text{ and } \int_0^\infty w'^2 dt \leq 1\} \tag{2.1}$$

and the variational problem (1.1), (1.2) transforms to

$$S^* = \sup_X \int_0^\infty e^{\beta w^2 - t} dt. \tag{2.2}$$

Clearly, Moser's inequality now states for (2.1), (2.2) that

$$\beta \leq 1 \Rightarrow S^* < K,$$

$$\beta > 1 \Rightarrow S^* = \infty$$

where K is a constant, which does not depend on β. Finally, if w is a maximizer, it is a solution of the problem

$$\text{(II)} \quad \begin{cases} w'' + \lambda w e^{\beta w^2 - t} = 0, & 0 < t < \infty, \\ w(0) = 0, \ w' > 0, \end{cases}$$

where $\lambda > 0$ is a Lagrangian multiplier, which ensures that $\int_0^\infty w'^2 dt = 1$.

Some preliminary observations can readily be made.

Proposition 1. *Suppose $\beta < 1$, then for any $w \in X$,*

$$\int_0^\infty e^{\beta w^2 - t} dt \leq \frac{1}{1-\beta}.$$

Proof. Since $w \in X$,

$$w(t) = \int_0^t w'(s) ds \leq \sqrt{t}\Big(\int_0^t w'^2 ds\Big)^{1/2} \leq \sqrt{t},$$

and so

$$\int_0^\infty e^{\beta w^2 - t} dt \leq \int_0^\infty e^{(\beta-1)t} dt = \frac{1}{1-\beta}.$$

Proposition 2. *Suppose $\beta > 1$. Then there exists a sequence $\{w_n\} \subset X$ such that*

$$\int_0^\infty e^{\beta w_n^2 - t} dt \to \infty \quad \text{as} \quad n \to \infty.$$

Proof. Define

$$w_n(t) = \begin{cases} t/\sqrt{n}, & 0 \leq t < n, \\ \sqrt{n}, & t \geq n. \end{cases}$$

Plainly, $w_n \in X$ for every $n \geq 1$, and

$$\int_0^\infty e^{\beta w_n^2 - t} dt > \int_n^\infty e^{\beta n - t} dt = e^{(\beta - 1)n} \to \infty$$

as $n \to \infty$.

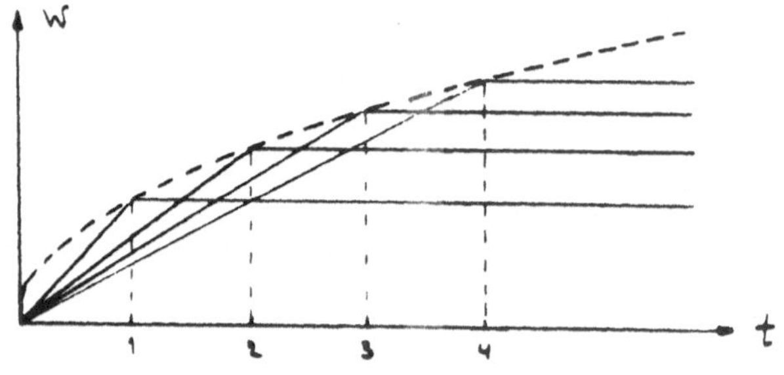

Fig.1. The sequence w_n.

3. Maximizers

If $\beta < 1$, the existence of a maximizer of (2.2) follows from a standard argument. Suppose $\{w_k\}$ is a maximizing sequence. Then, because $\{w_k\} \subset X$, and is therefore bounded on $H_0^1(\Omega)$, there exists a subsequence, also denoted by $\{w_k\}$, and a function $w_\beta \in H_0^1(\Omega)$, such that

$$w_k \rightharpoonup w_\beta \quad \text{in} \quad H_0^1$$

and also

$$w_k \to w_\beta \quad \text{in} \quad C(K),$$

where K is any compact set of $[0, \infty)$. Because

$$\int_0^\infty e^{\beta w_k^2 - t}\, dt < \frac{1}{1-\beta}$$

it follows by the dominated convergence theorem that

$$\int_0^\infty e^{\beta w_k^2 - t}\, dt \to \int_0^\infty e^{\beta w_\beta^2 - t}\, dt.$$

Finally, either

$$\int_0^\infty w_\beta'^2\, dt = 1$$

or

$$\int_0^\infty w_\beta'^2\, dt = 0.$$

However, since $w_\beta(0) = 0$, the second possibility implies that $w_\beta(t) \equiv 0$, and hence that

$$\int_0^\infty e^{\beta w_\beta^2 - t}\, dt = 1,$$

which can be ruled out.

To conclude the case $\beta < 1$, we note that

$$c_\beta = \lim_{t \to \infty} w_\beta(t) < \infty. \tag{3.1}$$

This follows from the differential equation for w, which states that

$$-w''(t) \le \lambda_\beta \sqrt{t} e^{(\beta - 1)t}.$$

Since $w' \in L^2(0,\infty)$ and $w'' < 0$, $w'(t) \to 0$ as $t \to \infty$, and hence

$$w'(t) = O(\sqrt{t}e^{(\beta-1)t}) \quad \text{as} \quad t \to \infty,$$

from which (3.1) follows.

Multiplying the equation by w_β and integrating it over $(0,\infty)$, we obtain for the Lagrangian multiplier λ_β,

$$\lambda_\beta = \left\{ \int_0^\infty w_\beta^2 e^{\beta w_\beta^2 - t} dt \right\}^{-1}, \tag{3.2}$$

and it can be shown that the right hand side of (3.2) is uniformly bounded with respect to $\beta \in (0,1)$.

We now turn to the case $\beta = 1$.

Let $\{\beta_k\} \subset (0,1)$ be a sequence, such that $\beta_k \to 1$ as $k \to \infty$. Then, taking suitable subsequences, which we denote again by $\{\beta_k\}$, we have as $k \to \infty$:

$$\begin{aligned} w_{\beta_k} &\rightharpoonup w_0 && \text{in} \quad H_0^1, \\ w_{\beta_k} &\to w_0 && \text{in} \quad C_{loc}, \\ \lambda_{\beta_k} &\to \lambda_0. \end{aligned}$$

As to the limits c_{β_k} there are two possibilities: *either* $c_{\beta_k} \to c_0 < \infty$ *(A), or* $c_{\beta_k} \to \infty$ *(B).*

Proposition 3. *Suppose Possibility (A) holds. Then w_0 is a maximizer.*

Proof. Because $w_\beta' \geq 0$ we have

$$w_\beta(t) \leq c_\beta \qquad \text{for all} \quad t \geq 0 \tag{3.3}$$

and similarly, because $w_0' \geq 0$, we have

$$w_0(t) \leq c_0 \qquad \text{for all} \quad t \geq 0. \tag{3.4}$$

Now let w be an element of X. Since, for any $\beta \in (0,1)$, w_β is a maximizer, we have

$$\int_0^\infty e^{\beta w^2 - t} dt \leq \int_0^\infty e^{\beta w_\beta^2 - t} dt. \tag{3.5}$$

Because (A) holds, it follows from (3.3) that the functions $w_{\beta_k}(t)$ are uniformly bounded with respect to the sequence $\{\beta_k\}$ and so, by the dominated convergence theorem,

$$\lim_{k\to\infty}\int_0^\infty e^{\beta_k w_{\beta_k}^2 - t}\,dt = \int_0^\infty e^{w_0^2 - t}\,dt. \tag{3.6}$$

Thus, if we set $\beta = \beta_k$ in (3.5) and let $k \to \infty$, we obtain

$$\int_0^\infty e^{w^2 - t}\,dt \le \int_0^\infty e^{w_0^2 - t}\,dt. \tag{3.7}$$

Since w was an arbitrary element in X, w_0 must be a maximizer.

Proposition 4. *Possibility (B) does not hold.*

Proof. The proof of Proposition 4 rests on two ingredients. The first one is an asymptotic estimate, which we formulate below but prove later.

Lemma 5 . *Suppose Possibility (B) holds. Then*

$$\lim_{\beta\uparrow 1}\int_0^\infty e^{\beta w_\beta^2 - t}\,dt = 1 + e,$$

where the limit is taken through the subsequence $\{\beta_k\}$.

Now let $w \in X$ and $\beta \in (0,1)$. Then, as before, because w_β is a maximizer,

$$\int_0^\infty e^{\beta w^2 - t}\,dt \le \int_0^\infty e^{\beta w_\beta^2 - t}\,dt.$$

If we let $\beta \to 1$ in this inequality through the sequence $\{\beta_k\}$, we obtain by Lemma 5

$$\int_0^\infty e^{w^2 - t}\,dt \le 1 + e. \tag{3.8}$$

The second ingredient is a test function due to Carleson and Chang:

$$\phi(t) = \begin{cases} \frac{1}{2}t, & 0 \le t < 2, \\ \sqrt{t-1}, & 2 \le t < T, \\ \sqrt{T-1}, & T \le t < \infty, \end{cases}$$

where $T = 1 + e^2$ has been chosen so that $\int_0^\infty \phi'^2 dt = 1$, and hence $\phi \in X$. The significant feature of this function is that

$$\int_0^\infty e^{\phi^2 - t} dt > 1 + e.$$

This contradicts (3.8) and thereby establishes Proposition 4.

We conclude with a sketch of the proof of Lemma 5. For the details we refer to [McLP].

Let $\beta \in (0,1)$, and set

$$\tau = t + \tau_\beta, \quad \tau_\beta = -\log \lambda_\beta,$$

and

$$y_\beta(\tau) = \sqrt{\beta}\, w_\beta(t).$$

Then, since w_β is a maximizer, y_β satisfies

$$\text{(III)} \quad \begin{cases} y'' + ye^{y^2 - \tau} = 0, & y > 0 \quad \text{for} \quad T(\gamma) < \tau < \infty \\ y(T(\gamma)) = 0, & y(\infty) = \gamma, \end{cases}$$

where

$$\gamma = \sqrt{\beta}\, c_\beta, \qquad T(\gamma) = \tau_\beta.$$

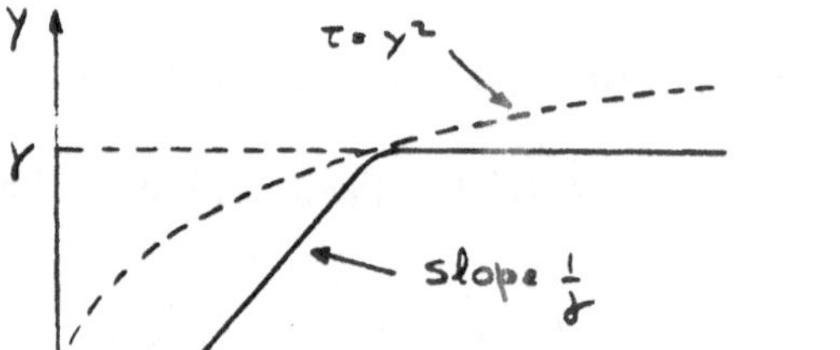

Fig.2. Characteristic graphs of $y(\tau, \gamma)$ for large γ.

For every $\gamma > 0$, Problem (III) has a unique solution $y(\tau, \gamma)$, its first zero being $T(\gamma)$. For large values of γ, the graph of y has a very characteristic piecewise linear slope [AP2]: except for a small transition region around $\tau = \gamma^2$, we have (see Fig.2)

$$y(\tau, \gamma) \sim \begin{cases} \dfrac{1}{\gamma}[\tau - T(\gamma)], & \tau \quad \text{small}, \\ \gamma, & \tau \quad \text{large}. \end{cases}$$

In addition, it was shown in [AP2,McLP] that

$$T(\gamma) = 2\log\gamma + 1 + o(1) \qquad \text{as} \quad \gamma \to \infty. \tag{3.9}$$

Write

$$\begin{aligned}
\int_0^\infty e^{\beta w_\beta^2 - t}\,dt &= e^{\tau_\beta} \int_{\tau_\beta}^\infty e^{y_\beta^2 - \tau}\,d\tau \\
&= e^{\tau_\beta}\left(\int_{\tau_\beta}^{\tau^*} + \int_{\tau^*}^\infty\right) e^{y_\beta^2 - \tau}\,d\tau \\
&= J_1 + J_2,
\end{aligned}$$

where $\tau^* = (1-\epsilon)\gamma^2$ and ϵ an arbitrary small positive number. Below we give an idea how J_1 and J_2 are estimated as $\gamma \to \infty$. As to J_1 we have

$$\begin{aligned}
J_1 &\sim \int_0^{\tau^* - T(\gamma)} e^{(\gamma^{-2}t - 1)t}\,dt \\
&\to \int_0^\infty e^{-t}\,dt \\
&= 1.
\end{aligned}$$

To estimate J_2, note that

$$y'(\tau^*) = \int_{\tau^*}^\infty y e^{y^2 - \tau}\,dt.$$

Therefore, since for every $\epsilon > 0$

$$y'(\tau^*) \to \frac{1}{\gamma} \quad \text{as} \quad \gamma \to \infty,$$

and

$$y(\tau^*) \to \gamma \quad \text{as} \quad \epsilon \to 0,$$

we have

$$\frac{1}{\gamma} \sim \gamma \int_{\tau^*}^\infty e^{y^2 - \tau}\,dt$$

as $\gamma \to \infty$ and $\epsilon \to 0$. It follows that

$$J_2 = e^{T(\gamma)} \int_{\tau^*}^\infty e^{y_\beta^2 - \tau}\,dt \sim \frac{e^{T(\gamma)}}{\gamma^2} \to e$$

as $\gamma \to \infty$, by (3.9). Thus, when $\beta \uparrow 1$,

$$\int_0^\infty e^{\beta w_\beta^2 - t}\,dt \to 1 + e,$$

as asserted.

4. The case $\alpha > 4\pi$

We now return to the elliptic boundary value problem (I). When $\Omega = B$, the unit ball in $\mathbf{R}^2$, and $0 < \lambda < \mu_1$ where μ_1 is the principal eigenvalue of $-\Delta$ on B, this problem was shown to have a (radial) solution for any $\alpha > 0$ [AP2]. Such a solution u corresponds to a solution w of (II) and a solution y of (III). Thus, if we write $u(0) = c$, we can conclude from (3.9) that

$$\begin{aligned}\lambda(c) &= \exp\{-T(c\sqrt{\beta})\}\\ &= \frac{4\pi}{\alpha e}\cdot\frac{1}{c^2}[1 + o(1)] \qquad \text{as} \quad c \to \infty.\end{aligned}$$

For general bounded domains Ω in $\mathbf{R}^2$, it was shown [S] that if

$$\text{(4.1)} \qquad \sup_M \int_\Omega e^{4\pi u^2} > 1 + e,$$

then there exists a number $\alpha^* > 4\pi$ such that for any $\alpha \in (4\pi, \alpha^*)$, there exists a function $u_\alpha \in M$ which locally maximizes $\int_\Omega e^{\alpha u^2}$ on M. In addition, there exists a number $\alpha_* \in (4\pi, \alpha^*]$ such that for almost every $\alpha \in (4\pi, \alpha_*]$ in the sense of Lebesgue measure, there exists a second critical point $u^\alpha \in M$ of $\int_\Omega e^{\alpha u^2}$ which is different from the first. For balls, (4.1) was shown to hold by the test function of Carleson and Chang, but also for domains which are close to a ball (4.1) can be shown to hold [S].

Finally, we mention a related problem studied by J.B. McLeod and K. McLeod [McL McL]:

$$\text{(IV)} \begin{cases} -\Delta u = \mu_1 u + (e^{u^p} - 1), & u > 0 \quad \text{in} \quad B_R \\ u = 0 & \qquad\quad \text{on} \quad \partial B_R, \end{cases}$$

where $B_R = \{x \in \mathbf{R}^2 : |x| < R\}$. They showed that if $p > 2$, there exists a number $\rho > 0$ such that (IV) has a solution if and only if $R \in (\rho, 1)$. This confirms a long standing conjecture that in $\mathbf{R}^2$ the rate of growth of the function e^{u^p} becomes supercritical if $p > 2$, much the same as on $R^N\,(N > 2)$ the rate of growth of the function u^p becomes supercritical when $p \geq (N+2)/(N-2)$.

Acknowledgements

This reseach was supported by grants from the Netherlands Organisation for Scientific Research (NWO), the Science and Engineering Research Council of the U.K., the National Science Foundation and the U.S. Army.

References

[AP1] Atkinson, F.V. & L.A. Peletier, Elliptic equations with nearly critical growth, J. Diff. Equ. 70 (1987), 349-365.

[AP2] Atkinson, F.V. & L.A. Peletier, Ground states and Dirichlet problems for $-\Delta u = f(u)$ in R^2, Arch. Rational Mech. Anal. 96 (1986), 147-165.

[BP] Brézis, H. & L.A. Peletier, Asymptotics for elliptic equations involving critical growth, Report Mathematical Institute, University of Leiden W 88-03, 1988.

[CC] Carleson, L. & S.-Y. A. Chang, On the existence of an extremal function for an inequality of J. Moser, Bull. Sc. Math. 2 110 (1986), 113-127.

[McLMcL] McLeod, J.B. & K.B. McLeod, Critical Sobolev exponents in two dimensions, Proc. Royal Soc. Edin., to appear.

[McLP] McLeod, J.B. & L.A. Peletier, Observations on Moser's inequality, Report Mathematical Institute, University of Leiden W 88-05, 1988; Arch. Rational Mech. Anal., to appear.

[M1] Moser, J., A sharp form of an inequality by N. Trudinger, Indiana U. Math. J. 20 (1971), 1077-1092.

[M2] Moser, J., On a nonlinear problem in differential geometry, Dynamical Systems (ed. M.M. Peixoto), N.Y. Academic press (1973), 273-280.

[S] Struwe, M., Critical points of embeddings of $H_0^{1,n}$ into Órlicz spaces, Report ETH–Zentrum, 8092 Zürich, Switzerland.

[T] Trudinger, N.S., On imbeddings into Orlicz spaces and some applications, J. Math. Mech. 17 (1967), 473-483.

J.B.M.: Department of Mathematics, University of Pittsburg, PA 15260, USA.

L.A.P.: Mathematical Institute, Leiden University, PB 9512, 2300 RA Leiden, The Netherlands

Evolution equations with discontinuous nonlinearities and non-convex constraints

A.Marino* C.Saccon*

Introduction

We consider here some semilinear parabolic problems with discontinuous nonlinearities, possibly subjected to some convex or even non-convex constraint conditions. We can find solutions of such problems following the abstract framework, which we feel quite simple, presented in section 4: the methods described there originated from the ideas proposed in [9], subsequently developed in [13] and [18]. In this setting, expecially in the constrained problems, the "potential " function f related to each problem plays a very important role (see 4.13)). We point out that in the case of the non convex constraint, which is made up by the intersection of a convex set K and a smooth hypersurface S, a key point is the non tangency between K and S (see (4.14)), which are studied in propositions (3.2) and (3.3). This fact resembles the situation in the regular setting: a sufficent condition for the intersection of two manyfolds M_1 and M_2 to be a manyfold is the non tangency between M_1 and M_2.

In general we remark that:

- since the nonlinearity is discontinuous the function f neither is ϕ-convex (see [13]) nor belongs to the classes considered in [18];
- we use some compactness arguments for the level sets of f, which are not required in the ϕ-convex context;
- we get existence theorems without uniqueness.

---

* Dipartimento di Matematica, Via Buonarroti 2, 56100 Pisa.

In the following we shall consider a real function $g : \mathbf{R} \to \mathbf{R}$, which will be allowed to be discontinuous. We remark here that all the results stated are formulated in such a way that they hold unchanged if g is replaced with an equivalent function.

We also consider a bounded open subset Ω of $\mathbf{R}^N$, a function $\varphi : \Omega \to \mathbf{R}$ (the obstacle) and and a real number $\rho > 0$.

We are interested in the following evolution problems: for $\mathcal{U} : I \to \mathrm{L}^2(\Omega)$ an absolutely continuous curve (I is an interval) we consider the problems

$$\text{(P.1)} \qquad \begin{cases} \mathcal{U}(t) \in \mathrm{H}^1_o(\Omega) & \forall t \text{ in } I \text{ and almost everywhere in } I: \\ \mathcal{U}'(t) = \Delta\mathcal{U}(t) + g(\mathcal{U}(t)); \end{cases}$$

$$\text{(P.2)} \qquad \begin{cases} \mathcal{U}(t) \in \mathrm{H}^1_o(\Omega), \mathcal{U}(t) \geq \varphi \quad \forall t \text{ in } I \text{ and a. e. in } I: \\ \mathcal{U}'(t) = \ \Delta\mathcal{U}(t) + g(\mathcal{U}(t)) \quad \text{in } \{x \mid \mathcal{U}(t)(x) > \varphi(x)\}, \\ \mathcal{U}'(t) = [\Delta\mathcal{U}(t) + g(\mathcal{U}(t))]^+ \text{ in } \{x \mid \mathcal{U}(t)(x) = \varphi(x)\} \end{cases}$$

(here the convex constraint $K = \{u \mid u \geq \varphi\}$ is involved);

$$\text{(P.3)} \qquad \begin{cases} \mathcal{U}(t) \in \mathrm{H}^1_o(\Omega), \mathcal{U}(t) \geq \varphi, \displaystyle\int_\Omega \mathcal{U}^2 \, \mathrm{d}x = \rho^2 \qquad \forall t \text{ in } I \\ \text{and there exists } \Lambda : I \to \mathbf{R} \text{ such that a.e. in } I: \\ \mathcal{U}'(t) = \ \Delta\mathcal{U}(t) + g(\mathcal{U}(t)) + \Lambda(t)\mathcal{U}(t) \quad \text{in } \{x \mid \mathcal{U}(t)(x) > \varphi(x)\}, \\ \mathcal{U}'(t) = [\Delta\mathcal{U}(t) + g(\mathcal{U}(t)) + \Lambda(t)\mathcal{U}(t)]^+ \text{ in } \{x \mid \mathcal{U}(t)(x) = \varphi(x)\}; \end{cases}$$

(here the non-convex constraint $\{u \mid u \geq \varphi, \int_\Omega u^2 \, \mathrm{d}x = \rho^2\}$ is involved; observe that Λ can be regarded as a "Lagrange multiplier" relative to the constraint $S_\rho = \{u \mid \int_\Omega u^2 \, \mathrm{d}x = \rho^2\}$).

We introduce now some notations and assumptions that will be often recalled later. We shall always require that

$$\text{(g.1)} \qquad |g(s)| \leq a + b|s| \qquad \forall s \in \mathbf{R} \qquad \text{for } a, b \text{ in } \mathbf{R}$$

This assumption however could be replaced with a more general growth condition, stating all the theorems in a local form. Sometimes we shall also consider the assumption:

$$\text{(g.2)} \qquad \begin{cases} \text{there exists } \tilde{g} \text{ equivalent to } g \text{ such that} \\ \tilde{g} \text{ is continuous almost everywhere.} \end{cases}$$

A meaningful situation in which (g.2) is satisfied is that of a function g with bounded variation.

If (g.1) is satisfied, we introduce the functions $\underline{g}, \bar{g} : \mathbf{R} \to \mathbf{R}$ defined by

$$\underline{g}(s) = \sup\{h(s) \mid h : \mathbf{R} \to \mathbf{R} \text{ is continuous}, h \leq g \text{ almost everywhere }\}$$
$$\bar{g}(s) = \inf\{h(s) \mid h : \mathbf{R} \to \mathbf{R} \text{ is continuous}, g \leq h \text{ almost everywhere }\}$$

and the sets

$$E^* = \{s \in \mathbf{R} \mid \underline{g}(s) = g(s) = \bar{g}(s)\} \quad , \quad E = \mathbf{R} \setminus E^*.$$

Observe that, if (g.1) holds, then meas(E)=0.

§1. The unconstrained case

In this section we study problem (P.1).

(1.1) THEOREM *Assume that (g.1) and (g.2) hold.*
Then for all u_o in $H^1_o(\Omega)$ there exists $\mathcal{U} : [0, +\infty[\to L^2(\Omega)$, an absolutely continuous curve, such that $\mathcal{U}(0) = u_o$,

$$(P.1)' \quad \begin{cases} \mathcal{U}(t) \in H^1_o(\Omega) & \forall t \geq 0 \text{ and for a.e. } t \geq 0: \\ \mathcal{U}'(t) = \Delta \mathcal{U}(t) + g(\mathcal{U}(t)) & \text{in } \{x \mid \mathcal{U}(t)(x) \notin E\}, \\ \mathcal{U}'(t) = \Delta \mathcal{U}(t) = 0 \in [\underline{g}(\mathcal{U}(t)), \bar{g}(\mathcal{U}(t))] & \text{in } \{x \mid \mathcal{U}(t)(x) \in E\}. \end{cases}$$

In particular, if, for instance, $g(s) \geq C > 0 \ \forall s$ or $g(s) \leq C < 0 \ \forall s$ (at least in a neighbourhood of E), then $\mathcal{U}$ solves (P.1) and for almost every t $\{x \mid \mathcal{U}(t)(x) \in E\}$ is a null set.

(1.2) THEOREM *(cfr. [1], [24], [25]) Assume that (g.1) holds.*
Then for all u_o in $H^1_o(\Omega)$ there exists $\mathcal{U} : [0, +\infty[\to L^2(\Omega)$, an absolutely continuous curve, such that $\mathcal{U}(0) = u_o$,

$$\begin{cases} \mathcal{U}(t) \in H^1_o(\Omega) & \forall t \geq 0 \text{ and for a.e. } t \geq 0: \\ \mathcal{U}'(t) = \Delta \mathcal{U}(t) + \beta(t)) & \text{where } \underline{g}(\mathcal{U}(t)) \leq \beta(t) \leq \bar{g}(\mathcal{U}(t)). \end{cases}$$

Theorems (1.1) and (1.2) can be deduced from theorems (4.11) and (4.9) (of section 4) respectively, by introducing the functional $f_1 : L^2(\Omega) \to \mathbf{R} \cup \{+\infty\}$ defined by:

$$f_1(u) = \begin{cases} \frac{1}{2} \int_\Omega |Du|^2 \, dx - \int_\Omega \int_0^u g(\sigma) \, d\sigma \, dx & \text{if } u \in H^1_o(\Omega), \\ +\infty & \text{if } u \in L^2(\Omega) \setminus H^1_o(\Omega), \end{cases}$$

and the operator $\mathcal{A}_1 : \mathrm{H}^1_o(\Omega) \to 2^{\mathrm{L}^2(\Omega)}$ defined by:

$$\alpha \in \mathcal{A}_1(u) \Leftrightarrow \begin{cases} \alpha = -\Delta u - \beta \text{ (in the distributional sense)} \\ \text{with } \underline{g}(u(x)) \leq \beta(x) \leq \bar{g}(u(x)) \text{ a.e. in } \Omega. \end{cases}$$

(1.3) REMARK *The operator appearing in the right hand side of (P.1)′ is precisely the minimal section of the multivalued operator $\mathcal{A}_1$ above (see remark (4.13)). The concrete expression of such minimal section is essential for solving (P.1), under the assumpion indicated in theorem (1.1).*

§2. The case with a convex constraint

In this section we study problem (P.2).

We shall consider the following assumption on the obstacle φ

$$(\varphi) \qquad\qquad \varphi \in \mathrm{W}^{2,2}(\Omega)$$

which could also be weakened, provided that all the following theorems were stated in the weak form of variational inequalities.

We also set

$$K = \{u \in \mathrm{L}^2(\Omega) \mid u(x) \geq \varphi(x) \text{ a.e. in } \Omega\}$$

and for u in K we define the "contact set" by

$$C(u) = \{x \in \Omega \mid u(x) = \varphi(x)\}.$$

(2.1) THEOREM *Assume that (g.1), (g.2) and (φ) hold.*

Then for all u_o in $\mathrm{H}^1_o(\Omega) \cap K$ there exists $\mathcal{U} : [0, +\infty[\to \mathrm{L}^2(\Omega)$, an absolutely continuous curve, such that $\mathcal{U}(0) = u_o$,

$$(P.2)' \quad \begin{cases} \mathcal{U}(t) \in \mathrm{H}^1_o(\Omega) \cap K \ \forall t \geq 0 \text{ and for a.e. } t \geq 0: \\ \\ \text{in } \{x \mid \mathcal{U}(t)(x) \notin E\} \\ \mathcal{U}'(t) = \begin{cases} \Delta\mathcal{U}(t) + g(\mathcal{U}(t)) & \text{in } \Omega \setminus C(\mathcal{U}(t)), \\ [\Delta\mathcal{U}(t) + g(\mathcal{U}(t))]^+ & \text{in } C(\mathcal{U}(t)), \end{cases} \\ \\ \text{in } \{x \mid \mathcal{U}(t)(x) \in E\} \\ \mathcal{U}'(t) = \Delta\mathcal{U}(t) = 0 \in \begin{cases} [\underline{g}(\mathcal{U}(t)), \bar{g}(\mathcal{U}(t))] & \text{in } \Omega \setminus C(\mathcal{U}(t)), \\ [\underline{g}(\mathcal{U}(t)), +\infty[& \text{in } C(\mathcal{U}(t)). \end{cases} \end{cases}$$

In particular, if, for instance, $g(s) \geq C > 0\ \forall s$ or $g(s) \leq C < 0\ \forall s$ (at least in a neighbourhood of E), then $\mathcal{U}$ solves (P.2). If $g(s) \geq C > 0\ \forall s$, then for almost every t $\{x \mid \mathcal{U}(t)(x) \in E\}$ is a null set.

(2.2) THEOREM *(cfr. [25]) Assume that (g.1)and (φ) hold. Then for all u_o in $H_o^1(\Omega) \cap K$ there exists $\mathcal{U} : [0, +\infty[\to L^2(\Omega)$, an absolutely continuous curve, such that $\mathcal{U}(0) = u_o$,*

$$\begin{cases} \mathcal{U}(t) \in H_o^1(\Omega) \cap K\ \forall t \geq 0 \text{ and for a.e. } t \geq 0: \\ \mathcal{U}'(t) = \Delta\mathcal{U}(t) + \beta(t)) \qquad \text{where} \\ \quad \underline{g}(\mathcal{U}(t)) \leq \beta(t) \leq \bar{g}(\mathcal{U}(t)) \text{ in } \Omega \setminus C(\mathcal{U}(t)), \\ \quad \underline{g}(\mathcal{U}(t)) \leq \beta(t) \qquad\qquad \text{in } C(\mathcal{U}(t)). \end{cases}$$

Theorems (2.1) and (2.2) can be deduced from theorems (4.11) and (4.9) (of section 4) respectively, by introducing the functional $f_1 : L^2(\Omega) \to \mathbf{R} \cup \{+\infty\}$ defined by:

$$f_2(u) = \begin{cases} f_1(u) & \text{if } u \in K, \\ +\infty & \text{if } u \in L^2(\Omega) \setminus K, \end{cases}$$

and the operator $\mathcal{A}_2 : H_o^1(\Omega) \cap K \to 2^{L^2(\Omega)}$ defined by:

$$\alpha \in \mathcal{A}_2(u) \Leftrightarrow \begin{cases} \alpha = -\Delta u - \beta & \text{(in the distributional sense) with} \\ \underline{g}(u) \leq \beta \leq \bar{g}(u) & \text{in } \Omega \setminus C(u), \\ \underline{g}(u) \leq \beta & \text{in } C(u). \end{cases}$$

(2.3) REMARK *The operator appearing in the right hand side of (P.2)' is precisely the minimal section of the multivalued operator $\mathcal{A}_2$ above (see remark (4.13)). The concrete expression of such minimal section is essential for solving (P.2), under the assumpion indicated in theorem (2.1).*

§3. The case with a non-convex constraint and some non-tangency conditions

In this section we study problem (P.3). We recall that

$$K = \{u \in L^2(\Omega) \mid u(x) \geq \varphi(x) \text{ a.e.}\},$$
$$C(u) = \{x \in \Omega \mid u(x) = \varphi(x)\}.$$

For $\rho > 0$ we set

$$\rho_K = \min\left\{ \left(\int_\Omega u^2\, dx\right)^{1/2} \mid u \in K \right\} = \left(\int_\Omega (\varphi^+)^2\, dx\right)^{1/2},$$

$$S_\rho = \left\{ u \in L^2(\Omega) \mid \int_\Omega u^2\, dx = \rho^2 \right\}.$$

(3.1) DEFINITION *We say that K and S_ρ are "tangent" at a point u_o of $K \cap S_\rho$, if the tangent plane $\{v \in L^2(\Omega) \mid \int_\Omega u_o(v - u_o)\, dx = 0\}$ to S_ρ at u_o is tangent to K at u_o.*

(3.2) PROPOSITION *(see [6]) Let $u_o \in K \cap S_\rho$. Then K and S_ρ are not tangent at u_o if and only if*

$$(N.T.u_o)\qquad \begin{cases} \int_\Omega u_o^2\, dx > \rho_K^2 \quad \text{and} \\ \text{meas}(\{x \mid \varphi(x) < u_o(x) < 0\} \cup \{x \mid 0 < u_o(x)\}) > 0. \end{cases}$$

Observe that, if $\varphi^+ \neq 0$, then the second condition is automatically fulfilled.

(3.3) PROPOSITION *(see [6]) If the conditions*

$$(N.T.)\qquad \begin{cases} \rho > \rho_K\ ,\ \varphi \in W^{1,2}(\Omega) \cap C(\Omega) \text{ and there is no open} \\ \text{subset } \Omega' \text{ of } \Omega \text{ such that } \varphi \in H_o^1(\Omega'), \varphi \geq 0 \text{ in } \Omega', \end{cases}$$

are satisfied, then K and S_ρ are not tangent at any u of $K \cap H_o^1(\Omega) \cap S_\rho$.

(3.4) THEOREM *Assume that (g.1), (g.2) and (φ) hold.*
Then for all u_o in $H_o^1(\Omega) \cap K \cap S_\rho$ such that $(N.T.u_o)$ holds there exist $T > 0$, $\mathcal{U} : [0, T[\to L^2(\Omega)$, an absolutely continuous curve and $\Lambda : [0, T[\to \mathbf{R}$, such that $\mathcal{U}(0) = u_o$,

$$(P.3)'\qquad \begin{cases} \mathcal{U}(t) \in H_o^1(\Omega) \cap K \cap S_\rho\ \forall t \text{ in } [0,T[\text{ and for a.e. } t \text{ in } [0,T[: \\[2ex] \text{in } \{x \mid \mathcal{U}(t)(x) \notin E\} \\ \mathcal{U}'(t) = \begin{cases} \Delta\mathcal{U}(t) + g(\mathcal{U}(t)) + \Lambda(t)\mathcal{U}(t) & \text{in } \Omega \setminus C(\mathcal{U}(t)), \\ [\Delta\mathcal{U}(t) + g(\mathcal{U}(t)) + \Lambda(t)\mathcal{U}(t)]^+ & \text{in } C(\mathcal{U}(t)), \end{cases} \\[2ex] \text{in } \{x \mid \mathcal{U}(t)(x) \in E\} \\ \mathcal{U}'(t) = \Delta\mathcal{U}(t) = 0 \quad \text{and} \\ \qquad 0 \in \begin{cases} [\underline{g}(\mathcal{U}(t)) + \Lambda(t)\mathcal{U}(t), \bar{g}(\mathcal{U}(t)) + \Lambda(t)\mathcal{U}(t)] & \text{in } \Omega \setminus C(\mathcal{U}(t)), \\ [\underline{g}(\mathcal{U}(t)) + \Lambda(t)\mathcal{U}(t), +\infty[& \text{in } C(\mathcal{U}(t)). \end{cases} \end{cases}$$

It can be shown that, assuming

$$0 \le g(s)s \le \sigma s^2 \qquad \forall s \text{ in } \mathbf{R}$$

with $\sigma < \inf \left\{ \frac{\int_\Omega |\mathrm{D}u|^2 \, dx}{\int_\Omega u^2 \, dx} \mid u \in \mathrm{H}^1_o(\Omega) \right\}$ *(the condition* $0 \le g(s)s$ *could be replaced by* $0 \le \underline{g}(s)s$ *and* $0 \le \bar{g}(s)s$ *for all s in E), then* $\mathcal{U}$ *solves (P.3) and for almost every t the set* $\{x \in \Omega \mid \mathcal{U}(t)(x) \in E\}$ *is a null set.*

Furthermore, if (N.T.) is assumed, then $T = +\infty$*.*

(3.5) THEOREM *Assume that (g.1)and* (φ) *hold.*
Then for all u_o *in* $\mathrm{H}^1_o(\Omega) \cap K \cap S_\rho$ *such that* (N.T.u_o) *holds, there exist* $T > 0$, $\mathcal{U} : [0, +T[\to \mathrm{L}^2(\Omega)$, *an absolutely continuous curve and* $\Lambda : [0, T[\to \mathbf{R}$ *such that* $\mathcal{U}(0) = u_o$,

$$\begin{cases} \mathcal{U}(t) \in \mathrm{H}^1_o(\Omega) \cap K \cap S_\rho \ \forall t \text{ in } [0,T[\text{ and for a.e. } t \text{ in } [0,T[: \\ \mathcal{U}'(t) = \Delta\mathcal{U}(t) + \beta(t)) + \Lambda(t)\mathcal{U}(t) \qquad \text{where} \\ \qquad \underline{g}(\mathcal{U}(t)) \le \beta(t) \le \bar{g}(\mathcal{U}(t)) \ \text{in } \Omega \setminus C(\mathcal{U}(t)), \\ \qquad \underline{g}(\mathcal{U}(t)) \le \beta(t) \qquad\qquad \text{in } C(\mathcal{U}(t)). \end{cases}$$

Furthermore, if (N.T.) holds, then $T = +\infty$*.*

Theorems (3.4) and (3.5) can be deduced from theorems (4.11) and (4.9) (of section 4) respectively, by introducing the functional $f_3 : \mathrm{L}^2(\Omega) \to \mathbf{R} \cup \{+\infty\}$ defined by:

$$f_3(u) = \begin{cases} f_1(u) & \text{if } u \in K \cap S_\rho, \\ +\infty & \text{elsewhere in } \mathrm{L}^2(\Omega), \end{cases}$$

and the operator $\mathcal{A}_3 : \mathrm{H}^1_o(\Omega) \cap K \to 2^{\mathrm{L}^2(\Omega)}$ defined by:

$$\alpha \in \mathcal{A}_3(u) \Leftrightarrow \begin{cases} \alpha = -\Delta u - \beta - \lambda u & \text{(in the distributional sense)} \\ \text{with } \lambda \in \mathbf{R} \text{ and} & \\ \underline{g}(u) \le \beta \le \bar{g}(u) & \text{in } \Omega \setminus C(u), \\ \underline{g}(u) \le \beta & \text{in } C(u). \end{cases}$$

(3.6) REMARK *The operator appearing in the right hand side of (P.3)′ is precisely the minimal section of the multivalued operator* $\mathcal{A}_3$ *above (see remark (4.13)). The concrete expression of such minimal section is essential for solving (P.3), under the assumpion indicated in theorem (3.4).*

§4. Abstract results

We present now some definitions and results, which are at the base of the proofs of the previous theorems. For the proofs of the results stated we refer the reader to [16] and [23].

We shall consider a Hilbert space H, with inner product $\langle\cdot,\cdot\rangle$ and norm $\|\cdot\|$ and a function $f : H \to \mathbf{R} \cup \{+\infty\}$.

We set $\mathcal{D}(f) = \{u \in H \mid f(u) < +\infty\}$. We recall the following definition (see [9]).

(4.1) DEFINITION *Let $u \in \mathcal{D}(f)$. We define the subdifferential $\partial^- f(u)$ of f at u as the set of all α's in H (possibly empty) such that:*

$$\liminf_{v \to u} \frac{f(v) - f(u) - \langle \alpha, v - u \rangle}{\|v - u\|} \geq 0.$$

We also recall the following density result (see [13] and [15]).

(4.2) THEOREM *Assume f to be lower semicontinuous in a neighbourhood of $\mathcal{D}(f)$. Then the set*

$$\{u \in \mathcal{D}(f) \mid \partial^- f(u) \neq \emptyset\}$$

is dense in $\mathcal{D}(f)$.

Theorem (4.2) makes possible the following definition.

(4.3) DEFINITION *Let f be lower semicontinuous in a neighbourhood of $\mathcal{D}(f)$. We define $p : \mathcal{D}(f) \to \mathbf{R} \cup \{+\infty\}$ by*

$$p(u) = \liminf_{\substack{v \to u \\ \partial^- f(v) \neq \emptyset}} \inf\{\|\alpha\| \mid \alpha \in \partial^- f(v)\}$$

(4.4) DEFINITION *We say that f is locally coercive, if for all u_o in $\mathcal{D}(f)$ there exists $\rho > 0$ such that the set*

$$\{v \in H \mid f(v) \leq C, \|v - u\| \leq \rho\}$$

is compact for all C in $\mathbf{R}$.

(4.5) DEFINITION *We say that f is ∇-continuous, if for all $u \in \mathcal{D}(f)$, for all sequences $\{u_n\}_n$ in $\mathcal{D}(f)$ and $\{\alpha_n\}_n$ in H we have*

$$\left.\begin{array}{lll} \lim_{n\to\infty} u_n = u & , & \sup_n f(u_n) < +\infty \\ \alpha_n \in \partial^- f(u_n)\ \forall n & , & \sup_n \|\alpha_n\| < +\infty \end{array}\right\} \Rightarrow \lim_{n\to\infty} f(u_n) = f(u)$$

Now we have the ingredients to state the main lemma, which can be proved (see [23]) in a quite simple way.

(4.6) LEMMA *Let $f : H \to \mathbf{R} \cup \{+\infty\}$ be a locally coercive function. Then for all u_o in $\mathcal{D}(f)$ there exist $T > 0$ and $\mathcal{U} : [0, T[\to H$, an absolutely continuous curve, such that $\mathcal{U}(0) = u_o$, $f(\mathcal{U}(t)) \leq f(u_o)$ $\forall t$ in $[0, T[$ and*

$$\text{(4.7)} \qquad \begin{cases} \exists \{\mathcal{U}_n\}_n, \{\alpha_n\}_n \text{ sequences in } L^2(0,T;H) \text{ such that:} \\ \alpha_n(t) \in \partial^- f(\mathcal{U}_n(t)) \text{ for a.e. } t \text{ in } [0,T[\text{ and} \\ \mathcal{U}_n \to \mathcal{U} \text{ uniformly on } [0,T[, \\ -\alpha_n \rightharpoonup \mathcal{U}' \text{ weakly in } L^2(0,T;H). \end{cases}$$

If furthermore f is ∇-continuous, then

$$\text{(4.8)} \qquad \begin{cases} f(\mathcal{U}(t_2)) - f(\mathcal{U}(t_1)) \leq \begin{cases} -\int_{t_1}^{t_2} \|\mathcal{U}'(t)\|^2 \, dt \\ -\int_{t_1}^{t_2} p(\mathcal{U}(t))^2 \, dt \end{cases} \\ \text{for almost all } t_1, \text{ for all } t_2 \text{ in } [0,T[\text{ with } t_1 \leq t_2. \end{cases}$$

(this implies in particular that $f \circ \mathcal{U}$ is equivalent to a non increasing function).

The following theorems follow in a very simple fashion from lemma (4.6). First we introduce some assumptions. We consider an operator $\mathcal{A} : \mathcal{D}(f) \to 2^H$ and the following conditions on $\mathcal{A}$.

$$(\mathcal{A}) \qquad \partial^- f(u) \subset \mathcal{A}(u) \qquad \forall u \in \mathcal{D}(f).$$

$$(\bar{\mathcal{A}}) \qquad \begin{cases} \text{for all } T > 0, \text{ for all } \mathcal{U}, \alpha \text{ in } L^2(0,T;H) \text{ and for all} \\ \{\mathcal{U}_n\}_n, \{\alpha_n\}_n \text{ sequences in } L^2(0,T;H), \text{ if} \\ \qquad \mathcal{U}_n \to \mathcal{U} \text{ uniformly on } [0,T[\quad , \quad \sup_n \sup_{[0,T[} f \circ \mathcal{U}_n < +\infty, \\ \qquad \alpha_n(t) \in \mathcal{A}(\mathcal{U}_n(t)) \quad \text{for a.e. } t \text{ in } [0,T[, \\ \qquad \alpha_n \rightharpoonup \alpha \quad \text{weakly in } L^2(0,T;H), \\ \text{then} \\ \qquad \alpha(t) \in \mathcal{A}(\mathcal{U}(t)) \quad \text{for a.e. } t \text{ in } L^2(0,T;H). \end{cases}$$

$$(\mathcal{A}')\quad \begin{cases} \forall T>0,\ \forall \mathcal{U}:[0,T[\to H, \text{ with } \mathcal{U} \text{ absolutely continuous and} \\ \qquad \sup\limits_{[0,T[} f\circ\mathcal{U} < +\infty \quad , \quad \mathcal{A}(\mathcal{U}(t)) \neq \emptyset \quad \text{a.e. in } [0,T[, \\ \text{one has for almost all } t \text{ in } [0,T[: \\ \qquad D_+(f\circ\mathcal{U})(t) \geq \langle \alpha, \mathcal{U}'(t)\rangle \quad \forall\alpha\in\mathcal{A}(\mathcal{U}(t)) \end{cases}$$

(here D_+ denotes the right lower derivative).

(4.9) THEOREM *Let $f : H \to \mathbf{R}\cup\{+\infty\}$ be a locally coercive function and $\mathcal{A}$ an operator satisfying $(\mathcal{A})$ and $(\bar{\mathcal{A}})$.*

Then for all u_o in $\mathcal{D}(f)$ there exist $T>0$ and $\mathcal{U} : [0,T[\to H$, an absolutely continuous curve, such that $\mathcal{U}(0) = u_o$, $f(\mathcal{U}(t)) \leq f(u_o)$ $\forall t$ in $[0,T[$, (4.7) holds and

$$(4.10)\qquad -\mathcal{U}'(t) \in \mathcal{A}(\mathcal{U}(t)) \quad \text{for a.e. } t \text{ in } [0,T[.$$

If furthermore f is ∇-continuous, then (4.8) holds.

The proof is an immediate consequence of the fact, which is evident, that any $\mathcal{U}$ satisfying (4.7) has the property (4.10).

(4.11) THEOREM *Let $f : H \to \mathbf{R}\cup\{+\infty\}$ be a locally coercive, ∇-continuous function and $\mathcal{A}$ an operator satisfying $(\mathcal{A})$, $(\bar{\mathcal{A}})$ and $(\mathcal{A}')$. For all u in $\mathcal{D}(f)$ with $\mathcal{A}(u) \neq \emptyset$ let $A(u)$ be a minimal section of $\mathcal{A}(u)$, namely an element in $\mathcal{A}(u)$ having minimal norm.*

Then for all u_o in $\mathcal{D}(f)$ there exist $T>0$ and $\mathcal{U} : [0,T[\to H$, an absolutely continuous curve, such that $\mathcal{U}(0) = u_o$, $f(\mathcal{U}(t)) \leq f(u_o)$ $\forall t$ in $[0,T[$, (4.7)and (4.8) hold and for almost every t in $[0,T[$ one has:

$$(4.12)\qquad \begin{cases} \mathcal{U}'(t) = -A(\mathcal{U}(t)), \\ (f\circ\mathcal{U})'(t) = -\|A(\mathcal{U}(t))\|^2 = -p(\mathcal{U}(t))^2. \end{cases}$$

The proof of theorem (4.11) is a consequence of the fact:

> if f is locally coercive, $(\mathcal{A})$, $(\bar{\mathcal{A}})$, and $(\mathcal{A}')$ hold, then any absolutely continuous curve $\mathcal{U}$ which satisfies (4.8) has the properties (4.12).

To prove this (see [16] it suffices to prove the relation

$$\|A(u)\| \leq p(u) \quad \forall u \text{ in } \mathcal{D}(f) \text{ with } \mathcal{A}(u)\neq\emptyset,$$

then use $(\mathcal{A}')$ to obtain $(f\circ\mathcal{U})'(t) = \langle A(\mathcal{U}(t)), \mathcal{U}'(t)\rangle = -\|A(\mathcal{U}(t))\|\,\|\mathcal{U}'(t)\|$ for almost every t in $[0,T[$.

(4.13) REMARK *We point out that to prove (4.12), where we have precisely the minimal section of $\mathcal{A}$ in the right hand side, we make an essential use of the functional f, namely of the variational nature of the problem.*

(4.14) REMARK *The non tangency condition, introduced in section 3, are essential to prove the inclusion $\partial^- f_3 \subset \mathcal{A}_3$, namely the assumption $\mathcal{A}$, for theorems (4.9) and (4.11) in the case of the functional f_3 to get theorems (3.5) and (3.4) respectively.*

REFERENCES

[1] **J.P.Aubin.** *Domaines de viabilité pour des inclusion différentielles opérationelles.* 1986.

[2] **J.P.Aubin-A.Cellina.** *Differential inclusions.* Springer Verlag (1984).

[3] **H.Brezis.** *Opérateurs maximaux monotone et semigroupes de contraction dans les espaces de Hilbert.* North Holland Mathematics Studies, 5, Notas de Matematica 50, Amsterdam, London 1973.

[4] **F.H.Clarke.** *Optimization and nonsmooth analysis.* John Wiley (1984).

[5] **G.Cŏbanov-A.Marino-D.Scolozzi.** *Evolution equations for the eigenvalue problem for the Laplace operator with respect to an obstacle.* to appear.

[6] **G.Cŏbanov-A.Marino-D.Scolozzi.** *Multiplicity of eigenvalues of the Laplace operator with respect to an obstacle and non tangency conditions.* Nonlinear Anal. Th. Meth. Appl., to appear.

[7] **G.Colombo-M.Tosques.** *Multivalued perturbations for a class of nonlinear evolution equations.* to appear.

[8] **E.De Giorgi-M.Degiovanni-A.Marino-M.Tosques.** *Evolution equations for a class of nonlinear operators.* Atti Accadem. Naz. Lincei Rend. Cl. Sci. Fis. Mat. Natur. (8) 75 (1983),1-8.

[9] **E.De Giorgi-A.Marino-M.Tosques.** *Problemi di evoluzione in spazi metrici e curve di massima pendenza.* Atti Accadem. Naz. Lincei Rend. Cl. Sci. Fis. Mat. Natur. (8) 68 (1980),180-187.

[10] **E.De Giorgi-A.Marino-M.Tosques.** *Funzioni (p,q)-convesse.* Atti Accadem. Naz. Lincei Rend. Cl. Sci. Fis. Mat. Natur. (8) 73 (1982),6-14.

[11] **M.Degiovanni-A.Marino-M.Tosques.** *General properties of (p,q)-convex functions and (p,q)-monotone operators.* Ricerche Mat. 32 (1983), 285-319.

[12] **M.Degiovanni-A.Marino-M.Tosques.** *Evolution equations associated with (p,q)-convex functions and (p,q)-monotone operators.* Ricerche Mat. 33 (1984), 81-112.

[13] **M.Degiovanni-A.Marino-M.Tosques.** *Evolution equations with lack of convexity.* Nonlinear Anal. The. Meth. Appl. 9, 12 (1985),1401-1443.

[14] **M.Degiovanni-M.Tosques.** *Evolution equations for (φ, f)-monotone operators.* Boll. Un. Mat. Ital. B (6) 5 (1986), 537-568.

[15] **M.Edelstein.** *On nearest points of sets in uniformly convex Banach spaces.* J. London Math. Soc., 43 (1968), 375-377.

[16] **M.Frigon-C.Saccon.** . to appear.

[17] **A.Marino.** *Evolution equations and multiplicity of critical points with respect to an obstacle.* Contributions to modern calculus of variations,(Bologna 1985) Res. Notes in Math. , 148, Pitman, London - New York, 1987, 123-144.

[18] **A.Marino - C.Saccon - M.Tosques.** *Curves of maximal slope and parabolic variational inequalities on non convex constraints.* to appear in Ann. Scuola Norm. Sup. Pisa.

[19] **A.Marino-D.Scolozzi.** *Punti inferiormente stazionari ed equazione di evoluzione con vincoli unilaterali non convessi.* Rend. Sem. Mat. e Fis. Milano, 52 1982, 393-414.

[20] **A.Marino-D.Scolozzi.** *Autovalori dell'operatore di Laplace ed equazioni di evoluzione in presenza di ostacolo.* (Bari 1984) Problemi differenziali e teoria dei punti critici , Pitagora, Bologna, 1984, 137-155.

[21] **A.Marino-M.Tosques.** *Curves of maximal slope for a certain class of non regular functions.* Boll. Un. Mat. Ital. B (6) 1 (1982), 143-170.

[22] **C.Saccon.** *Some parabolic equations on non convex constraints.* Boll. Un. Mat. Ital. , to appear

[23] **C.Saccon.** . to appear.

[24] **Shi Shuzhong.** *Théorèmes de viabilité pour les inclusion aux dérivée partielles.* C.R.A.S. , 1986.

[25] **Shi Shuzhong.** *Nagumo type condition for partial differential inclusions.* Nonlinear Anal. Th. Meth. Appl., 12 (1988), 951-967.

Nonlinear Variational Two-Point Boundary Value Problems

J. MAWHIN

1. Historical Introduction

Let us consider the two-point boundary value problem

$$u''(x) = f(x, u(x)), \quad x \in I, \qquad u(0) = u(\pi) = 0 \tag{1}$$

where $I = [0, \pi]$, $f: I \times \mathbf{R} \to \mathbf{R}$ is a Caratheodory function. As early as in 1915, Lichtenstein [9] observed that if the function $F: I \times \mathbf{R} \to \mathbf{R}$ defined by

$$F(x, u) = \int_0^u f(x, v)dv$$

is such that

$$F(x, u) \geq A \tag{2}$$

for some real number A and all $(x, u) \in I \times \mathbf{R}$, then the corresponding action integral

$$\varphi: u \mapsto \int_I [(1/2)(u'(x))^2 + F(x, u(x))]\, dx$$

is bounded below on the set $C_0^1(I)$ of functions u of class C^1 on I which vanish at 0 and π, and hence φ has an infimum d on $C_0^1(I)$. Lichtenstein introduced then a minimizing sequence (u_n) such that $\varphi(u_n) \to d$ as $n \to \infty$, namely the one coming from the associated Ritz method. Writing

$$u_n(x) = \sum_{j=1}^{\infty} (u_{n,j}/j) \sin jx,$$

so that

$$\sum_{j=1}^{\infty} u_{n,j} \sin jx$$

is the Fourier series of u'_n, Lichtenstein proved the existence of a subsequence (u_{n_k}) such that, for each $j \in \mathbb{N}^*$, $(u_{n_k,j})$ converges to some u_j as $k \to \infty$. He then showed that the corresponding function u defined by

$$(3) \qquad u(x) = \sum_{j=1}^{\infty} (u_j/j) \sin jx$$

belongs to $C^2(I)$, vanishes at 0 and π, minimizes φ over $C_0^1(I)$ and hence is a solution of (1). To realize this program, Lichtenstein observed that $(\|u'_n\|_{L^2})$ is bounded, so that the sequence $(\sum_{j=1}^{\infty} n_{n,j}^2)$ is bounded, and a diagonal process provides the existence of convergent subsequences $(u_{n_k,j})$ and (u_{n_k}). Lichtenstein shows then that (u_{n_k}) converges uniformly on I to the function u given by (3), and hence

$$(1/2) \int_I (u'_{n_k}(x))^2 dx \to d - \int_I f(x, u(x)) dx = d - \Delta.$$

This easily implies that

$$(4) \qquad (\pi/4) \sum_{j=1}^{\infty} u_j^2 \leq d - \Delta,$$

and a contradiction argument based upon the definition of d shows that the equality holds in (4). From the fact that the infimum d of φ is reached by u, Lichtenstein deduces the relation

$$(\pi/2) u_1 + l^{-1} \int_I f(x, u(x)) \sin lx \, dx = 0, \qquad l \in \mathbb{N}^*,$$

and

$$(5) \qquad (\pi/2) \sum_{l=1}^{\infty} u_l v_l + \int_I f(x, u(x)) v(x) dx = 0,$$

for all $v(x) = \sum_{j=1}^{\infty} (v_j/j) \sin jx$ such that $\sum_{j=1}^{\infty} v_j^2$ converges. Choosing for $v(x)$ the Green function $G(x_0, x)$ of the associated linear problem, he deduces from (5) that

$$u(x_0) = - \int_I G(x_0, x) f(x, u(x)) dx, \qquad x_0 \in I,$$

so that $u \in C_0^2(I)$ and is a solution of (1) such that $\varphi(u) = d$.

The reader will easily detect in the above arguments, possibly sometimes in an implicit form, the basic ingredients of the present setting for the direct method of the calculus of variations for the study of (1), namely:

1) the obtention of a minimizing sequence (u_k);

2) the use of the boundedness of (u_k) in the Sobolev space $H_0^1(I)$ to obtain a subsequence converging weakly in $H_0^1(I)$ and uniformly on I to some u;

3) the use of the lower semicontinuity of the square of the H_0^1 norm to show that u minimizes φ;

4) the use of the Euler equation to show that the (weak) solution u is indeed a classical one.

2. Sharp Nonresonance Conditions by Variational Methods

If we consider the linear problem

$$\begin{aligned} u''(x) + \lambda u(x) &= h(x), \quad x \in I, \\ u(0) = u(\pi) &= 0 \end{aligned} \tag{6}$$

for which $F(x,u) = h(x)u-(\lambda/2)u^2$, we see that the Lichtenstein condition (2) will hold if and only if $\lambda < 0$. Thus (2) is not sharp as it is well known that (6) is solvable for each $h \in L^1(I)$ when $\lambda < 1$. Now, condition (2) is only used by Lichtenstein, together with the trivial inequality

$$\int_I (u'(x))^2 dx \geq 0, \tag{7}$$

to prove that φ is bounded below. If one replaces (7) by the Poincaré's inequality

$$\int_I (u'(x))^2 dx \geq \int_I u^2(x) dx, \tag{8}$$

which holds on $H_0^1(I)$, we easily see that φ is still bounded from below if (2) is replaced by the more general condition

$$F(x,u) \geq (B/2)u^2 + A \tag{9}$$

for some $A \in \mathbb{R}$, $B > -1$ and all $(x,u) \in I \times \mathbb{R}$. This was noticed in 1938 by Cinquini [1] (without reference to Lichstenstein paper but rather on Tonelli's basic work in the calculus of variations), although the existence of a solution for (1) when F satisfies (9) had already been obtained in 1930 by Hammerstein [8]. Hammerstein's approach was however much more complicated that the above method, being based upon a treatment of the integral equation equivalent to (1) by a Galerkin method, the corresponding finite-dimensional Galerkin equations being solved by a variational argument. Applied to (6), condition (9) obviously gives the sharp condition $\lambda < 1$. We refer to [10] for a complete modern treatment of the above mentioned results.

The next step in weakening (9) was made in [12] (and extended to the partial differential case in [11]) and consisted in replacing (9) by the condition

$$\liminf_{|u|\to\infty} 2u^{-2}F(x,u) \geq b(x) \tag{10}$$

uniformly in $x \in I$ where the function $b \in L^1(I)$ satisfies some conditions generalizing $B > -1$ in the constant case. The following lemma, synthetizing some results of [5] and [6], is the corresponding substitute for the Poincaré inequality.

Lemma 1. *Let $b \in L^1(I)$. Then the following three conditions are equivalent.*

(a) *The quadratic form*

$$Q: H_0^1(I) \to \mathbf{R}, \quad u \mapsto \int_I [(u'(x))^2 + b(x)u^2(x)]dx$$

is positive definite.

(b) *There exists $\delta > 0$ such that, for each $u \in H_0^1(I)$, one has*

$$Q(u) \geq \delta \, ||u||^2_{H_0^1}.$$

(c) *The smallest eigenvalue of the problem*

$$\begin{aligned} -u''(x) + b(x)u(x) &= \lambda u(x) \\ u(0) = u(\pi) &= 0 \end{aligned} \tag{11}$$

is positive.

Proof. We shall use the variational characterization

$$\lambda_1 = \min_{||u||_{L^2}=1} Q(u),$$

for the smallest eigenvalue of (11). If (b) holds, then, by Poincaré's inequality, we have

$$\lambda_1 \geq \delta \min_{\|u\|_{L^2}=1} \int_I (u'(x))^2 dx \geq \delta > 0$$

and (c) is satisfied. If (c) holds, then, for all $u \in H_0^1(I) \setminus \{0\}$, we have

$$Q(u) \geq \lambda_1 \int_I u^2(x) dx > 0$$

and (a) is satisfied. Finally, if (a) holds and not (b), there will exist a sequence (u_n) in $H_0^1(I)$ such that $\|u_n\|_{H_0^1} = 1$ and

$$Q(u_n) \to 0 \quad \text{if} \quad n \to \infty. \tag{12}$$

Then, taking if necessary a subsequence, we can assume that $u_n \to u$ weakly in $H_0^1(I)$ and uniformly on I. Therefore, by the lower semi-continuity of the square of the norm in $H_0^1(I)$, we get from (12) $Q(u) \leq 0$ and hence, by (a), $u = 0$. But then (12) implies that $\|u_n\|_{H_0^1} \to 0$ as $n \to \infty$, a contradiction.

We have now the following generalization of the Lichtenstein- Hammerstein- Cinquini theorem, which is a synthesis of some results of [5] and [6].

Theorem 1. *Assume that there exists $b \in L^1(I)$ with the property that for every $\varepsilon > 0$ there exists $c_\varepsilon \in L^1(I)$ and $a_\varepsilon \in L^1(I)$ such that*

$$2F(x,u) \geq (b(x) - \varepsilon)u^2 + c_\varepsilon(x)|u| + a_\varepsilon(x) \tag{13}$$

for $(x,u) \in I \times \mathbb{R}$. If b satisfies one of the three equivalent conditions of Lemma 1, then (1) has a solution which minimizes φ on $H_0^1(I)$.

Proof. It is easily checked that the functional φ is weakly lower semicontinuous over $H_0^1(I)$. Let $\delta > 0$ given by Lemma 1 and fix $0 < \varepsilon < \delta$. Then, by (13) and Lemma 1,

$$\begin{aligned}\varphi(u) &\geq (1/2) \int_I [(u'(x))^2 + b(x)u^2(x) - \varepsilon u^2(x) + c_\varepsilon(x)|u(x)| + a_\varepsilon(x)]dx \\ &\geq (1/2)(\delta - \varepsilon)\|u\|^2_{H_0^1} - \|c_\varepsilon\|_{L^1}\|u\|_{H_0^1} - \|a_\varepsilon\|_{L^1},\end{aligned}$$

so that φ is coercive and has a minimum.

The following result, proved in [12] (see also [11]), gives an explicit condition upon b which implies that the conditions of Lemma 1 are satisfied.

Lemma 2. *Assume that $b = b_0 + b_1$ where b_0 and b_1 belong to $L^1(I)$, $b_0(x) \geq -1$ for* a.e. *$x \in I$, $b_0(x) > -1$ on a subset of I with positive measure. Then there exists $\delta_1 > 0$ such that if $\|b_1\|_{L^1} < \delta_1$, b satisfies the conditions of Lemma* 1.

Other nonresonance conditions have been recently introduced by de Figueiredo and Gossez [2] in the situation where

$$\liminf_{|u|\to\infty} 2u^{-2}F(x,u) \geq -1 \tag{14}$$

uniformly in $x \in I$. Under a supplementary growth condition

$$|f(x,u)| \leq C|u| + d(x)$$

for some $C \geq 0$ and $d \in L^1(I)$, they assume the existence of some set $J \subset I$ with meas $J > 0$ and some $B > -1$ such that the set

$$\bigcap_{x\in J} \{s \in \mathbb{R} \setminus \{0\} : 2s^{-2}F(x,s) \geq B\}$$

has positive density at $+\infty$ and at $-\infty$. A measurable subset E of $\mathbb{R}$ is said to have positive density at $+\infty$ when

$$\liminf_{r\to+\infty} r^{-1}\text{meas}(E \cap [0,r]) > 0,$$

and similarly at $-\infty$. By a delicate analysis based upon the theory of rearrangements of functions, they are able to show that, under those conditions, the functional φ is still coercive.

3. A Sharp Nonresonance Condition Using Topological Degree

If we apply the results of the previous section to the special case

$$\begin{aligned} u''(x) &= g(u(x)) - h(x), \quad x \in I, \\ u(0) &= u(\pi) = 0, \end{aligned} \tag{15}$$

with $g: \mathbb{R} \to \mathbb{R}$ continuous and $h \in L^1(I)$, the conditions of Section 2 reduce to one of the following assumptions, with $G(u) = \int_0^u g(v)dv$:

$$\liminf_{|u|\to\infty} 2u^{-2}G(u) > -1$$

or

$$\liminf_{|u|\to\infty} 2u^{-2}G(u) = -1, \quad |g(u)| \leq C|u| + D$$

and there exists some $B > -1$ such that set

$$\{s \in \mathbf{R} \setminus \{0\}: 2s^{-2}G(s) \geq B\}$$

has positive density at $+\infty$ and $-\infty$.

We want now to describe some recent results of Fernandes, Omari and Zanolin [3], motivated by earlier ones of Fernandes–Zanolin for periodic solutions [4], which essentially ensure the solvability of (15) under the assumption that $h \in L^\infty(I)$ and that

$$\limsup_{u\to-\infty} 2u^{-2}G(u) > -1 \quad \text{and} \quad \limsup_{u\to+\infty} 2u^{-2}G(u) > -1. \tag{16}$$

One surprising feature of this result is that although its assumptions are made only upon the potential G of g, the only known proof uses topological degree arguments and it remains an open problem to prove it by critical point theory.

As usual with topological degree arguments, we associate to (15) the family of problems

$$(17_s) \qquad \begin{aligned} u''(x) &= s[g(u(x)) - h(x)], \quad x \in I \\ u(0) &= u(\pi) = 0, \end{aligned}$$

where $s \in]0,1[$, and we introduce, following [3], the following properties, with $R \in \mathbf{R}$:

(P_R^-) For each $s \in]0,1[$ and each possible solution u of (17_s), one has

$$\min_I u \neq -R;$$

(P_R^+) For each $s \in]0,1[$ and each possible solution u of (17_s), one has

$$\max_I u \neq R.$$

The first lemma is an easy consequence of degree theory. Its proof, a special case of the theory of curvature bound sets [7], consists in showing that if $\Gamma = \{u \in C_0^0(I): -A < u(x) < B\}$ no possible solution u of (17_s) belongs to $\partial\Gamma$.

Lemma 3. *Assume that there exist positive constants A, B such that P_A^- and P_B^+ are satisfied. Then problem* (16) *has at least one solution u such that $-A < u(x) < B$ for all $x \in I$.*

The second lemma gives sufficient conditions in order that $(P_R^{\pm})$ holds for some $R > 0$.

Lemma 4. *Assume that $h \in L^\infty(I)$ and let $R > 0$. If*

$$g(-R) + \|h\|_{L^\infty} < 0,$$

then (P_R^-) holds. If

$$g(R) - \|h\|_{L^\infty} > 0,$$

then (P_R^+) holds.

The proof follows easily from a maximum principle argument.
The crucial lemma is the following one.

Lemma 5. *Suppose that g is bounded below on $]-\infty, 0]$ (resp. bounded above on $[0, +\infty[$) and that*

$$\limsup_{v \to -\infty} 2v^{-2} G(v) > -1$$

$$(\textit{resp.} \ \limsup_{v \to +\infty} 2v^{-2} G(v) > -1).$$

Then (P_R^-) (resp. (P_R^+)) holds for an unbounded set of R's.

Proof (sketched). We consider P_R^+, the other case being similar. Without loss of generality (up to a modification of h) we can assume that $g(v) < 0$ for all $v \in [0, \infty[$. Let $\rho > 0$ be such that

$$\limsup_{v \to +\infty} 2v^{-2} G(v) > -\rho > -1, \tag{18}$$

and let $H(x) = \int_0^x h(s)ds$. As a consequence of (18), there exists an increasing sequence (w_n) of real numbers such that $w_n \to +\infty$, $w_n > L = 2\pi\|H\|_{L^\infty}$ for all n and

$$(\rho/2)v^2 + G(v) < (\rho/2)w_n^2 + G(w_n)$$

for all $v \in [0, w_n[$. By a careful study of the time map associated to the problem, one shows then that if $n \in \mathbb{N}^*$, $s \in]0,1[$ and u is a solution of (17_s) such that $\max_I u = w_n$, then

$$\tau(n) = 2\int_0^{w_n - L} \frac{dv}{2\|H\|_{L^\infty} + (\rho(w_n^2 - v^2))^{1/2}} \leq \pi.$$

Now,

$$\tau(n) \to \pi/\rho^{1/2} > \pi \quad \text{as } n \to \infty,$$

and hence condition $P^+_{w_n}$ will hold for all sufficiently large n, which completes the proof.

Theorem 2. *Assume that $h \in L^\infty(I)$ and that condition* (16) *holds. Then problem* (15) *has at least one solution.*

Proof. Either there exists $A > 0$ such that $g(-A) + \|h\|_{L^\infty} < 0$, and then P^-_A holds by Lemma 4, or $g(v) \geq -\|h\|_{L^\infty}$ for all $v \leq 0$, in which case g is bounded from below on $]-\infty, 0]$ and, because of the corresponding condition (16), (P^-_R) holds for an unbounded set of R's. In the same way, P^+_B holds for some $B > 0$ and Lemma 3 implies the existence result.

REFERENCES

[1] S. Cinquini, *Sopra i problemi di valori al contorno per equazioni differenziali non lineari,* Boll. Un. Mat. Ital. (1) **17** (1938), 99–105.

[2] D.G. De Figueiredo and J.P. Gossez, *Nonresonance below the first eigenvalue for a semilinear elliptic problem,* Math. Ann. **281** (1988), 589–610.

[3] M.L.C. Fernandes, P. Omari and F. Zanolin, *On the solvability of a semilinear two-point BVP around the first eigenvalue,* J. Differential and Integral Equations, to appear.

[4] M.L.C. Fernandes and F. Zanolin, *Periodic solutions of a second order differential equation with one-sided growth restrictions on the restoring term,* Archiv Math. (Basel) **51** (1988), 151–163.

[5] A. Fonda and J.P. Gossez, *Semicoercive variational problems at resonance,* to appear.

[6] A. Fonda and J. Mawhin, *Quadratic forms, weighted eigenfunctions and boundary value problems for nonlinear second order ordinary differential equations,* Proc. Royal Soc. Edinburgh **112A** (1989), 145–153.

[7] R.E. Gaines and J. Mawhin, *Coincidence Degree and Nonlinear Differential Equations,* Lecture Notes in Math., **568**, Springer, Berlin, 1977.

[8] A. Hammerstein, *Nichtlineare Integralgleichungen nebst Anwendungen,* Acta Math. **54** (1930), 117–176.

[9] L. Lichtenstein, *Uber einige Existenzprobleme der Variationsrechnung. Methode der unendlichvielen Variablen,* J. Reine Angew. Math. **145** (1915), 24–85.

[10] J. Mawhin, *Problèmes de Dirichlet variationnels non linéaires,* Sémin. Math. Sup. **104**, Presses Univ. Montréal, Montréal, 1987.

[11] J. Mawhin, J. Ward and M. Willem, *Variational methods and semilinear equations,* Arch. Rat. Mech. Anal. **95** (1986), 269–277.

[12] J. Mawhin and M. Willem, *Variational methods and boundary value problems for vector second order differential equations and applications to the pendulum equation,* in "Nonlinear Analysis and Optimization," Vinti ed., Lecture Notes in Math., **1107**, Springer, Berlin, 1984, 181–192.

Université de Louvain
Institut Mathématique
Chemin du Cyclotron, 2
B-1348 Louvain-la-Neuve,
Belgique

SOME RELATIVE ISOPERIMETRIC INEQUALITIES

AND APPLICATIONS TO NONLINEAR PROBLEMS

by

Filomena Pacella (*)

(*) Dipartimento di Matematica - Universitá di Roma I "La Sapienza" - Piazzale A. Moro 2 - 00185 Roma , Italy.

INTRODUCTION

The importance of isoperimetric inequalities in many problems in Analysis as well as in Differential Geometry or in Mathematical Phisics is well known. ([3] , [11] , [22] , [25] , [4] , [12]). In particular the classical isoperimetric inequality, which states that among all sets of fixed finite measure the ball has the smallest perimeter (in the sense of De Giorgi [8]), plays a fondamental role in the symmetrization of Dirichlet problems ([3] , [27] , [28]), in studying Sobolev inequalities ([22] , [26]) and, more generally, in deriving geometrical properties for solutions of partial differential equations ([16] , [19] , [25]).

However, if the problems one deals with are not Dirichlet problems, for instance if we study partial differential equations with Neumann or mixed boundary conditions, the classical isoperimetric inequality is, in general, no longer appropriate. Instead, in these situations it would be useful to have some kind of "relative" (in a suitable sense that will be explained in the next sections) isoperimetric inequalities that when applied, for example, to the level sets of the solutions of such problems take account of the fact that these functions may not vanish on the boundary of the set where they are defined.

In this lecture we describe recent results obtained in [10] , [20] , [21] and [23] about some relative isoperimetric inequalities suitable to treat mixed boundary problems and indicate some applications to semilinear elliptic equations involving critical Sobolev exponents.

The subject is divided in three sections. In the first one we present an isoperimetric inequality for convex cones. In section 2 we describe a way of symmetrizing functions which do not have compact support in their domain. Finally in section 3 we apply the results of section 1 and 2 to Sobolev inequalities and semilinear elliptic equations with mixed boundary conditions.

1. ISOPERIMETRIC INEQUALITIES FOR CONVEX CONES

Let Ω be an open set in $\mathbf{R}^n$ and E a measurable set. The De Giorgi perimeter of E relative to Ω is defined as ([8])

$$P_\Omega(E) = \sup \{ \, | \int_E \operatorname{div} \psi \, dx \, | \; , \; \psi \in [C_0^\infty(\Omega)]^n \; , \; |\psi| \leq 1 \}$$

that is $P_\Omega(E)$ is the total variation on Ω of the characteristic function of the set E. In other words $P_\Omega(E)$ measures the part of the boundary of E which lies in Ω. If $\Omega = \mathbf{R}^n$, $P_{\mathbf{R}^n}(E)$ is usually denoted by P(E) and gives a way of measuring the whole boundary of E. Several properties of $P_\Omega(E)$ are listed in [1], [8] and [22].

Denoting by $|E|$ the Lebesgue measure of E and assuming $|E| < +\infty$, by the classical isoperimetric inequality we have

$$(1.1.) \qquad P(E) \geqslant n C_n^{1/n} |E|^{\frac{n-1}{n}}$$

where C_n is the measure of the unit ball in $\mathbf{R}^n$. As it is well known the equality in (1.1.) holds iff E is a ball in $\mathbf{R}^n$.

We are now interested in obtaining some kind of "relative isoperimetric inequalities", that is to prove that the perimeter of E relative to some fixed open set Ω also satisfies some isoperimetric inequality of type (1.1.). In case this is possible we also would like to find the optimal sets. As it can be easily understood the problem of getting such relative isoperimetric inequalities is not trivially solvable and its solution depends strongly on the geometrical properties of the set Ω.

A class of open sets Ω for which it is possible to derive some relative isoperimetric inequalities is that of the convex cones.

Let Σ_α be an open cone in $\mathbf{R}^n$, $n \geq 2$, with vertex at the origin and solid angle $\alpha \in \,]0, \omega_{n-1}[$, where ω_{n-1} is the (n-1)-dimensional Hausdorff measure of the unit sphere S^{n-1}. To be more precise, if A_α is a subset of S^{n-1} with $H_{n-1}(A_\alpha) = \alpha$ then $\Sigma_\alpha = \{ \lambda x , x \in A_\alpha , \lambda \in]0,+\infty[\}$.

We denote by $\Sigma(\alpha,R)$ the open sector in $\mathbf{R}^n$ with solid angle α and radius $R>0$, that is $\Sigma(\alpha,R) = \Sigma_\alpha \cap B_R$ where B_R is the ball in $\mathbf{R}^n$ with center at the origin and radius $R>0$. By the symbol α_n we mean the measure of any unitary sector $\Sigma(\alpha,1)$ with solid angle α.

THEOREM 1.1. Let Σ_α be a convex cone in $\mathbf{R}^n$, $n \geq 2$. Then

$$P_{\Sigma_\alpha}(E) \geq n\alpha_n^{1/n}|E|^{\frac{n-1}{n}} \tag{1.2.}$$

for any measurable set $E \subset \Sigma_\alpha$ with $|E| < +\infty$. Moreover if $\partial\Sigma_\alpha \setminus \{P\}$ ($P =$ vertex of Σ_α) is smooth, equality in (1.1.) holds if and only if E is a convex sector $\Sigma(\alpha,R)$ homothetic to Σ_α.

The isoperimetric inequality (1.2.) is proved in [20], by adapting a proof of the classical isoperimetric inequality based on Brunn-Minkowski's inequality ([11], [15]). If $n=2$, (1.2.) is an easy consequence of the classical isoperimetric inequality in the plane (see [3]). Another proof of (1.2.), for a particular type of convex cone is given in [21].

As remarked in [20] the relative isoperimetric inequality (1.2.) does not hold without any assumption on the cone Σ_α. Easy counterexamples of nonconvex cones for which (1.2.) is not satisfied are constructed in [20]. Anyway the problem of characterizing completely the cones sharing the isoperimetric property (1.2.) is still open.

2. α - SYMMETRIZATION

Let Ω be a bounded connected open set in $\mathbf{R}^n$, whose boundary is lipschitz continuous and is made of two manifolds Γ_o and Γ_1. According to [23] to any such Ω it is possible to associate an isoperimetric constant by defining

$$Q(\Gamma_1,\Omega) = \sup_E \frac{|E|^{\frac{n-1}{n}}}{P_\Omega(E)} \tag{2.1.}$$

where the supremum is taken over all measurable subsets E of Ω such that $\partial E \cap \Gamma_o$ does not contain any set of positive $(n-1)$-dimensional Hausdorff measure. Since $Q(\Gamma_1,\Omega)$ depends mainly on Γ_1, it is called "the isoperimetric constant of Ω relative to Γ_1".

Several examples about the way of computing or estimating $Q(\Gamma_1,\Omega)$ are shown in [23]. We recall here some of the main properties

PROPOSITION 2.1. - We have

i) $H_{n-1}(\Gamma_o) = H_{n-1}(\partial\Omega) \Longrightarrow Q(\Gamma_1,\Omega) = (nC_n^{1/n})^{-1}$

ii) $H_{n-1}(\Gamma_1) = H_{n-1}(\partial\Omega) \Longrightarrow Q(\Gamma_1,\Omega) = +\infty$

iii) $H_{n-1}(\Gamma_1) > 0 \Longrightarrow Q(\Gamma_1,\Omega) \geq (n\,(C_n/2)^{1/n})^{-1}$

iv) if $\Omega = \Sigma(\alpha,R)$ is a convex sector and $\tilde{\Gamma}_o = \{ x \in \partial\Sigma(\alpha,R) ,\ |x| = R \}$, $\tilde{\Gamma}_1 = \partial\Sigma(\alpha,R) \setminus \Gamma_o$ then the supremum in (2.1.) is a maximum and $Q(\tilde{\Gamma}_1,\Sigma(\alpha,R)) = (n\alpha_n^{1/n})^{-1}$.

<u>Proof.</u> If $H_{n-1}(\Gamma_0) = H_{n-1}(\partial\Omega)$ then the sets E considered in the definition (2.1.) are entirely contained in Ω so that $P_\Omega(E) = P(E)$. Thus i) follows from the classical isoperimetric inequality (1.1.). Property ii) is obvious since in (2.1.) we can take sets $E \subset \Omega$ with measure near $|\Omega|$ and perimeter relative to Ω arbitrarily small. Regarding to iii) we just have to consider the sets $E = B_\rho(x_0) \cap \Omega$ where $B_\rho(x_0)$ is a ball centered in a point $x_0 \in \Gamma_1$ and radius $\rho>0$ and observe that E approximates $\Sigma(\omega_{n-1}/2,\rho)$ as $\rho \to 0$. Finally iv) is a consequence of Theorem 1.1.
//

REMARK 2.1. - The problem of determining $Q(\Gamma_1,\Omega)$ for a fixed bounded open set Ω with $\Omega = \Gamma_0 \cup \Gamma_1$ has some similarity with the so called partitioning problem in Geometric Measure Theory (see [12]). There the problem is to minimize $P_\Omega(F)$ among all sets $F \subset \Omega$ with $|F| = \sigma|\Omega|$, $\sigma \in]0,1[$; here instead we restrict the class of the competing sets E to those which do not intersect Γ_0 (up to a set of zero (n-1)-dimensional Hausdorff measure). Therefore our problem is a kind of "partitioning problem with obstacle" where the obstacle is represented by Γ_0. //

If $H_{n-1}(\Gamma_1) > 0$ and $Q(\Gamma_1,\Omega) < +\infty$, by iii) $Q(\Gamma_1,\Omega) \in [(n(\omega_n/2)^{1/n})^{-1}, +\infty[$ Therefore there exists a number $\alpha_n \in]0, \omega_n/2]$, such that $Q(\Gamma_1,\Omega) = (n\alpha_n^{1/n})^{-1}$ and, obviously, α_n can be thought of as the measure of a unitary sector $\Sigma(\alpha,1)$. Consequently it makes sense to define the class E_{α_n} of all open sets Ω of the type considered above such that $Q(\Gamma_1,\Omega) = (n\alpha_n^{1/n})^{-1}$. From iv) of Proposition 2.1. we immediately have that any convex sector $\Sigma(\alpha,R)$ such that $R^{-n}|\Sigma(\alpha,R)| = \alpha_n$, belongs to E_{α_n}.

As we will see below the classes E_{α_n} are suitable to study the symmetrization of measurable functions defined in Ω and vanishing on Γ_0.

Let us define

$$V^p(\Omega) = \{ u \in H^{1,p}(\Omega) \ , \ u \equiv 0 \text{ on } \Gamma_o \} \quad , \ p>1 \, .$$

For $\Omega \in E_{\alpha_n}$ we denote by $C_\alpha(\Omega)$ a convex sector in E_{α_n} such that $|\Omega| = |C_\alpha(\Omega)|$. Then to any $u \in V^p(\Omega)$ we associate the unique radial decreasing function $C_\alpha(u)$ defined in $C_\alpha(\Omega)$, having the same distribution function as u, that is

$$|\{ x \in \Omega : |u(x)| > t \}| = |\{ x \in C_\alpha(\Omega) : |C_\alpha(u)(x)| > t \}| \qquad \forall t>0 \, .$$

Another way of defining $C_\alpha(u)$ is to set $C_\alpha(u)(x) = u^*(\alpha_n |x|^n)$, where u^* is the decreasing rearrangement of u : $u^*(s) = \sup \{ t \text{ such that } \mu(t) > s \}$, $\mu(t) = |\{ x \in \Omega : |u(x)| > t \}|$.

The transformation $u \to C_\alpha(u)$ is called α-symmetrization according to [3] where it was defined in connection with another class of open sets, in the case n=2. Of course, the very definition implies

$$\text{(2.2.)} \qquad \int_\Omega |u(x)|^s dx = \int_{C_\alpha(\Omega)} |C_\alpha(u)(x)|^s dx \qquad \forall s>0$$

Moreover, if $u \in V^p(\Omega)$ and $\Omega \in E_{\alpha_n}$ then

$$\text{(2.3.)} \qquad \int_\Omega |Du|^p dx \geq \int_{C_\alpha(\Omega)} |DC_\alpha(u)|^p dx \qquad \forall p>1$$

Let us notice explicitely that property (2.3.) holds because both Ω and $C_\alpha(\Omega)$ belong to E_{α_n}. In fact, if $\Omega \in E_{\alpha_n}$ (i.e. $Q(\Gamma_1,\Omega) = (n\alpha_n^{1/n})^{-1}$) by (2.1.) we have the isoperimetric inequality

(2.4.) $$P_\Omega(E) \geq n\alpha_n^{1/n}|E|^{\frac{n-1}{n}}$$

for any set admissible in the definition (2.1.).

Then, if we apply (2.4.) to the level sets of a function $u \in V^p(\Omega)$ and use the co-area formula ([11]) , after some computations we obtain (2.3.) as in [26] , Lemma 1.

Morever, if Ω is a convex sector $\Sigma(\alpha,R)$, (2.4.) becomes an equality for any convex sector $E = \Sigma(\alpha,r) \subset \Sigma(\alpha,R)$ $(r<R)$ homothetic to $\Sigma(\alpha,R)$. Therefore $\Sigma(\alpha,R)$ is symmetrized in itself and $C_\alpha(v) = v$ for any radial decreasing function $v \in V^p(\Sigma(\alpha,R))$. Hence, in the α-symmetrization the convex sectors play the same role as the ball in the Schwartz symmetrization.

Thus one could think of using the α-symmetrization in the study of partial differential equation with mixed boundary conditions, in the same way as the Schwartz symmetrization has been used in the study of Dirichlet problems ([3] , [16] , [25] , [27] , [28]). A model problem for the case of uniformly elliptic operators was considered in [23] where actually the constant $Q(\Gamma_1,\Omega)$ and the class E_{α_n} were introduced with the aim of extending some 2-dimensional results of C.Bandle about the symmetrization of mixed boundary problems to higher dimensions and weaker regularity assumption.

3. APPLICATIONS

Let Ω be a bounded domain in $\mathbf{R}^n$, n>2 , of the same type considered in the previous section.

Here we examine the existence of positive solutions of a semilinear elliptic equation with mixed boundary conditions in the following form

$$(3.1.)\qquad \begin{cases} -\Delta u = u^{\frac{n+2}{n-2}} & \text{in } \Omega \\ u = 0 & \text{on } \Gamma_o \\ \frac{\partial u}{\partial \nu} = 0 & \text{on } \Gamma_1 \end{cases}$$

where ν is the outer normal to $\partial\Omega$ and $\frac{n+2}{n-2}+1$ is the critical Sobolev exponent for the embedding $V^p(\Omega) \hookrightarrow L^s(\Omega)$.

By using the well-known Pohozaev identity ([24]) we immediately derive some negative results for (3.1.) . In fact for a solution u of (3.1.) Pohozaev's identity becomes

$$\frac{n-2}{2n} \int_{\Gamma_1} u^{2n/(n-2)} (x\cdot\nu)\, d\sigma = \frac{1}{2} \int_{\Gamma_1} \left(\frac{\partial u}{\partial \nu}\right)^2 (x\cdot\nu)\, d\sigma - \frac{1}{2} \int_{\Gamma_o} \left(\frac{\partial u}{\partial \nu}\right)^2 (x\cdot\nu)\, d\sigma$$

which implies the nonexistence of positive solutions to (3.1.) whenever $x\cdot\nu > 0$ a.e. on Γ_o and $x\cdot\nu = 0$ a.e. on Γ_1 . This happens, for instance if Ω is a sector $\Sigma(\alpha,R)$ and $\Gamma_o = \{ x \in \partial\Sigma(\alpha,R) ,\ |x| = R \}$, $\Gamma_1 = \partial\Sigma(\alpha,R) \setminus \Gamma_o$.

Our approach to problem (3.1.) is via minimization of the functional

$$J(u) = \frac{\int_\Omega |Du|^2 dx}{\left(\int_\Omega |u|^t dx\right)^{2/t}} \quad , \quad t = \frac{2n}{n-2}$$

Of course this procedure fails if one deals with the Dirichlet problem (case $\Gamma_0 \equiv \partial\Omega$) unless there are some lower order perturbation as shown by Brezis and Nirenberg in [7].

As we will see later the situation is indeed quite different if we deal with the mixed boundary problem. To this purpose we investigate, in general, the best Sobolev constant $S^p(\Omega)$, for the embedding $V^p(\Omega) \hookrightarrow L^q(\Omega)$, $q = \frac{np}{n-p}$, $1<p<n$. That is

$$S^p(\Omega) = \inf_{\substack{u \in V^p(\Omega) \\ u \neq 0}} J^p(u) \quad , \quad J^p(u) = \frac{\int_\Omega |Du|^p dx}{\left(\int_\Omega |u|^q dx\right)^{p/q}}$$

in the case $p=2$ we set $S^2(\Omega) = S(\Omega)$ and $J^p = J$.

In particular we would like to answer the following questions

i) how does $S^p(\Omega)$ depend on Ω ?

ii) is $S^p(\Omega)$ achieved ?

Of course the second question is particularly relevant when $p=2$ in connection with problem (3.1.). In the case of the space $H^{1,p}(\Omega)$ we already know the answers to the above questions, that is the best Sobolev constant does not depend on Ω and it is never achieved. In general, for the space $V^p(\Omega)$ not only does $S^p(\Omega)$ depend on Ω but it is not even possible to find a positive constant which bounds from below all the numbers $S^p(\Omega)$. This is easily seen taking Ω as a sector $\Sigma(\alpha,R)$ and letting the amplitude α go to zero ([21]). Obviously this depends on the fact that the functions we are considering do not vanish on the whole boundary Ω.

By using the isoperimetric constant $Q(\Gamma_1,\Omega)$, defined in section 2 we can get some estimates for $S^p(\Omega)$

THEOREM 3.1. - If $\Omega \in E_{\alpha_n}$ then

(3.2.) $\quad S^p(\Omega) \geq S^p(\alpha_n)$

where $\quad S^p(\alpha_n) = \left(\dfrac{B^{1/q}}{n \alpha_n^{1/n}} \right)^{-1}$ and B is a constant which depends only on p and q.

Moreover the estimate (3.2.) is "the best possible" as long as Ω varies in E_{α_n} since $S^p(\alpha_n) = S^p(\Sigma(\alpha,R))$ for any convex sector $\Sigma(\alpha,R)$ in E_{α_n}.

Proof. See [21] Theorem 2.1. //

From (3.1.) we get at once

$$\|Du\|_p \geq S^p(\alpha_n)\, \|u\|_q \qquad \text{for all } u \in V^p(\Omega).$$

Actually, by using again the α-symmetrization and the same procedure as in [7] it is possible to prove (see [10]) that

(3.3.) $$\|Du\|_p \geq S^p(\alpha_n)\, \|u\|_q^p + \lambda\, \|u\|_s^p \qquad \text{for all } u \in V^p(\Omega)$$

for a positive constant $\lambda = \lambda(\alpha_n, s)$ depending on the class E_{α_n} and s, with $s \in \left]0, \frac{(p-1)n}{n-p}\right[$.

From (3.3.) we immediately obtain that $S^p(\Omega)$ is not achieved whenever $S^p(\Omega) = S^p(\alpha_n)$, $\Omega \in E_{\alpha_n}$. In particular by Theorem 3.1. $S^p(\Sigma(\alpha,R))$ is not achieved for any convex sector $\Sigma(\alpha,R)$. However these are not the only cases when it is possible to check the condition $S^p(\Omega) = S^p(\alpha_n)$. For instance in [10] the following geometrical condition is derived

PROPOSITION 3.1.- If Ω belongs to the class $E_{C_n/2}$, then $S^p(\Omega) = S^p(C_n/2)$ and therefore $S^p(\Omega)$ is not achieved.

This is the case for example of the open set contained between two concentric spheres with Γ_1 denoting the boundary of the interior ball (see [23] , Example 3.3.).

Now in order to understand when $S^p(\Omega)$ is achieved let us examine the behaviour of the minimizing sequences. As a consequence of the concentration-compactness principle of P.L.Lions we have ([21] , [17] , [18])

PROPOSITION 3.2. - Let u_n be a minimizing sequence for $S^p(\Omega)$, then either u_n is relatively compact or there exists a subsequence u_{n_k} such that $u_{n_k} \to 0$ in $V^p(\Omega)$ and $|u_{n_k}|^q \rightharpoonup \delta_{x_0}$, $|\nabla u_{n_k}|^p \rightharpoonup S^p(\Omega)\, \delta_{x_0}$ in the sense of measures for some point $x_0 \in \overline{\Gamma}_1$.

Hence if there is a lack of compactness for the minimizing sequence a concentration phenomenon occurs near Γ_1. This implies that the value of $S^p(\Omega)$ is the same in any neighborhood of x_0 , that is $S^p(\Omega) = S^p(\Omega \cap B_\varepsilon)$ where B_ε is a ball centered in x_0 with radius $\varepsilon > 0$ sufficiently small and $V^p(\Omega \cap B_\varepsilon) = \{u \in H^{1,p}(\Omega \cap B_\varepsilon)$, $u \equiv 0$ on $\partial B_\varepsilon \cap \Omega \}$. In particular, if Γ_1 is regular and $S^p(\Omega)$ is not achieved we have that $S^p(\Omega) = S^p(\mathbf{R}^n_+) = S^p(C_n/2)$, according to our notation , that is $S^p(\Omega)$ is the best Sobolev constant for the embedding $V^p(\Sigma(\omega_{n-1}/2,R)) \hookrightarrow L^q(\Sigma(\omega_{n-1}/2,R))$. Therefore if we know that $S^p(\Omega) < S^p(C_n/2)$ that is if it is possible to construct functions u in $V^p(\Omega)$ which make the infimum lower than the value one would get with functions which concentrate near points belonging to $\overline{\Gamma}_1$, then $S^p(\Omega)$ is achieved. Of course this depends on the geometry of Ω and the major question is to understand what geometrical characteristics Ω should have in order to be able to construct such functions.

As the next propositions indicate it is conceivable that the "isoperimetric" properties of Ω (relatively to Γ_1 in the sense explained in section 2) play an important role in the existence of a minimizing function for $S^p(\Omega)$ and consequently to obtain a positive solution of problem (3.1.), in the case $p = 2$.

PROPOSITION 3.3. - Let $Q(\Gamma_1,\Omega) > [n(C_n/2)^{1/n}]^{-1}$ and suppose that Γ_1 is smooth. Then there exists a number $1 < \bar{p}(\Omega) < n$ such that $S^p(\Omega)$ is achieved foe every $1 < p < \bar{p}(\Omega)$.

Proof. The result immediately follows from the continuity of $S^p(\Omega)$ with respect to $p \in [1, n[$ and the fact that $[Q(\Gamma_1,\Omega)]^{-1}$ is actually the best constant for the embedding $V^1(\Omega) \hookrightarrow L^{\frac{n}{n-1}}(\Omega)$, $V^1(\Omega)$ being defined as the space $V^1(\Omega) = \{ u \in BV, u \equiv 0 \text{ on } \Gamma_0\}$ (see also [21]). //

PROPOSITION 3.4. - Let us suppose that $\partial\Omega$ is smooth. Then there exists $\varepsilon > 0$ such that $S^p(\Omega)$ is achieved whenever $\operatorname{diam}(\Gamma_0) < \varepsilon$.

Proof. Suppose that the diameter of Γ_0 is smaller than a positive number ε. Thus Γ_0 is contained in a ball B_ε with radius ε. Then the p-capacity of B_ε with respect to a ball B_R with fixed radius $R > \varepsilon$, cap (B_ε, B_R), tends to zero as $\varepsilon \to 0$ (see [22], section 2.2.). Therefore, if ε is sufficiently small, by the definition of capacity it is possible to construct a function $u \in H_0^{1,p}(B_R)$ which is identically 1 in a neighborhood of B_ε and such that $\int_{B_R} |Du|^p dx$ is very small. By considering the function $v = 1 - u$ and extending it to 1 in $\Omega \setminus B_R$ we can define a new function $\phi \in V^p(\Omega)$ which coincides with v in $B_R \cap \Omega$. Consequently $\int_\Omega |D\phi|^p dx$ is very small, which implies that $S^p(\Omega) < S^p(C_n/2)$ and hence the infimum is achieved.

//

The last proposition allows to construct many domains Ω for which $S^2(\Omega)$ is achieved. It also shows that if the mixed boundary problem (3.1.) is "close" to a homogeneous Neuman problem then a positive solution of (3.1.) exists with energy, obviously, very small.

Let us observe that for the domains Ω satisfying the hypotheses of the previous proposition the isoperimetric constant $Q(\Gamma_1, \Omega)$ is very large. Therefore, even if not mentioned explicitely, the sufficient condition given in Proposition 3.4. is again a condition on the isoperimetric properties of Ω with respect to Γ_1.

Finally we would like to point out that the two previous propositions have been stated for smooth domains Ω only for semplicity. Similar results apply in the case of open sets with lipschitz continuous boundary after suitable modifications.

We close by remarking that another result which enphasizes the deep connection between the "relative" isoperimetric properties of Ω and the solutions of mixed boundary problems has been obtained in [5] . The result obtained there is of different nature since it does not deal with an existence problem but gives a description of the geometrical structure of positive solutions of semilinear elliptic equations in the general form $\Delta u + f(u) = 0$ with mixed boundary conditions of (3.1.) type. In particular it is shown that all positive solutions of such equations in some convex sectors are spherically symmetric. This symmetry result is strictly related to the relative isoperimetric inequality (1.2.) in the same way as the well-known symmetry theorem of Gidas, Ni and Nirenberg ([13], see also [19]) is connected with the classical isoperimetric inequality (1.1.); we refer the reader to [5] for a precise statement of the previous result and for further comments on the subject.

REFERENCES

[1] G.Anzellotti, M.Giaquinta, U.Massari, G.Modica and L.Pepe, Note sul problema di Plateau, Pisa, Editrice Tecnico Scientifica (1974).

[2] T.Aubin, Equations différentielles non linéaires et probléme de Yamabe concernant la courbure scalaire, J. Math. Pures Appl. 5 (1976) ,269-293.

[3] C.Bandle, Isoperimetric inequalities and applications, Pitman , London (1980).

[4] P.H.Bérard, From vanishing theorems to estimating theorems : the Bochner technique revisited, Bull. Americ. Math. Soc. (to appear).

[5] H.Berestycki and F.Pacella , Symmetry properties for positive solutions of elliptic equations with mixed boundary conditions, Journ. Funct. Anal. (to appear).

[6] H.Brezis, Some variational problems with lack of compactness,Proceedings of Symposia in Pure Mathematics 45 (1986) , 165-201.

[7] H.Brezis and L.Nirenberg, Positive solutions of nonlinear elliptic equations involving critical Sobolev exponents, Comm. Pure Appl. Math. XXXVI (1983) , 437-477.

[8] E.De Giorgi, Su una teoria generale della misura (r-1)-dimensionale in uno spazio a r dimensioni, Ann. Mat. Pura Appl. , (4) 36 (1954) , 191-213.

[9] E.De Giorgi, Sulla proprietá isoperimetrica della ipersfera nella classe degli insiemi aventi frontiera orientata di misura finita, Mem. Accad. Naz. Lincei (8) 5 (1958) , 33-44.

[10] H.Egnell,F.Pacella and M.Tricarico, Some remarks on Sobolev inequalities, Nonlinear Analysis T.M.A. (to appear).

[11] H.Federer, Geometric measure theory, Springer-Verlag, New York (1969).

[12] S.Gallot, Some links between isoperimetry, spectrum and topology, (in preparation).

[13] B.Gidas,W.M.Ni and L.Nirenberg, Symmetry and related properties via the maximum principle, Comm. Math. Phys. 68 (1979) 209-243.

[14] M.Grüter, Boundary regularity for solutions of a partitioning problem, Arch. Rat. Mech. Anal., (3) 97 (1987), 261-270.

[15] H.Hadwiger, Vorlesungen über inhalt, oberflache und isoperimetric, Springer-Verlag, Berlin (1957).

[16] B.Kawohl, Rearrangements and convexity of level sets in P.D.E., Lectures Notes in Mathematics, 1150, Springer-Verlag, Berlin (1985).

[17] P.L.Lions, The concentration-compactness principle in the calculus of variations. The locally compact case (part 1 and 2), Ann. I.H.P. Anal. Nonlin., (1984), 109-145 and 223-283.

[18] P.L.Lions, The concentration-compactness principle in the calculus of variations. The limit case (part 1 and 2), Riv. Mat. Iberoamericana 1 (1985), 145-201 and 45-121.

[19] P.L.Lions,Two geometrical properties of solutions of semilinear problems, Appl. Anal. 12 (1981), 267-272.

[20] P.L.Lions and F.Pacella, Isoperimetric inequalities for convex cones, (to appear).

[21] P.L.Lions, F.Pacella and M.Tricarico, Best constants in Sobolev inequalities for functions vanishing on some part of the boundary and related questions, Indiana Univ. Math. Journ. (2) 37 (1988).

[22] V.Maz'ja, Sobolev spaces, Springer-Verlag, Berlin (1985).

[23] F.Pacella and M.Tricarico, Symmetrization for a class of elliptic equations with mixed boundary conditions, Atti Sem. Mat. Fis. Univ. Modena, XXXIV (1985-1986), 75-94.

[24] S.Pohozaev, Eigenfunction of the equations $\Delta u + f(u) = 0$, Soviet Math. Doklady 6, (1965), 1408-1411.

[25] G.Polya and G.Szegö, Isoperimetric inequalities in mathematical physics, Ann. Math. Studies, Princeton (1951).

[26] G.Talenti, Best constant in Sobolev inequalities, Ann. Mat. Pura Appl. 110 (1976), 353-372.

[27] G.Talenti, Elliptic equations and rearrangements, Ann. Scuola Norm. Sup. Pisa 3 (1976), 697-718.

[28] H.F.Weinberger, Symmetrization in uniformly elliptic problems, Studies in Mathematical Analysis, Stanford Univ. Press (1962).

PART II

Partial Differential Equations and Problems in Geometry

Approximation in Sobolev Spaces between two manifolds and homotopy groups

FABRICE BETHUEL

Abstract

We consider two compact manifolds M^n and N^k and the Sobolev spaces $W^{1,p}(M^n, N^k)$, for $1 < p < n = \dim M^n$. We give a necessary and sufficient condition for smooth maps between M^n and N^k to be dense in $W^{1,p}(M^n, N^k)$. This condition can be simply stated in terms of homotopy groups, and is $\pi_{[p]}(N^k) = 0$. In cases where such a condition does not hold, we show that we can approximate maps in $W^{1,p}(M^n, N^k)$ by maps smooth except on a singular set which has a simple shape. We consider also the problem of the weak density of smooth maps.

0. Introduction

The problem of density of smooth maps between two compact manifolds M^n and N^k was first considered by Eells and Lemaire ([EL]). Schoen and Uhlenbeck [SU1], [SU2] have proved that smooth maps are dense in $W^{1,p}(M^n, N^k)$ in the case $p \geq n = \dim M$. They also gave an example of non density of smooth maps; they showed that $C^\infty(B^3, S^2)$ is not dense in $H^1(B^3, S^2)$: for instance the radial projection π from B^3 to S^2 defined by $\pi(x) = \frac{x}{|x|}$ cannot be approximated by smooth maps. White [W1] gave interesting results concerning the closure of $C^\infty(M^n, N^k)$ in $W^{1,p}(M^n, N^k)$. In [BZ], Zheng and the author showed that smooth maps are dense in $W^{1,p}(M^n, S^k)$, when $p < k$. They also generalized the counterexample of Schoen and Uhlenbeck, and proved that smooth maps are not dense in $W^{1,p}(M^n, N^k)$ when $p < n$ and $\pi_{[p]}(N^k)$ is not trivial. Escobedo [E] extended some of the previous results to the Sobolev spaces $W^{r,p}(M^n, N^k)$, with $r \geq 1$.

In cases where we do not have density of smooth maps in $W^{1,p}(M^n, N^k)$, it seems nevertheless interesting to be able to approximate maps in $W^{1,p}(M^n, N^k)$ by maps smooth except on a singular set of low dimension which has a simple shape. Such results have been obtained in [BZ]: for instance, maps of $H^1(B^3, S^2)$, smooth except on a finite number of point singularities are dense in $H^1(B^3, S^2)$. In another direction in [Bel] we were able to characterize the maps of $H^1(B^3, S^2)$ which can be approximated by smooth maps, and to prove that $C^\infty(B^3, S^2)$ is sequentially dense in $H^1(B^3, S^2)$ for the weak topology.

We consider in this paper two compact riemannian manifolds M^n and N^k of dimension n and k respectively. N^k is isometrically embedded in $\mathbf{R}^\ell (\ell \in \mathbf{N}^*)$, M^n may have a boundary, but not N^k. For $1 < p < n$ we consider the Sobolev space $W^{1,p}(M^n, N^k)$ defined by :

$$W^{1,p}(M^n, N^k) = \{u \in W^{1,p}(M^n, \mathbf{R}^\ell); u(x) \in \mathbf{N}^k \ a.e.\}.$$

and $[p]$ represents the largest integer less than or equal to p.

The following theorem is our main result, and gives a necessary and sufficient condition for smooth maps to be dense in $W^{1,p}(M^n, N^k)$.

Theorem 1. *Let $1 < p < n$. Smooth maps between M^n and N^k are dense in $W^{1,p}(M^n, N^k)$ if and only if $\pi_{[p]}(N^k) = 0$.*

The fact that this condition is necessary is proved in [BZ], (Theorem 2). This paper is devoted to the proof of the fact that this condition is sufficient. Theorem 1 settles the problem of density of smooth maps in $W^{1,p}(M^n, N^k)$ for every p, since for $p \geq n$ we have the result of [SU2].

Modifying the proof of Theorem 1 we have the following:

Theorem 1 bis. *Let $1 < p \leq n$, and assume $\pi_{[p]}(N^k) = 0$ and $\partial M^n \neq \emptyset$. Let u be in $W^{1,p}(M^n, N^k)$, such that u restricted to ∂M is in $W^{1,p}(\partial M, N^k) \cup C^0$ (resp. $C^\infty(\partial M, N^k)$). If there is a map in $C^0(M^n, N^k)$ such that $u = v$ on ∂M^n, then u can be approximated in $W^{1,p}(M^n, N^k)$ by maps in $W^{1,p}(M^n, N^k) \cap C^0$ (resp. $C^\infty(M^n, N^k)$) which coincide with u on ∂M^n.*

When $\pi_{[p]}(N^k) \neq 0$, by Theorem 1, smooth maps are not dense in $W^{1,p}(M^n, N^k)$. In this case, we are nevertheless able to approximate maps in $W^{1,p}(M^n, N^k)$ by maps regular except on a singular set of low dimension. More precisely, we consider the class R_p^0 (resp. R_p^∞) of maps in $W^{1,p}(M^n, N^k)$ defined in the following way: $u \in W^{1,p}(M^n, N^k)$ is in R_p^0 (resp. R_p^∞) if and only if u is continuous (resp. smooth) except on a singular set $\sum(u)$, where $\sum(u) = \bigcup_{i=1}^r \sum_i$ $r \in \mathbf{N}$, where for $i = 1$ to r, $\sum_i$ is a subset of a submanifold of M^n of dimension $n - [p] - 1$, and the boundary of $\sum_i$ is a union of subsets of submanifolds of M^n of dimension $n - [p] - 2$. If $p \geq n - 1$, $\sum_i$ is a point.

Then we have the following theorem.

Theorem 2. *For every $1 < p < n$, R_p^0 (resp. R_p^∞) is dense in $W^{1,p}(B^n, N^k)$.*

We have also the following (which is similar to Theorem 1 bis).

Theorem 2 bis. *Assume $1 < p < n$, and $\partial M^n \neq \emptyset$. Let u be in $W^{1,p}(M^n, N^k)$ such that u restricted to ∂M^n is in $W^{1,p}(\partial M^n, N^k) \cap C^\infty$ (resp. $C^\infty(\partial M^n, N^k)$).*

If there is a map v in $C^\infty(M^n, N^k)$ such that $u = v$ on ∂M^n, then u can be approximated in $W^{1,p}(M^n, N^k)$ by maps in R_p^0 (resp. R_p^∞), which coincide with u on ∂M^n.

When $\pi_{[p]}(N^k) \neq 0$, we also consider the problem of the density of smooth maps for the weak topology in $W^{1,p}(M^n, N^k)$, induced by the weak topology in $W^{1,p}(M^n, \mathbf{R}^\ell)$. We have the following theorem:

Theorem 3. *If $\pi_{[p]}(N^k) \neq 0$, and if p is not an integer, then smooth maps are not sequentially dense for the weak topology in $W^{1,p}(M^n, N^k)$. Moreover, a map which is a weak limit of a sequence of smooth maps is in the strong closure of smooth maps.*

This theorem is useful when one tries to minimize a functional in a certain class of maps in $W^{1,p}(M^n, N^k)$, for instance $C^\infty(M^n, N^k)$.

When $\pi_{[p]}(N^k) \neq 0$, and p is an integer, the problem of sequential density of smooth maps for the weak topology is more difficult and we are not able to answer that question. Nevertheless we have the following theorem.

Theorem 4. *If p is an integer and $\pi_{[p]}(N^k) \neq 0$, then smooth maps between M^n and N^k are dense in $W^{1,p}(M^n, N^k)$ for the weak topology.*

Note that here we are not able to prove that smooth maps are sequentially dense for the weak topology. Adopting the method of [Be] we are able to prove this result in the special case $N^k = S^k$. We have:

Theorem 5. *If p is an integer, $p \leq n - 1$, smooth maps between M^n and S^p are sequentially dense in $W^{1,p}(M^n, S^p)$ for the weak topology.*

The paper is divided as follows. Here, we shall consider only the case $M^n = C^n = [0,1]^n$, and $n - 1 < p < n$. We first prove that we are able to approximate each map in $W^{1,p}(C^n, N^k)$, by maps in R_p^∞, that is, maps in $W^{1,p}(C^n, N^k)$ smooth except at most at a finite number of point singularities. This in fact is a general result (see Theorem 2 for $n-1 < p$) and holds even if $\pi_{[p]}(N^k) \neq 0$. Then we conclude using the fact that $\pi_{n-1}(N^k) = 0$ and the following lemma, which is proved in [BZ]:

Lemma 1. *Let $1 < p < n$ and assume $\pi_{n-1}(N^k) = 0$. Let u be a map in $W^{1,p}(M^n, N^k)$ such that u is continuous except at a finite number of point singularities. Then u can be approximated in $W^{1,p}(M^n, N^k)$ by smooth maps between M^n and N^k.*

We recall some usual notations: For $n > 1$, B^n is the open unit ball in $\mathbf{R}^n$ and S^n is the boundary of B^{n+1}. $B^n(y, \delta)$ is the open ball of radius

δ, centered at y. We set $C^n = [0,1]^n$, $C'^n = [-\frac{1}{2}, \frac{1}{2}]^n$, and $C'^n(\mu) = [-\frac{1}{2\mu}, \frac{1}{2\mu}]^n$ for $\mu \in \mathbf{R}^+$. For $q \in \mathbf{N}^*$, $\pi_q(\mathbf{N}^k)$ is the q-th homotopy group of N^k. For $x = (x_1, x_2, \dots, x_i, \dots, x_n) \in \mathbf{R}^n$ we set $\|x\| = \text{Max}\{|x_i|, 1 \leq i \leq n\}$. For $\delta > 0$ sufficiently small, and for $y_0 \in N^k$, we set $\widetilde{B}_\rho(y_0, \delta) = N^k \cap B^\ell(y_0, \delta)$.

We consider an open neighborhood θ of N^k in $\mathbf{R}^\ell$, such that the nearest point projection $\bar{\pi} : \theta \to N^k$ is a smooth fibration.

For u in $W^{1,p}(M^n, N^k)$, we set $E(u) = \int_{M^n} |\nabla u|^p \, dx$. If W is a subset of M^n we set $E(u; W) = \int_W |\nabla u|^p \, dx$. If Σ is a submanifold of M^n we set $\underline{E}(u, \Sigma) = \int_\Sigma |\nabla u|^p \, d\sigma$, where $d\sigma$ is the volume measure on Σ induced by the volume measure on M^n.

I. Proof of Theorem 1 when $M^n = C^n$ and $n-1<p<n$:

We assume throughout this section that $M^n = C^n$, $n-1 < p < n$ and $\pi_{n-1}(N^k) = 0$. Let u be in $W^{1,p}(C^n, N^k)$. We are going to approximate u by maps u_m which are continuous except at most at a finite number of point singularities (the conclusion then follows from Lemma 1 applied to u_m and the assumption $\pi_{n-1}(N^k) = 0$). In order to construct our approximation sequence u_m, we divide, in a convenient way, the cube C^n in $(m+1)^n$ little cubes C_r having an edge of length $\frac{1}{m}$, and we divide our cubes into two categories. The "good cubes" are the cubes such that the energy of u restricted to these cubes, and the energy of u restricted to the boundary of these cubes are small. For these cubes, the image of u is mainly located in some small geodesic ball in N^k, and we can approximate u using a standard mollifying technique. The "bad cubes" are the other ones, on which we approximate u by maps having point singularities. Next we present the basic method for dividing C^n into little cubes C_r, in a convenient way.

I.1. The method for dividing C^n into small cubes C_r

Without loss of generality, we may assume that u restricted to ∂C^n is in $W^{1,p}(\partial C^n, N^k)$ and thus continuous by Sobolev's Embedding Theorem. For $k = 1, \dots, n$, we set $e_k = (0, \dots, 1, \dots, 0) \in \mathbf{R}^n$ (where $k-1$ zeros preceed 1). For $1 \leq k \leq n$, and $a \in 0,1$, we denote $P(a,k)$ the restriction to C^n of the hyperplane passing through the point $A_k(a) = ae_k$ and orthogonal to e_k. For $m \in \mathbf{N}^*$ and for $\alpha \in [\frac{1}{4m}, \frac{3}{4m}]$ we consider the hyperplanes $P(\alpha + \frac{j}{m}, k)$ for $0 \leq j \leq m-1$ and the union of these hyperplanes $W(m, \alpha, k) = \bigcup_{j=0}^{m-1} P(\alpha + \frac{j}{m}, k)$. For almost every $\alpha \in [\frac{1}{4m}, \frac{3}{4m}]$ u restricted to $W(m, \alpha, k)$ is in $W^{1,p}$ and thus continuous by Sobolev's Embedding Theorem. Moreover, we have clearly

$$\int_{\frac{1}{4m}}^{\frac{3}{4m}} E(u; W(m, \alpha, k))\ d\alpha \leq E(u).$$

Thus there is some α_k in $[\frac{1}{4m}, \frac{3}{4m}]$ such that u restricted to $W(m, \alpha, k)$ is in $W^{1,p} \hookrightarrow C^0$ and such that

$$E(u, W(m, \alpha_k, k)) = \sum_{j=0}^{m-1} E(u; P(\alpha_k + \frac{j}{m}, k)) \leq 2mE(u). \tag{1}$$

Considering now the "slicings" of C^n by the set $W(m, \alpha_k, k)$ obtained by the method above when we change the direction k, we see that we have divided C^n in $(m+1)^n$ small cubes that we denote by $C_1, C_2, C \ldots, C_{(m+1)^n}$. The cubes which are not in contact with the boundary have edges of length $\frac{1}{m}$ (and are translates of $[0, \frac{1}{m}]^n$). The cubes which are in contact with the boundary are diffeomorphic to $[0, \frac{1}{m}]^n$ by linear maps f_r such that $|\nabla f_r| \leq 4$, $|\nabla f_r^{-1}| \leq 4$ (these inequalities are due to the fact that $\alpha_k \in [\frac{1}{4m}, \frac{3}{4m}]$). Inequality (1) then gives us

$$\sum_{r=1}^{(m+1)^n} E(u; \partial C_r) \leq K_1' m E(u) + E(u; \partial C) \leq K_1 m E(u) \tag{2}$$

for m large enough. For every small cube C_r we define the scaled energy $\widetilde{E}_m(u; C_r)$ by: $\widetilde{E}_m(u; C_r) = E(\widetilde{u}_{m,r}; C'^n)$ where $\widetilde{u}_{m,r}$ is the map from C'^n to N^k defined by $\widetilde{u}_{m,r}(x) = u(\frac{x}{m} + x_r)$, where x_r is the barycenter of C_r for the cubes which are not in contact with the boundary. ($\widetilde{u}_{m,r}$ is a "blow-up" map of u restricted to C_r).

We also set $\widetilde{E}_m(u; \partial C_r) = E(\widetilde{u}_{m,r}, \partial C'^n)$. We have the following scaling equalities

$$\begin{cases} \widetilde{E}_m(u; C_r) &= m^{n-p} E(u; C_r) \\ \widetilde{E}_m(u; \partial C_r) &= m^{n-p-1} E(u; \partial C_r). \end{cases} \tag{3}$$

Next we are going to divide our cubes C_r into two categories, the "good" and the "bad" cubes.

I.2. Definition of the "good" and of the "bad" cubes

Let $\varepsilon > 0$ be small (to be fixed later). We consider first the cubes C_r (for $r = 1, \ldots, (m+1)^n$), such that $\widetilde{E}_m(u; \partial C_r) \geq \varepsilon$. We denote by $P_{1,m}$ the union of cubes C_r which verify this condition, and $I_{1,m}$ the set of indexes r of these cubes (that means $C_r \subset P_{1,m}$ if and only if $r \in I_{1,m}$; $P_{1,m} = \bigcup_{r\in I_{1,m}} C_r$). We consider also the cubes C_r such that $\widetilde{E}_m(u; C_r) \geq \varepsilon m^{-\nu}$ where ν is some positive constant, which is fixed and small. We denote by $P_{2,m}$ the union of these cubes and $I_{2,m}$ the set of indexes for these cubes. We have $P_{2,m} = \bigcup_{r\in I_{2,m}} C_r$. We set $P_m = P_{1,m} \cup P_{2,m}$, $I_m = I_{1,m} \cup I_{2,m}$. P_m is the union of the "bad cubes." We consider also $Q_m = \overline{C^n \setminus P_n}$ and $J_m = \{(1, \dots, (m+1)^n\} \setminus P_m$ (the set of indexes for the cubes in Q_m). Q_m is the union of the "good" cubes. Next, we are going to show that the volume of P_m is small. Indeed, using relation (2) and the scaling equalities (3), it is easy to see that

$$\# I_{1,m} \leq K_2 m^{n-p} E(u) \varepsilon^{-1}.$$

Likewise the equality

$$\sum_{r=1}^{(m+1)^n} E(u; C_r) = E(u),$$

and the scaling equality (3) gives

$$\# I_{2,m} \leq K_2 m^{n-p+\nu} E(u) \varepsilon^{-1}.$$

Hence we see that $(\# I_m) m^{-n} = \mathrm{vol}(P_m) \to 0$ when $m \to +\infty$, and $E(u; P_m) \to 0$ when $m \to +\infty$.

We are going to approximate u in different ways on P_m and Q_m. Since $E(u; P_m)$ tends to zero, we do not need to approximate u very closely on P_m; we only have to construct u_m on P_m in such a way that $u_m = u$ on ∂P_m and $E(u_m; P_m) \leq C E(u; P_m)$ and u_m is continuous except at most at a finite number of point singularities. This is the purpose of the next construction.

I.3. Construction of the approximation map u_m on P_m

For the construction of the map u_m on each cube C_r included in P_m we are going to use the following lemma:

Lemma 2. *Let $n-1 < p < n$, $\mu > 0$ and v be a map in $W^{1,p}(C^n(\mu), N^k)$ such that u restricted to $\partial C^n(\mu)$ is in $W^{1,p}(\partial C^n(\mu), N^k) \hookrightarrow C^0$. There is an absolute constant K_3 depending only on p and n, and some map w in $W^{1,p}(C^n(\mu), N^k)$ continuous except at most at a finite number of point singularities, such that $w = v$ on ∂C^n and $E(w) \leq K_3 E(v)$.*

Before we give the proof of Lemma 2, we complete the construction of u_m on P_m. Defining u_m on each cube C_r in P_m as the map w obtained by Lemma 2 for $\mu = \frac{1}{m}$, $C^n(\mu) = C_r$, and $v = u$, we see that u_m is continuous on P_m except at most at a finite number of point singularities, u_m and u coincide on ∂C_r and thus on ∂P_m, and we have:

$$E(u_m; P_m) \leq K_3 E(u; P_m) \to 0 \quad \text{when } n \to +\infty. \tag{4}$$

Next we give the proof of Lemma 2.

Proof of Lemma 2. By a simple scaling argument, we may assume without loss of generality that $C^n(\mu) = C^n = [0,1]^n$. We are going to use the method of section I.1.1 for dividing C^n into little cubes. Thus, for every $s \in \mathbb{N}^*$, we may divide C^n in $(s+1)^n$ little cubes C_ℓ of length $\frac{1}{s}$ (except those in contact with the boundary, which are linearly diffeomorphic to $[-\frac{1}{2s}, \frac{1}{2s}]^n$), such that

$$\sum_{\ell=1}^{(s+1)^n} E(v; \partial C^\ell) \leq K_1 s E(v)$$

for s large. On each cube C_ℓ (for $\ell = 1, \ldots, (s+1)^n$), we define a map w_s by

$$w_s(x) = v\left(\frac{x - x_\ell}{2s\|x - x_\ell\|} + x_\ell\right)$$

where x_ℓ is the barycenter of C_ℓ, and if C_ℓ is not in contact with the boundary, and in a similar way if C_ℓ is in contact with the boundary (using the fact that, in this case C_ℓ is diffeomorphic to $[-\frac{1}{2s}, \frac{1}{2s}]^n$ by a linear map f_ℓ such that $|\nabla f_\ell| \leq 4$, $|\nabla f_\ell|^{-1} \leq 4$). It is then easy to show that w_s is in $W^{1,p}(C^n, N^k)$, is continuous except at the point x_ℓ, $w_s = v$ on ∂C^n, and for each cube C_ℓ we have

$$E(w_s; C^\ell) \leq K_4 s^{-1} E(v; \partial C^\ell). \tag{5}$$

Adding these inequalities for $\ell = 1, \ldots, (s+2)^n$, and combining with (5) we obtain:

$$E(w_s) \leq \frac{K_4}{s} \sum_{\ell=1}^{(s+1)} E(v; \partial C^\ell) \leq K_4 K_1 E(v)$$

for s large enough. Thus for s large enough, we set $w = w_s$, and w satisfies the conditions of Lemma 2. This completes the proof of Lemma 2. Next we are going to construct the approximation map u_m on Q_m.

I.4. Construction of the approximation map

On Q_m we are going to approximate u by maps u_m which are continuous. First we construct a map w_m such that w_m is in $W^{1,p}(C^n, N^k)$, and such that for each cube C_r in Q_m the image of w_m on C_r lies in a small geodesic ball of N^k (then it is easy to construct on Q_m a map u_m continuous, using a simple mollifying argument). For this purpose we need two technical lemmas:

Lemma 3. *Let $\delta > 0$ be small. Let $q \in \mathbf{N}^*$ and $p > q$. We consider the cube C^q and a map v in $W^{1,p}(C^q, N^k)$. Then there is some $\varepsilon_0(\delta, q, p, N^k)$ such that if $E(v) \leq \varepsilon_0(\delta, q, p, N^k)$, then the image of C^q by v (which is continuous by the Sobolev Embedding Theorem) lies in some domain $\widetilde{B}_\rho(y, \delta)$ for some $y \in N^k$.*

Note that Lemma 3 is a simple consequence of Sobolev's Embedding Theorem, since $W^{1,p}(C^q, N^k) \hookrightarrow C^0(C^q, N^k)$. For technical reasons (which will become clear in the sequel), we choose δ_0 such that for every $y \in N^k$, $B(y, 4n\delta_0)$ lies in θ. We choose also $\varepsilon = \varepsilon_0(\frac{\delta_0}{2n}, p, n-1, N^k)$ (recall that ε is the constant needed for the definition of P_m and Q_m). We need also the following result:

Lemma 4. *For $\delta > 0$ small enough there is some constant $\widetilde{K}(\delta)$, depending only on N^k and δ, such that there is some smooth map $\varphi(y, \delta)$, for every y in N^k, from N^k to $\widetilde{B}_\rho(y, \delta)$ such that $|\nabla\varphi(y, \delta)| \leq \widetilde{K}(\delta)$ and $\varphi(y, \delta) = \mathrm{Id}$ on $\widetilde{B}_\rho(y, \delta)$.*

We come back now to the construction of u_m on Q_m. For each cube C_r in Q_m, we have (by the definition of Q_m), $\widetilde{E}(u; \partial C_r) \leq \varepsilon$. If we apply Lemma 4 to C_r, $q = n - 1$, p and each face of ∂C_r we see that, (after a "blow-up" of the cube) the image of ∂C_r by u (which is continuous on ∂C_r by the construction of C_r) lies in some domain $\widetilde{B}_\rho(y_r, \delta_0)$, for some $y_r \in N^k$.

Then we define a map $w_m \in W^{1,p}(C_r, N^k)$ in the following way:

$$w_m = \varphi(y_r, 2\delta_0) \circ u, \quad \text{on } C_r.$$

Since φ is Lipschitz, by the chain rule deviation for maps in $W^{1,p}$, w_m is clearly in $W^{1,p}(C_r, \widetilde{B}_\rho(y_r, 2\delta_0))$. Moreover, since $\varphi(y_r, 2\delta_0) = \mathrm{Id}$ on $\widetilde{B}_\rho(y_r, 2\delta_0)$, and since the image of u restricted to ∂C_r is in $\widetilde{B}_\rho(y_r, \delta_0)$ we have $w_m = u$ on ∂C_r. Thus defining w_m in such a way for each C_r in Q_m, we see that $w_m \in W^{1,p}(Q_m, N^k)$ and $w_m = u$ on ∂Q_m. We now are going to show that w_m approximates u on Q_m. For $C_r \subset Q_m$, consider the set:

$$U_{m,r} = \{x \in C_r \mid u(x) \neq w_m(x)\}.$$

We have

$$(6)\qquad \int_{C_r} |\nabla(u-w_m)|^p\,dx = \int_{U_{m,r}} |\nabla(u-w_m)|^p\,dx$$
$$\le K_5 \int_{U_{m,r}} (|\nabla u|^p + |\nabla w_m|^p)\,dx$$
$$\le K_5(1+K(2\delta_0)^p)\int_{U_{m,r}} |\nabla u|^p\,dx.$$

Thus, in order to prove that w_m approximates u on Q_m, we only have to show that (mes $U_{m,r}$) $\times\ m^n$ is small. Since it is easier to argue on the "blow-up" maps, consider the maps $\widetilde{u}$ and $\widetilde{w}_m$ defined on C'^n by

$$\widetilde{u}\ (x) = u(\frac{x}{m}+x_r);\ \widetilde{w}_m(x) = w(\frac{x}{m}+x_r)$$

where x_r is the barycenter of C_r; ($\widetilde{u}$ and $\widetilde{w}_m$ are the "blow-up" maps of u and w_m respectively). We consider the set

$$A_{m,r} = \{x \in C'^n \mid \widetilde{u}(x) \notin B(y_r, 2\delta_0)\}.$$

It is easy to see that

$$(7)\qquad m^r \text{ mes } U_{m,r} \le \text{mes } A_{m,r}.$$

Since if x is in $U_{m,r}$, $u(x) \neq w_m(x)$ and thus $u(x) \notin B(y_r, 2\delta_0)$ (the factor m^n in (7) being a scaling factor). We are going to estimate mes $A_{m,r}$. We claim that

$$(8)\qquad \text{mes } A_{m,r} \le m^{-\nu}\varepsilon \setminus \varepsilon_0(\delta_0, N^k).$$

Proof of the Claim: For $a \in [-\frac{1}{2}, \frac{1}{2}]$ we consider the hyperplanes $P(a,1)$ defined in section I.1.1. We have, by Fubini's theorem, and since $C_r \subset Q_m$ implies that $E(\widetilde{u}; C'^n) \le \varepsilon m^{-\nu}$:

$$\int_{-1/2}^{1/2} E(\widetilde{u}; P(a,1))\,da \le E(\widetilde{u}; C'^n) \le \varepsilon m^{-\nu}$$

it follows that

$$(9)\ \text{mes}\{a \in [-\frac{1}{2}, \frac{1}{2}] | E(\widetilde{u}; P(a,1)) \le \varepsilon_0(\delta_0, N^k)\} \ge 1 - m^{-\nu}\varepsilon\setminus\varepsilon_0(\delta_0, N^k)$$

where $\varepsilon_0(\delta_0, N^k)$ is the constant arising in Lemma 4. For every a such that $E(\widetilde{u}; P(a,1) \le \varepsilon_0(\delta_0, N^k)$, we may apply Lemma to $P(a,1)$ which is a translate of C'^{n-1}, p, $q = n - 1$; since $u(\partial C_r) \subset \widetilde{B}_\rho(y_r, \delta_0)$, Lemma 4 then shows that $\widetilde{u}(P(a,1)) \subset \widetilde{B}_\rho(y_r, 2\delta_0)$, and hence $A_{m,r} \cap P(a,1) = \emptyset$. Inequality (8) follows from this relation and (9). Next we complete the estimate on $\int_{Q_m} |\nabla(u - w_m)|^p\, dx$.

Adding relation (7), for all cubes C_r in Q_m we obtain

(10)

$$\int_{Q_m} |\nabla(u - w_m)|^p\, dx \le K_5(1 + \widetilde{K}(2\delta_0)^p) \int_{\cup r \in I_m U_{m,r}} |\nabla u|^p\, dx$$
$$\le K_5(1 + \widetilde{K}(2\delta_0)^p) E(u;_{r\in I_m} U_{m,r}).$$

Relations (7) and (8) show that

$$\operatorname{mes}\left(\bigcup_{r\in J_m} U_{m,r}\right) \le \sum_{r\in J_m} m^{-n} \operatorname{mes} A_{m,r} \le \# J_m m^{-\nu} m^n \varepsilon \setminus \varepsilon_0(\delta_0, N^k).$$

Since $\# J_m \le (m+1)^n$, we see that $\operatorname{mes}(\cup_{r\in I_m} U_{m,r}) \to 0$ when $m \to +\infty$ and thus by (10)

(11) $$\int_{Q_m} |\nabla(u - w_m)|^p\, dx \to 0 \quad \text{when } m \to +\infty.$$

Hence w_m approximates u in $W^{1,p}$ on Q_m. Next we have to "smoothen" w_m.

For this purpose we are going to use the following lemma.

Lemma 5. *Let $\mu > 0$, let $p > 1$, and let $\widetilde{B}_\rho(y, \delta)$ be some ball in N^k such that $\widetilde{B}_\rho(y,\delta) \subset \theta$. Let v be in $W^{1,p}(C^n(\mu), \widetilde{B}_\rho(y,\delta))$ such that v restricted to ∂C^n is in $W^{1,p}(\partial C^n, \widetilde{B}_\rho(y,\delta)) \cap C^0$. Then v can be approximated in $W^{1,p}(C^n(\mu), N^k)$ by maps v_m in $W^{1,p}(C^n(\mu), N^k) \cap C^0$, which coincide with v on $\partial C^n(\mu)$.*

We apply Lemma 5 to each cube C_r in Q_m and $\widetilde{B}(y_r, 2\delta_0)$. The lemma gives us a map u_m such that $u_m = w_m = u$ on ∂Q_m and such that

(12) $$\int_{Q_m} |\nabla(w_m - u_m)|^p\, dx \le \frac{1}{m}.$$

Combining (12) and (11) we see that

(13) $$\int_{Q_m} |\nabla(u - u_m)|^p\, dx \to 0 \quad \text{when } m \to +\infty.$$

This completes the construction of u_m on Q_m.

I.5. Proof of Theorem 1 completed in the case $M^n = C^n$ and $n-1<p<n$:

In section I.1.3 we have constructed u_m on P_m such that u_m is continuous except at a finite number of point singularities, and such that $u_m = u$ on ∂P_m. In section I.1.4 we have constructed u_m on Q_m such that u_m is in $W^{1,p}(Q_m; N^k) \cap C^0$ and $u_m = u$ on ∂Q_m. Thus, since $u_m = u$ on $\partial P_m \cap \partial Q_m$, and since $P_m \cup Q_m = C^n$, the map u_m is in $W^{1,p}(C^n, N^k)$, and moreover is continuous except at a finite number of points. We have:

$$\int_{C^n} |\nabla(u-u_m)|^p \, dx \leq \int_{Q_m} |\nabla(u-u_m)|^p \, dx + K_5[E(u; P_m) + E(u_m; P_m)].$$

Thus combining (5) and (13) we see that

$$\int_{C^n} |\nabla(u-u_m)|^p \, dx \to 0.$$

This proves that we can approximate u by maps in $W^{1,p}(C^n, N^k)$ continuous except at a finite number of point singularities. Theorem 1 then follows, in the case considered here, from the hypothesis $\pi_{n-1}(N^k) = 0$ and Lemma 1.

REFERENCES

[Be] F. Bethuel, *A characterization of maps in $H^1(B^3, S^2)$ which can be approximated by smooth maps*, to appear.

[BZ] F. Bethuel and X. Zheng, *Density of smooth functions between two manifolds in Sobolev spaces*, J. Funct. Analysis **80** (1988), 60-67.

[E] M. Escobedo, to appear.

[EL] J. Eells and L. Lemaire, Bull. London Math. Soc. **10** (1978), 1-68.

[SU1] R. Schoen and K. Uhlenbeck, *A regularity theory for harmonic maps*, J. Diff. Geom. **17** (1982), 307-335.

[SU2] R. Schoen and K. Uhlenbeck, *Approximation theorems for Sobolev mappings*, preprint.

[W1] B. White, *Infima of Energy Functions in homotopy classes*, J. Diff. Geom. **23** (1986), 127-142.

[W2] B. White, *Homotopy classes in Sobolev spaces and the existence of energy minimizing maps*, Acta Mathematica **160** (1988), 1-17.

Fabrice Bethuel
CERMA ENPC
La Courtine
93167 Noisy le Grand Cedex, France

The "magic" of Weitzenböck formulas

Jean-Pierre BOURGUIGNON

Introduction

In recent years, problems of geometric origin have attracted the attention of many analysts. In some instances, Geometry was pointing to the most subtle case of an analytic problem, e.g., the limiting case of Sobolev inequalities in the Yamabe problem or in Yang-Mills theory. Moreover, phenomena of a geometric nature were appearing in a P.D.E. context like the "bubbling off" phenomenon in the harmonic map problem. In many of the geometric situations considered, the problem could be reduced to solving an elliptic scalar equation, most often a non-linear one.

One of our main points in writing this survey is to show evidence that, although legitimate, such an emphasis on scalar equations is indeed a reduction, i.e., that by restricting the considerations to that special case some crucial information is lost. The theme of this talk will therefore be *to make propaganda for studying systems of Partial Differential Equations*, and their relationships to scalar equations.

To be more specific, in the study of Systems of P.D.E.s, a tool that, over the years, geometers have learned to use is coined under the name of *Weitzenböck formulas.* These formulas relate Laplace operators that one can universally define on sections of vector bundles. This points to the fact that to grasp all the information available one should form *all* possible Laplacians, and study how they are related. This problem is of course meaningless for scalar equations, since then only the Laplace-Beltrami is operating. This pushes us into the realm of another (a priori unrelated) domain of mathematics, namely *representation theory*.

The (may be strange sounding) title of this talk aims at suggesting two different, and partly contradictory, ideas. *In the study of "harmonic" objects, Weitzenböck formulas* (that we define in great generality in Section II) *are indeed a powerful tool which can be developed quite systematically*†, but *the reason why they work so well is not yet completely clear.*

Another aim for this report may also be to convince more analysts that, as soon as they are dealing with operators with variable coefficients, they should not be afraid of using geometric tools, e.g., shift from open sets of Euclidean spaces to manifolds. As a metaphore, we should say that, as a magnetic field quite often lifts the indeterminacies in a physical experiment, and reveals a finer structure, geometric objects appear, and become useful, as soon as one has left Euclidean spaces, and objects invariant under translations.

Throughout this article we will work over Riemannian manifolds, hence assume a basic familiarity with standard objects connected to that topic such as the Levi-Civitå connection, properties of the Riemann curvature tensor, etc. (A possible reference is [4].) Most often, thanks to the metric, we will identify 1-forms and vectors.

This article is organized as follows.

In Section I, we recall some standard situations where Analysis and Geometry have come together.

In Section II, after having recalled the first use of Weitzenböck formulas as made by S. Bochner more than fifty years ago, we give a fairly general definition of them, and explain their main uses.

In Section III, we show a number of different instances where passing to a system, and using a Weitzenböck formula proves useful. Examples that we work out include deriving eigenvalue estimates for scalar operators from an appropriate Weitzenböck formula, and subelliptic estimates for the norms of sections satisfying a system of equations.

In Section IV, we present two recent developments related to Weitzenböck formulas. The first one gives improved estimates for the norm of the gradient of the length of a harmonic section of a bundle (a so-called *Kato inequality*) using again elementary representation theory. The second one, due to A. Polombo, deals with sections of an endomorphism bundle for which Weitzenböck formulas appear when computing the Laplacian of its eigenvalues.

Acknowledgements: The author thanks the organizers of the meeting *"Problèmes variationnels non-linéaires"* held in Paris in June 1988 for having given him the opportunity of addressing the conference on the subject matter of this article.

† Other reports on these formulas are available such as [27] which is quite comprehensive, and [3] which is geared toward the most recent analytic results drawn from these formulas.

I. Reducing geometric problems to solving an elliptic equation

a) *The Yamabe problem.*

The Yamabe problem was originally stated in [28] as a theorem. It is by now solved thanks to successive contributions (cf. [24], [1], [18], [19], and also [2] for another proof). An interesting survey discussion of this problem can be found in [14]. The problem is, "*to find metrics on a compact manifold with constant scalar curvature in a given conformal class of Riemannian metrics*".

On a compact manifold M of dimension $n \geq 3$, if one gives oneself a metric g, it is well known that the scalar curvature $s_{\tilde{g}}$ of the conformally related metric $\tilde{g} = \varphi^{\frac{4}{n-2}} g$ where φ is a positive function has the expression†

$$s_{\tilde{g}} = \varphi^{-\frac{n+2}{n-2}} \left(4 \frac{n-1}{n-2} \Delta_g + s_g \, \varphi \right) \tag{1}$$

where, of course, s_g denotes the scalar curvature of g and Δ_g the Laplace-Beltrami operator acting on functions.

Hence, looking for conformally related metrics with constant scalar curvature σ amounts to solving the *non-linear eigenvalue problem*

$$L_g \, \varphi = \sigma \, \varphi^{\frac{n+2}{n-2}} \tag{2}$$

for the elliptic operator $L_g \equiv 4\frac{n-1}{n-2} \Delta_g + s_g$ which is often called the *conformal Laplacian* of the metric g because of its equivariance under conformal changes of metrics.

Notice that this eigenvalue problem is variational, i.e., is the Euler-Lagrange equation of the *total scalar curvature* functional $\tilde{g} \mapsto \int_M s_{\tilde{g}} v_{\tilde{g}}$ (where v_g denotes the volume element of a metric g).

Uniqueness of the metric with constant scalar curvature in a conformal class can easily be established by a maximum principle argument when, in the class, there is a metric with negative scalar curvature. (This condition is in fact equivalent to requiring the functional to have a negative minimum). For the positive case, uniqueness is not in general guaranteed. In section III, we shall describe an instance where uniqueness can be established.

---

† Notice that the choice we made of parametrizing the conformal change by introducing the power $\frac{4}{n-2}$ had the effect of concentrating the non-linearity in the zeroth order term. For surfaces (i.e., when $n = 2$), the appropriate parametrization is $\tilde{g} = e^{2\varphi} g$, and leads to a Liouville type eigenvalue problem.

b) Kähler-Einstein metrics

The search for *Kähler-Einstein metrics* is another instance where a geometric problem can be converted into an analytic one involving an elliptic scalar equation.

Let M be a complex manifold of real dimension $n = 2m$ on which one fixes a Kähler class $[\omega]$. We look for a Kähler metric $\tilde{\omega}$ cohomologous to the positive form ω of type $(1,1)$, i.e., as is classical, we consider all forms $\tilde{\omega} = \omega + \partial\overline{\partial}\varphi$ which are positive. Such a function φ is referred to as a *Kähler potential* with respect to ω.

If we denote the Ricci curvature of the Kähler metric ω viewed as a $(1,1)$-form by ρ_ω, the geometric condition we want to impose is the so-called *Einstein condition* $\rho_{\tilde{\omega}} = k\,\tilde{\omega}$. Because of the possible interpretation of the Ricci curvature as the curvature of the canonical bundle, hence its expression in a local holomorphic system of coordinates (z^α) as

$$(3)\quad (\rho_{\tilde{\omega}})_{\alpha\overline{\beta}} = \frac{1}{2\pi i}\,\frac{\partial^2 \log\det(g_{\alpha\overline{\beta}})}{\partial z^\alpha\,\partial \overline{z}^\beta} \quad \text{(here, we write } \omega = i\sum_{\alpha,\beta=1}^{n} g_{\alpha\overline{\beta}}dz^\alpha d\overline{z}^\beta\text{)}\,,$$

it is possible to express the geometric condition by a partial differential equation, namely

$$\det\frac{\tilde{\omega}^m}{\omega^m} = e^{k\varphi}\,e^f\,,$$

or in more classical terms an equation of Monge-Ampère type

$$(4)\qquad \det\left(g_{\alpha\overline{\beta}} + \frac{\partial^2\varphi}{\partial z^\alpha\partial\overline{z}^\beta}\right) = e^{k\varphi}\,e^f$$

where the function f can be calculated explicitely from ω and the Ricci curvature ρ_ω.

The geometry of the situation is quite rich. As an example, in order to solve the equation, the first Chern class must vanish or be definite. If the first Chern class is negative, the equation has a unique solution as shown by T. Aubin and S.T. Yau. If it vanishes, uniqueness is lost, but existence persists. In fact the equation reduces to that of the Calabi conjecture solved by S.T. Yau. The full solution of the positive case is still open, and the geometric obstructions known today are all connected with the presence of holomorphic vector fields on the complex manifold M. They have not yet been properly tied to the analytic formulation of the problem.

c) Hodge-de Rham theory of harmonic forms

To understand the topology of a compact manifold, it is useful to describe its cohomology classes as equivalence classes of closed differential forms modulo exact forms. This approach due to de Rham is also well adapted to an analytic treatment (cf. [26]). Indeed, if a is a closed exterior differential k-form, hence defines a cohomology class $\alpha = [a]$ in the cohomology group $H^k(M,\mathbf{R})$, all cohomologous forms can be written as $a + db$ where b is an exterior differential $(k-1)$-form. Then, a Riemannian metric g being given on M, one can look for the form $\tilde{a} \in [\alpha]$ with smallest possible L^2-norm. One then has the well known

THEOREM (W. HODGE-G. DE RHAM).– *Let (M,g) be a compact Riemannian manifold and $\alpha \in H^k(M,\mathbf{R})$. There exists a unique exterior differential k-form a_0 called the harmonic representative of α with critical L^2-norm. It has minimal norm in its class, and satisfies the elliptic system*

$$\begin{cases} da_0 = 0 \\ d^*a_0 = 0 \end{cases}$$

where d^* denotes the codifferential, i.e., the formal adjoint of d for the global scalar product defined by g.

In the preceding statement, the two equations in the system have different statutes. The fact that a_0 is *closed* defines the set on which the variational problem is posed, hence warranties its geometric significance. The fact that a_0 is *coclosed* expresses that a_0 is a critical point of the chosen functional.

This first order system has the disadvantage of involving forms of different degrees. As is well known, it is possible to transform it into a second-order system involving only exterior differential k-forms by forming the *Hodge-de Rham Laplacian*† $\frac{1}{2}\,dd^* + d^*d$ having also the space of harmonic forms as kernel. (This follows from integrating over M the image of a harmonic form under the Laplacian against the form itself, and using that d and d^* are adjoint operators.)

It is quite fascinating to realize that this dissymetry between the two parts of the equation exactly parallels the one appearing in electromagnetism in the Maxwell equations for the vacuum when the electromagnetic field ω is (as it should be) viewed as an exterior differential 2-form on space-time. The only (crucial) difference there is that space-time is endowed with a *Lorentzian* metric, hence the system of second order P.D.E. one builds with it is *hyperbolic*, hence propagates waves. (Hopefully so, if one likes to listen to the radio!)

† The coefficient $\frac{1}{2}$ may look unusual. It is necessary to fit our definition of a Laplacian, and our conventions for d and the inner products on spaces of exterior forms.

II. The general use of Weitzenböck formulas

In 1946, S. Bochner used a formula due to Weitzenböck (1923) to refine the preceding discussion in the case of harmonic 1-forms.

If D denotes the Levi-Cività covariant derivative associated with the metric g (and D^* its formal adjoint), one indeed has

Weitzenböck formula for 1-forms

$$\frac{1}{2}\,d^*d + dd^* = D^*D + r \tag{5}$$

where the Ricci curvature term r is to be understood as a linear map on 1-forms by using the metric g to raise one of its indices.

THEOREM (S. BOCHNER [5]).– *A compact Riemannian manifold with positive Ricci curvature has no non-trivial harmonic 1-forms, hence its first Betti number vanishes.*

PROOF.– Thanks to (5), the Hodge-de Rham Laplacian $\frac{1}{2}\,dd^* + d^*d$ appears as the sum of a non-negative operator D^*D (by integrating again by parts against the form itself !) and an operator r which is positive by assumption, hence is positive. It can therefore have no kernel. ■

This suggests a general pattern which will be one of our threads through these developments

$$\left\{\begin{array}{l}\bullet \textit{ Solution of a system (of geometric origin)}\\ \bullet \textit{ Strict inequality imposed on anappropriate geometric quantity}\end{array}\right\}$$

$$\Downarrow$$

$$\{\bullet \textit{ Solution vanishes identically}\}$$

This scheme can be refined if one weakens the assumption on the Ricci curvature, and supposes only that it is non-negative. We then have

THEOREM.– *On an n-dimensional compact Riemannian manifold M with non-negative Ricci curvature any harmonic 1-form is parallel. In particular, the first Betti number of M is at most n.*

PROOF.– Since $r \geq 0$, any harmonic 1-form must be annihilated by D^*D, hence by D after integrating by parts against itself.

But there cannot exist more than n parallel 1-forms on an n-dimensional manifold, since any linear combination of parallel 1-forms which vanishes at a point vanishes everywhere. ∎

The scheme we introduced earlier can therefore be refined to cover that case too

$$\left\{\begin{array}{l}\bullet\ \textit{Solution of a system (of geometric origin)}\\ \bullet\ \textit{border case of an inequality on a geometric quantity}\end{array}\right\}$$

$$\Downarrow$$

$$\left\{\begin{array}{c}\bullet\ \textit{Solution of the original system is also solution of a}\\ \textit{``stronger'' system}\end{array}\right\}$$

We are then ready to do things a little more systematically. For that purpose, we wander a bit into the theory of natural bundles.

Over any Riemannian manifold (M,g), a ***natural Riemannian vector bundle*** $\mathcal{E}$ gives rise† to a genuine Riemannian vector bundle $\mathcal{E}M$ with the property that, for any isometry φ between two Riemannian manifolds (M,g) and (N,h), a bundle isometry $\mathcal{E}\varphi : \mathcal{E}M \mapsto \mathcal{E}N$ covers φ.

Typical examples are given by the tangent bundle, and all tensor bundles built on it.

A ***natural differential operator*** $\mathcal{A}$ between two natural Riemannian vector bundles $\mathcal{E}_1$ and $\mathcal{E}_2$ is then defined as a collection of differential operators $\mathcal{A}M$ mapping sections of $\mathcal{E}_1M$ to sections of $\mathcal{E}_2N$ whenever (M,g) is a Riemannian manifold with the property that, if φ is an isometry from (M,g) to (N,h) and σ a section of $\mathcal{E}_1M$, then

$$\mathcal{E}_2\varphi \circ \mathcal{A}M(\sigma) \circ \varphi^{-1} = \mathcal{A}N(\mathcal{E}_1\varphi \circ \sigma \circ \varphi^{-1}).$$

A typical example of such a natural Riemannian operator is given by the Levi-Cività connection ∇ between any tensor bundle and its tensor product with the cotangent bundle.

We will not go any further along this axiomatic approach, and abandon these cumbersome notations although they are somewhat inevitable if one is to *prove* theorems about natural objects.

The key fact which takes our previous discussion into the realm of representation theory is the following basic theorem.

† It is a typical instance where one should speak of a *functor*.

THEOREM ([22]).– *Any natural Riemannian differential operator $\mathcal{A}$ of order 1 between sections of two natural Riemanniann bundles $\mathcal{E}_1$ and $\mathcal{E}_2$ associated with the bundle of orthonormal frames has an O_n-equivariant principal symbol, i.e., on a Riemannian manifold (M,g), for $A = \mathcal{A}M$, $E_1 = \mathcal{E}_1 M$, and $E_2 = \mathcal{E}_2 M$, σ_A maps $E_1 \otimes T^*M$ into E_2, and one has $A = \sigma_A \circ D$.*

If one wants to detect all possible first natural differential operators between bundles E_1 and E_2, one therefore has to determine all O_n-invariant maps from $TM \otimes E_1$ to E_2, a purely algebraic problem completely determined by the representations ρ_1 and ρ_2 by which E_1 and E_2 are respectively associated with the bundle of orthonormal frames.

This point of view turns out to be a little too restrictive in that one would also like to deal with spinor bundles (both for mathematical and physical reasons). The difficulties involved are purely topological. The manifold M to be considered must have vanishing second Stiefel-Whitney classes (a specific class in $H^2(M, \mathbf{Z}/2\mathbf{Z})$). The previous developments then go through with only minor modifications, the major ones being that one should start with the bundle of spinor frames (once a spin structure has been chosen, a well defined two-fold covering of the bundle of orthonormal frames), and use representations of the group Spin_n (instead of O_n or SO_n).

Before defining Weitzenböck formulas in some generality, one needs to single out a class of operators which they involve.

DEFINITION.– *A second order differential operator A mapping sections of a natural bundle E to themselves is called a Laplacian if its principal symbol $\sigma_A \in \mathrm{Hom}(S^2T^*M \otimes E, E)$ satisfies, for $\xi \in T^*M$ and $\epsilon \in E$,*

$$\sigma_A(\xi \otimes \xi)\,\epsilon = -\,g(\xi,\xi)\,\epsilon\ .$$

One can then set the following.

DEFINITION.– *A Weitzenböck formula is an identity relating two natural Laplacians acting on sections of the same bundle.*

Typical examples of second order natural differential operators mapping sections of a natural bundle $\mathcal{E}$ to themselves are compositions with their adjoints of first-order natural differential operators defined on sections of $\mathcal{E}$, and their linear combinations. Moreover, in case of a convex combination, one is sure that such an operator is non-negative by a mere integration by parts.

The following crucial general fact was precisely what Bochner exploited in the case of differential 1-forms.

PROPOSITION.– *The difference between two natural Laplacians acting on sections of a bundle of tensors of a given rank is an operator of order zero involving only the curvature of the bundle on which they are defined.*

PROOF.– Since the two operators have the same principal symbol, their difference is a natural first-order differential operator. Its symbol must therefore be an O_n-invariant map from $TM \otimes E$ to E. But, according to a classical theorem in invariant theory, there is no such map between the source and the target since their ranks as tensor representations differ only by one.

Their difference must then be of order zero.

To prove that this term involves only the curvature requires a more thorough analysis of the space of germs of metrics for which we refer to [22]. ∎

The game one ought to play is then the following† :

- *determine all natural Laplacians acting on sections of a given bundle* E ;
- *write down explicitely the Weitzenböck formulas relating the Laplacian one is interested in with all other Laplacians ;*
- *determine which ones involve favourable curvature conditions.*

We close this section by giving one more example due to A. Lichnerowicz of use of a Weitzenböck formula which, in the hands of M. Gromov and H.B. Lawson, turned out to be at the heart of the solution of a significant geometric problem, namely "*determine when a manifold admits a metric with positive scalar curvature*".

We will be working with the bundle of spinors on a (necessarily spin) manifold M. This bundle naturally inherits a connection from the Levi-Cività connection that we still denote by D. Since, thanks to Clifford multiplication at each point, spinors are acted upon by tangent vectors, one can define a special first-order differential operator, the *Dirac operator* $\mathcal{D}$, by setting, for a spinor field ψ,

$$\mathcal{D}\psi = \sum_{i=1}^{n} e_i \cdot D_{e_i}\psi \tag{6}$$

where (e_i) is an orthonormal basis of T_mM and $\cdot$ denotes Clifford multiplication. A spinor field ψ satisfying $\mathcal{D}\psi = 0$ is called a *harmonic spinor*.

---

† To the author's knowledge, the first instances where this idea was applied are [6] and [13]

We then have the *Lichnerowicz formula*

Weitzenböck formula for spinors

$$(7) \qquad \mathcal{D}^2 = D^*D + \frac{1}{4}s\,,$$

from which, by applying our scheme, we derive two important results.

THEOREM.– *Any compact spin manifold with positive scalar curvature has no harmonic spinor.*

Since the Dirac operator is elliptic and since, in dimensions $n = 4k$, there is a topological invariant, the so-called A-genus $\widehat{A}(M)$, measuring its index, i.e., the balance between positive harmonic spinors and negative ones, one then obtains

COROLLARY.– *Any $4k$-dimensional compact spin manifold with non-vanishing $\widehat{A}$-genus has no metric with positive scalar curvature.*

The refined scheme can also be applied there to the effect that

THEOREM.– *Any non trivial harmonic spinor on a compact spin manifold with non-negative scalar curvature is parallel, and the metric has vanishing scalar curvature.*

They are of course many other situations where one can apply our scheme. The most famous ones are connected to *complex analytic geometry* where Weitzenböck formulas applied to the operator $\overline{\partial}^*\overline{\partial}$ are used to show that *positive* bundles have no holomorphic sections (cf. [12]).

All the instances of applications of Weitzenböck formulas we encountered so far made use of the comparison between the Laplacian given to us by the geometry and D^*D, often called the *rough Laplacian* (although *plain* Laplacian would probably be a better denomination). As a result, in all these examples, we could draw a conclusion only in the case where the curvature of the bundle we considered was *positive*.

A case of application of the more elaborate game we discussed before, i.e., looking for all possible natural Laplacians built out of natural first order differential operators can be found in [6] and [13].

There, a rigidity result is proved for Einstein metrics with negative curvature by showing that the linearized Einstein equation (i.e., the system that an infinitesimal deformation of an Einstein metric by Einstein metrics must satisfy) has no solutions. This is done by comparing his Laplacian via a Weitzenböck formula to the appropriate operator which is positive on the objects we consider (and not to D^*D), but then the curvature appears with the opposite sign, allowing us to draw a conclusion in a case of negative curvature.

Before going to more elaborate uses of Weitzenböck formulas, let us notice that all previous uses of the Weitzenböck formulas concentrating on vanishing theorems can be transformed into estimating the lowest eigenvalue of the Laplacian studied. The most obvious way is by introducing a uniform lower bound on the curvature term. More refined ways are by now available, calling for integral norms of the curvature. (Reference [3] emphasizes that aspect).

III. Some less well known uses of Weitzenböck formulas

a) Eigenvalue bounds and symmetries

The first fact we want to draw attention to is that *"symmetries of a geometric situation are always expressed as solutions of a system of Partial Differential Equations"*. (A possible reference is [11].)

i) An infinitesimal isometry X, a so-called *Killing field*, is a solution of the system $\mathcal{L}_X g = 0$, which, when expressed in terms of covariant derivatives, becomes $\varsigma(DX) = 0$ where ς denotes the operation of symmetrization of a linear map. The first-order operator having precisely the map $\frac{1}{2}\varsigma$ as principal symbol is traditionally denoted by δ^*, so that the equation for a Killing field X is $\delta^* X = 0$.

One then has

Another Weitzenböck formula† for 1-forms

(8) $$\frac{1}{4}\, d^*d + dd^* = \delta\delta^* + r$$

where we have denoted by δ the formal adjoint of δ^*. (It is just the covariant divergence which appears in Elasticity theory.)

From this formula, one can derive the following

THEOREM.– *A compact Riemannian manifold with negative Ricci curvature has no non-trivial Killing fields.*

PROOF.– Any non-trivial Killing field X would lie in the kernel of the operator $\frac{1}{4}d^*d + dd^* - r$ which is positive when the Ricci curvature is negative, hence a contradiction. ∎

† Although the geometrically motivated operators appearing on both sides of formulas (8) and (8') are second order elliptic operators, they are not Laplacians, but differ from a Laplacian by the same amount, hence our speaking of Weitzenböck formulas.

Using a previously described procedure, in the border case, the preceding statement can be refined as follows.

COROLLARY.– *Any Killing field on a compact Riemannian manifold with non-positive Ricci curvature is parallel.*

Provided the Ricci curvature is controlled, formula (8) allows us also to control the L^2-norm of the hessian of a function f with the L^2-norm of its Laplacian. (Once again, for that purpose, we take the inner product of both sides with df, and integrate against df.)

ii) An infinitesimal conformal transformation Y is a solution of the system $\mathcal{L}_Y g + \frac{1}{n} d^*Y\, g = 0$ (Y must preserve the metric density $g \otimes (v_g)^{-\frac{1}{n}}$).

If we denote by C the differential operator acting on vector fields that is so defined, one easily sees that $C(Y) = \varsigma_0(DY)$ where ς_0 denotes the trace-free part of the symmetric part of DY.

We then have

Yet another Weitzenböck formula on 1-forms

$$\frac{1}{4} d^*d + \frac{n-1}{n} dd^* = C^*C + r \tag{8'}$$

which can be used in a more subtle way than previous ones.

Suppose f is an eigenfunction for the Laplacian on functions, say $\Delta f = \lambda f$. Then, unless f is a constant, df is a non-trivial eigenform for all Laplacians of the type $\Delta_{a,b} = a\, dd^* + b\, d^*d$ since $\Delta_{a,b}(df) = a\, \lambda\, df$.

As a consequence, we obtain the following result

THEOREM (A. LICHNEROWICZ-M. OBATA [15]).– *On a compact Riemannian manifold* (M, g) *with Ricci curvature bounded from below by* $k > 0$, *the lowest non-zero eigenvalue* λ_1 *of the Laplacian satisfies*

$$\lambda_1 \geq \frac{n}{n-1} k \tag{9}$$

with equality if and only if (M, g) *is isometric to the standard sphere of radius* $k^{-1/2}$.

PROOF.– The differential 1-form df is an eigenform of $\frac{n-1}{n} dd^* + d^*d$ for the eigenvalue $\frac{n-1}{n} \lambda_1$. Since $\frac{n-1}{n} dd^* + d^*d = C^*C + r$, from $r \geq k\, g$, one gets $\frac{n-1}{n} \lambda_1 \geq k$, hence the desired inequality.

Equality can only occur if df lies in the kernel of C, i.e., if the gradient of f is an infinitesimal conformal transformation. This fact (which implies that the conformal group is non compact) forces (M, g) to be isometric to a standard sphere. ∎

There is yet another system involving the operator C which has some interesting applications to a variational problem. Namely, if one considers the trace free part of the Ricci curvature $\tilde{r}_0$ of a metric $\tilde{g}$ related conformally to a metric g by $\tilde{g} = \varphi^{-2} g$, then

$$\tilde{r}_0 = r_0 + \frac{1}{n-2} \varphi^{-1}(C(d\varphi)) . \tag{10}$$

This formula can be related to the differential of the scalar curvature by the second Bianchi identity, and the following conclusion drawn (for details, see [9]).

PROPOSITION.– *If a conformal class contains an Einstein metric, this metric is up to diffeomorphism the only metric of constant scalar curvature.*

iii) On a complex manifold, a vector field Z is said to be *holomorphic* if its components (Z^α) in a holomorphic system of coordinates (z^α), i.e., such that one has $Z = \sum Z^\alpha \partial/\partial z^\alpha$, are holomorphic functions. On a Kähler manifold it is convenient to consider the form ζ of type $(0,1)$ associated to Z by the duality defined by the Kähler form ω, and to translate the condition on Z into a condition on ζ. One gets that $\nabla''\zeta = 0$ where ∇'' denotes the $(0,2)$ part of the 2-form $\nabla\zeta$.

In this context the basic Weitzenböck formula is

$$\overline{\partial}\,\overline{\partial}^* + \overline{\partial}^*\overline{\partial} = \nabla''^*\nabla'' + \rho_\omega . \tag{11}$$

It has two main applications. We first give the one related to eigenvalue bounds in the same spirit as the Obata-Lichnerowicz theorem. It is due to A. Lichnerowicz.

THEOREM.– *If the Ricci form ρ_ω of a Kähler metric ω satisfies $\rho_\omega \geq \kappa\omega$ with $\kappa > 0$, then the lowest non-zero eigenvalue λ_1 of the Laplacian on functions satisfies*

$$\lambda_1 \geq \kappa , \tag{12}$$

with equality if and only if, denoting by f a corresponding eigenfunction, $\overline{\partial} f$ is a (necessarily non isometric) holomorphic vector field.

The proof is obtained in a manner similar to the previous one by applying the Weitzenböck formula (11) to $\overline{\partial} f$.

The second application provides an obstruction to the existence of Kähler metrics with constant scalar curvature, in particular of Kähler-Einstein metrics, hence to solving (4). We state it in the Kähler-Einstein case.

THEOREM (Y. MATSUSHIMA [16]).– *On a positive Kähler-Einstein manifold, the Lie algebra of Killing vector fields is a real form of the Lie algebra of holomorphic vector fields, hence reductive.*

From the previous theorem, one derives examples of compact Kähler manifolds with positive first Chern class having no Kähler-Einstein metrics because their holomorphic vector fields do not form a reductive Lie algebra, e.g., complex manifolds obtained from complex projective spaces $\mathbf{C}P^m$ by blowing up one or two points.

b) Subelliptic estimates

Weitzenböck formulas also proved useful in Yang-Mills theory. This variational problem is defined over the space $\mathcal{C}$ of G-connections for a given compact Lie group G over a vector G-bundle $\pi : E \mapsto M$. For a connection ∇, its curvature, denoted by F^∇, is a 2-form on M with values in the bundle $\mathcal{G}_E$ of infinitesimal G-automorphisms of E which is associated to the bundle of G-frames of E by the adjoint representation on the Lie algebra of G. At a connection ∇, the *Yang-Mills functional* $\mathcal{YM}$ is then defined as half the L^2-norm of its curvature, i.e.,

$$\mathcal{YM}(\nabla) = \frac{1}{2}\int_M \|F^\nabla\|^2 v_g\,,$$

the norm used involves a Riemannian metric g on the base manifold M, and an appropriate invariant inner product along the fibres of $\mathcal{G}_E$.

The system of Yang-Mills equations describing the critical points of $\mathcal{YM}$ is

$$\begin{cases} d^\nabla F^\nabla = 0 \\ d^{\nabla *} F^\nabla = 0 \end{cases} \tag{13}$$

hence states that the curvature F^∇ of the connection ∇ is harmonic. This system is therefore a nice elliptic system in the curvature. Notice that the notion of exterior differential extends to the case of vector-valued exterior differential forms such as F^∇ to the expense that it involves a connection on that bundle (in our case, we of course used the connection ∇, hence the notation d^∇). This makes the system non linear in ∇. Here, we point out that the system is far more complicated as a system in the connection because of the presence of an infinite dimensional group of invariance, the *gauge group*, i.e., the group of sections of the automorphism bundle of E.

Notice further that the two parts of the system have different statutes, as we already pointed out when we commented on the system of Maxwell equations. The first one $d^\nabla F^\nabla = 0$ is a universal identity valid for any connection ∇, the so-called *second Bianchi identity*. The second one is the true *field equation* (as physicists would say).

To study the regularity of the system, one reduces it to a subelliptic estimate by the following token. One starts with the Weitzenböck formula for the Hodge-de Rham Laplacian on 2-forms

$$d^\nabla d^{\nabla *} + d^{\nabla *} d^\nabla = \nabla^* \nabla + \text{ curvature term}$$

that one applies to the curvature F^∇. The curvature term here refers to the bundle $\wedge^2 T^*M \otimes \mathcal{G}_E \mapsto M$ equipped with the tensor product connection, hence splits into two terms since the curvature is a derivation, one involving the curvature of the Riemannian metric g on M, the other involving the curvature of the bundle $\mathcal{G}_E$, i.e., in fact F^∇ itself. This introduces a non-linear dependance in F^∇ into the problem, and this fact has far reaching consequences for the study of the possible singularities of the problem. By taking the scalar product of the previous equation with F^∇, one then has

$$(14) \qquad (\nabla^* \nabla F^\nabla, F^\nabla) = Q_2(F^\nabla) + Q_3(F^\nabla)$$

where Q_2 et Q_3 are respectively quadratic and cubic functions in F^∇. The final step needed to get the subelliptic estimate is to make appear the Laplacian of $\|F^\nabla\|$. This follows from the identity :

$$(15) \qquad (\nabla^* \nabla F^\nabla, F^\nabla) = \frac{1}{2} \Delta \|F^\nabla\|^2 + \|\nabla F^\nabla\|^2 .$$

Since $\Delta \|F^\nabla\|^2 = 2\,\|F^\nabla\|\,\Delta \|F^\nabla\| - 2\|d\|F^\nabla\|\|^2$, using (15), we can transform (14) into the inequality

$$(16) \qquad \Delta \|F^\nabla\| \leq c_2 \|F^\nabla\| + c_3 \|F^\nabla\|^2$$

(where c_2 ans c_3 are positive constants, independent of ∇) as soon as one has noticed that

$$(17) \qquad \|d\|F^\nabla\|\|^2 \leq \|\nabla F^\nabla\|^2 ,$$

the so-called *Kato inequality* which can be reduced to a Cauchy-Schwarz inequality after simple algebraic manipulations. This estimate is crucial (cf. [25]) in proving that a Yang-Mills has no point singularity, hence does not blow up in the neighbourhood of a point.

c) Results requiring a more careful analysis of the curvature term.

In this paragraph we quote a few results using Weitzenböck formulas, but requiring a more precise description of the curvature term. We will be more sketchy, and refer most of the time to the original articles. The point we want to make here is the following. Although Weitzenböck formulas go back to the golden days of tensor calculus, hence are often presented with plenty of indices, it is sometimes crucial to understand the curvature contributions in intrinsic terms.

We begin by a slight (but crucial) modification of the Weitzenböck formula for 2-forms involving the Hodge-de Rham Laplacian. Indeed, on an oriented 4-dimensional manifold, the bundle of exterior 2-forms $\wedge^2 T^*M$ splits as the direct sum of two 3-dimensional bundles, the bundle of *positive* (or *self-dual*) 2-forms $\wedge^2 T^*M$ and the bundle of *negative* (or *anti-self-dual*) ones $\wedge^2 T^*M$ which are the eigenbundles of the Hodge linear map $*$ sending a decomposable 2-form viewed as an oriented 2-plane to its orthogonal oriented complement.

Connections on a bundle over an oriented compact 4-manifold for which the curvature is self-dual (or anti-self-dual) are absolute minima of the Yang-Mills functional (cf. [10]) as one easily sees since $d^* = * \circ d \circ *$.

On some special manifolds, to construct the special solutions of the Yang-Mills equations, that self-dual solutions are, C. Taubes (cf. [23]) starts by producing an almost self-dual connection by an explicit geometric construction. He then goes on improving this connection by an iteration scheme. To control the steps of his scheme, it is needed to bound the size of the correction term. It is convenient to take this term as $d^*\alpha$ where α is an anti-self-dual 2-form. This 2-form at stage n, called α_n, is determined as the solution of the system

$$P_- \circ d^{\nabla_{n-1}} \circ (d^{\nabla_{n-1}})^* \alpha_n = P_-(F^{\nabla_{n-1}}) \tag{18}$$

where P_- denotes projection onto negative forms. One then wants a priori estimates on some Sobolev norms of α_n. This is achieved by the Weitzenböck formula for the (elliptic) operator $P_- \circ d \circ d^*$ acting on anti-self-dual 2-forms.

$$P_- \circ d \circ d^* = \nabla^*\nabla + \ell_1(R_g) + \ell_2(P_-(F^\nabla))$$

where ℓ_1 and ℓ_2 are linear maps from $\wedge^- T^*M$ to itself depending linearly respectively on R_g and $P_-(F^\nabla)$. The extra piece of information, here, is that the curvature of the bundle enters only *via its anti-self-dual part.*

The representation theoretical fact behind this phenomenon is the following elementary lemma (one must say, a well kept secret).

LEMMA.– *As a tensor product of* SO_4*-modules,* $\wedge^+\mathbf{R}^4 \otimes \wedge^-\mathbf{R}^4$ *is isomorphic to the* SO_4*-module of traceless symmetric 2-tensors on* $\mathbf{R}^4$.

This fact is pertinent to our discussion since the self-dual part of the curvature $P_+(R^\nabla)$ acting on an anti-self-dual 2-form will produce a symmetric 2-tensor, hence does not contribute to the anti-self-dual 2-form that will appear after applying the curvature term.

This is of course crucial to keep the approximation scheme going since we want correction terms to get smaller and smaller.

To close this section let us mention two examples taken from [7] of further information which can be derived from a proper structuring of the curvature term in a Weitzenböck formula :

i) the formula for the Hodge-de Rham Laplacian *on forms of degree half the dimension of the manifold* involves only the scalar curvature and the Weyl conformal curvature tensor of the Riemannian metric. *The tracefree part of the Ricci curvature has no effect.* This makes this formula particularly attractive for conformally flat manifolds (i.e., those for which the Weyl conformal curvature tensor vanishes).

ii) for exterior differential forms taking their values in bundles of exterior forms, e.g., the Riemann curvature tensor when viewed as a 2-form taking its values in 2-forms, or a linear map viewed as a 1-form with values in 1-forms, the two terms of the Weitzenböck formula for the vector-valued Hodge-de Rham Laplacian behave very differently with respect to symmetries (or skew symmetries) between the form and bundle variables. This has interesting geometric consequences on the interplay between vector-valued harmonic forms exhibiting these symmetries, and the curvature of the Riemannian metric.

IV. Some more recent developments

We present here two recent developments connected to Weitzenböck formulas.

a) An improved Kato inequality

Although elementary, the Kato inequality (17) was a basic ingredient in using Weitzenbčk formulas, notably in establishing subelliptic estimates. Let us recall that it says that, for any tensor field t, one has the following inequality

$$\|d\|t\|\|^2 \leq \|Dt\|^2 \ . \tag{19}$$

The key observation is that this inequality can be phrased as

$$\|(t, Dt)\|^2 \leq \|t\|^2 \|Dt\|^2$$

where (t, Dt) means the 1-form obtained by contracting all indices of t against those of Dt not involving differentiation.

In many geometric situations (as we saw), one has to study tensor fields of a certain type which are annihilated by a natural first order differential operator, say A. As we pointed out, this means that Dt lies in the kernel of the symbol map σ_A. In this situation it is therefore natural to study the following finite dimensional variational problem where the type of the tensor t is fixed (we denote this space of tensors by E)

$$\text{(20)} \qquad \textit{for } t \in E,\ T \in \ker \sigma_A \subset \mathbf{R}^n \otimes E,\ \textit{find} \max_{\substack{\|t\|^2=1 \\ \|T\|^2=1}} \|(t,T)\|^2 .$$

The solution to this variational problem can be easily achieved by the Lagrange multiplier technique. Whereas the plain Kato inequality, based on the Cauchy-Schwarz inequality, proposes 1 as maximum, under the refined assumption that t satisfies a first-order differential condition, one can improve this constant.

The reason for this is the following. The value 1 would be achieved for $T = \xi \otimes t$ which does not belong to the kernel of σ_A. (Tensors of the type $\xi \otimes t$ do never belong to any irreducible component of $\mathbf{R}^n \otimes E$.) The specific value of the maximum has to be worked out for each specific case of a tensor type E and a fisrt order differential operator A. (For details, see [8]). Here, we give one specific example which was our starting point in this investigation.

PROPOSITION.– *If h is a traceless symmetric 2-tensor field satisfying the Codazzi equation (i.e., h being closed as a TM-valued 1-form), then*

$$\|d\|h\|\|^2 \leq \frac{n}{n+2} \|Dh\|^2 .$$

Such an h appears naturally as the second fondamental form of a minimal hypersurface in Euclidean space. Minimality is responsible for h being traceless, and the Codazzi equation relates the first and second fundamental forms of any hypersurface in Euclidean space.

Such an improved Kato inequality is basic in studying the regularity of the minimal submanifold equation (see [21] for a use of a weaker version of this improved Kato inequality, but there with a much less transparent proof).

b) New types of Weitzenböck formulas

One of the main drawbacks of Weitzenböck formulas is that, as we saw, they work when the curvature has a fixed sign, most often positive.

In [17], A. Polombo has been able to circumvent this difficulty for fields of symmetric endomorphisms ϵ of natural bundles by establishing Weitzenböck formulas for their eigenvalues. (A typical geometric example of such an object is the curvature viewed as a map on 2-forms). Indeed, if one introduces the spectral decomposition $\epsilon = \sum \lambda_i \omega_i \otimes \omega_i$, it is possible to compute $\Delta\lambda_i$ outside the bifurcation set of the eigenvalues provided ϵ is harmonic in an appropriate sense. One then gets a formula of the type

$$\Delta\lambda_i = \sum (\lambda_j - \lambda_i)\left(c_{ij} + \|(D\omega_i, \omega_j)\|^2\right) \tag{21}$$

where c_{ij} is a curvature term.

The flexibility gained in this way lies in the possibility of choosing the functions of the eigenvalues to be considered. The usual tensorial Weitzenböck formulas were of course dealing with the tensorial function $\sum_i \lambda_i^2 = |\epsilon|^2$. The new idea here is to venture into non invariantly defined functions of the eigenvalues such as $\sup_i \lambda_i$. If this function is chosen cleverly, one can prove that the function considered is subharmonic under negative curvature assumptions, and finally get vanishing theorems. In fact, what counts is more the fact that eigenvalues are not too dispersed.

There is of course a price to pay for daring to leave the safe banks of tensor territory. The functions to be considered are *not everywhere differentiable.* Hopefully, Analysis comes to the rescue since it is possible to prove that harmonicity of the field ϵ ensures that the "bad set" has codimension at least 2, hence that the subharmonic function can be extended across it.

We close by stating one specific application of the scheme we just described

THEOREM.– *Let M be a compact 4-dimensional Kähler-Einstein manifold with non-positive scalar curvature.*

If the eigenvalues of the anti-self-dual part of the Weyl tensor $\lambda_1 \leq \lambda_2 \leq \lambda_3$ satisfy the dispersion condition $-1.08\,\lambda_1 \leq \lambda_2 \leq -2\,\lambda_1$, then M has constant holomorphic sectional curvature, hence is a quotient of the standard disk in $\mathbf{C}^2$.

This result is an improved version of a result due to Y.T. Siu and P. Yang (cf. [21]), the understanding of which was Polombo's starting point.

References

[1] T. AUBIN, *Equations différentielles non linéaires et problème de Yamabe concernant la courbure scalaire*, J. Maths Pures Appl. **55** (1976), 269-296.

[2] A. BAHRI, à paraître.

[3] P. BÉRARD, *From vanishing theorems to estimating theorems; the Bochner technique revisited*, Bull. Amer. Math. Soc. **19** (1988), 371-406.

[4] M. BERGER, P. GAUDUCHON, E. MAZET, *Le spectre d'une variété riemannienne*, Lect. Notes in Math. **194**, Springer, Berlin-New York (1971).

[5] S. BOCHNER, *Vector fields and Ricci curvature*, Bull. Amer. Math. Soc. **52** (1946), 776-797.

[6] J.-P. BOURGUIGNON, Exposé n°XVI, *in Géométrie riemannienne de dimension* 4, Séminaire A. Besse, CEDIC-Nathan, Paris (1981).

[7] J.-P. BOURGUIGNON, *Les métriques à courbure harmonique sur une variété compacte à signature non nulle sont d'Einstein*, Inventiones Math. **63** (1981), 263-286.

[8] J.-P. BOURGUIGNON, *Une amélioration de l'inégalité de Kato et applications géométriques* (à paraître).

[9] J.-P. BOURGUIGNON, J.-P. EZIN, *Fonctions courbure interdites dans une classe conforme et transformations conformes*, Trans. Amer. Math. Soc. **301** (1987), 723-736.

[10] J.-P. BOURGUIGNON, H.B. LAWSON, *Stability and isolation phenomena for Yang-Mills fields*, Commun. Math. Phys. **79** (1981), 189-230.

[11] S. KOBAYASHI, *Transformation groups in differential geometry*, Erg. Math. **70**, Springer, Berlin-Heidelberg-New York (1972).

[12] K. KODAIRA, J. MORROW, *Complex manifolds*, Holt-Rinehart-Wilson, New York (1971).

[13] N. KOISO, *Non-deformability of Einstein metrics*, Osaka J. Math. **15** (1978), 419-433.

[14] J.M. LEE, T. PARKER, *The Yamabe problem*, Bull. Amer. Math. Soc. **17** (1987), 37-91.

[15] A. LICHNEROWICZ, *Géométrie des groupes de transformations*, Dunod, Paris (1958).

[16] Y. MATSUSHIMA, *Remarks on Kähler-Einstein manifolds*, Nagoya Math. J. **46** (1972), 161-173.

[17] A. Polombo, *Condition d'Einstein et courbure négative en dimension 4*, C.R. Acad. Sci. Paris **307** (1988), 667-670.

[18] R. Schoen, *Conformal deformation of a Riemannian metric to constant scalar curvature*, J. Differential Geom. **20** (1984), 479-495.

[19] R. Schoen, C.I.M.E. Lecture Notes Series (1988).

[20] R. Schoen, L. Simon, S.T. Yau, *Curvature estimates for minimal hypersurfaces*, Acta Math. **134** (1975), 479-495.

[21] Y.T. Siu, P. Yang, *Compact Kähler surfaces of non-positive bisectional curvature*, Inventiones Math. **64** (1981), 471-487.

[22] P. Stredder, *Natural differential operators on Riemannian manifolds and representations of the orthogonal and special orthogonal groups* , J. Differential Geom. **10** (1976), 647-660.

[23] C.H. Taubes, *Self-dual Yang-Mills connections on non selfdual 4-manifolds*, J. Differential Geom. **17** (1982), 139-170.

[24] N. Trudinger, *Remarks concerning the conformal deformation of Riemannian structures on compact manifolds*, Ann. Scuola Norm. Sup. Pisa **22** (1968), 265-274.

[25] K. Uhlenbeck, *Removable singularities for Yang-Mills fields*, Commun. Math. Phys. **83** (1982), 11-30.

[26] F.W. Warner, *Foundations of differentiable manifolds and Lie groups*, Universitext **94**, Springer, New York-Heidelberg-Berlin (1983).

[27] H. Wu, *The Bochner technique in differential geometry*, Math. Reports, London (1987).

[28] H. Yamabe, *On a deformation of Riemannian structures on compact manifolds*, Osaka Math. J. **12** (1960), 21-37.

Jean Pierre BOURGUIGNON

Centre de Mathématiques
Ecole Polytechnique
U.R.A. du C.N.R.S. D 0169
F-91128 PALAISEAU Cedex
(France)

A Remark on Minimal Surfaces with Corners

MICHAEL GRÜTER

Here, I want to report on joint work with Leon Simon [GS].

In the Calculus of Variations one is basically confronted with two problems: the *existence* of minimizers or stationary points, and—since one usually obtains weak solutions—the question of *regularity*. We want to make a small contribution to the latter problem. Although we consider the classical case of two-dimensional minimal surfaces in $\mathbb{R}^3$ we shall not use the parametric approach but instead work in the context of Geometric Measure Theory (henceforth called GMT) which turns out to have some advantages even in this case.

We are interested in minimal surfaces with corners, for example minimal surfaces in $\mathbb{R}^3$ bounded by a polygon or having a partially free boundary. If a polygonal boundary Γ is given and one looks at the Douglas–Radó solution of the Plateau problem for Γ it is in general only possible to prove Hölder continuity of the solution up to the boundary. This is because the parametrization is a conformal map and thus preserves angles. Furthermore, there may exist branch points at a corner. For a discussion of minimal surfaces in the parametric setting the reader is referred to the fundamental monograph [NJ], the papers [DG], [JJ], [GM1], and the references given there.

In the approach via GMT we can only handle minimizing solutions in the class of (integer multiplicity) rectifiable currents. See Section 1 for the relevant definitions. In [GS] we treat the case of a fixed boundary having a corner, the partially free boundary problem—on which we concentrate here—as well as an application to minimal graphs in $\mathbb{R}^{n+1}$. All these applications are based on a simple but important observation which can be phrased as:

"A unique measure theoretic tangent cone contained in a hyperplane and regularity off the corner imply regularity up to the corner."

In Section 2 we describe in some detail the application of this principle to the partially free boundary problem. We hardly use the language of GMT and try to stay in the setting of submanifolds.

Our result gives that in a neighbourhood of the corner the minimizing current is given by the graph of a $C^{1,\alpha}$-function (compare the end of Section 2) defined over a simply connected domain in a plane. For further information one can then use non-parametric methods.

A fundamental tool in the proof is an abstract regularity theorem for varifolds, originally proved by Allard [AW1] in the case of interior regularity. The analogous result for the fixed smooth boundary is also due to Allard [AW2] while the case of a free boundary was treated by Jost and the author [GJ]. Although the exact statements of these theorems are rather technical and use the language of GMT they really describe smooth surfaces (in arbitrary dimension and codimension) using only terms that make sense for such weak objects as varifolds. For smooth n-dim minimal surfaces in $\mathbb{R}^{n+k}$ they say, e.g. in the interior case, that if the n-area of the surface in a ball of radius ρ centered at some point on the surface is not much larger than the area of the corresponding n-dim ball, then the surface can be written as a graph over an n-dim ball B of radius $\beta\rho$, $\beta = \beta(n,k)$. Actually, we will use a more technical version of this theorem which allows a-priori (under some additional condition) to determine the n-plane containing B.

1. Terminology

Let $U \subset \mathbb{R}^{n+k}$ be an open set. We define

$$\mathcal{D}^n(U) = \{C^\infty\text{-}n\text{-forms } \omega, \operatorname{spt}\omega \subset U\}$$

with the usual topology of uniform convergence of all derivatives on compact subsets. Its dual space is denoted by $\mathcal{D}_n(U)$. The elements of $\mathcal{D}_n(U)$ are called *n-currents* in U. If $T \in \mathcal{D}_n(U)$ and $W \subset U$ is open the *mass* of T in W is defined by

$$\mathbf{M}_W(T) := \sup\{T(\omega) : \omega \in \mathcal{D}^n(U), \operatorname{spt}\omega \subset W, |\omega| \le 1\} \le \infty.$$

The *boundary* of T is the $(n-1)$-current ∂T in U given by (d = exterior derivative)

$$\partial T(\omega) := T(d\omega).$$

The n-currents can be thought of as generalized n-dim oriented submanifolds of locally finite $\mathcal{H}^n$-measure ($\mathcal{H}^n = n -$ dim Hausdorff measure) as one sees from the following example.

Let $M \subset U$ be an n-dim oriented C^1-submanifold such that for every compact set K

$$\mathcal{H}^n(M \cap K) < \infty.$$

Then $[[M]] \in \mathcal{D}_n(U)$ is defined by (oriented integration)

$$[[M]](\omega) = \int_M \omega\,;$$

for the mass we get

$$\mathbf{M}_W([[M]]) = \mathcal{H}^n(M \cap W).$$

If also ∂M is C^1 and has locally finite $\mathcal{H}^{n-1}$-measure we have by Stokes' formula

$$(\partial[[M]])(\omega) = [[M]](d\omega) = \int_M d\omega = \int_{\partial M} \omega = [[\partial M]](\omega),$$

i.e.

$$\partial[[M]] = [[\partial M]].$$

A set $M \subset \mathbb{R}^{n+k}$ is called *countably n-rectifiable* if M is $\mathcal{H}^n$-measurable and if

$$M \subset \bigcup_{j=0}^{\infty} M_j$$

where $\mathcal{H}^n(M_0) = 0$ and for $j \geq 1$ M_j is an n-dim C^1-submanifold of $\mathbb{R}^{n+k}$. A basic fact about countably n-rectifiable subsets M is that they possess $\mathcal{H}^n$-a.e. an *approximate tangent space* T_xM (in the measure theoretic sense).

An n-current $T \in \mathcal{D}_n(U)$ is called *(integer multiplicity) rectifiable*, if

$$T(\omega) = \int_M \langle \omega, \xi \rangle \theta \, d\mathcal{H}^n, \qquad \omega \in \mathcal{D}^n(U),$$

where $M \subset U$ is countably n-rectifiable, $\theta \geq 0$ a locally $\mathcal{H}^n$-integrable integer valued function, and $\xi(x) = \tau_1 \wedge \ldots \wedge \tau_n$, for $\mathcal{H}^n$-a.e. $x \in M$, $\{\tau_1, \ldots, \tau_n\}$ being an orthonormal basis of the approximate tangent space T_xM.

We use the notation $\mu_T = \mathcal{H}^n \llcorner \theta$ and

$$T = \tau(M, \theta, \xi).$$

Forgetting the orientation ξ we get an (*integer multiplicity*) *rectifiable n-varifold*

$$V = \mathbf{v}(M, \theta).$$

A general *varifold* in U is a Radon measure on $U \times G(n+k, n)$. For a rectifiable n-varifold $V = \mathbf{v}(M, \theta)$ and any function $f \in C_c^0(U \times G(n+k, n), \mathbf{R})$ we have

$$V(f) = \int_M f(x, T_x M)\theta(x) d\mathcal{H}^n(x).$$

The other notation that will be used is hopefully self-explaining.

For more information on GMT the reader is referred to [FH] and [SL].

2. Minimizing Rectifiable Currents with Partially Free Boundary

In the partially free boundary problem one looks at the following situation (locally).

Suppose, we are given a supporting C^2-surface S in $\mathbf{R}^3$ and a C^2-curve γ which lies (locally) on one side of S such that near the origin $\gamma \cap S = \{0\}$. Furthermore, we assume that γ meets S transversely.

Let us consider a rectifiable (integer multiplicity) 2-current T in $B_1(0) \subset \mathbf{R}^3$ having the properties:

$$\operatorname{spt}\left[(\partial T - [[\gamma]]) \llcorner B_1(0)\right] \subset S, \tag{1}$$

$$\mathbf{M}_W(T) \leq \mathbf{M}_W(T + X) \tag{2}$$

for every open $W \subset\subset B_1(0)$, and any rectifiable 2-current X such that $\operatorname{spt} X \subset W$ as well as $\operatorname{spt} \partial X \subset S$.

Note, that the existence of such a solution T to the global partially free boundary problem (S connected, endpoints of γ contained in S) is easily deduced from quite general compactness theorems (compare [FH] 5.1.6(1), and for a more detailed argument [GM2] 1.8). What can be said about the regularity of T?

In a neighbourhood of each $x \in \operatorname{spt} T \sim \operatorname{spt} \partial T$ we have that T is given by (m-times, $m \in \mathbf{N}$) integration over an embedded minimal surface. This follows from the classical interior regularity theory developed by DeGiorgi, because (near x) T can be decomposed into area-minimizing oriented frontiers ($n = 2 < 7$). The regularity near $x \in \gamma \sim \{0\}$, i.e. at the fixed smooth part of the boundary away from the corner 0, follows from the work of Hardt and Simon [HS] (no dimensional restriction in the codimension one case). Finally, the regularity at the free part of the boundary—away from the corner 0—was shown in [GM3], see also [GM4]. Here, again it is important that $n = 2 < 7$. Furthermore, it turns out that $\operatorname{spt} T$ intersects S orthogonally off the corner.

Therefore, from the existing regularity theories, we conclude that $M := \operatorname{spt} T$ is a connected embedded minimal $C^{1,\alpha}$-hypersurface away from the origin, and that its boundary and S intersect orthogonally. From now on we assume that T has multiplicity one and lies on one side of S.

Note, that for the three cases considered above (interior, fixed boundary, free boundary regularity) one has abstract regularity results in the setting of rectifiable varifolds (arbitrary dimension and codimension) by [AW1], [AW2], and [GJ] respectively. We shall make good use of them later.

Let us first consider the case where γ is perpendicular to S. In this situation we can reflect across S and use the method of [HS] to show regularity including the corner.

It remains to consider the case where the smaller angle α between T_0S and t (tangent plane to S at 0, tangent to γ at 0 respectively) satisfies $0 < \alpha < \pi/2$.

The first step in the proof is to derive a monotonicity formula valid at the corner.

If S were a plane and γ a ray one could simply use the usual vector field

$$X(x) = f(|x|)x$$

in the first variation formula

$$\int_M \operatorname{div}_M X \, d\mathcal{H}^2 = 0$$

for vector fields X with compact support that are tangent to S and γ respectively.

In the curved case one has to make a slight modification in the definition of X to get that

$$e^{c\kappa r}\mathcal{H}^2(M \cap B_r(0))r^{-2}$$

is an increasing function of r. Here, c is an absolute constant and κ some curvature bound for S and γ near 0.

From the monotonicity formula one deduces as usual the existence of a tangent cone C to M at 0 such that the Radon measures

$$\mathcal{H}^2 \llcorner (\lambda_i^{-1} M)$$

corresponding to the blown-up minimal surfaces $\lambda_i^{-1}M$ converge (in the sense of Radon measures) to μ_C as $\lambda_i \downarrow 0$ for some sequence $\{\lambda_i\}$. Note that C is a rectifiable 2-current in $\mathbf{R}^3$ with density $\theta_C = 1$ a.e.

Furthermore, C is an area-minimizing solution of the partially free boundary problem with fixed boundary t and supporting surface the plane T_0S.

Thus we may (again) apply the above-mentioned regularity theories to C to get the following:

(3) C is a cone that intersects T_0S orthogonally,

(4) the fixed boundary of C is t,

(5) C is a smooth manifold away from the origin.

From this we deduce that

$$\partial(C \llcorner B_1(0)) - \partial C \llcorner B_1(0)$$

is supported in the unique great circle ($0 < \alpha < \pi/2$!) determined by t that is perpendicular to T_0S.

Since M lies on one side of S and M has density one (implies $\theta^2(M,a) = 1/2$ for $a \in \gamma \sim \{0\}$) we see that

$$\partial(C \llcorner B_1(0)) - \partial C \llcorner B_1(0) = [[\beta]]$$

where β is one of the two geodesic arcs determined by $t \cap S^2$ which connect perpendicularly to $T_0S \cap S^2$ while staying on one side of T_0S.

This means, that there are only two possible tangent cones at the corner each of them a conical region in a plane. But this obviously implies the *uniqueness* of C because—as can easily be seen—in the case of non-uniqueness one must necessarily have a whole continuum of different tangent cones.

Set $U := \operatorname{spt} C$ and $M_\lambda := \lambda^{-1}M$.

Note that by uniqueness

(6) $$\mathcal{H}^2 \llcorner M_\lambda \to \mathcal{H}^2 \llcorner U \quad \text{as} \quad \lambda \downarrow 0.$$

The next step is to improve the above measure theoretic convergence to get convergence of M_λ to U in the Hausdorff-distance.

From (6) we get for compact sets $K \subset \mathbb{R}^3$

$$\lim_{\lambda \downarrow 0}\left[\sup_{x \in K \cap U} \operatorname{dist}(x, M_\lambda)\right] = 0.$$

On the other hand by the various monotonicity formulae (in the interior, at the fixed and free boundary) we see that e.g.

$$\lim_{\lambda \downarrow 0}\left[\sup_{x \in M_\lambda \cap A} \operatorname{dist}(x, U)\right] = 0$$

where $A = B_{3/2}(0) \sim B_{1/4}(0)$. Set $A' = B_1(0) \sim B_{1/2}(0)$. We thus have that

$$M_\lambda \to U \quad \text{in} \quad A$$

with respect to Hausdorff-distance.

We want to apply the varifold regularity theorems by Allard and Jost–Grüter to M_λ on A'.

Since M_λ has mean curvature zero, intersects S_λ $(= \lambda^{-1}S)$ orthogonally along the free boundary, and has density one (away from the boundary) the only hypothesis that remains to be checked is the following:

$$\mathcal{H}^2(M_\lambda \cap B_\rho(x)) \le (1+\epsilon)\pi\rho^2, \tag{7}$$

if x is in the interior of $M_\lambda \cap A$; and

$$\mathcal{H}^2(M_\lambda \cap B_\rho(x)) \le (1+\epsilon)\pi\rho^2/2, \tag{8}$$

if x is in the (free or fixed) boundary $(\partial M_\lambda) \cap A$. Here $\epsilon > 0$ is given, and $\rho > 0$ is some radius.

But because of the measure theoretic convergence (6) we have

$$\lim_{\lambda \downarrow 0} \mathcal{H}^2(M_\lambda \cap B_\rho(x)) = \mathcal{H}^2(U \cap B_\rho(x)). \tag{9}$$

Actually, this argument is not quite correct since $x \in M_\lambda$. Instead of using some fixed x one has to work with a sequence $\{x_\lambda\}$, $x_\lambda \in M_\lambda$, and $x_\lambda \to x_0 \in U$.

Having established (7) and (8) for λ small enough the varifold regularity theorems then imply that $M_\lambda \cap A'$ can *locally* be written as the graph of a $C^{1,\alpha}$-function. Of course, we knew that already but we shall now show that $M_\lambda \cap A'$ can be written *globally* as the graph of a $C^{1,\alpha}$-function defined on a domain in the plane H containing U.

To achieve this we argue as follows.

One more technical version of the varifold regularity theorems implies that the varifold in question can be written as a graph over some domain in a fixed plane H if the excess with respect to H is small.

The exact definition of the excess is a bit technical so we don't repeat it here. Let it suffice to say that the excess measures the mean L^2-deviation of the approximate tangent spaces to the rectifiable varifold from the fixed plane H.

Furthermore, the excess can be estimated by the normalized height of M_λ over H (similar to the Caccioppoli-inequality).

But, because of the convergence of M_λ to U with respect to Hausdorff-distance we get

$$\mathcal{H}\text{-dist}(M_\lambda \cap A, U \cap A) < \eta \tag{10}$$

for $\lambda \leq \lambda_0(\eta)$ and any given $\eta > 0$.

That is, the height is uniformly arbitrarily small. The precise argument is the following.

For fixed small $\rho > 0$ and $0 < \epsilon \ll \rho$ we cover $U^\epsilon \cap A$ ($U^\epsilon = \epsilon$-neighbourhood of U) by a finite number of balls $\{B_i\}$ of radius ρ_i, $\rho' \leq \rho_i \leq \rho$ such that the following conditions are fulfilled.

(i) The balls concentric with $\{B_i\}$ having radius $\rho_i - 2\epsilon$ still cover $U^\epsilon \cap A$.

(ii) For each ball B_i we either have (a_i = centre (B_i))

$$\mathcal{L}^2(B_i \cap U) = \pi\rho_i^2, \quad \text{if } a_i \in \text{int } U$$

or

$$\mathcal{L}^2(B_i \cap U) = \pi/2\rho_i^2, \quad \text{if } a_i \in \partial U.$$

Denote by N the number of balls $\{B_i\}$. Note, that for fixed $\{\rho_i\}$ and $\epsilon' < \epsilon$ we can use the same balls $B_i^{(\epsilon)}$ for ϵ', i.e. N can be chosen independent of ϵ. Now choose $\lambda_0(\epsilon)$ small enough such that (10) holds for $\eta = \epsilon$ and such that γ_λ is close enough in C^1 to t (one boundary ray of U), and $S_\lambda \cap A$ is C^1-close enough to $T_0 S \cap A$ ($\lambda \leq \lambda_0(\epsilon)$).

Next, choose $\lambda_1(\epsilon) \leq \lambda_0(\epsilon)$ such that for $i = 1, \ldots, N$ and for $\lambda \leq \lambda_1(\epsilon)$ we have (7) respectively (8) for the balls B_i with centre a_i and the corresponding radii ρ_i. This is possible because of (ii) and (9).

From (10) with $\eta = \epsilon$ we can then pick (for fixed $\lambda \leq \lambda_1(\epsilon)$) for each i a point $x_i^{(\lambda)} \in M_\lambda$ with $|a_i - x_i^{(\lambda)}| < \epsilon$.

The balls $B_i^{(\lambda)} = B_{\rho_i - \epsilon}(x_i^{(\lambda)})$ then satisfy (because of (ii) and the choice of $\lambda_1(\epsilon)$)

$$\begin{aligned}\mathcal{H}^2(M_\lambda \cap B_i^{(\lambda)}) &\le \mathcal{H}^2(M_\lambda \cap B_i) \le (1+\epsilon)\pi\rho_i^2 \\ &= \pi(\rho_i - \epsilon)^2 \left[(1+\epsilon)\left(1+\frac{\epsilon}{\rho_i - \epsilon}\right)^2\right] \\ &\le (1+\epsilon')\pi(\rho_i-\epsilon)^2.\end{aligned}$$

Here $\epsilon' = \epsilon'(\epsilon, \rho')$ and $\epsilon' \to 0$ as $\epsilon \to 0$.

Since $\partial M_\lambda \cap A \subset (\gamma_\lambda \cap A) \cup (S_\lambda \cap A)$, and γ_λ and S_λ are C^0-close to t and $T_0 S$ respectively, we see that for a ball B_i with $\mathcal{L}^2(B_i \cap U) = \frac{\pi}{2}\rho_i^2$ we can pick $x_i^{(\lambda)} \in \partial M_\lambda$ such that $|a_i - x_i^{(\lambda)}| < \epsilon$.

Thus, for these i, we get ($B_i^{(\lambda)} = B_{\rho_i - \epsilon}(x_i^{(\lambda)})$)

$$\mathcal{H}^2(M_\lambda \cap B_i^{(\lambda)}) \le \frac{1}{2}(1+\epsilon')\pi(\rho_i - \epsilon)^2.$$

Furthermore, by (i) and (10) we see that $\{B_i^{(\lambda)}\}$ still covers $M_\lambda \cap A$.

Together with the height estimate (10) and the C^1-closeness of γ_λ and S_λ to t and $T_0 S$ respectively we conclude that for $\epsilon > 0$ small enough ($\epsilon' \to 0$ as $\epsilon \to 0$) we can write $M_\lambda \cap A'$ (A' to avoid the boundary $\partial A \cap M_\lambda$) as the graph of a $C^{1,\alpha}$-function w_λ defined on a domain in H. The varifold regularity theorems give the additional information that

$$|Dw_\lambda|_{L^\infty} \le C\epsilon'. \tag{11}$$

But now the regularity up to the corner is immediate. We can take e.g. the sequence $\lambda_k = 2^{-k}$ for k large enough and piece together the different graphs in the annuli

$$A_m = B_{2^{-m}}(0) \sim B_{2^{-(m+1)}}(0).$$

Thus, we get a simply connected domain $W \subset H$ and a function $w \in C^{1,\alpha}(\overline{W} \sim \{0\})$, such that for some $R > 0$ fixed,

$$\overline{M} \sim \{0\} \cap B_R(0) = \operatorname{graph} w$$

and (because of (11))

$$\lim_{\substack{x \to 0 \\ x \in W}} |Dw(x)| = 0.$$

REFERENCES

[AW1] W.K. Allard, *On the first variation of a varifold*, Ann. Math. **95** (1972), 417–491.

[AW2] W.K. Allard, *On the first variation of a varifold: boundary behavior*, Ann. Math. **101** (1975), 418–446.

[DG] G. Dziuk, *Über quasilineare elliptische Systeme mit isothermen Parametern an Ecken der Randkurve*, Analysis **1** (1981), 63–81.

[FH] H. Federer, *Geometric Measure Theory*, Springer, 1969.

[GJ] M. Grüter and J. Jost, *Allard type regularity results for varifolds with free boundaries*, Ann. Sc. Norm. Sup. Pisa (4) **13** (1986), 129–169.

[GM1] M. Grüter, *The monotonicity formula in geometric measure theory, and an application to a partially free boundary problem*, in: Partial Differential Equations and Calculus of Variations. Eds. S. Hildebrandt, R. Leis; Lect. Notes in Math. **1357**, Springer, 1988.

[GM2] M. Grüter, *Regularität von minimierenden Strömen bei einer freien Randbedingung*, Habilitationsschrift, Düsseldorf, 1985.

[GM3] M. Grüter, *Optimal regularity for codimension one minimal surfaces with a free boundary*, Man. Math. **58** (1987), 295–343.

[GM4] M. Grüter, *Regularity results for minimizing currents with a free boundary*, J. Reine Angew. Math. **375/376** (1987), 307–325.

[GS] M. Grüter and L. Simon, *Regularity of minimal surfaces near corners*, in preparation.

[HS] R. Hardt, and L. Simon, *Boundary regularity and embedded solutions for the oriented Plateau problem*, Ann. Math. **110** (1979), 439–486.

[JJ] J. Jost, *Continuity of minimal surfaces with piecewise smooth free boundaries*, Math. Ann. **276** (1987), 599–614.

[NJ] J.C.C. Nitsche, *Vorlesungen über Minimalflächen*, Springer, 1975.

[SL] L. Simon, *Lectures on geometric measure theory*, Proc. Centre for Mathematical Analysis, Australian National University, Canberra, **3**, 1983.

Fachbereich Mathematik
Universität des Saarlandes
D-6600 Saarbrücken
West Germany

Extremal Surfaces of Mixed Type in Minkowski Space $\mathbf{R}^{n+1}$

GU CHAOHAO (C. H. GU)*

Abstract

A connected 2-dimensional submanifold in Minkowski space is called a surface of mixed type if it contains a space-like part and a time-like part simultaneously. In the present paper we consider the extremal surfaces of mixed type in Minkowski space $\mathbf{R}^{n+1}$.

Suppose that the surface is C^3 and the gradient of the square of the area density does not vanish on the light-like points of the surface, then we obtain the general explicit expression of the surface and prove that

(a) The time-like part and space-like part are separated by a null-curve.

(b) The surface is analytic not only on the space-like part but also in some mixed region.

(c) There is an explicit algorithm for the construction of all these extremal surfaces of mixed type globally, starting from given analytic curves in $\mathbf{R}^n$.

The same results for 3-dimensional Minkowski space were obtained earlier [G2], [G3].

1. Introduction

Extremal surfaces in the Minkowski spaces $\mathbf{R}^{n+1}$ are 2-dimensional C^2-submanifolds with vanishing mean curvature vectors. A connected surface is called of mixed type if it contains a space-like part and a time-like part simultaneously. There were a series of works devoted to the purely space-like or purely time-like extremal surfaces in Minkowski space

* The work was supported by the French Univ. council and the Chinese Fund of Natural Science.

or Lorentzian manifolds (e.g. [B][CY][Ch][Q][Ba][M][G1]), the author of this present paper considered extremal surfaces of mixed type in 3-dimensional Minkowski space $\mathbf{R}^{2+1}$ [G2], [G3]. The following results were obtained:

(a) The bordline of the space-like part and the time-like part is a null curve.

(b) The surface is analytic not only on the space-like part where the equation is elliptic but also on some mixed region near the bordline in spite of the fact that the equation is hyperbolic on the time-like part.

(c) Starting with an analytic plane curve there is an "explicit algorithm" for constructing all generic extremal surfaces of mixed type globally.

For the case $n = 2$ these results were obtained under the assumption that the extremal surfaces are C^2 and free of flat points ($K \neq 0$). The main approach was using the Legendre transformation to linearize the partial differential equation $H = 0$ which is quasilinear and of mixed type. For the case of higher dimensions we have to solve a system of quasilinear partial differential equations of mixed type. The Legendre transformation does not work. It is found that the above-mentioned results hold in $\mathbf{R}^{n+1}$ by considering the existence and the behavior of isothermal coordinates. The only assumption is that at all light-points, the gradient of the square of the area density (see §2) does not vanish. In the case of $n = 2$, this assumption is a consequence of $K \neq 0$.

In §2 some notations and definitions are presented. §3 is devoted to the existence and behavior of isothermal coordinates. In §4 the general expressions and properties (a), (b) are obtained, and the "explicit algorithm" for global construction of extremal surfaces of mixed type is described in §5 together with two examples.

The main results were obtained when the author was visiting the University Paris VI, Laboratory of Relativistic Mechanics. The author is very grateful to Prof. Y. Choquet-Bruhat for her hospitality and for the attentions to this work. He is indebted to Prof. H. Brezis and the organizers of the International Symposium on Variational Problems, held in Paris, 1988 June.

2. Notations and definitions

Let $X = (x_1, \ldots, x_n, x_{n+1})$ be the position vector of a point in the Minkowski space $\mathbf{R}^{n+1}$. The scalar product of two vectors X and Y is

$$X \cdot Y = x_1 y_1 + x_2 y_2 + \cdots + x_n y_n - x_{n+1} y_{n+1} \tag{2.1}$$

in particular,

$$X^2 = x_1^2 + x_2^2 + \cdots + x_n^2 - x_{n+1}^2 \tag{2.2}$$

The Lorentz metric of the space $\mathbf{R}^{n+1}$ is

$$ds^2 = dx_1^2 + dx_2^2 + \cdots + dx_n^2 - dx_{n+1}^2 \tag{2.3}$$

Let $z = (x_3, x_4, \ldots, x_{n+1})$, The local equation of a surface S_2 in $\mathbf{R}^{n+1}$ in a suitable coordinate system can be written as

$$z = f(x_1, x_2) \qquad (\text{or } z_i = f_i(x_1, x_2), \quad i = 3, 4, \ldots, n+1) \tag{2.4}$$

We also use the notation $x = x_1, y = x_2$ and

$$p = f_x, \quad q = f_y, \quad r = f_{xx}, \quad s = f_{xy}, \quad t = f_{yy} \tag{2.5}$$

Here p, q, r, s, t have $n-1$ components. Moreover, we use the notations

$$\begin{gathered} p^2 = p_3^2 + \cdots + p_n^2 - p_{n+1}^2, \qquad q^2 = q_3^2 + \cdots + q_n^2 - q_{n+1}^2 \\ p.q = p_3.q_3 + \cdots + p_n.q_n - p_{n+1}.q_{n+1} \end{gathered} \tag{2.6}$$

The induced metric or the first fundamental form of S_2 is

$$ds_1^2 = dx_1^2 + dx_2^2 + df^2 = (1+p^2)\,dx^2 + 2pq\,dx\,dy + (1+q^2)\,dy^2 \tag{2.7}$$

Let

$$w = (1+p^2)(1+q^2) - (p.q)^2 \tag{2.8}$$

If $w > 0$ at $P \in S_2$ then S_2 is space-like at P and if $w < 0$ at $P \in S_2, S_2$ is time-like at P. The metric is Riemannian or Lorentzian if the surface is space-like or time-like.

A connected surface is of mixed type if it contains both a space-like part and a time-like part.

The area element of S_2 is

$$dA = |w|^{1/2}\,dx\,dy \tag{2.9}$$

If S_2 is a C^2 surface and satisfies

$$\delta \int dA = 0 \tag{2.10}$$

then the surface is called extremal. It is well-known that S_2 is extremal if and only if its mean curvature vector H vanishes or the system of partial differential equations

$$(1+p^2)t - 2p.qs + (1+q^2)r = 0 \tag{2.11}$$

is satisfied. It is seen that the system is elliptic or hyperbolic if $w > 0$ or $w < 0$ respectively. Since we consider extremal surfaces of mixed type, we have to consider this system in a mixed region.

In this paper we assume that S_2 is C^2 and

$$dw \neq 0 \tag{2.12}$$

at each point $P \in S_2$ where $w = 0$. From this assumption it is easily seen that the time-like region and space-like region are separated by C^1-curves.

3. Isothermal coordinates

Isothermal coordinates (θ_1, θ_2) are special local coordinates of a surface such that the metric ds_1^2 is conformally flat, i.e.,

$$ds_1^2 = \begin{cases} A(\theta_1,\theta_2)(d\theta_1^2 + d\theta_2^2) & \text{(space-like part)} \\ A(\theta_1,\theta_2)(d\theta_1^2 - d\theta_2^2) & \text{(time-like part)} \end{cases} \tag{3.1}$$

The existence of isothermal coordinates near a space-like point or a time-like point is well-known. The problem is to construct isothermal coordinates near a light-like point with some continuity.

As in the case of Euclidean space [O] the isothermal coordinates for the extremal surface take a more explicit form

$$\theta_1 = x + F(x,y), \qquad \theta_2 = y + G(x,y) \tag{3.2}$$

where F and G are determined by the following completely integrable system of equations

$$\partial F/\partial x = \pm(p^2+1)/\left|w\right|^{1/2} \qquad \partial F/\partial y = \pm p.q/\left|w\right|^{1/2} \tag{3.3}$$

and

$$\partial G/\partial x = +p.q/\left|w\right|^{1/2}, \qquad \partial G/\partial y = +(q^2+1)/\left|w\right|^{1/2} \tag{3.4}$$

Here the plus and minus signs in (3.3) stand for the space-like case and time-like case respectively.

Let C be a curve defined by $w = 0$ and be the bordline of space-like and time-like regions. Around a point $P \in C$ there is a local parametrization (t, w) of S_2

$$x = x(t, w) \qquad y = y(t, w) \tag{3.5}$$

such that the curve $w = 0$ is the local representation of the bordline C. Let (t, w) be defined on a region Ω containing P and Ω_+, Ω_- be the space-like part and time-like part of Ω respectively.

Equation (3.3) can be written in the form

$$\begin{aligned} \partial F/\partial t &= \pm |w|^{-1/2} \alpha(t, w), \qquad & \partial F/\partial w &= \pm |w|^{-1/2} \beta(t, w) \\ \partial G/\partial t &= + |w|^{-1/2} \gamma(t, w), \qquad & \partial G/\partial w &= |w|^{-1/2} \delta(t, w) \end{aligned} \tag{3.6}$$

Here α, β, γ, δ are C^1-functions. From the condition of integrability it is seen that $\alpha(t, w) = w\alpha_1(t, w)$, $\gamma(t, w) = w\gamma_1(t, w)$ here α_1 and γ_1 are C^1 in Ω_+ and Ω_-.

Let P be an arbitrary point at C^1 and assume that and $F(P) = 0$, $G(P) = 0$. By integration we have

$$F(t, w) = \begin{cases} \int_0^w w^{-1/2} \beta(t, w)\, dw & (w > 0) \\ 0 & (w = 0) \\ -\int_0^w (-w)^{-1/2} \beta(t, w)\, dw & (w < 0) \end{cases} \tag{3.7}$$

$$G(t, w) = \begin{cases} \int_0^w w^{-1/2} \delta(t, w)\, dw & (w > 0) \\ 0 & (w = 0) \\ \int_0^w (-w)^{-1/2} \delta(t, w)\, dw & (w < 0) \end{cases} \tag{3.8}$$

Considering the limits of derivatives of $|w|^{-1/2} F$ and $|w|^{-1/2} G$ when $w \to 0$, we have:

Lemma 1. *There exist functions*

$$\begin{aligned} \theta_1 &= x(t, w) + |w|^{1/2} a(t, w) \\ \theta_2 &= y(t, w) + |w|^{1/2} b(t, w) \end{aligned} \tag{3.9}$$

*such that (θ_1, θ_2) are isothermal coordinates of S_2 on Ω_+ and Ω_-. Here $a(t, w)$ and $b(t, w)$ are C^1-functions for $\theta_2 \geq 0$ or $\theta_2 \leq 0$.**

---

* This lemma was also proved by Miss Zhang Lin independently.

Let (θ_1, θ_2) be valued in a region Σ and the image of $\Omega_+ \cup C$ be denoted by Σ_+. There exists a conformal mapping $(\theta_1, \theta_2) \to (\phi_1, \phi_2)$ such that the image of curve C lies on the line $\phi_2 = 0$ and that of Σ_+ lies in $\phi_2 > 0$. Evidently, we have

$$\phi_1 = l(t, 0) + w^{1/2} c(t, w), \qquad \phi_2 = w^{1/2} h(t, w) \tag{3.10}$$

here l, c and h are C^1 when $w > 0$ and continuous up to $w = 0$. Evidently the conformal mapping $(\theta_1, \theta_2) \to (\phi_1, \phi_2)$ can be extended to the time-like part continuously such that (ϕ_1, ϕ_2) are isothermal coordinates of the time-like part, i.e.

$$\phi_{1\theta_1} = \phi_{2\theta_2} \qquad \phi_{2\theta_1} = \phi_{1\theta_2} \tag{3.11}$$

are satisfied. It is seen that the time-like part corresponds to a region in $\phi_2 < 0$ and ϕ_1, ϕ_2 can be written as

$$\phi_1 = l(t, 0) + (-w)^{1/2} c_1(t, w), \qquad \phi_2 = (-w)^{1/2} h_1(t, w) \tag{3.12}$$

Hence we have

Lemma 2. *The behavior of the isothermal coordinates near the bord curve C is shown by* (3.10) *and* (3.12).

We need the following:

Lemma 3. *Under the condition* (2.12)

$$\lim_{\phi_2 \to 0\pm} A(\phi_1, \phi_2) = 0 \tag{3.13}$$

Proof. For $\phi_2 > 0$

$$A\, d\phi_1 \wedge d\phi_2 = w^{1/2}\, dx \wedge dy \tag{3.14}$$

From (3.2), (3.3) and (3.4) we see that

$$d\theta_1 \wedge d\theta_2 = (2 + w^{-1/2}(2 + p^2 + q^2))\, dx \wedge dy \tag{3.15}$$

and it is clear that the Jacobian $\partial(\phi_1, \phi_2)/\partial(\theta_1, \theta_2)$ is bounded. Hence

$$A(2 + w^{-1/2}(2 + p^2 + q^2)) = w^{1/2} \tilde{F}(x, y) \tag{3.16}$$

Here $\tilde{F}(x,y)$ is a bounded function. Along the line C, $2+p^2+q^2 \neq 0$. In fact we have $1+p^2>0$, $1+q^2>0$ in Ω_+. If along C, $2+p^2+q^2=0$ we must have $1+p^2=0$, $1+q^2=0$ along C. Then we have $p.q=0$ along C since $w=0$ and hence $w_x=w_y=0$ along C. This is a contradiction. Then (3.16) implies (3.13) for the case $\phi_2 \to 0_+$. For $\phi_2<0$ we have

$$d\theta_1 \wedge d\theta_2 = (2+|w|^{-1/2}(q^2-p^2))\, dx \wedge dy$$

and (3.16) becomes

$$A(2+|w|^{-1/2}(q^2-p^2)) = |w|^{1/2}\tilde{F}(x,y) \qquad (3.17)$$

Since the surface is C^2 we obtain (3.13) for the case $\phi_2 \to 0_-$. We need the following:

Lemma 4.

$$\lim_{w\to 0_+} c = 0, \qquad \lim_{w\to 0_-} h \neq 0 \qquad (3.18)$$

$$\lim_{w\to 0_-} c_1 = 0, \qquad \lim_{w\to 0} h_1 \neq 0 \qquad (3.19)$$

Proof. For $w>0$ from

$$\begin{aligned} A\, d\phi_1 \wedge d\phi_2 &= A\{((l_t + w^{1/2}c_t))(w^{-1/2}h/2 + w^{1/2}h_w) \\ &\quad - (w^{-1/2}c/2 + w^{1/2}c_w)w^{1/2}h_t\}\, dt \wedge dw \\ &= w^{1/2} dx \wedge dy \end{aligned}$$

and (3.16) we see that $\lim_{w\to 0_+} h \neq 0$, since $l_t \neq 0$. Moreover, from

$$ds_1^2 = A(w^{-1}(c^2+h^2)\, dw^2/4 + c(t,w)w^{-1/2}\, dt\, dw + \cdots)$$

and (3.17) we see that $\lim_{w\to 0_+} c(t,w) = 0$, since the coefficients of ds_1^2 in the coordinates (t,w) should be C^1. (3.18) is proved. (3.19) is proved in a similar way. Q.E.D.

4. General expressions for extremal surfaces of mixed type

As the Euclidean case in the isothermal coordinates (ϕ_1,ϕ_2) an extremal surface $X(\phi_1,\phi_2)$ satisfies the following equations

$$\begin{aligned} X &= X_{\phi_1\phi_1} + X_{\phi_2\phi_2} = 0 \qquad \text{(space-like part)} \\ X &= X_{\phi_1\phi_1} - X_{\phi_2\phi_2} = 0 \qquad \text{(time-like part)} \end{aligned} \qquad (4.1)$$

Hence the general expression of the space-like part of an extremal surface is

$$X = \operatorname{Re} F(\sigma) \tag{4.2}$$

Here $\sigma = \phi_1 + i\phi_2$ and F are an analytic functions of σ. Moreover, $F(\sigma)$ satisfies the condition

$$F'(\sigma)^2 = 0 \tag{4.3}$$

which means that the parameters ϕ_1, ϕ_2 are isothermal. Without loss of generalities we may set locally

$$F_{n+1}(\sigma) = \int b(\sigma)\, d\sigma \qquad (b(\sigma) \neq 0) \tag{4.4}$$

and then (4.4)

$$\begin{aligned} X_{n+1} &= \operatorname{Re} \int b(\sigma)\, d\sigma \\ X_i &= \operatorname{Re} \int b(\sigma)\alpha_i(\sigma)\, d\sigma \end{aligned} \tag{4.5}$$

where $b(\sigma)$ and $\alpha_i(\sigma)$ are analytic functions with

$$\sum \alpha_i^2(\sigma) = 1 \tag{4.6}$$

Similarly the time-like part of the extremal surface has the expression

$$\begin{aligned} X_{n+1} &= \int f(\lambda)\, d\lambda + \int g(\mu)\, d\mu \\ X_i &= \int f(\lambda)\beta_i(\lambda)\, d\lambda + \int g(\mu)\gamma_i(\mu)\, d\mu \end{aligned} \tag{4.7}$$

Here

$$\lambda = \phi_1 + \phi_2, \qquad \mu = \phi_1 - \phi_2 \tag{4.8}$$

and the condition of that ϕ_1, ϕ_2 are isothermal is

$$\sum \beta_i^2 = 1, \qquad \sum \gamma_i^2 = 1 \tag{4.9}$$

Let

$$b(\sigma) = u + iv, \qquad \alpha_i(\sigma) = l_i + im_i \tag{4.10}$$

where u, v, l_i, m_i are real. Then (4.5) becomes

$$\begin{aligned} X_{n+1} &= \int u\,d\phi_1 - v\,d\phi_2 \\ X_i &= \int (ul_i - vm_i)\,d\phi_1 - (um_i + vl_i)\,d\phi_2 \end{aligned} \tag{4.11}$$

Here

$$X_{n+1\phi_1} = u, \qquad X_{i\phi_1} = (ul_i - vm_i) \tag{4.12}$$

and

$$A = \sum_i (ul_i - vm_i)^2 - u^2 = (u^2 + v^2)\sum_i m_i^2 \tag{4.13}$$

since (4.6) means

$$\sum_i l_i^2 - \sum_i m_i^2 = 1, \qquad \sum_i l_i m_i = 0$$

From Lemma 3 we see that $m_i = 0$ along $\phi_2 = 0$. Then along $\phi_2 = 0$. we have

$$X_{\phi_1} = \begin{pmatrix} ul_i \\ u \end{pmatrix} \qquad X_{\phi_2} = \begin{pmatrix} -vl_i \\ v \end{pmatrix} \tag{4.14}$$

and

$$X_w = (w^{-1/2}c/2 + w^{1/2}c_w)X_{\phi_1} + (w^{-1/2}h/2 + w^{1/2}h_w)X_{\phi_2} \tag{4.15}$$

X_w is C^1 and linearly independent of X_{ϕ_1}. From Lemma 4 we see that $X_{\phi_2} \to 0$ as $\phi_2 \to 0_+$. Hence $v(\phi_1, 0) = 0$. Consequently the analytic functions $b(\sigma)$, $\alpha_i(\sigma)$ are real-valued along the line $\phi_2 = 0$. From the reflection principle of analytic functions we see that

$$u(\phi_1, 0), \qquad l_i(\phi_1, 0)$$

are real analytic functions of ϕ_1.

For the region Ω_- we have

$$A = 2fg(\sum \beta_i \gamma_i - 1) \tag{4.16}$$

From Lemma 3 we have

$$\sum \beta_i \gamma_i = 1$$

along $\phi_2 = 0$. Considering the fact $\sum \beta_i^2 = 1$, $\sum \gamma_i^2 = 1$ we have

$$\beta_i = \gamma_i \tag{4.17}$$

Now

$$X_{\phi_1} = \begin{pmatrix} (f+g)\beta_i \\ (f+g) \end{pmatrix}, \qquad X_{\phi_1} = \begin{pmatrix} (f-g)\beta_i \\ (f-g) \end{pmatrix} \tag{4.18}$$

along C. We obtain $f = g$ along $\phi_2 = 0$ similarly. From the continuity of X_{ϕ_1} we have

$$f(\phi_1) = g(\phi_1) = u(\phi_1, 0)/2, \qquad \beta_i(\phi_1) = \gamma_i(\phi_1) = l_i(\phi_1, 0) \tag{4.19}$$

Thus we obtain the following theorem:

Theorem. *The general expressions of extremal surfaces of mixed type in* $\mathbf{R}^{n+1}$ *with grad* $w \neq 0$ *along the bordline* $w = 0$ *are*

$$\left.\begin{aligned} X_{n+1} &= \operatorname{Re} \int b(\sigma)\, d\sigma \\ X_i &= \operatorname{Re} \int b(\sigma)\alpha_i(\sigma)\, d\sigma \end{aligned}\right\} \quad (\phi_2 \geq 0) \tag{4.20}$$

and

$$\begin{aligned} X_{n+1} &= \int f(\lambda)\, d\lambda + \int f(\mu)\, d\mu \\ X_i &= \int f(\lambda)\beta_i(\lambda)\, d\lambda + \int f(\mu)\beta_i(\mu)\, d\mu \qquad (\phi_2 < 0) \\ &(\lambda = \phi_1 + \phi_2, \quad \mu = \phi_1 - \phi_2) \end{aligned} \tag{4.21}$$

Here $b(\sigma)$, $\alpha_i(\sigma)$ *are analytic functions defined on some region of* $\phi_2 > 0$ *and real-valued along* $\phi_2 = 0$. *Moreover* $\sum \alpha_i^2 = 1$, $f(\phi_1) = b(\phi_1)/2$, $\beta_i(\phi_1) = \alpha_i(\phi_1)$.

From this theorem we have the following consequences.

Consequence 1. The bordline between the space-like part and time-like part is a null curve.

In fact, the curve has the expression

$$X_{n+1} = \int u(\phi_1, 0)\, d\phi_1, \qquad X_i = \int u(\phi_1, 0) l_i(\phi_1, 0)\, d\phi_1$$

with $\sum l_i^2(\phi_1, 0) = 1$

Consequence 2. Near the bordline C the extremal surface is analytic.

Because $u(\phi_1, 0)$, $l_i(\phi_1, 0)$ are real analytic functions the surface is analytic in the time-like part near the curve C. It is easily verified that the surface is analytic along the curve C, since X can be expressed via parameters (ϕ_1, ϕ) analytically where $\phi = \pm\phi_2^2$ as $\phi_2 > 0$ or $\phi_2 < 0$

5. Global construction of extremal surfaces of mixed type

Let

$$X_i(s) = a_i(s) \qquad (i = 1, 2, \ldots, n) \tag{5.1}$$

be an analytic curve in $\mathbf{R}^n$. Here s is the arc length. Hence $a_i(s)$ are analytic functions of s and

$$\sum a_i^2(s) = 1 \tag{5.2}$$

The "algorithm of the constructions" is

(a) Lift the given curve to obtain a null analytic curve in $\mathbf{R}^{n+1}$

$$X_i = a_i(s), \qquad X_{n+1} = s \tag{5.3}$$

(b) Time-like extension

Choose a real analytic functions $s = s(\phi_1)$ and construct the surface defined by

$$X_i = \{a_i(s(\lambda)) + a_i(s(\mu))\}/2, \qquad X_{n+1} = \{s(\lambda) + s(\mu)\}/2 \tag{5.4}$$

Here $\lambda = \phi_1 + \phi_2$, $\mu = \phi_1 - \phi_2$ $(\phi_2 < 0)$

(c) Space-like extension

$$X_i = \mathrm{Re}\{a_i(s(\phi_1 + i\phi_2))\}, \qquad X_{n+1} = \mathrm{Re}\{s(\phi_1 + i\phi_2)\} \tag{5.5}$$

The analytic continuation is possible since a_i and s are analytic.

The formulas (5.4) and (5.5) are equivalent to the formulas (4.21) and (4.22) respectively.

In time-like extension or space-like extension a new bordline may appear, then we do a time-like or space-like extension again. Thus we will obtain a whole extremal surface of mixed type.

In case $n = 2$ we have described two examples [G2,G3]. If the plane curve is a circle we obtain a helicoid. If the plane curve is defined by $y = \cos hx$, we obtain a rule surface $z = ythx$. It is a complete graph.

We consider examples for $n = 3$.

Example 1. Take the normal cubic curve

$$X_1 = t, \qquad X_2 = t^2, \qquad X_3 = t^3 \tag{5.6}$$

as a starting curve. The lift in $\mathbf{R}^{3+1}$ is defined by (5.6) and the elliptic integral

$$X_4 = \int_0^t \sqrt{1 + 4t^2 + 9t^4}\, dt = E(t) \tag{5.7}$$

The time-like extension is

$$\begin{aligned} X_1 &= (\lambda + \mu)/2 = \phi_1 \\ X_2 &= (\lambda^2 + \mu^2)/2 = \phi_1^2 + \phi_2^2 \\ X_3 &= (\lambda^3 + \mu^3)/2 = \phi_1^3 + 3\phi_1\phi_2^2 \\ X_4 &= (E(\lambda) + E(\mu))/2 = J(\phi_1, \phi_2^2) \end{aligned}$$

The space-like extension is

$$X_1 = \phi_1 \tag{5.8}$$

$$
\begin{aligned}
X_2 &= \phi_1^2 - \phi_2^2 \\
X_3 &= \phi_1^3 - 3\phi_1\phi_2^2 \\
X_4 &= \{E(\lambda) + E(\mu)\}/2 = J(\phi_1, -\phi_2^2)
\end{aligned}
\tag{5.9}
$$

Let

$$
\tilde{\sigma} = \begin{cases} -\phi_2^2 & \phi_2 > 0 \\ \phi_2^2 & \phi_2 < 0 \end{cases}
\tag{5.10}
$$

There is a unified analytic expression for the two parts of the surface

$$
X_1 = \phi_1, \qquad X_2 = \phi_1^2 + \tilde{\sigma}, \qquad X_3 = \phi_1^3 + 3\phi_1\tilde{\sigma}, \qquad X_4 = J(\phi_1, \tilde{\sigma})
$$

or

$$
X_3 = X_1^3 + 3X_1(X_2 - X_1^2), \qquad X_4 = J(X_1, X_2 - X_1^2)
\tag{5.11}
$$

Example 2. Take the analytic curve

$$
X_1 = \sin t, \qquad X_2 = (\sin^2 t)/2, \qquad X_3 = (t - (\sin 2t)/2)/2
\tag{5.12}
$$

in $\mathbf{R}^3$. The lift in $\mathbf{R}^{3+1}$ is defined by

$$
X_4 = t
\tag{5.13}
$$

The time-like extension is

$$
\begin{aligned}
X_1 &= (\sin\lambda + \sin\mu)/2, & X_2 &= (\sin^2\lambda + \sin^2\mu)/4 \\
X_3 &= (\lambda+\mu)/4 - (\sin 2\lambda + \sin 2\mu)/8, & X_4 &= (\lambda+\mu)/2
\end{aligned}
\tag{5.14}
$$

One can write down the parametric representation of the space-like as well. However, from (5.14) we can obtain the unified equation of the surface including the space-like part. Eliminating λ and μ we obtain the equations for the extremal surface

$$
\begin{aligned}
(4X_2 - 1)\sin^2 X_4 + \cos 2X_4(2X_1^2 - \sin^2 4X_4) &= 0 \\
(2X_3 - X_4)\sin X_4 + \cos X_4(2X_1^2 - \sin^2 X_4) &= 0
\end{aligned}
\tag{5.15}
$$

Because of the analyticity they are the equations for the entire extremal surface of Mixed type.

BIBLIOGRAPHY

[B] Bartnik, R. and Simon, L., *Space-like hypersurfaces with prescribed boundary values and mean curvature*, Commun. Math. Phys. **87**(1982), 131–152.

[Ba] Barbashov, B.M., Nesternko, V. V. and Chervyakov, A. M., *General solutions of nonlinear equations in the geometric theory of the relativistic string*, Commun. Math. Phys. **34**, 1982, 471–481.

[Ch] Choquet-Bruhat, Y., *Maximal submanifolds with constant mean extrinsic curvature of a Lorentzian Manifold*, Ann. d. Scuda Norm. Sup. Pisa **3**(1976), 361–376.

[CY] Cheng, S.Y. and Yau, S.T., *Maximal space-like surfaces in Lorentz-Minkowski spaces*, Ann. Math. **104**(1976) 407–419.

[G1] Gu, C.H., *On the motion of strings in a curved space-time*, Proc. of 3rd Grossmann meeting on general relativity.

[G2] Gu, C.H., *The extremal surfaces in the 3-dimensional Minkowski spaces*, Acta Math. Sinica, New series 1985, Vol. **1**, No. 2, 173–180.

[G3] Gu, C.H., *A Global study of extremal surfaces in 3-dimensional Minkowski spaces*, Proc. DD6 symposium, Lect. Notes in Math. **1255**, Berlin-Heidelberg-New York 1985, 26–33.

[M] Milnor, T., *Harmonically immersed Lorentz surfaces*, Geometrodynamics, Proc. 1985 Corenza, 201–213. preprint, 1986.

[O] Ossermann, R., *A survey of minimal surfaces*, 1969.

[Q] Quien, N., *Plateau's problem in Minkowski space*, Analysis **5**, (1985) 43–60.

Gu Chaohao
University of Science and Technology
Hefei Anhui China
and
Inst. of Math
Fudan University
Shanghai, China

Convergence of Minimal Submanifolds to a Singular Variety

ROBERT GULLIVER

Abstract: A sequence of minimal hypersurfaces M_h is considered, whose varifold limit V is not a density-one smooth hypersurface. Six geometrical problems are outlined, with the idea of studying asymptotic behavior as $h \to \infty$ in terms of additional structures on V. Variational limits for the Dirichlet integral are presented in some detail; the examples involve homogenization of manifolds.

What is the limit of a sequence of minimal hypersurfaces in $\mathbb{R}^{n+1}$?

This question has too few answers, and too many answers. Too *few* answers, in the sense that it has been successfully studied in the context of only a limited number of problems. Too *many* answers, in the sense that various choices of what should be preserved after passage to the limit lead to a wide variety of distinct structures to be considered as part of the asymptotic objects. We shall indicate this multiplicity of structures in a list of six problems; give two related examples of a sequence of minimal surfaces M_h in $\mathbb{R}^3$, tending toward objects which are singular in some sense; and state two new theorems regarding the last two problems in the list.

The problems

(Ar). One context in which a successful theory has been developed is as follows. If one wishes to preserve the local AREA (mass, n-dimensional Hausdorff measure) of a sequence of minimal surfaces M_h, then the convergence may be understood in terms of Almgren's theory of varifolds and rectifiability. Recall that a varifold is a Radon measure on the Grassmann bundle of n-planes over $\mathbb{R}^{n+1}$; the area measure of a C^1 hypersurface is lifted to the Grassmann bundle to define the associated varifold. If the area of M_h in a fixed compact set $K \subset \mathbb{R}^{n+1}$ is bounded independently of h, then a subsequence of the M_h converges weakly to a varifold V, whose area in K is the limit of the area of M_h as $h \to \infty$ [A, S1]. If M_h minimizes area inside K relative to its boundary on ∂K, then V is the varifold associated to a hypersurface with a singular set of dimension $\leq n-7$ (see [S1, p. 221]).

(2FF). Much less attention has been paid to the SECOND FUNDAMENTAL FORM of the hypersurfaces M_h. As $h \to \infty$ and the M_h converge weakly to a varifold with singularities, essential information about the second fundamental form may be lost (see Example II below). This phenomenon is partly responsible for the necessity of a regularity hypothesis on one tangent cone in Simon's result on the uniqueness of tangent cones [S2]. Our ongoing work with M. Grüter has yielded preliminary results on this problem.

(Geo B). For the intrinsic problem of finding the Riemannian distance between two points of M_h, or the length of the shortest GEODESIC with prescribed BOUNDARY CONDITIONS, the useful notion of Hausdorff convergence appears in the work of Gromov and his collaborators. Admissible objects in this context are locally compact metric spaces. Under the hypotheses of bounded topology, bounded diameter, bounded sectional curvature and a lower bound on volume, the limiting metric space is a Riemannian manifold [G–L–P].

(Geo I). A superficially related problem of the intrinsic geometry of M_h is to find the GEODESIC with prescribed INITIAL CONDITIONS. The appropriate structure, however, appears to be completely different from (Geo B) above. For example, it would be appropriate to consider the geodesic flow as a group of measure-preserving transformations of the unit tangent bundle. For the case of a sequence of manifolds M_h of unbounded topology, there is no work yet that we are aware of. It would also be of interest to pursue the relationship with problem (2FF) above.

(Spec). Substantial work has been done on determining the SPECTRUM of the Laplace operator on a Riemannian manifold, by Colin de Verdiére among others. Recent work of Anné and Fukaya has illuminated certain interesting cases with singular limit spaces. For a complete minimal hypersurface M_h in $\mathbb{R}^{n+1}$, one should restrict attention to a compact subset $K \subset \mathbb{R}^{n+1}$ and pose homogeneous Dirichlet boundary conditions on $M_h \cap \partial K$, for example. With a sequence M_h of unbounded topology, an appropriate notion of convergence would be the convergence of each eigenvalue, or convergence in the uniform operator topology of the resolvent away from the spectrum. The relevant structures to be considered as part of the limit object are the same as in (Dir), below.

(Dir). Finally, one may consider variational convergence of the Riemannian DIRICHLET INTEGRAL of M_h:

$$D_h(u) := \int_{M_h} |du|^2 \, \mathrm{dVol}_{M_h}.$$

Note that a varifold also has a well-defined Dirichlet integral (see [H]). Since D_h is defined on $L^2(M_h \cap K)$, taking values in $[0, \infty]$, a useful notion of convergence will begin with embeddings of this sequence of function spaces in a fixed Hilbert space. We shall take the direct approach of embedding $L^2(M_h \cap K)$ in $L^2(\Omega)$, for some singular manifold Ω, by composition with an a.e. bi-Lipschitz mapping $\Phi_h : \Omega \setminus E_h \to M_h$, where the E_h are open subsets of Ω. With respect to these embeddings, we say that a functional $D_0 : L^2(\Omega) \to [0, \infty]$ is the $\underline{\Gamma - \text{limit}}$ of D_h if (1) for every $u \in L^2(\Omega)$ and every L^2-convergent $u_h \to u$, $D_0(u) \leq \liminf D_h(u_h \circ \Phi_h^{-1})$; and (2) for each $u \in L^2(\Omega)$ there is an L^2-convergent sequence $u_h \to u$, such that $D_0(u) = \lim D_h(u_h \circ \Phi_h^{-1})$. Γ-convergence is strong enough to imply convergence of solutions to boundary-value problems, and weak enough that compactness holds in very general situations [dG–F]. Observe that there is no requirement that D_0 be of the same form as D_h; nor that it be local: $D_0(u + v) = D_0(u) + D_0(v)$ for u and v with disjoint supports; nor that it be regular: $C_0^\infty(\Omega)$ be D_0-dense in the domain of D_0. In particular, the Γ-limit of a sequence of smooth manifolds need not be a manifold with singularities. One possibility is the following functional on a Riemannian manifold Ω with singular set of measure 0:

$$D_0(u) = \int_\Omega |du|^2 b(x) \, \mathrm{dVol}_\Omega + \int_{\bar\Omega \times \bar\Omega} (u(x) - u(y))^2 J(\mathrm{dVol}_\Omega(x) \, \mathrm{dVol}_\Omega(y)),$$

where J is a nonnegative Borel measure on $\bar\Omega \times \bar\Omega$, not necessarily locally finite. J is called the "jumping measure" because of its interpretation for Brownian motion [Fuk]. An interesting special case is when jumping occurs only between pairs $x \in \Omega$ and $Tx \in \Omega$ where $T : \Omega \to \Omega$ is an isometry of order two. In this case, the functional D_0 becomes

$$D_0(u) = \int_\Omega |du|^2 b(x) \, \mathrm{dVol}_\Omega + \int_{\bar\Omega} (u(x) - u(Tx))^2 d\nu(x), \tag{1}$$

for some nonnegative Borel measure ν on $\bar\Omega$ such that sets of vanishing capacity have measure zero. The second term of (1) must be evaluated using a quasi-continuous representative of $u \in H^1(\Omega, b)$. A functional of the form (1) may be thought of as representing temperature conduction on a manifold which is arranged in two layers, so that Tx and x are adjacent, with a varying degree of thermal contact between the layers: perfect thermal contact on a set E if $\nu(A) = +\infty$ for every $A \subset E$ of positive capacity, complete thermal insulation where $\nu = 0$, and partial thermal contact on an open set E with $0 < \nu(E) < \infty$. We call a functional D_0 of the form (1) the Dirichlet integral of a *relaxed Riemannian manifold* $\{\Omega, g, b, \nu\}$.

We should note here that Γ-convergence and the problem (Dir) are intimately related to stable convergence of the stopping measure for Brownian motion on the manifolds M_h [B–dM–M].

The examples

Two-dimensional minimal surfaces in $\mathbf{R}^3$ have the special advantage that they may be represented explicitly in terms of two meromorphic functions on a Riemann surface. It appears likely that examples analogous to the unstable minimal surfaces given below exist as hypersurfaces in $\mathbf{R}^{n+1}$ for all n, although existence has not been proved; one might expect minimax methods to be appropriate here. Even in these low dimensions, and with simple singularity structure on the limit varifold, interesting structures appear.

Example I. Consider the blowdown of the second Scherk surface:

$$M_h = \{x \in \mathbf{R}^3 : \cos(hx_3) = \sinh(hx_1)\sinh(hx_2)\}.$$

From far away, the surface M_h looks like the union M_∞ of the (x_1, x_3)-plane and the (x_2, x_3)-plane. On closer inspection, one sees that M_h features a "tower" of horizontal handles connecting the opposite quadrants $x_1x_2 > 0$ at altitudes $2k\pi/h$ and connecting the other pair of quadrants $x_1x_2 < 0$ at altitudes $(2k+1)\pi/h$ for integers k (compare the computer drawing of M_1 on p. 87 of [K]).

(Ar). As $h \to \infty$, M_h converges in the weak-varifold sense to M_∞ considered as a varifold. Note that for any neighborhood V of the singular set $\mathrm{Sing}(M_\infty)$, which is the x_3-axis, $M_h \cap V$ is unstable for large h. However, M_∞ is stable: this illustrates the necessity to redefine stability of varifolds as part of problem (2FF) above. For problem (2FF), one associates to M_∞ the additional structure of a varifold on the Grassmann bundle, which in this example has infinite measure over each interval of $\mathrm{Sing}(M_\infty)$. For problem (Geo B), one may consider M_∞ as a metric space; it is not difficult to show that M_∞ is the Hausdorff limit of M_h as $h \to \infty$.

Problem (Geo I) is more interesting in this example. Consider a geodesic Γ on M_h ($1 << h < \infty$) which starts at $x_1 = 1$, $x_3 = 0$ with x_2 close to 0, and with initial velocity $x_1' = -1$, x_2' close to 0 and $x_3' = v_3$ arbitrary. As Γ approaches the x_3-axis, it may sweep along the side of one of the handles to enter the portion of M_h close to the (x_2, x_3)-plane with velocity roughly $\pm e_2$, e.g. when $v_3 = 0$ since M_h meets the (x_1, x_2)-plane in two geodesics. Here we write $e_2 = (0, 1, 0)$, etc. as usual. Or Γ may slip between two handles and cross to the opposite portion $\{x_1 < 0\}$ of the (x_1, x_3)-plane, asymptotically speaking as $h \to \infty$. Or again, it may wrap around one of the handles, or a consecutive sequence of handles, several times before leaving the vicinity of the x_3-axis in a less and less predictable direction as $h \to \infty$. In fact, for $h = \infty$ the geodesic flow must be considered to be nondeterministic. Considering the time-one map as an L^1-norm-preserving transformation of positive functions on the unit tangent bundle $T_1 M_\infty$ of M_∞ leads to a measure on $T_1 M_\infty \times T_1 M_\infty$ as the kernel. It appears likely that this measure, when restricted to distinct components of $M_\infty \setminus \mathrm{Sing}(M_\infty)$, is absolutely continuous with respect to the natural invariant measure, in contrast to the atomic measures which appear in the smooth case.

Both problems (Spec) and (Dir) may be treated in terms of a mapping $\Phi_h : \Omega \setminus E_h \to M_h$, an involutive isometry $T : \Omega \to \Omega$, and a nonnegative Borel measure ν_h on $\bar{\Omega}$, as in formula (1) above. We choose Ω to be the *disjoint* union of $M_\infty \cap \{x_1 \geq x_2\}$ and of $M_\infty \cap \{x_1 \leq x_2\}$ so that Ω is isometric to two copies of the Euclidean plane $\{\mathbb{R}^2, \mathrm{can}\}$. We choose $T : \Omega \to \Omega$ to be the reflection $T(x_1, x_2, x_3) = (x_2, x_1, x_3)$ in the plane $\Pi = \{x \in \mathbb{R}^3 : x_1 = x_2\}$, but interchanging the two copies of the x_3-axis. Similarly, we define an involutive isometry $S_h : M_h \to M_h$ by the same formula, $S_h(x_1, x_2, x_3) = (x_2, x_1, x_3)$: M_h will be "cut apart" along $M_h \cap \Pi$, the fixed-point set of S_h. For each $y \in \Omega$, which we associate to $\bar{y} \in M_\infty$, consider the line orthogonal to Π and passing through $\bar{y}$: if this line meets M_h before crossing Π, we preliminarily define $\Phi_h(y) \in M_h$ to be the point where this line meets M_h; otherwise, we say $y \in E_h$. For $\bar{y} \in M_\infty \cap \Pi = \mathrm{Sing}(M_\infty)$, Φ_h may be extended by continuity. Thus $\Phi_h^{-1} : M_h \to \Omega$ is a parallel projection, double-valued on ∂E_h, where Φ_h further fails to be Lipschitz-continuous. We now make a small change in Φ_h^{-1} and in the set E_h, by adding a constant multiple $\epsilon > 0$ of the normal vector N to M_h to the projection vector $\pm(e_1 - e_2)$. Here N is oriented so that it points toward the x_3-axis along $M_h \cap \Pi$. Pulling back the metrics g_h of M_h to Ω, we may write

$$\Phi_h^*(g_h) = g_{ij}^h(y)\, dy^i\, dy^j$$

in terms of Euclidean coordinates on $\Omega_h := \Omega \setminus \bar{E}_h$. Since Φ_h and Φ_h^{-1}

satisfy Lipschitz bounds uniformly in h, there are positive constants $\lambda < \Lambda$ so that the metrics $(g_{ij}^h - \lambda\delta_{ij})$ and $(\Lambda\delta_{ij} - g_{ij}^h)$ remain positive definite. For any Borel set B, we write ∞_B for the measure defined by $\infty_B(A) = +\infty$ if $A \cap B$ has positive capacity, and $\infty_B(A) = 0$ otherwise. We choose $\nu_h = \infty_{\partial E_h}$. Then for any $u \in L^2(M_h)$, it may be shown that

$$(2)\ D_h(u) = \int_{\Omega_h} g_h^{ij}(y)\partial_i v\, \partial_j v \sqrt{\det(g_{ij}^h(y))}dy + \int_{\bar{\Omega}_h} (v(y) - v(Ty))^2 d\nu_h(y),$$

where $v := u \circ \Phi_h \in L^2(\Omega_h)$. Because of our choice of ν_h, the second integral has the effect of "glueing together" Ω_h along ∂E_h, thereby re-establishing the geometry of M_h in the context of the manifold Ω. Observe that T is an isometry for $\Phi_h^*(g_h)$. It now follows from Theorem 1 below that $\bar{\Omega}$ carries a uniformly positive metric $a_{ij}(y) \in L^\infty(\Omega)$ and a nonnegative Borel measure ν, so that the functional

$$(3) \qquad D_0(v) = \int_\Omega a^{ij}(y)\partial_i v\, \partial_j v\, dy + \int_{\bar{\Omega}} (v(y) - v(Ty))^2 d\nu(y)$$

is the Γ-limit of $D_h(v \circ \Phi_h^{-1})$. That is, a certain relaxed Riemannian manifold $\{\Omega, a_{ij}(y), 1, \nu\}$ is the Γ-limit of the blowdown sequence M_h for the second Scherk surface. Moreover, in this (relatively simple) case, it may be shown with capacity arguments that $a_{ij}(y) = \delta_{ij}$ and $\nu = \infty_B$, where B consists of the two copies in Ω of the x_3-axis. Note that the relaxed Riemannian manifold $\{\Omega, \mathrm{can}, 1, \infty_B\}$ corresponds to the set $M_\infty \subset \mathbf{R}^3$ as a varifold. Further, the additional hypotheses of Theorem 2 below are satisfied, and we may conclude that the spectrum of $M_h \cap K$ converges to the spectrum of $M_\infty \cap K$.

Example II. The second Scherk surface of Example I may be deformed, while retaining the symmetry planes, so that the new sequence M_h of unstable minimal surfaces has a Γ-limit in problems (Dir) and (Spec) which is no longer a varifold.

The family of deformed surfaces is described parametrically by the formula

$$(4) \qquad x(\omega, \alpha) = 2\sin 2\alpha\, \mathrm{Re} \int_1^\omega \frac{(z^{-1} - z, iz^{-1} + iz, 2)}{z^2 + z^{-2} - 2\cos 2\alpha} \frac{dz}{z} + c_2 e_2,$$

where $0 < \alpha < \pi/2$. For $\omega \in \mathbb{C} \cup \{\infty\} \setminus \{e^{i\Theta} : \alpha \leq |\Theta| \leq \pi - \alpha\}$, formula (4) describes a minimal surface in $\mathbf{R}^3$ which meets the horizontal planes $x_3 = 0$, $x_3 = \pi$ and $x_3 = -\pi$ at right angles along the four arcs of the circle $|\omega| = 1$ between the poles $\omega = \pm e^{i\alpha}$, $\omega = \pm e^{-i\alpha}$. We may extend this minimal surface by reflection in the planes $x_3 = (2k+1)\pi$ $(k \in \mathbb{Z})$, to form a complete minimal surface Σ_α (see pp. 86–88 in [K]). Note that Σ_α also has a vertical plane $\Pi = \{x : x_1 = 0\}$ of reflective symmetry (where $\omega \in \mathbb{R}$) and a plane $x_2 =$ const. where $\omega \in i\mathbb{R}$. We choose c_2 so that $X(i\mathbb{R})$ lies in the (x_1, x_3)-plane. It may be shown via long computations that $\Sigma_{\pi/4}$ is obtained from the surface M_1 of Example I after a $\pi/4$-rotation about the x_3-axis. Viewed from far away, Σ_α looks like the union of the two planes $x_1 = \pm x_2 \tan\alpha$; near the x_3-axis, Σ_α is a tower of handles, with the same topological structure as the second Scherk surface $\Sigma_{\pi/4}$. As $\alpha \to 0$, the handles connecting the regions $|x_1| > |x_2| \tan\alpha$ shrink to small necks around the lines $x_2 = 0$, $x_3 = (2k+1)\pi$ $(k \in \mathbb{Z})$. In fact, the "waist" curve σ corresponds to $\omega \in i\mathbb{R} \cup \{\infty\}$ traversed once; its curvature equals the positive principal curvature κ, given everywhere on Σ_α by

$$\kappa = \frac{|\omega^2 + \omega^{-2} - 2\cos 2\alpha|}{\sin 2\alpha \big(|\omega| + |\omega^{-1}|\big)^2},$$

which satisfies $\cos^2\alpha \le \kappa \sin 2\alpha \le 1$ along σ. Thus as $\alpha \to 0$, σ is C^2-close to a circle of radius $\sin 2\alpha$ in the (x_2, x_3)-plane. The "waist" curve of a handle in the opposite direction ($\omega \in \mathbb{R} \cup \{\infty\}$) flattens out toward a double covering of the line segment $[(2k-1)\pi, (2k+1)\pi]$ of the x_3-axis.

We shall choose a sequence $\alpha_h \to 0$, to be specified more precisely below, and define

$$M_h := \frac{1}{h}\Sigma_{\alpha_h}.$$

Regarding problem (Ar), the varifold limit of M_h is the (x_2, x_3)-plane Π, covered twice. For problem (2FF), one finds again a second fundamental form having infinite weight over each interval of the x_3-axis. For problem (Geo B), the Hausdorff limit of M_h consists on two copies of Π identified along the x_3-axis. The limit in problem (Geo I), however, consists of two *disjoint* copies of Π. Namely, as $h \to \infty$, the probability that a given geodesic will be significantly deflected by the curved area near one of the handles tends to zero.

The problems (Dir) and (Spec) may be treated in a manner similar to Example I. The surface M_h is "cut apart" along the plane Π. Each piece is represented by projection in the direction $\pm e_1 + \epsilon N(x)$ from $x \in \overline{M_h \text{ to } \Phi_h^{-1}(x)} \in M_{\infty,h} := \{y \in \mathbb{R}^3 : y_1 = \pm y_2 \tan\alpha_h\}$. The reference manifold Ω is again two disjoint copies of $(\mathbb{R}^2, \text{can})$ which is identified with $\big(M_{\infty,h} \cap \{y_1 \ge 0\}\big) \cup \big(M_{\infty,h} \cap \{y_1 \le 0\}\big)$. An open set $E_h \subset \Omega$ corresponds to those points of $M_{\infty,h}$ not on any of the projection lines. A Lipschitz mapping $\Phi_h : \Omega \setminus E_h \to M_h$ is thereby uniquely defined. As before, ν_h is chosen to be $\infty_{\partial E_h}$. As in Example I, the hypotheses of Theorem 1 and 2 are satisfied, so that there is a relaxed Riemannian manifold, which may be shown to have the form $\{\Omega, \text{can}, 1, \nu\}$, which is the Γ-limit of the manifold M_h. Moreover, the spectrum of M_h converges to the spectrum of $\{\Omega, \text{can}, 1, \nu\}$. It is not difficult to show that the measure ν given by Theorem 1 is supported on the set $B \subset \Omega$ consisting of both copies of the x_3-axis. Moreover, ν is x_3-translation invariant. It follows that ν must be of the form $\nu = c\, dx_3 \mid B$ for some $0 \leq c < \infty$, or that $\nu = \infty_B$, which we indicate by $c = \infty$. It may be shown by means of arguments regarding capacities that

$$c = \lim_{h \to \infty} \frac{2\pi h}{|\log \sin 2\alpha_h|},$$

provided that this limit exists. In other words, when $\alpha_h \to 0$ at a rate proportional to

$$\alpha_h \sim e^{-bh}$$

the Γ-limit consists of two planes which are neither totally insulated from each other nor in perfect contact.

The Theorems

We would like to state two theorems, obtained recently in collaboration with Gianni dal Maso and Umberto Mosco, in the context of problems (Dir) and (Spec) above. Our reliance on an involutive structure, as in formula (2) above, substantially limits the range of applicability of such theorems. On the other hand, since the measures involved are not Radon measures, we cannot expect to find such general theorems as in the book of Fukushima [Fuk]. Earlier results of the present type require that the topological modifications take place in a set of asymptotically negligible capacity [C–F].

Suppose that a Riemannian manifold-with-boundary $\bar{\Omega}$ (the "reference manifold") is diffeomorphic to a bounded Lipschitz domain in $\mathbf{R}^n$, $n \geq 2$: this hypothesis allows us to apply results from the literature of homogeneization theory. Consider a sequence of Riemannian manifolds $\{M_h, g_h\}$ which may be represented using uniformly Lipschitz-continuous mappings $\Phi_h : \Omega_h \to M_h$, where $\Omega_h \subset \Omega$ is strongly connected: there exist (e.g. harmonic) extensions $\Pi_h : H^1(\Omega_h, \Phi_h^* g_h) \to H^1(\Omega)$ such that for all $u \in H^1(\Omega_h, \Phi_h^* g_h)$,

$$\|\Pi_h u\|_{H^1(\Omega)} \leq c_0 \|u\|_{H^1(\Omega_h, \Phi_h^* g_h)}. \tag{5}$$

We further assume that Φ_h has local inverse mappings satisfying a uniform Lipschitz condition. Finally, assume that the volume forms $b_h(x)dx$ of

$\Phi_h^* g_h$ converge weak* in $L^\infty(\Omega)$ to a bounded, uniformly positive $b(x)dx$. Then there is a uniformly positive L^∞ metric $a_{ij}(x)dx^j dx^i$ on Ω so that the Dirichlet integrals of $\{\Omega_h, \Phi_h^* g_h\}$ Γ-converge in $L^2(\Omega)$ to

$$G(u) := \int_\Omega a^{ij}(x) D_j u\, D_i u\, b(x)\, dx.$$

On the other hand, a specific functional of the form of G may be shown to be the Γ-limit provided that $g_h^{ij}(x) b_h(x) 1_{\Omega_h}$ converges in measure in Ω to $a^{ij}(x)b(x)$.

Note that Ω_h does not yet reflect the topology of M_h, as in the example above, so that the Dirichlet integral of $\{\Omega_h, \Phi_h^* g_h\}$ will not be equivalent to the Dirichlet integral D_h of M_h. We require that Ω carry an involutive bi-Lipschitz homeomorphism $T : \Omega \to \Omega$, which is simultaneously an isometry for all the metrics $\Phi_h^* g_h$. Each mapping $\Phi_h : \Omega_h \to M_h$ is assumed to be one-to-one except on a compact T-invariant subset Σ_h of $\partial\Omega_h$, where it is two-to-one.

Theorem 1 [dM–G–M]. *Under the hypothesis above, there exists a quasi T-invariant Borel measure ν on $\bar{\Omega}$, such that after passing to a subsequence, the functional*

$$F(u) := G(u) + \int_{\bar{\Omega}} (u(x) - u(Tx))^2 d\nu(v) \tag{6}$$

is the Γ-limit in $L^2(\Omega)$ of $D_h(u \circ \Phi_h^{-1})$.

A measure ν is quasi-T-invariant if for all $u \in H^1(\Omega)$, $u(x)^2$ and $u(Tx)^2$ have the same integral with respect to ν. Theorem 1 states that Riemannian manifolds satisfying certain uniformity conditions enjoy compactness in terms of Γ-convergence, provided that the category under consideration is expanded to allow relaxed Riemannian manifolds. These uniformity conditions are not related to curvature bounds, to lower bounds on injectivity radius, or to bounds on topological type, as illustrated by Examples I and II. The hypothesis that $b_h \to b$ in weak*-L^∞, where $b \geq$ const. > 0, implies that not too much volume is being cut away from Ω. The strong connectivity hypothesis (5) prevents substantial regions from being nearly separated in Ω_h. The proof begins by defining $\nu_h = \infty_{\Sigma_h}$ and showing that M_h is equivalent to the relaxed Riemannian manifold $\{\Omega_h, \Phi_h^* g_h, 1, \nu_h\}$. Complementary to this compactness theorem are results which allow one to show that a specific functional of the form (6) is the Γ-limit of $D_h(u \circ \Phi_h^{-1})$ by computing capacitary densities [Bu–dM–M].

Theorem 2 [dM–G–M]. *In addition to the hypotheses of Theorem 1, assume that* $b_h 1_{\Omega_h} \rightarrow b$ *in measure on* Ω. *Then the* jth *eigenvalue of the subsequence* M_h, *counted with multiplicities, converges to the* jth *eigenvalue of the functional* (6) *with the weight* $b(x)$.

Remarks 1. It will be apparent to the reader that our discussion of minimal hypersurfaces of $\mathbf{R}^{n+1}$ has in fact focussed on their *intrinsic* geometry, except in problem (2FF). In terms of Riemannian geometry, the emphasis on minimal submanifolds here serves to direct attention to an as yet undefined class of manifolds, which readily exhibit unbounded topological type, which tend to have handles which are not too thin, and to which most recent theorems do not apply.

2. Mathematicians familiar with the ideas related to Young measures may note strong analogies in a quite different context [T].

We would like to thank SFB 256 and the University of Bonn for hospitality during the development of this research.

REFERENCES

[A] Allard, W., *First variation of a varifold.* Annals of Math. **95**, 1972, 417–491.

[B–dM–M] Baxter, J., G. dal Maso and U. Mosco, *Stopping times and* Γ*-convergence.* Trans. Amer. Soc. **303**, 1987, 1–38.

[Bu–dM–M] Buttazzo, G., G. dal Maso and U. Mosco, *A derivation theorem for capacities with respect to a Radon measure.* J. Functional Anal. **71**, 1987, 263–278.

[C–F] Chavel, I. and E. Feldman, *Diffusion on manifolds with small handles.* Indiana U. Math. J. **34**, 1985, 449–461.

[dM–G–M] dal Maso, G., R. Gulliver and U. Mosco, *Asymptotic spectrum of manifolds of increasing topological type.* preprint, Bonn, 1989.

[dM–M] dal Maso, G. and U. Mosco, *Wiener criteria and energy decay for relaxed Dirichlet problems.* Arch. Rat. Mech. Anal. **95**, 1986, 345–387.

[dG–F] de Giorgi, E. and T. Franzoni, *Su un tipo di convergenza variazionale.* Atti Accad. Naz. Lincei Fis. Mat. **58**, 1975, 842–850.

[Fuk] Fukushima, M., *Dirichlet Forms and Markov Processes.* North-Holland, Amsterdam, 1980.

[G–L–P] Gromov, M., J. Lafontaine and P. Pansu, *Métriques sur les Variétés Riemanniennes.* Cedic/Fernand Nathan, Paris, 1981.

[H] Hutchinson, J., *Second fundamental form for varifolds and the existence of surfaces minimizing curvature,* Indiana U. Math. J. **35**, 1986, 45–71.

[K] Karcher, H., *Embedded minimal surfaces derived from Scherk's examples.* Manuscripta Math. **62**, 1988, 83–114.

[S1] Simon, L., *Lectures on Geometric Measure Theory.* Proc. Centre Math. Analysis, Australian Nat. Univ., vol. 3, Canberra, 1983.

[S2] Simon L., *Asymptotics for a class of non-linear evolution equations, with applications to geometric problems.* Annals of Math. **118**, 1983, 525–571.

[T] Tartar, L., *Compensated compactness and applications to partial differential equations.* pp. 136–212 in Nonlinear Analysis and Mechanics, Heriot-Watt Symposium IV, Pitman, 1979.

Department of Mathematics
University of Minnesota
127 Vincent Hall
206 Church Street S.E.
Minneapolis, Minnesota 55455

Harmonic Diffeomorphisms between Riemannian Manifolds

FREDERIC HELEIN

In this lecture I will speak principally about a joint work with J.-M. Coron[C H] in which we studied the following problem : let M and N be two Riemannian manifolds with or without boundary, which are diffeomorphic, and let u be a harmonic C^1 - diffeomorphism between M and N. In the case where M and N have nonempty boundaries, we assume that the restriction of u to ∂M is a diffeomorphism between ∂M and ∂N. Then we want to know if u is or is not a minimizing harmonic map, i.e. if u minimizes the energy functional among the maps which have the same boundary data as u and which are homotopic to u. In the case where M and N have empty boundaries, the answer is generally no because of the counterexample of the identity map from S^3 to S^3: this is a harmonic diffeomorphism but the infimum of the energy in its homotopy class is zero (see [E S]).

Let us consider first a simple example with boundary: the source manifold is the unit ball B^n and, N is a Riemannian manifold diffeomorphic to B^n. We suppose too that the harmonic diffeomorphism u is of class C^1. Then it is interesting to represent the manifold N using the chart u^{-1}: $N \to B^n$. Or equivalently we assume that N is equal to the ball B^n equipped with the pull-back image g_{ij} of the metric on N by the map u: $N = (B^n, g_{ij})$. We reduce then our problem to the case of the identity map from (B^n , c) to (B^n , g_{ij}) where c is the canonical Euclidian metric. Here since u is of class C^1 , the g_{ij} are of class C^0 on B^n.

Let us suppose for the moment that the g_{ij} are of class C^1 . Then it is easy to find the Euler equation of our map. For a harmonic map $\varphi : (B^n, c) \to (B^n, g_{ij})$ this is :

$$\forall\, k,\ \Delta \varphi^k + \Gamma^k_{ij} (\varphi)\, \partial_\alpha\varphi^i\, \partial_\alpha\varphi^j = 0$$

where Γ^k_{ij} are the Christoffel symbols and with the usual summation conventions for indices, and where $\partial_\alpha = \partial / \partial x^\alpha$. But here for $Id : (B^n, c) \to (B^n, g_{ij})$, this becomes :

$$\forall k,\ \Gamma^k_{ii} = 0$$

or

$$(1)\quad \forall k,\ \partial_k(g_{ii}) = 2\,\partial_i(g_{ik}).$$

In fact, in the general case where the g_{ij} are of class C^0, one can show that the equation (1) holds in a distribution sense.

One can consider a slightly more general situation in the case where the dimension n is greater than or equal to 3: we can replace B^n by B^n minus a finite collection of points $\{ a_1, ..., a_p \}$, and u is a harmonic diffeomorphism between $B^n_* = B^n \setminus \{ a_1, ..., a_p \}$ and N. For example if $B^n_* = B^n \setminus \{ a_1, a_2 \}$ one can consider manifolds N with the following shape :

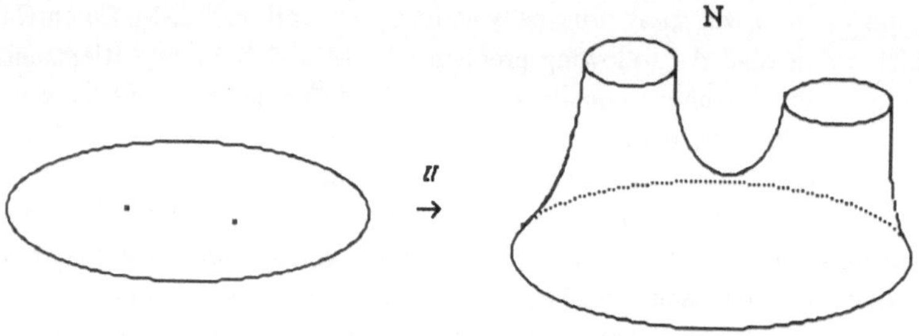

diffeomorphic to the ball minus two holes. The metric coefficients g_{ij} which we obtain by this way are not continuous on the a_q's. However, if we suppose that these g_{ij} satisfy some technical conditions which express roughly that N is the interior of a compact manifold with boundary, we can show that equation (1) still holds in the distribution sense. In the following this allows us to give some results about harmonic maps with point singularities.

To summarize, our hypothesis is (1) and we want to know if $Id : (B^n, c) \to (B^n, g_{ij})$ is a minimizing harmonic map, i.e. for any map φ in $H^1_*(B^n, B^n)$

$$(2)\quad \begin{cases} E(\varphi) := (1/2)\int_{B^n} g_{ij}[\varphi(x)]\,\partial_\alpha\varphi^i(x)\,\partial_\alpha\varphi^j(x)\,dx \\[2ex] \qquad \geq (1/2)\int_{B^n} g_{ii}(x)\,dx = E(Id). \end{cases}$$

Here $H^1_*(B^n, B^n)$ is the set of the maps φ in $H^1(B^n, B^n)$ which agree with Id on the boundary and such that $E(\varphi)$ has a sense and is finite.

First let us consider a different but simpler situation. Let $\pi : B^n \to (\mathbf{R}^2, \lambda\, \delta_{ij})$ be a map which associates to each x in B^n the two first coordinates (x^1, x^2) of x in an orthogonal basis of $\mathbf{R}^n$. $\mathbf{R}^2$ is equipped with a metric $\lambda(z)\,\delta_{ij}$ conformal to the Euclidian metric of $\mathbf{R}^n$. Then it is not difficult to see that Π is a minimizing harmonic map. This can be shown by using an area estimate argument or by using the coarea formula of Federer. Let φ be a smooth map from B^n into $(\mathbf{R}^n, \lambda\, \delta_{ij})$ which agrees with Π on ∂B^n, then

$$E(\varphi) = (1/2)\int_{B^n} |\nabla\varphi(x)|^2 \lambda[\varphi(x)]\,dx$$
$$\geq \int_{B^n} |d\varphi^1(x) \wedge d\varphi^2(x)|\,\lambda[\varphi(x)]\,dx,$$

and it follows from the coarea formula that this last expression is equal to

$$\int_{B^n} \mathcal{H}^{n-2}[\varphi^{-1}(z)]\,\lambda(z)\,dz,$$

where $\mathcal{H}^{n-2}[\varphi^{-1}(z)]$ is the $(n-2)$ - dimensionnal Hausdorff measure of the subset $\varphi^{-1}(z)$ of B^n. A consequence of Sard's theorem is that for almost every z in B^2, $\varphi^{-1}(z)$ is a smooth submanifold of B^n the boundary of which is the same as the boundary of $\Pi^{-1}(z)$, since φ agrees with Id on ∂B^n. Also $\Pi^{-1}(z)$ is obviously an $\mathcal{H}^{n-2}$-minimizing submanifold. This leads to the inequality

$$\mathcal{H}^{n-2}[\varphi^{-1}(z)] \geq \mathcal{H}^{n-2}[\Pi^{-1}(z)] \quad \text{for almost every z in } B^2.$$

Hence

$$E(\varphi) \geq \int_{B^n} \mathcal{H}^{n-2}[\Pi^{-1}(z)]\,\lambda(z)\,dz = E(\Pi).$$

From this inequality it follows by density that for any φ in $H^1_*(B^n, (B^2, \lambda\, \delta_{ij}))$

$$(3) \quad \int_{B^n} \gamma_{ij}[\varphi(x)]\,\partial_\alpha\varphi^i(x)\,\partial_\alpha\varphi^j(x)\,dx \geq \int_{B^n} \gamma_{ii}(x)\,dx$$

where $\gamma_{ij}(x) = \langle \Pi(e_i), \Pi(e_j) \rangle\, \lambda[\Pi(x)]$. Now for any R in SO (n) we can define a similar map $\Pi_R : B^n \to (B^2, \lambda_R\, \delta_{ij})$ by $\Pi_R(x) = (\langle Rx, e_1 \rangle, \langle Rx, e_2 \rangle)$. Then we can obtain an inequality (3_R) similar to (3) where $\gamma_{ij}(x)$

is replaced by $\gamma^R_{ij}(x) = \langle \Pi_R(e_i), \Pi_R(e_j)\rangle \lambda_R[\Pi_R(x)]$. And if there exists a real positive function f on $SO(n) \times \mathbf{R}^2$ such that $f(R, z) = \lambda_R(z)$ for almost every (R, z) in $SO(n) \times \mathbf{R}^2$, then we can sum the inequalties (3_R) over $SO(n)$. So if

$$(4)\quad h_{ij}(x) = \int_{SO(n)} \langle \Pi_R(e_i), \Pi_R(e_j)\rangle f(R, \Pi_R(x))\, d\sigma(R),$$

then h_{ij} satisfies an inequality, similar to (3), where γ_{ij} is replaced by h_{ij}.

Hence we have shown that for every metric h_{ij}, if there exists a positive function f such that (4) holds, then there corresponds to this metric h_{ij} a harmonic minimizing diffeomorphism. So one way to prove that a given harmonic diffeomorphism $Id : (B^n, c) \to (B^n, g_{ij})$ is minimizing is to prove that (4) holds with a positive f. This method was first used by J.-M. Coron and by R. Gulliver in [C G] for related problems. Since it seems difficult to say if (4) is true in the general case we have studied the case of $SO(n)$ - equivariant maps and manifolds. Then we have to consider the identity map from (B^n, c) to B^n equipped with the metric

$$ds^2 = g_{//}(r)[(dx)_{//}]^2 + g_{\perp}(r)[(dx)_{\perp}]^2$$

where $(dx)_{//} = \langle dx, x/|x|\rangle$, $(dx)_{\perp} = dx - (dx)_{//}\, x/|x|$, $r = |x|$, and $g_{//}$ and $g_{\perp}$ are positive continuous maps on $[0, 1)$. Without loss of generality it suffices to show that there exists a map φ from $[0, +\infty)$ to $(0, +\infty)$ such that

$$g_{ij}(x) = \int_{SO(n)} \langle \Pi_R(e_i), \Pi_R(e_j)\rangle\, \varphi(|\Pi_R(x)|)\, d\sigma(R)$$

to obtain (4).

In this special situation hypothesis (1) becomes

$$(5)\quad (d/dr)[r^{2(n-1)} g_{//}(r)] = (n-1)\, r^{2(n-2)}\, (d/dr)[r^2 g_{\perp}(r)],$$

and (4) is equivalent to

$$(6)\quad \begin{cases} r^2 g_{\perp}(r) = (n-2)/(n-1) \displaystyle\int_0^r (2 - t^2/r^2)(1 - t^2/r^2)^{(n-4)/2}\, t\, \varphi(t)\, dt \\ r^2 g_{//}(r) = (n-2) \displaystyle\int_0^r (t^2/r^2)\, (1 - t^2/r^2)^{(n-4)/2}\, t\, \varphi(t)\, dt. \end{cases}$$

We now have to study the solvatibility of (6) with the compatibility equation (5). The main difficulty is to be sure that the function φ which satisfies (6) is a nonnegative function. Because of this requirement, the results which follow are actually partial.

Theorem 1. *If* $n = 3$, *if (5) is satisfied, and if* $r^2 g_\perp(r)$ *is a nondecreasing function, then there exists a nonnegative function* φ *such that (6) holds, and hence the identity map from* (B^n, c) *to* (B^n, g_{ij}) *is a minimizing map.*

Let us remark that this condition has a simple geometric interpretation. Suppose that the target manifold N is a hypersurface of revolution in $\mathbf{R}^4$ with the metric induced by the embedding of N in $\mathbf{R}^4$. Then $r^2 g_\perp(r)$ represents the square of the distance of a point of N to the axis of revolution of N. This condition is always true if for example N is the graph of a SO (3) - invariant map from B^3 to **R**. If $n = 4$, the results are not as good as in dimension 3 :

Theorem 2. *If* $n = 4$, *and if (5) is satisfied, there exists a nonnegative function* φ *such that (6) holds if and only if* $r^4 g_{//}(r)$ *is a nondecreasing function. Hence in this case the identity map from* (B^n, c) *to* (B^n, g_{ij}) *is a minimizing map.*

If $n \geq 5$, when the dimension increases, the results become weaker. In dimension 3, the condition that $r^2 g_\perp(r)$ is nondecreasing is not optimal. On one hand there are examples where $r^2 g_\perp(r)$ is decreasing over a non negligeable set and where the relation (6) holds with a nonnegative φ (see [C H]). On the other hand there exist SO (3) - equivariant harmonic diffeomorphisms such that the map φ which satisfies the relation (6) is negative somewhere. To see this it suffices to consider the example with $g_{//}(r) = (1 - r^4) / (1 + r^4)^2$ and $g_\perp(r) = 1 / (1 + r^4)^2$. Then we have a harmonic SO (3) - equivariant diffeomorphism such that $2g_\perp(r) - g_{//}(r)$ is decreasing and one can show that if relations (6) were true with a nonnegative φ then $2g_\perp(r) - g_{//}(r)$ would be nondecreasing.

Analogous results are proved in the case of maps u with singularities. In this case the source manifold is $B^n \setminus \{0\}$ and N is diffeomorphic to $B^n \setminus \{0\}$. Now let us consider some examples of applications of the above results.

Dimension 3. The ellipsoid

Let $\mathcal{E}_a^n = \{ (x, x_{n+1}) \in \mathbf{R}^n \times \mathbf{R} : (x)^2 + (x_{n+1})^2 / a^2 = 1 \}$ where a belongs to $(0, +\infty)$. Let us consider the set $H^1_*(B^n, \mathcal{E}_a^n)$ of maps φ in $H^1(B^n, \mathbf{R}^n)$ which satisfy $\varphi(x) \in \mathcal{E}_a^n$ for almost every x in B^n and such that $\varphi(x) = (x, 0)$ on ∂B^n. Then in the case where $n = 3$, we show that

- If $a^2 \geq 8$, $u_*(x) = (x / |x|, 0)$ is minimizing.
- If $a^2 < 8$, there exists a smooth SO (3) - equivariant minimizing map which is a diffeomorphism between B^3 and the upper hemisphere of $\mathcal{E}_a^3$, $\mathcal{E}_a^{3+} = \mathcal{E}_a^3 \cap \{ (x, x_4) : x_4 > 0 \}$.

Let us recall the results which were known about this problem in dimension n greater or equal than 3. It was first solved for $n \geq 3$ and $a = 1$ in [J K], then for $n \geq 7$ and $a \geq 1$ in [B], and for $n \geq 3$ and $a \leq 1$ in [Hé 1]. Furthermore A. Baldes showed in [B] that if $n \geq 3$, $u_*(x) = (x / |x|, 0)$ could be minimizing only when $a^2 \geq 4(n-1)/(n-2)^2$. One can summarize all these papers by the following conjecture for $n \geq 3$:

- If $a^2 \geq 4(n-1)/(n-2)^2$, u_* is minimizing.
- If $a^2 < 4(n-1)/(n-2)^2$, there exists a smooth SO (n) - equivariant minimizing diffeomorphism between B^n and $\mathcal{E}_a^{n+}$.

Graph of a SO (3) - invariant real valued function

If N is the graph of a SO (3) - invariant map f from $\overline{B^3}$ to $\mathbf{R}$, then in the set of the maps φ from B^3 to the graph of f which satisfies $\varphi(x) = (x, f(x))$ on ∂B^3, there exists a smooth SO (3) - equivariant map minimizing map u if f is of class C^1 on $\overline{B^3}$, even if N is very sharp.

Example of R. Gulliver and B. White

This is a harmonic map u from B^3 to a SO (3) - equivariant Riemannian manifold Σ which has roughly the shape of a hyperboloid of revolution. u is regular on B^3 except at 0, and the blow - up sequence at 0, $u_\lambda(x) = u(\lambda x)$ tends to its homogeneous tangent map more slowly than any positive power of λ when λ tends to 0 (see [G W]). In [C H] we have shown that this map is harmonic minimizing.

Dimension 4

In this case, one can show that another example due to R. Gulliver and B. White with the same property is minimizing.

Now let us turn to the case of Riemannian manifolds of dimension 2. Here we will obtain optimal results:

Theorem 3. *Any harmonic diffeomorphism between Riemannian surfaces is minimizing (by minimizing we mean minimizing in its homotopy class; the homotopies are required to agree with the considered map on ∂M).*

First let us see how to prove it in the case where the source manifold is the closed unit disc $\overline{B^2}$ (the target manifold N is diffeomorphic to $\overline{B^2}$). Let Id : (B^2, c) $\rightarrow$ (B^2, g_{ij}) be such a harmonic diffeomorphism. The Euler equation (1) can be integrated for any dimension but for n = 2, the solution is particularly interesting. Indeed from (1) we get :

$$\begin{bmatrix} g_{11} g_{21} \\ g_{12} g_{22} \end{bmatrix} = \begin{bmatrix} \lambda \, 0 \\ 0 \, \lambda \end{bmatrix} + \begin{bmatrix} -\operatorname{Re}\varphi & \operatorname{Im}\varphi \\ \operatorname{Im}\varphi & \operatorname{Re}\varphi \end{bmatrix}$$

where λ is a continuous positive real - valued function on $\overline{B^2}$, and φ is a holomorphic map from B^2 to **C** such that $\lambda^2 > |\varphi|^2$ (this last condition expresses that the metric g_{ij} is positive definite). Let $\omega : \overline{B^2} \rightarrow (0, +\infty)$ be defined by

$$\omega(x) = [2 - (x^1)^2 - (x^2)^2]^{-4}.$$

Since $\lambda^2 > |\varphi|^2$ on $\overline{B^2}$, there exists $\varepsilon > 0$ such that $\lambda^2 > |\varphi|^2 + \varepsilon\omega$ on $\overline{B^2}$. Using ω we make the following decomposition of the g_{ij} by

$$(g_{ij}) = (g'_{ij}) + (g''_{ij})$$

where

$$(g'_{ij}) = \begin{bmatrix} \lambda - \sqrt{|\varphi|^2 + \varepsilon\omega} & 0 \\ 0 & \lambda - \sqrt{|\varphi|^2 + \varepsilon\omega} \end{bmatrix}$$

and

$$(g''_{ij}) = \begin{bmatrix} \sqrt{|\varphi|^2 + \varepsilon\omega} & 0 \\ 0 & \sqrt{|\varphi|^2 + \varepsilon\omega} \end{bmatrix} + \begin{bmatrix} -\operatorname{Re}\varphi & \operatorname{Im}\varphi \\ \operatorname{Im}\varphi & \operatorname{Re}\varphi \end{bmatrix}.$$

Now we consider the two harmonic diffeomorphisms Id : $(B^2, c) \to (B^2, g')$, and Id : $(B^2, c) \to (B^2, g'')$. The first map is conformal and hence minimizing. Moreover a straightforward computation shows that the Gauss curvature of (B^2, g'') is negative everywhere. It is possible to extend the metric g" to all of $\mathbf{R}^2$ by a metric σ such that $(\mathbf{R}^2, \sigma)$ is complete and has negative Gauss curvature everywhere too. But a well-known result of Hartman says that any harmonic map into a complete Riemannian surface with negative Gauss curvature everywhere is unique with respect to its own boundary conditions or in its homotopy class [Ha]. This implies that Id : $(B^2, c) \to (B^2, g'')$ is minimizing. Hence Id is minimizing from (B^2, c) to $(B^2, g' + g'')$.

Now let us present the proof of the generalization of this result to the case of two diffeomorphic Riemannian surfaces (M, h) and (N, g) with or without boundary. We use isothermal charts on M such that the metric h is given by $\rho (z) dzd\bar{z}$ where ρ is a positive map. The condition which expresses that $u : (M, h) \to (N, g)$ is harmonic is that the pull-back image $u_* (g)$ of the metric g by u is of the form

$$\lambda (z) dzd\bar{z} - \operatorname{Re} [\varphi (z) (dz)^2]$$

where $\varphi (z) (dz)^2$ defines a holomorphic two-form on M. Here we remark that since any holomorphic two-form on S^2 must vanish, any harmonic map from M is conformal provided that M is diffeomorphic to S^2, and hence is minimizing. Therefore we will consider the case where M has a non empty boundary or where M has a nonnegative genus. Then there exists h_0 on M which is conformal to h and such that (M, h_0) has a nonnegative Gauss curvature everywhere. Let suppose that in our isothermal chart, h_0 has the expresssion $\sqrt{\omega} (z) dzd\bar{z}$. Then, choosing ε small enough we can write

$$u^* g = h' + h''$$

where $h' = (\lambda - \sqrt{|\varphi|^2 + \varepsilon\omega}) dzd\bar{z}$

and $h'' = \sqrt{|\varphi|^2 + \varepsilon\omega}\, dzd\bar{z} - \operatorname{Re} [\varphi (z) (dz)^2]$.

We set $g' = (u^{-1})_* h'$ and $g'' = (u^{-1})_* h''$. Then $g = g' + g''$. To conclude we remark that $u : (M, h) \to (N, g')$ is conformal and thus minimizing, and that $u : (M, h) \to (N, g'')$ is minimizing too because it is harmonic and (N, g'') has a nonpositive Gauss curvature.

To conclude let us speak about a related result which is proved in [Hé 2] .

Theorem. *Let* $u : (M, h) \to (N, g)$ *be a continuous map between Riemannian surfaces which satisfies*

(i) *u is a homeomorphism between* M *and* N,

(ii) *u is quasiconformal, i.e.* $|u_{\bar z}| < k\,|u_z|$ *with* $k < 1$,

(iii) $\varphi(z)(dz)^2 = (|u_x|^2 - |u_y|^2 - 2i\langle u_x, u_y\rangle)(dz)^2$ *is a holomorphic two-form on M.*

Then u is harmonic regular (thus minimizing).

The idea of the proof is based on the following formal - quite non rigorous - observation : the condition (iii) can be intepreted by saying that the pull-back image of the metric on N by u (which is not a metric because of the lack of regularity of u) is of the form

$$\lambda(z)\,dzd\bar z - \mathrm{Re}[\varphi(z)(dz)^2]$$
$$= (\lambda(z) - |\varphi(z)|)\,dzd\bar z + [\,|\varphi(z)|\,dzd\bar z - \mathrm{Re}[\varphi(z)(dz)^2]\,]$$

The first term looks like a metric conformal to h. Moreover on any simply connected open subset Ω of M on which $\varphi \neq 0$, one can write

$$|\varphi(z)|\,dzd\bar z - \mathrm{Re}[\varphi(z)(dz)^2] = |d\alpha(z)|^2$$

where $\alpha : \Omega \to \mathbf{R}$ is a scalar harmonic function. These two remarks imply that the restriction of u to any Ω is minimizing and hence regular. Regularity can therefore be proved on M minus the points where φ is zero. But these points are isolated and by a result of J. Sacks and K. Uhlenbeck [S U], u is regular everywhere.

REFERENCES

[A] S.I. AL'BER, *On n-dimensional Problems in the Calculus of Variations in the Large,* Soviet. Math. Dokl. **5** (1964), p. 700-704.

[B] A. BALDES, *Stability and uniqueness properties of the equator map from a ball into an ellipsoid*, Math. Z. **185** (1984), p. 505 - 516.

[C H] J. - M. CORON, F. HELEIN, *Harmonic diffeomorphisms, minimizing harmonic maps and rotational symmetry*, to appear in Comp. Math.

[C G] J. - M. CORON, R. GULLIVER, p - *harmonic maps into spheres*, preprint.

[E S] J. EELLS, J. H. SAMPSON, *Harmonic mappings of Riemannian manifolds*, Amer. J. Math. **86** (1964), p. 109 - 160.

[G W] R. GULLIVER, B. WHITE, *On convergence rates of Harmonic maps near points of discontinuity*, to appear in Math. Ann.

[Ha] P. HARTMAN, *On homotopic harmonic maps*, Canad. J. Math. **19** (1967), p. 547 - 570.

[Hé 1] F. HELEIN, *Regularity and uniqueness of harmonic maps into an ellipsoid*, Manuscripta Math. **60** (1988), p. 235 - 257.

[Hé 2] F. HELEIN, *Homéomorphismes quasiconformes entre variétés Riemanniennes*, to appear in C. R. Acad. Sci. Paris.

[J K] W. JÅGER, H. KAUL. *Rotationally symmetric harmonic map from a ball into a sphere and the regularity problem for weak solutions of elliptic systems*, J. Reine Angew. Math. **343** (1983), p. 146 - 161.

[S U] J. SACKS, K. UHLENBECK, *The existence of minimal immersions of two spheres*, Ann. Math. **113** (1981), p. 1 - 24.

Frédéric Hélein,
Centre de Mathématiques,
Ecole Polytechnique, Palaiseau Cedex
FRANCE

SURFACES OF MINIMAL AREA SUPPORTED BY A GIVEN BODY IN $\mathbb{R}^3$

G. Mancini
Dipartimento di Matematica, Università di Bologna
R.Musina
SISSA, Trieste

Existence theory in the "Plateau problem" is mostly concerned with surfaces spanning some given boundary configuration ([A], [Cou], [HiN], [GJ], [S], to quote a few).

Here we consider minimal surfaces "supported" by a given body Ω in $\mathbb{R}^3$: sphere-like surfaces might be expected in this case, with free boundaries arising where the surface touches the "obstacle" Ω .

In case Ω is smooth, this problem fits in the more general framework of minimal spheres in Riemannian manifolds. Basic existence results, for compact manifolds without boundary, have been obtained by Sacks and Uhlenbeck [SU] (see also [MiMo] for an estimate of the Morse index of such objects and application to the topology of manifolds).

The main difficulty in this kind of variational problems is some lack of compactness due to invariance with respect to the action of the non compact group of conformal tranformations.

In their celebrated paper, Sacks and Uhlenbeck tackle this difficulty solving first approximated variational problems and then passing to the limit. The limiting procedure is based upon estimates for the solutions of the associated Euler-Lagrange equations, which show that the only obstruction to convergence is the existence of minimal spheres. In case of manifolds with boundary one has, however, to deal with inequalities rather than with equations.

Here we present, in a very special case, a direct approach, based on the analysis of the behaviour, with respect to weak convergence, of some integrands related to integral homotopic invariants.

As an application, we also give an existence result for disk-type minimal surfaces with obstructions belonging to prescribed homotopy classes (see [MM]).

1. S^2-type surfaces of minimal area supported by a given body

Let $M: = \mathbb{R}^3 \setminus \Omega$, Ω being a bounded open subset of $\mathbb{R}^3$. We assume, in this section, Ω connected, and for simplicity, that M is connected and $0 \in \Omega$. We will call sphere in M any L^∞ map from S^2 into M, having finite Dirichlet integral. We consider

Problem 1. Find a "smooth" sphere with minimal area in the class of non contractible spheres in M.

Since Problem 1 is a free boundary problem, the expected optimal smoothess is hölder continuity of the first derivatives.

A Dirichlet Principle. If ∂M is smooth, with second fundamental form b, we say that a $\mathcal{C}^{1,\alpha}$ map U from S^2 into M is a minimal sphere if

$$\Delta U = \chi_{U^{-1}(\partial\Omega)} b(U)\,(dU,dU)\,\nu(U) \qquad \text{a.e.}$$

Here Δ denotes the Laplace-Beltrami operator on S^2, χ is the characteristic function of the "coincidence set" and ν stands for the inner unite normal vector field of ∂M relative to M (see [D]).

By elliptic regularity theory (see [H1], [Du], [F], etc.) we can regard minimal spheres as stationary points of the Dirichlet integral. Since it can be easily seen that a minimal sphere is conformal, and taking into account the ε–conformality principle of Morrey, Problem 1 reduces to minimizing the Dirichlet integral in the class of noncontractible spheres. This is in fact a variational problem with a (homotopically invariant) integral constraint.

Let

$$X(M): = \{U \in L^\infty(\mathbb{R}^2, \mathbb{R}^3): U(z) \in M \text{ for a.e. } z \in \mathbb{R}^2, \text{ and } \int |\nabla U|^2 < +\infty\}$$

$$V(U): = \int \det(U, \nabla U), \qquad U \in X(M),$$

$$p\xi: = \xi/|\xi|, \qquad \xi \in \mathbb{R}^3 \setminus \{0\}.$$

Notice that $\nabla(pU) \in L^2$ $\forall U \in X(M)$ and $(4\pi)^{-1} V(pU) \in \mathbb{Z}$ is the "degree" of pU (see [BC]). Also, denoted by Π: $S^2 \to \mathbb{R}^2$ the stereographic projection, and assuming $U \circ \Pi \in \mathcal{C}(S^2, M)$, then $U \circ \Pi$ is a non contractible sphere in M, provided $V(pU) \neq 0$.

After setting $X^e = X^e(M) := \{U \in X(M) : V(pU) \neq 0\}$, we are led to consider

Problem 1'. $$\min_{X^e} \int |\nabla U|^2 \ (:= I_\infty)$$

As noticed above, if a minimizer exists, it is a non contractible minimal sphere, and hence a solution to Problem 1. We state this fact explicitely (see [MM] for details):

Theorem 1. Assume $\Omega := \{F<0\}$, with $F \in \mathcal{C}^2(\mathbb{R}^3, \mathbb{R})$, $\nabla F \neq 0$ on $\partial\Omega$. Let $U_\infty \in X^e(M)$ be such that $\int |\nabla U_\infty|^2 = I_\infty$. Then U_∞ solves Problem 1'.

Thus a solution to Problem 1 follows from

Theorem 2. Assume M is a Lipschitz neighborhood retract. Then I_∞ is achieved.

Sketch of the proof of Theorem 2. Minimizing sequences can be assumed to be weakly convergent: if $U_n \in X^e$, $\int |\nabla U_n|^2$ is bounded, then there is $U \in L^\infty$ s.t. (for some subsequence) $\nabla U_n \to \nabla U$ weakly in L^2 and $U_n \to U$ a.e.. But, in general, the weak limit will be a constant, so that the existence of a minimizer is not a trivial fact. The following Lemma describes in a crucial way the behaviour of minimizing sequences.

Lemma 1. Let $U_n \in X^e$ be minimizing, with weak limit U. If $U \notin X^e$, then U is a constant, and, either

(i)there is (exactly one) $a \in \mathbb{R}^2$ s.t. $$\int_{|z-a| \leq r} |\nabla U_n|^2 \to I_\infty \quad \forall r>0$$

or

(ii) $U_n \to 0$ in H^1_{loc}.

Using the concentration function of $|\nabla U_n|^2$ (see [LPL]), one can build, from U_n, and by means of translations and dilations, a new minimizing sequence whose convergence in X^e will follow from the above Lemma.

Sketch of the proof of Lemma 1. It relies on a basic inequality, which we now describe.

Let $U_n \in X^e$, $\nabla U_n \to \nabla U$ weakly in L^2, $U_n \to U$ a.e.. If $U \notin X^e$, then, eventually replacing U_n by $U_n^*(z) := U_n(z|z|^{-2})$, there is $a \in \mathbb{R}^2$ s.t.

$$(1) \qquad \liminf \int_{|z-a| \leq r} |\nabla U_n|^2 \geq \int_{|z-a| \leq r} |\nabla U|^2 + I_\infty \quad \forall r>0.$$

In particular,

$$(2) \qquad \liminf \int_{\mathbb{R}^2} |\nabla U_n|^2 \geq \int_{\mathbb{R}^2} |\nabla U|^2 + I_\infty.$$

While we don't derive here Lemma 1 from (1), we present the basic arguments to prove (1). The main tool is a Lemma on "separation of spheres", first observed by Wente

Lemma 2 (see [BCL]). Let $\int |\nabla V_n|^2$ be uniformly bounded, and $V_n \to V$ a.e. Then, there is a finite set of points $a_i \in \mathbb{R}^2$, integers d_i such that

$$(3) \qquad \det(V_n, \nabla V_n) \to \det(V, \nabla V) + 4\pi \ \Sigma\, d_i\, \delta_{a_i} \ \text{ weakly in the sense of measures}$$

Eventually replacing U_n by U_n^*, we can assume $V_n = pU_n$ satisfies (3) with some $d_i \neq 0$. Now, to get (1), we "close" the surface $U_n|_{D_s(a_i)}$, with a second surface given by $k_n^s := h_n^s(a_i + (z-a_i)\,|z-a_i|^{-2}s^2)$, $|z-a_i| \geq s$, where

$$\int_{|z-a_i| \leq s} |\nabla h_n^s|^2 = \inf\{ \int_{|z-a_i| \leq s} |\nabla v|^2 : v \in H^1(D_s(a_i), M),\ v - U_n \in H^1_0(D_s)\}$$

Denoted by V_n^s such an extension, notice that

$$(4) \quad V(pV_n^s) = 4d_i\pi + o(1) + \int_{|z-a_i| \leq s} \det(pU, \nabla pU) - \int_{|z-a_i| \leq s} \det(ph_n^s, \nabla ph_n^s)$$

Using suitable comparison functions, and assuming M is a Lipschitz neighborhood retract, we can prove the following estimate

$$(5)\quad \liminf \int_{|z-a_i|\le s} |\nabla h_n^s|^2 \le \int_{|z-a_i|\le s} |\nabla U|^2 \quad \text{for a.e. } s.$$

From (5), (4) one easily gets, for a.e. s,

$$I_\infty \le \limsup \int_{\mathbb{R}^2} |\nabla V_n^s|^2 \le \limsup \int_{|z-a_i|\le s} |\nabla U_n|^2 + \int_{|z-a_i|\le s} |\nabla U|^2$$

Finally, for r fixed, and passing eventually to a subsequence, the above argument yields

$$\liminf \int_{|z-a_i|\le r} |\nabla U_n|^2 \ge I_\infty - \int_{|z-a_i|\le s} |\nabla U|^2 + \liminf \int_{s\le|z-a_i|\le r} |\nabla U_n|^2$$

for a.e. s, and hence (1).

Remark. The above techniques go over to general three dimensional Riemannian manifolds M, after suitably replacing the volume integral V (work in progress).

2. S^2 - type minimal surfaces enclosing many obstacles in $\mathbb{R}^3$

Several questions arise in connection with Problem 1: non existence of branch points, embeddedness, different topological type, etc... Difficult existence and regularity problems arise in the case of a (partly) thin obstacle Ω (see [De G] for an approach via geometric measure theory).

A simpler question to be considered arises when dropping the connectivity assumption on Ω. More precisely, assume $\Omega = \Omega_1 \cup .... \cup \Omega_p$, where Ω_i are disjoint open bounded sets and let $\xi_i \in \Omega_i$. Denoted $p_i \xi = (\xi - \xi_i) / |\xi_i - \xi|$, we consider

Problem 2.

$$\min_{\{U \in X(M) : V(p_i U) \ne 0 \ \forall i\}} \int |\nabla U|^2$$

Differently from the case of a single obstacle, here minimizing sequences do not converge, in general, even up to translations and dilations: they might weakly converge to non contractible minimal spheres "missing" some of the Ω_i's. Sacks and Uhlenbeck give in their paper a very precise description in terms of "splitting" of the energy, in the case of a general Riemannian manifold M without boundary.

Lemma ([SU]). Let Γ be a connected component of $C(S^2, M)$. Then either min $\{\int |dU|^2 : U \in \Gamma\}$ is achieved, or $\forall\, \delta > 0\ \exists\, \Gamma_1\ \Gamma_2$ connected components of $C(S^2, M)$ such that

$$\inf\{\int |dU|^2 : U \in \Gamma_1\} + \inf\{\int |dU|^2 : U \in \Gamma_2\} \leq \delta + \inf\{\int |dU|^2 : U \in \Gamma\}$$

Using such a description, they are able to prove the existence of many minimal spheres in M. Following these ideas, and in order to prevent the "dichothomy" described above, one is led to introduce

A Courant-Douglas type condition

(CD) $$\inf_{V(pU_j)\neq 0\ i=1...p} \int |\nabla U|^2 < \min_{\mathcal{P}} \left(\sum_j I^{\sigma_j}\right)$$

where $(\sigma_1,..., \sigma_2) \in \mathcal{P}$ is any partition of $\{1,...,p\}$ and

$$I^\sigma = \inf_{V(pU_j)\neq 0 \forall j \in \sigma} \int |\nabla U|^2, \text{ if } \sigma \subset \{1,...,p\}.$$

Using similar estimates as in the previous section, one can prove

Theorem 3 ([M]). Assume $M = \overline{\mathbb{R}^3 \setminus \Omega}$ is a Lipschitz neighborhood retract. Then

$$\inf_{U \in X(M), V(pU_i)\neq 0 \forall i} \int |\nabla U|^2 \text{ is achieved provided (CD) holds true.}$$

3. Pairs of solutions of the Plateau Problem for disk type minimal surfaces with obstructions.

Given Ω as above and a Jordan curve $\Gamma \subset C$, let

$X_\Gamma := \{u \in H^1(D,M) : u|_{\partial D}$ is a continuous weakly monotone parametrization of Γ satisfying a three point boundary condition$\}$

We assume $X_\Gamma(M)$ is not empty.

It is well known (see [T], [Hi]) that $\inf \int |\nabla u|^2$ is attained in X_Γ at some u which, assuming $\partial\Omega \in C^2$, satisfies

(i) $u \in C^{1,\alpha}(D) \cap C(\bar{D})$, $\Delta u = 0$ in $\{u \notin \bar{\Omega}\}$

(ii) $|u_x|^2 - |u_y|^2 = u_x u_y = 0$ in D

Furthermore $u|_{\partial D}$ is a homeomorphism and u minimizes the area functional in the class X_Γ.

Definition 1. Let $u \in X_\Gamma \cap C(D)$. We say that u, together with u, encloses Ω, if there is a change of parameters $g_u \in C(\dot{D}, \dot{D})$ such that

$$(6) \quad U(z) := \begin{cases} u(z) & \text{if } |z| \le 1 \\ u\,(g_u(z|z|^{-2})) & \text{if } |z| \ge 1 \end{cases}$$

is continuous and not homotopically trivial as a map from S^2 into M.

For a given u in X_Γ, one can prove (at least whenever $u \in C^1(\dot{D})$, which is the case if $\Gamma \cap \partial\Omega = \emptyset$, see [Hi2]) that $V(pu) = V(pu\, g_u)$, where $g_u(r,\theta) = r\, u^{-1}(u(1,\theta))$. Thus the following condition

$$(7) \quad V(pu) \neq V(pu) \quad (\text{here } V(u) = \int_D \det(u, \nabla u)\,)$$

implies that u, together with u, encloses Ω in the sense of Definition above (cpr. [BC]). Accordingly, we consider

Problem 3. To minimize $\int |\nabla u|^2$ in the class
$X_\Gamma^e := \{u \in X_\Gamma(M) : V(pu) \neq V(pu)\}$.

While Problem 2 is always well posed (i.e. X_Γ^e is not empty), it does not have, in general, any solution (e.g. if Γ reduces to a point). More generally:

$$(8) \quad I_\Gamma := \inf \{\int |\nabla u|^2 : u \in X_\Gamma^e\} \le I_\infty + \int |\nabla u|^2$$

and, whenever equality occurs, there are minimizing sequences which do not converge in X_Γ^e. Nevertheless, we have the following result.

Theorem 4. I_Γ is achieved provided strict inequality holds in (8).

Such a minimizer will enclose, together with the small solution u, the obstacle Ω, and,in case $\partial\Omega$ is smooth, will satisfy the minimal surface equation far away from the obstacle. Furthermore it is area minimizing in its class.

To prove Theorem 3, we first reduce to a Dirichlet problem.

Lemma 3. Let $\int |\nabla u_n|^2 \to I_\Gamma$, $V(pu_n) \neq V(p\,u)$, $u_n \to u$ weakly in H^1.Then

$$I_\Gamma = \inf\{\int |\nabla v|^2 : v-u \in H^1_o,\ V(pv) \neq V(p\,u)\}.$$

Finally, since strict inequality in (8) implies $I_\infty + \int |\nabla u|^2 > \inf\ \{\int |\nabla v|^2 : v-u \in H^1_o,\ V(pv) \neq V(p\,u)\}$, Theorem 3 follows from Lemma 3 and (cpr.(2))

Lemma 4. Let $u_n - u \to 0$ weakly in H^1_o. If $V(pu_n) \neq V(pu)$ while $V(pu) = V(pu)$, then

$$\liminf \int_D |\nabla u_n|^2 \geq I_\infty + \int_D |\nabla u|^2 .$$

REFERENCES

[A] **Alt, H.W.**: Die Existenz einer Minimalflache mit freimen Rand vorgeschriebener Lange, Arch. Rat. Mech. Anal. 51 (1973), 304-320

[BC] **Brezis,H., Coron,J.M.**: Large solutions for harmonic maps in two dimensions, Comm. Math. Phys. 92, 203-215 (1983)

[BCL] **Brezis, H., Coron, J.M., Lieb,E.H.**: Harmonic maps with defects, Comm. Math. Phys. 107, 649-705 (1986)

[Cou] **Courant, R.**: Dirichlet principle, conformal mappings and minimal surfaces, Interscience, New York, 1950.

[DeG] **De Giorgi E.**: Problemi di superfici minime con ostacoli: forma non cartesiana, B.U.M.I., 8, 80-88(1973).

[Du] **Duzaar, F.**: Variational inequalities and harmonic mappings, J. Reine Angewandte Math. 374, 39-60 (1987)

[F] **Fuchs, M.**: Variational inequalities for vector-valued functions with nonconvex obstacles, Analysis 5, 223-238 (1985)

[GJ] **Gruter, M., J.Jost**: On embedded minimal disks in convex bodies, A.I.H.P. Analyse non lineaire, 3, 5, 1986, 345-390

[Hi1] **Hildebrandt, S.**: On the regularity of solutions of two-dimensional variational problems with obstructions, Comm. Pure Appl. Math. 25, 479-496 (1972)

[Hi2] **Hildebrandt, S.**: Boundary behaviour of minimal surfaces, Arc. Rat. Mech. Anal. 35, 47-82 (1969).

[HiN] **Hildebrandt, S., J.C.C. Nitsche**: Minimal surfaces with free boundaries, Acta Matematica, 143 (1979) 251-272

[LPL] **Lions, P.L.**: The concentration-Compactness principle in the Calculus of Variations: the limit case, Rev. Mat. Iberoamericana 1, 145-201 vol. 1 and 45-121 vol. 2 (1985)

[M] **Musina R.**: S^2-type minimal surfaces enclosing many obstacles in $\mathbb{R}^3$, preprint.

[MM] **Mancini G., R. Musina**: Surfaces of minimal area enclosing a given body in $\mathbb{R}^3$.

[MiMo] **Micaleff M.J., J.D. Moore**: Minimal two spheres and the topology of manifolds with positive curvature on totally isotropic two planes, Ann. of Math., 198-227(1988).

[Ni] **Nitsche, J.C.C.**: Vorlesungen uber minimalflachen, Springer, Berlin 1975

[SU] **Sacks J., K.Uhlenbeck**: The existence of minimal immersions of 2-spheres, Ann. of Math., 113, 1-24(1981)

[S] **Struwe, M.**: On a free boundary problem for minimal surfaces, Invent. Math., 75 (1984), 547-560

[T] **Tomi, F.**: Minimal surfaces and surfaces of prescribed mean curvature spanned over obstacles, Math. Ann. 190, 248-264 (1971)

[W] **Wente, H.**: Large Solutions to the Volume Constrained Plateau Problem, Arch. Rat. Mech. and Analysis, 75, 59-77 (1980).

Calibrations and New Singularities in Area-minimizing Surfaces: A Survey

FRANK MORGAN

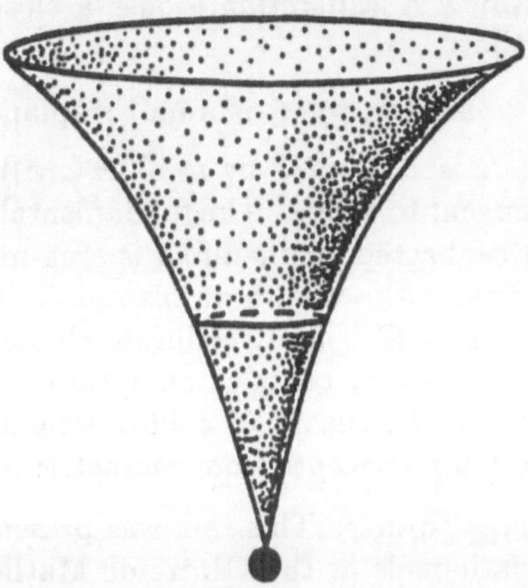

Contents

1. Introduction

In the past ten years the theory of calibrations has shed much light on the behavior of m-dimensional area-minimizing submanifolds of n-dimensional spaces. (*Area-minimizing* means that no other submanifold with the same boundary, or alternatively in the same homology class, has less m-dimensional area.) Area-minimizing submanifolds often have interesting singularities. An early obstacle to understanding was the difficulty of establishing that any singular example actually was area-minimizing. Calibrations produce a rich variety of such examples, as indicated in §3.

Lately the theory of calibrations has developed into a fine theoretical tool. For example, the Angle Criterion (see §4) tells at precisely what angles two sheets of an area-minimizing submanifold can cross. G. Lawlor (see Section 5) has completely classified which cones over products of spheres are area-minimizing.

A recent application (see §6) treats minimizers of integrands more general than area.

1.1 Calibrations. A *calibration* is just a closed differential m-form φ normalized so that

$$\sup\{\langle \xi, \varphi \rangle : \xi \text{ is a unit } m\text{-plane}\} = 1.$$

A submanifold S is *calibrated* by φ if $\langle \xi, \varphi(x) \rangle = 1$ whenever ξ is the oriented unit tangent to S at x. The fundamental Theorem of Calibrations 2.1 says that a calibrated submanifold is area-minimizing.

Section 3 indicates the wealth of examples. It begins with the powers of the Kahler form on $\boldsymbol{C}^n$. They calibrate all complex analytical varieties and thus prove that every complex analytic variety is area-minimizing.

Other surveys of the theory of calibrations are provided by [Ha] and [M1]. For underlying concepts from geometric measure theory, see [M2].

1.2 Acknowledgements. This talk was presented at the conference on Problèmes Variationnels at the Centre de Mathématiques Appliquées de l'Ecole Normale Superiéure in Paris in June 1988. An improvement of the results of §6 from $\boldsymbol{R}^6$ to $\boldsymbol{R}^4$ was obtained at that time while I was visiting J. M. Coron at the Université de Paris-Sud, Centre d'Orsay. This work was partially supported by a National Science Foundation grant.

2. The fundamental theorem of calibrations

2.1. Fundamental Theorem of Calibrations (cf. [HaL, Introduction]). *Let φ be a* <u>*calibration*</u>*, i.e., a closed differential m-form on an n-dimensional complete Riemannian manifold M normalized so that*

$$\sup\{\langle \xi, \varphi(x) \rangle : x \in M, \xi \text{ is a tangent } m\text{-plane to } M \text{ at } x\} = 1. \quad (1)$$

Let S be a C^1 compact oriented submanifold with or without boundary *calibrated by φ:* i.e., $\langle \xi, \varphi(x) \rangle = 1$ *whenever ξ is a tangent plane to S at x. Then S is* homologically area-minimizing: i.e., *if $S' \sim S$, then area $S' \geq$ area S.*

Remarks. S may be allowed singularities. Indeed the theorem holds in the category of the rectifiable currents of geometric measure theory (cf. [M2, Chapter 4]).

When M is $\boldsymbol{R}^n$ (or any manifold with trivial degree-m homology), S minimizes area among all m-dimensional surfaces (rectifiable currents) with the same boundary. For general, usually compact M, a closed calibrated surface S provides a least-area representative of its homology class.

The calibration φ may also be allowed singularities (cf. §3.7).

Proof. Since $\langle \xi, \varphi(x) \rangle = 1$ whenever ξ is a tangent plane to S at x, $\int_S \varphi =$ area S. By (1), for any surface S', $\int_{S'} \varphi \leq$ area S'. Since φ is closed, if $S' \sim S$, then $\int_{S'} \varphi = \int_S \varphi$. Assembling these three equalities and inequalities yields

$$\text{area } S' \geq \int_{S'} \varphi = \int_S \varphi = \text{ area } S,$$

as desired.

3. Examples of calibrations

3.1. The Kahler calibration. The first nontrivial calibrations were the normalized powers $\varphi = \omega^p/p!$ of the Kahler 2-form ω, defined on $\boldsymbol{C}^n \cong \boldsymbol{R}^{2n}$ by

$$\omega = dx_1 \wedge dy_1 + dx_2 \wedge dy_2 + \ldots + dx_n \wedge dy_n.$$

By Wirtinger's inequality [W], cf. [F1, §1.8.2]), $\langle \xi, \varphi \rangle \leq 1$, with equality if and only if the real $2p$-plane ξ in $\boldsymbol{R}^{2n} \cong \boldsymbol{C}^n$ is a complex p-plane. Consequently, by the fundamental theorem 2.1, portions of complex analytic varieties are always area-minimizing.

This first instance of the principle of calibrations was implicit in Wirtinger [W, 1936], explicit for complex submanifolds in De Rham [DeR, 1957], and applied to singular complex varieties in the context of rectifiable currents by Federer [F3, §4, 1965].

The results obtain not only in $\boldsymbol{C}^n$, but in any complex manifold with closed Kahler form, i.e., any Kahler manifold.

3.2. The quaternionic calibration. On n-dimensional quaternionic space $\boldsymbol{H}^n \cong \boldsymbol{R}^{4n}$, multiplication by i, j, and k give three complex structures and three associated Kahler forms $\omega_i, \omega_j, \omega_k$. The quaternionic 4-form q is defined as

$$q = \frac{1}{3}\left(\omega_i^2/2 + \omega_j^2/2 + \omega_k^2/2\right).$$

Marcel Berger [B1, Theorem 6.6] proved an analogue of Wirtinger's inequality for suitably normalized powers $\varphi = c_p q^p$ of the quaternionic form q : $\langle \xi, \varphi \rangle \leq 1$, *with equality if and only if the real* $4p$*-plane* ξ *in* $\boldsymbol{R}^{4n} \cong \boldsymbol{H}^n$ *is a quaternionic* p*-plane.* Consequently, by the fundamental theorem 2.1, portions of quaternionic varieties are always area-minimizing.

These results obtain not only in $\boldsymbol{H}^n$, but in any quaternionic manifold (with a closed quaternionic form), sometimes called "quaternionic-Kahler." Actually in $\boldsymbol{H}^n$ the results are not so interesting, because every quaternionic variety is complex analytic (e.g., for the Kahler form ω_i). In quaternionic manifolds there are interesting results. For example, in quaternionic projective space $\boldsymbol{H}P^n$, each $\boldsymbol{H}P^k (1 \leq k < n)$ is a quaternionic submanifold and therefore is area-minimizing in its homology class.

Other quaternionic symmetric spaces include the real Grassmannians $G(4, \boldsymbol{R}^n)$ of oriented 4-planes in $\boldsymbol{R}^n$ and the complex Grassmannians $G(2, \boldsymbol{C}^n)$ of complex 2-planes in $\boldsymbol{C}^n$. The interesting example of $G(4, \boldsymbol{R}^6) \cong G(2, \boldsymbol{C}^4)$ is analyzed in detail in [M3].

As far as I know, Berger was the first to see the principle of calibrations underlying the special fact that complex analytic varieties are area-minimizing and apply that principle to other examples such as quaternionic varieties [B2, §6, last paragraph].

3.3 The Lie bracket 3-form. On any Lie algebra $\mathfrak{g}$ consider the 3-form

$$\varphi(X, Y, Z) = \langle [X, Y], Z \rangle.$$

Here $[X, Y]$ is the Lie bracket and the inner product $\langle\ \ \rangle$ is with respect to the negative definite Killing form. Clearly φ is alternating in X and Y. That φ is alternating follows from the identity

$$\langle [X, Y], Z \rangle = \langle [Y, Z], X \rangle = \langle [Z, X], Y \rangle.$$

This form calibrates a homologically area-minimizing $SU_2 \cong \boldsymbol{S}^3$ in any compact irreducible Lie group G (except SO_2 and SO_3). It is totally geodesic and has constant sectional curvature κ, where κ is the maximum sectional curvature of G. It may be obtained by exponentiating the space of the highest restricted root $\bar{\delta}$ (which always has multiplicity 2). For example, the standard $SU_2 \subset SU_n$ is homologically area-minimizing.

S. Helgason [He, Theorem VII.11.1, Exercise VII.7] discovered these $\boldsymbol{S}^3$'s from a different perspective, independent of area-minimization. Given Helgason's description, it is actually quite easy so show them area-minimizing. Since the above 3-form φ is invariant under the adjoint actions of G on $\mathfrak{g}$, φ defines a bi-invariant form on G. Suppose X, Y, Z give an orthonormal basis for the Lie algebra SU_2. Since the group

SU_2 is totally geodesic, SU_2 is a Lie triple system, and $[X, Y] \in SU_2$. Since $\langle [X,Y], X\rangle = \langle [X,Y], Y\rangle = 0$, $[X,Y]$ is a multiple of Z. Since the sectional curvature in a Lie group is given by

$$K(X,Y) = -\frac{1}{4}\langle [X,Y], [X,Y]\rangle$$

for X, Y orthonormal, the fact that SU_2 has maximal sectional curvature implies that φ attains its maximum on SU_2. Therefore by the fundamental theorem SU_2 is homologically area-minimizing.

The Lie bracket 3-form φ was first used as a calibration by D. C. Thi [Thi, Corollary 5.4] to prove that SU_2 is homologically area-minimizing in SU_n. A. T. Fomenko [Fo, Cor. 8] had earlier proved that SU_2 is homologically area-minimizing in any simply connected Lie group by different methods.

3.4. The special Lagrangian calibration. On $\mathbf{C}^n \cong \mathbf{R}^n \oplus i\mathbf{R}^n$ define the special Lagrangian form by

$$\varphi = \operatorname{Re}\left((dx_1 + i dy_1) \wedge \ldots \wedge (dx_n + i dy_n)\right).$$

Reese Harvey and Blaine Lawson [HaL, Theorem III.1.10] proved that for any unit n-plane ξ, $\langle \xi, \varphi\rangle \le 1$, with equality if and only if ξ is the $x_1 x_2 \ldots x_n$ plane ξ_0 or more generally ξ belongs to the SU_n orbit of ξ_0. This SU_n orbit is called the set of special Lagrangian planes, because the U_n orbit of ξ_0 is the set of Lagrangian planes. The fundamental theorem now asserts that if the tangent planes to a submanifold M of $\mathbf{C}^n$ are all special Lagrangian, then M is area-minimizing. Such submanifolds are called special Lagrangian.

Harvey and Lawson [HaL, §III.3] gave many interesting examples of special Lagrangian submanifolds with singularities, including a cone over a 2-torus in $\mathbf{R}^6$ which fails to be real-analytic.

Harvey and Lawson [HaL, see Introduction] were the first to systematically propound the principle of calibrations. It was they who coined the term *calibration*. Their foundational paper discovered other interesting new calibrations, including the so-called associative and Cayley forms.

3.5 Nance calibrations. D. Nance [N] generalized the special Lagrangian calibration on $\mathbf{C}^n$ by replacing the n appearances of the imaginary number i by imaginary unit quaternions $u_1, \ldots, u_n$.

$$\varphi = \operatorname{Re}\left((dx_1 + u_1 dy_1) \wedge \ldots \wedge (dx_n + u_n dy_n)\right).$$

These calibrations play a central role in the proof of the Angle Criterion 4.1.

3.6. Grassmann calibrations. H. Gluck, F. Morgan, and W. Ziller [GMZ] used certain new calibrations on the Grassmannian $G(k, \mathbf{R}^p)$ of oriented k-planes in $\boldsymbol{R}^p$ for k even in order to prove that each subgrassmannian $G(k, \boldsymbol{R}^q)$ for $k < q < p$ is area-minimizing in its homology class. H. Tasaki [Tas] generalized these results to quaternionic subgassmannians

$$G(k, \boldsymbol{H}^q) \subset G(k, \boldsymbol{H}^p)$$

for any positive integers $k < q < p$. (These Grassmannians of quaternionic k-planes in $\boldsymbol{H}^p$ turn out not to be quaternionic manifolds, so that the quaternionic calibrations of §3.2 does not apply.)

3.7 Coflat calibrations. H. Federer [F2, §6] showed that one can allow calibrations with rather bad singularities, called *coflat calibrations.* Following work of H. B. Lawson [Ln, §5], he used a calibration with an isolated singularity at the origin to prove that the cone over $\boldsymbol{S}^n \times \boldsymbol{S}^n$ in $\boldsymbol{R}^{2n+2}$ is area-minimizing for $n \geq 3$. This cone has an interesting singularity itself at the origin.

Later B. Cheng [C] used coflat calibrations to prove for example that the cone $\boldsymbol{C}$ over the unitary group

$$U_n \subset \{n \times n \text{ complex matrices}\} \cong \boldsymbol{C}^{n^2}$$

is area-minimizing for $n \geq 4$. In this case the dimension of the singular set of the calibration $(2n^2 - 2)$ far exceeds the dimension of the calibrated cone $(n^2 + 1)$.

3.8 Vanishing calibrations. From early considerations of calibrations supported in a neighborhood of the calibrated surface ("vanishing calibrations"), G. Lawlor [L2] developed a new theory which provides sufficient curvature criteria for a submanifold to be area-minimizing. This method seems powerful. For example, Lawlor completely classified area- minimizing cones over products of spheres (see §5). Moreover, Lawlor's theory applies also to unoriented submanifolds and provides the first examples of nonorientable area-minimizing cones. S. Pedersen [P] obtained other applications.

4. Pairs of planes and the angle criterion

Here is one of the most basic questions about possible singularities in area-minimizing submanifolds. When is a pair of oriented m-planes P, Q which intersect at the origin area-minimizing? It may be best to think of 2 unit m-discs bounded by 2 circles or spheres and ask whether there is another oriented surface with the same boundary and less area. A pair of distinct 2-planes in $\boldsymbol{R}^3$ is never area-minimizing, as Figure 4.0.1 suggests.

The answer depends on the m angles $0 \leq \alpha_1 \leq \alpha_2 \leq \ldots \leq \alpha_m$ which characterize the geometric relationship between the planes. α_1 is the smallest angle between any unit vectors v_1 in P and w_1 in Q. α_2 is the smallest angle between $v_2 \perp v_1$ and $w_2 \perp w_1$. Similarly each α_k is

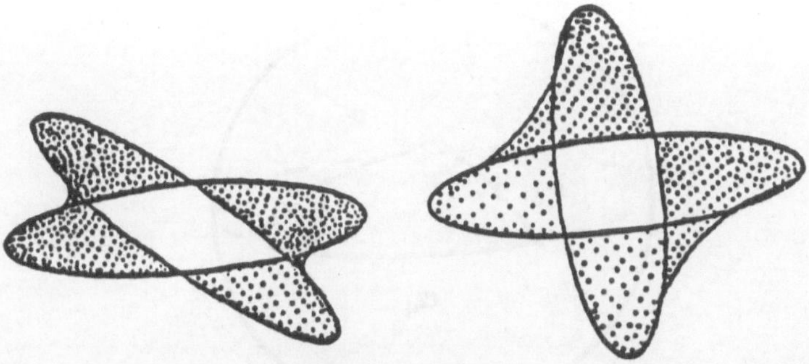

Figure 4.0.1
A pair of discs in $\mathbf{R}^3$ is never area-minimizing, as shown by these comparison surfaces of less area.

the smallest angle orthogonal to what has gone before. Finally α_m is the angle between the vectors v_m and w_m which complete the oriented orthonormal bases for P and Q. While $0 \leq \alpha_1 \leq \ldots \leq \alpha_{m-1} \leq \pi/2$, it is sometimes necessary to take $\alpha_m > \pi/2$ in order to get the orientations right.

For a pair of 2-planes in $\mathbf{R}^3$, $\alpha_1 = 0$ because they have a direction in common. α_2 is the usual dihedral angle.

The following criterion was conjectured by F. Morgan [M5, Introduction] and proved by complementary work of G. Lawlor [L1] and D. Nance [N].

4.1. The angle criterion. *A pair of oriented m-planes through the origin in $\mathbf{R}^n$ is area- minimizing if and only if their characterizing angles $0 \leq \alpha_1 \leq \alpha_2 \leq \ldots \leq \alpha_m$ satisfy the angle criterion*

$$\alpha_m \leq \alpha_1 + \alpha_2 + \ldots + \alpha_{m-1}.$$

Nance proved the sufficiency of the angle criterion with an elegant combination of calibrations, spherical geometry, and quaternions. Suppose P, Q is a pair of m-planes satisfying the angle criterion. P and Q lie in some $\mathbf{R}^{2m} \cong \mathbf{C}^m \cong \mathbf{R}^m \oplus i\mathbf{R}^m$ with orthonormal basis $e_1, \ldots, e_m, ie_1, \ldots, ie_m$. We may assume that

$$P = e_1 \wedge e_2 \wedge \ldots \wedge e$$
$$Q = e^{i\alpha_1} e_1 \wedge e^{i\alpha_2} e_2 \wedge \ldots \wedge e^{i\alpha_m} e_m.$$

Since $\alpha_m \leq \alpha_1 + \ldots + \alpha_{m-1}$, one can draw a geodesic m-gon on the unit 2-sphere with sidelengths $\alpha_1, \ldots, \alpha_m$, as in Figure 4.1.1. Each side of the length α_j is an arc of a great circle centered at the "polar" point u_j.

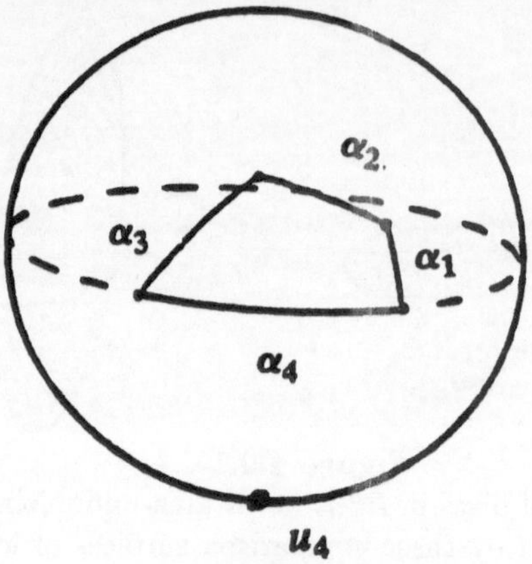

Figure 4.1.1
If the angles satisfy $\alpha_m \leq \alpha_1 + \ldots + \alpha_{m-1}$, they form the sidelengths of a spherical polygon. u_j is the exterior pole of the arc of length α_j. For example, if α_4 lies on the equator and the polygon lies above the equator then u_4 is the south pole.

Now view the 2-sphere as the set of imaginary unit quaternions, and consider the Nance calibration

$$\varphi = \operatorname{Re}\left((dx_1 + u_1 dy_1) \wedge \ldots \wedge (dx_m + u_m dy_m)\right).$$

One can show in a few lines [N, p. 163] that $\langle \xi, \varphi \rangle \leq 1$, with equality if ξ is P or Q. Therefore by the fundamental theorem 2.1, the pair of planes is area-minimizing.

5. Cones over products of spheres

Some of the first known examples of singularities in area-minimizing submanifolds were certain cones C over products of spheres:

$$C = \left\{ tx : 0 \leq t \leq 1,\ x \in S_{r_1}^{m_1} \times \ldots \times S_{r_k}^{m_k} \right\},$$

where $S_{r_j}^{m_j} = \{x \in \mathbf{R}^{m_j+1} : |x| = r_j\}$. C is a cone of dimension $m + 1 = \Sigma m_j + 1$ in R^{m+k}. C is *minimal* (i.e., the first variation of the area is 0) or equivalently has 0 mean curvature if the radii satisfy $r_j^2 = m_j/m$. The question is when such a C is area-minimizing.

For $m \leq 5$, all such cones are unstable and hence not area-minimizing [Sn, Theorem 6.1.1, p. 102].

In 1969 E. Bombieri, E. De Georgi, and E. Giusti [BDG] proved that the cone over $S^3 \times S^3$ in R^8 is area-minimizing. In 1972 H. B. Lawson [Ln, Theorem 4] proved that the cone over a pair $S^{m_1} \times S^{m_2}$ of spheres is area-minimizing if $m \geq 7$. In 1974 H. Federer [F2, §6.3] showed how to obtain these results with coflat calibrations (see 3.7) with isolated singularities at the origin. The same year P. Simoes [S] completed the classification for cones over pairs of spheres by showing that the cone over $S^2 \times S^4$ is area-minimizing, while the cone over $S^1 \times S^5$ is not. In 1978 D. Bindschadler [Bi] used coflat calibrations to prove that the cone over a k-fold product $(S^{m_1})^k$ of spheres is area-minimizing if $k = 3$ and $m_1 \geq 3$, or $k = 4$ and $m_1 \geq 5$, or $k \geq 5$ and $m_1 \geq k - 1$. Recently, B. Cheng (unpublished) extended Bindschadler's results to include $S^{m_1} \times S^{m_2} \times S^{m_2}$ whenever $m_1, m_2 \geq 4$.

Now G. Lawlor has completely classified the area-minimizing cones over products of spheres.

5.1. Theorem [L2, Theorem 4.1.1]. *Let C be an $(m+1)$-dimensional minimal cone over a product of spheres*

$$S_{r_1}^{m_1} \times \ldots \times S_{r_k}^{m_k}$$

$(m = \Sigma m_j,\ r_j^2 = m_j/m)$.

(1) *If $m \leq 5$, C is unstable and hence not area-minimizing.*
(2) *If $m \geq 7$, C is area-minimizing.*
(3) *If $m = 6$, then C is stable, and C is area-minimizing if and only if none of the spheres is a circle (i.e., $m_j > 1$).*

The proof uses Lawlor's curvature criterion from vanishing calibrations (see §3.8).

6. General elliptic integrands

The energy of a soap film is almost exactly proportional to its area. The surface energy of a salt crystal, however, depends on the direction, too: certain directions expose more bonds than others.

For oriented surfaces S of dimension $n-1$ in R^n, the energy $\Phi(S)$ is determined by an *integrand* $\Phi : S^{n-1} \to R$:

$$\Phi(S) = \int_S \Phi(n),$$

where n is the oriented unit normal to S. One expects good regularity results for Φ-minimizing surfaces if Φ is "elliptic," which guarantees that flat hyperplanes are uniquely Φ-minimizing. Technically, Φ is *elliptic* if the radial plot of $1/\Phi$ is uniformly convex. Φ is elliptic if and only if the associated Euler-Lagrange equation is an elliptic partial differential equation.

Hypersurfaces which minimize *area* are free of singularities up through $\boldsymbol{R}^7$ ([Sn, Theorem 6.2.1]; see [M2, §8.2, §10], [F1, Theorem 5.4.15]). The strongest known regularity results for hypersurfaces minimizing for a general smooth elliptic integrand guarantee regularity only up through $\boldsymbol{R}^3$ [AlmSS, Theorem II.7]. However, until recently, there were no examples of singularities below $\boldsymbol{R}^8$. Now there is a new example in $\boldsymbol{R}^4$.

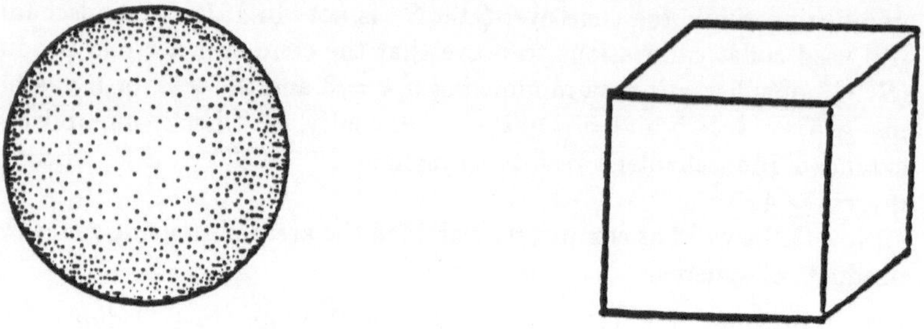

Figure 6.0.1
Soap bubbles and salt crystals minimize different kinds of surface energies.

6.1 Theorem [M4, Theorem 6.2]. *There is a real-analytic integrand* Φ : $S^3 \to \boldsymbol{R}$ *such that the cone over* $\boldsymbol{S}^1 \times \boldsymbol{S}^1 \subset \boldsymbol{R}^4$ *is* Φ*-minimizing.*

The integrand Φ is a function of $\Theta = \tan^{-1} \frac{\sqrt{x_3^2+x_4^2}}{\sqrt{x_1^2+x_2^2}}$ alone. Its radial plot appears in Figure 6.1.1.

6.2 Proof by calibration. As pointed out by H. Federer [F2, see p. 351-352] the fundamental theorem of calibrations extends to show surfaces S Φ-minimizing for general elliptic integrands Φ. The revised hypothesis on the calibration Φ requires that for every $(n-1)$-plane ξ,

$$\langle \xi, \Phi(x) \rangle \leq \Phi({}^*\xi), \tag{1}$$

with equality whenever ξ is a tangent plane to S at x. Here the Hodge dual ${}^*\xi$ is just the oriented unit normal to ξ. Unfortunately this revised hypothesis is harder to verify. Indeed, Theorem 6.1 seems to be the first application of calibrations to general elliptic integrands.

The required 3-form φ comes from a modification of the 7-form φ_1 of Federer's proof [F2, §6.3] that the cone over $\boldsymbol{S}^3 \times \boldsymbol{S}^3$ is area-minimizing (cf. §3.7). Fortunately this form φ_1, as well as the integrand Φ and the cone over $\boldsymbol{S}^3 \times \boldsymbol{S}^3$ itself, is $SO_2 \times SO_2$ invariant.

Lemma 6.4 below provides the means for verifying hypothesis (1). It shows how to associate an integrand G with φ such that (1) holds if

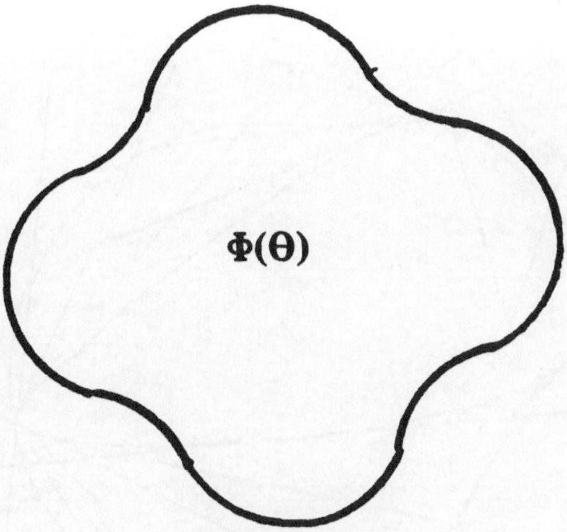

Figure 6.1.1
The cone over $S^1 \times S^1 \subset \mathbf{R}^4$ is Φ-minimizing for this plotted elliptic integrand Φ. Φ is about as far from the area integrand as possible.

$G \le W(\Phi)$, where $W(\Phi)$ defines the famous Wulff crystal associated with Φ.

6.3. The Wulff crystal (cf. [Tay] for further background and references). For any integrand Φ, there is a unique (convex) region C of least surface energy $\Phi(\partial C)$ for its volume, called the Wulff crystal. The soap bubble and salt crystal of Figure 6.0.1 give a couple of examples.

There is a classical Wulff construction for obtaining the Wulff crystal from a radial plot $\{r = \Phi(n)\}$ of the integrand. As in Figure 6.3.1, in each direction n, go out a distance $\Phi(n)$, and construct the hyperplane normal to the radial vector. The Wulff crystal is enveloped by these hyperplanes. In other words, the Wulff crystal is the intersection of all the halfspaces

$$H_n = \{w : w \bullet n \le \Phi(n)\}.$$

Its boundary is the radial plot of the function

$$W(\Phi)(w) \equiv \inf\{n \bullet w)^{-1}\Phi(n) : n \bullet w > 0\}.$$

6.4. Lemma [M4, Lemma 4.2]. *Let Φ be an integrand on $U \subset \mathbf{R}^n$. Let φ be a differential $(n-1)$-form. Let G be the associated integrand $G(w) = \sup\{|\varphi(z)| : w$ is the oriented unit normal to the $(n-1)$-plane $\Phi(x)^*\}$. If $G \le W(\Phi)$, then for all $x \in U$ and unit $(n-1)$-planes ξ,*

$$\langle \xi, \varphi(x) \rangle \le \Phi(^*\xi).$$

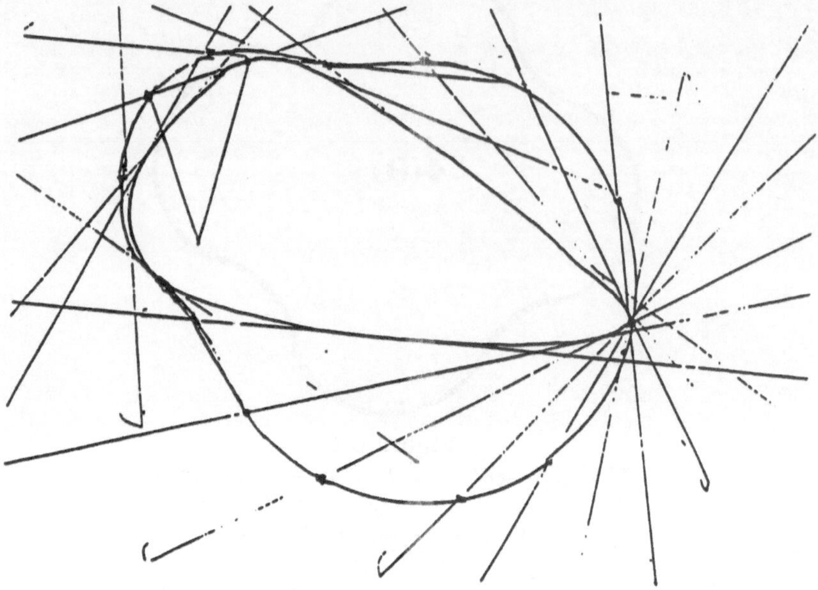

Figure 6.3.1
This Wulff crystal $G = W(\Phi)$ is enveloped by lines through points in the radial plot of Φ, perpendicular to the radial directions.

Remarks. This is the lemma which establishes the essential hyupothesis 6.2(1). For the purposes of this lemma, it is convenient to define $\sup \emptyset = 0$.

Proof. We may assume $\langle \xi, \Phi(x) \rangle > 0$. In other words, if w is the oriented unit normal to $\varphi(x)^*$, $^*\xi \bullet w > 0$.

$$\begin{aligned}\langle \xi, \varphi(x) \rangle &= (^*\xi \bullet w)|\varphi(x)| \le (^*\xi \bullet w)G(w) \\ &\le (^*\xi \bullet w)W(\Phi)(w) \le \Phi(^*\xi)\end{aligned}$$

by the definition of $W(\Phi)$.

REFERENCES

[AlmSS] Almgren, Jr., F. J. , R. Schoen, and L. Simon, *Regularity and singularity estimates on hypersurfaces minimizing elliptic variational integrals*, Acta Math. **139** (1977), 217-265.

[Be1] Berger, M., *Du côté de chez Pu*, Ann. Scient. Ec. Norm. Sup. **4** (5) (1972), 1-44.

[Be2] Berger, M., *Quelques problèmes de géométrie Riemannienne ou Deux variations sur les espaces symétriques compacts de rang un*, L'Enseignement Mathematique **XVI** (1970), 73-96.

[Bi] Bindschadler, David, *Invariant soutions to the oriented Plateau problem of maximal codimension*, Trans. Amer. Math. Soc. **261** (1980), 439-462.

[BDG] Bombieri, E., E. De Giorgi, and E. Guisti, *Minimal cones and the Bernstein problem*, Inven. Math (1969), 243-268.

[C] Cheng, Benny, *Area-minimizing equivariant cones and coflat calibrations*, Ph.D. Thesis, MIT, 1987.

[DeR] De Rham, G., *On the area of complex manifolds*, Notes for the Seminar on Several Complex Variables, Inst. for Adv. Study, Princeton 1957-58.

[F1] Federer, Herbert, *Geometric Measure Theory*, Springer-Verlag, New York, 1969.

[F2] Federer, Herbert, *Real flat chains, cochains, and variational problems*, Ind. U. Math. J. **24** (1974), 351-407.

[F3] Federer, Herbert, *Some theorems on integral currents*, Trans. Am. Math. Soc. **117** (1965), 43-67.

[Fo] Fomenko, A. T., *Minimal compacta in Riemannian manifolds and Reifenberg's conjecture*, Izv. Akad. Nauk SSSR Serv. Mat. **36** (1972); English translation Math. USSR Izv. **6** (1972), 1037-1066.

[GMZ] Gluck, Herman, Frank Morgan and Wolfgang Ziller, *Calibrated geometries in Grassmann manifolds*, Comm. Math. Helv., **64** (1989), 256–268.

[Ha] Harvey, Reese, *Calibrated geometries*, Proc. Int. Cong. Math., 1983.

[HaL] Harvey, Reese, and H. Blaine Lawson, Jr., *Calibrated goemetries*, Acta Math. **148** (1982), 47-157.

[He] Helgason, Sigurdur, *Differential Geometry, Lie Groups, and Symmetric Spaces*, Academic Press, New York, 1978.

[L1] Lawlor, Gary, *The angle criterion*, Inventiones Math. **95** (1989), 437-446.

[L2] Lawlor, Gary, *The curvature criterion*, Ph.D Thesis, Stanford U., 1988.

[Ln] Lawson, Jr., H. Blaine, *The equivariant Plateau problem and interior regularity*, Trans. Amer. Math. Soc. **173** (1972), 231-247.

[M1] Morgan, Frank, *Area-minimizing surfaces, faces of Grassmannians, and calibrations*, Am. Math. Monthly **95** (1988), 813-822.

[M2] Morgan, Frank, *Geometric Measure Theory: a Beginner's Guide*, Academic Press, 1988.

[M3] Morgan, Frank, *Least-volume representatives of homology classes in* $G(2,\boldsymbol{C}^4)$, Annales Sci. de l'Ecole Normale Sup. **22** (1988), 127-135.

[M4] Morgan, Frank, *A sharp counter-example on the regularity of* Φ*-minimizing hypersurfaces*, preprint (See, Bull. Amer. Math. Soc. **22** (1990), 295-299.).

[M5] Morgan, Frank, *On the singular structure of three-dimensional area-minimizing surfaces*, Trans. Amer. Math. Soc. **276** (1983), 137-143.

[N] Nance, Dana, *Sufficient conditions for a pair of n-planes to be area-minimizing*, Math. Ann. **279** (1987), 161-164.

[P] Pedersen, Sharon, *Optimal vector fields on spheres*, Ph.D. Thesis, U. of Penn., 1988.

[S] Simoes, Plinio, *On a class of minimal cones in* $\boldsymbol{R}^n$, Bull. Amer. Math. Soc. **80** (1974), 488-489.

[Sn] Simons, James, *Minimal varieties in Riemannian manifolds*, Ann. Math. **88** (1968), 62-105.

[Tas] Tasaki, Hiroyuki, *Calibrated geometries in quaternionic Grassmannians*, preprint, (1987).

[Tay] Taylor, Jean E., *Crystalline variational problems*, Bull. Am. Math. Soc. **84** (1987), 568-588.

[Thi] Thi, Dao Trong, *Minimal real currents on compact Riemannian manifolds*, Izv. Akad. Nauk. SSSR (Ser. Mat) **41** (1977); English translation in Math. USSR Izv. **11** (1977), 807-820.

[W] Wirtinger, W., *Eine Determinantenidentitat und ihre Anwendung auf analytische Gebilde und Hermitesche Massbestimmung*, Monatsh. f. Math. u. Physik **44** (1936), 343-365.

Frank Morgan
Dept. of Mathematics
Williams College
Williamstown, MA 01267

Harmonic Maps with Free Boundaries

KLAUS STEFFEN

In this note we describe some examples of minimizing harmonic maps between Riemannian manifolds with singularities at a free boundary and we discuss the regularity results for minimizing harmonic maps at a free boundary which we have obtained jointly with Frank Duzaar.

We consider a Riemannian manifold Ξ of dimension $\nu \geq 2$ with non-empty boundary Σ and mappings $u: \Xi \to X$ into another Riemannian manifold X of dimension $n \geq 1$ such that $u(\Sigma)$ is contained in a given closed submanifold S of X. There is no restriction on the dimension m of S, $0 \leq m \leq n$. We call Σ the *free boundary* and S the *supporting manifold* for the free boundary values. We then want to study the regularity of mappings $u: \Xi \to X$ which are energy minimizing subject to the *free boundary* condition $u(\Sigma) \subset S$. By the regularity results of Schoen & Uhlenbeck [SU1], [SU2] this is an open problem only at the free boundary Σ. Note that the free boundary condition is nonlinear even for a Euclidean space X. In suitable coordinates on X it is equivalent to $n - m$ conditions of Dirichlet type $u_{i|\Sigma} = 0$, $m < i \leq n$.

Existence of energy minimizing mappings $u: \Xi \to X$ satisfying a free boundary condition can be obtained with the direct method of the calculus of variations if one works in the Sobolev space $H^{1,2}(\Xi, X)$ of $H^{1,2}$-mappings $u: \Xi \to \mathbb{R}^{n+k}$ such that $u(\xi) \in X$ for μ_Ξ almost all $\xi \in \Xi$ where μ_Ξ denotes the Riemannian measure on Ξ. As usual, we have assumed here that X is isometrically embedded in some Euclidean space $\mathbb{R}^{n+k}$ as a closed submanifold. (One can give an intrinsic definition of $H^{1,2}(\Xi, X)$ using the notion of approximate differentiability of mappings $u: \Xi \to X$, for example. But for the methods we use in the proof of our main regularity results it is essential to be able to work in an ambient space $\mathbb{R}^{n+k}$ for X.) For $u \in H^{1,2}(\Xi, X)$ the *energy* is then defined by

$$\mathcal{E}(u) := \frac{1}{2} \int_\Xi |Du(\xi)|^2 d\mu_\Xi(\xi), \ |Du(\xi)|^2 := \text{trace}[Du(\xi)^\dagger Du(\xi)]$$

where $Du(\xi)^\dagger$ denotes the adjoint of the differential $Du(\xi) \in \mathrm{Hom}(T_\xi\Xi, \mathbb{R}^{n+k})$ with respect to the Riemannian metric on Ξ and the Euclidean metric on $\mathbb{R}^{n+k}$. Since each $u \in H^{1,2}(\Xi, X)$ has a trace $u_{|\Sigma}: \Sigma \to \mathbb{R}^{n+k}$ defined μ_Σ almost everywhere on Σ we may interpret the free boundary condition $u(\Sigma) \subset S$ for $H^{1,2}$ mappings in the sense of traces. The energy functional is weakly lower semi-continuous with respect to weak convergence in $H^{1,2}(\Xi, \mathbb{R}^{n+k})$ and $H^{1,2}(\Xi, X)$ as well as $\{u \in H^{1,2}(\Xi, \mathbb{R}^{n+k}): u(\Sigma) \subset S\}$ are weakly closed in $H^{1,2}(\Xi, \mathbb{R}^{n+k})$. Thus, the direct method can be applied to prove e.g. the following:

Proposition 1. *Suppose $\Xi \cup \partial\Xi$ is a compact Riemannian manifold with boundary, $\Sigma \neq \emptyset$ is open in $\partial\Xi$ and $\Gamma \subset \partial\Xi \setminus \Sigma$ has positive boundary measure $\mu_{\partial\Xi}(\Gamma) > 0$. Define the class $\mathcal{A} \subset H^{1,2}(\Xi, X)$ of admissible mappings $u: \Xi \to X$ by the free boundary condition $u(\Sigma) \subset S$ and an additional 'fixed boundary condition' $u_{|\Gamma} = g$ where $g: \Gamma \to X$ are given Dirichlet boundary data. If $\mathcal{A}$ is nonempty then $\mathcal{A}$ contains a mapping of minimal energy.*

Note that the proposition is also true for a purely free boundary condition, i.e. $\Sigma = \partial\Xi$ and $\Gamma = \emptyset$, provided S is compact. However, the minimizing mappings u are trivial in this case, $u \equiv \mathrm{const} \in S$. It is an interesting problem, not addressed here, to prove by some topological method in the calculus of variations the existence of nontrivial harmonic mappings $u: \Xi \to X$ which are locally on $\Xi \cup \partial\Xi$ minimizing for the energy functional with respect to a purely free boundary condition. Since our regularity results are of a local nature we could apply them to such mappings to infer partial regularity at the free boundary.

$H^{1,2}$ mappings $u: \Xi \to X$ are discontinuous, in general. In fact, it may be impossible to approximate u by smooth mappings from Ξ to X in $H^{1,2}$ norm (see [SU2], [BZ]). This phenomenon also occurs for energy minimizing mappings subject to boundary conditions and cannot be circumvented by working in the strong closure of the set of smooth mappings in $H^{1,2}(\Xi, X)$ satisfying the same boundary conditions. The reason is that the strong closure of this set of smooth mappings is generally not weakly closed in $H^{1,2}(\Xi, X)$ and hence not suitable for the direct method of the calculus of variations. Moreover, in some situations there are no smooth mappings in $H^{1,2}(\Xi, X)$ at all which satisfy the prescribed boundary conditions while non-smooth mappings in $H^{1,2}(\Xi, X)$ do exist which fulfill these conditions. This will be clear from the examples below.

Since one cannot expect regularity of energy minimizing harmonic maps $u \in H^{1,2}(\Xi, X)$ in general, the problem arises to prove optimal partial regularity, i.e. to give an optimal estimate for the size of the possible singular set. This has been done by Schoen & Uhlenbeck [SU1], [SU2] in the interior Ξ and at a fixed boundary $\Gamma \subset \partial\Xi$ as above with the result that the singular set is of codimension at least 3 in Ξ (discrete in case $\dim \Xi = 3$) and one has complete regularity near Γ, if smooth boundary data are prescribed there. Concerning the regularity at a free boundary Σ it seems reasonable to expect the same estimate on the size of the singular set as in the interior Ξ. In contrast to the fixed boundary case, one cannot hope for more regularity at the free boundary than in the interior. This is made evident by the following examples which exhibit singularities of energy minimizing harmonic mappings at a free boundary. While the first two of these examples are based on a well-known harmonic map with an isolated interior singularity the remaining two examples present some new singularity phenomena for harmonic maps. In particular, we exhibit classical harmonic vector functions with singularities at a free boundary.

Example 1. We let $\Xi = \Delta_+^\nu$ the open unit half ball in $\mathbb{R}^{\nu-1} \times \mathbb{R}_+$, $\Sigma = \Delta^\nu \cap \mathbb{R}^{\nu-1} \times \{0\}$ its bounding equatorial $\nu-1$ disc and $\Gamma = \partial\Xi \setminus \Sigma$ the spherical part of the boundary. For X, S we take the n sphere $S^n \subset \mathbb{R}^{n+1}$ with its equatorial $n-1$ sphere $S^{n-1} \times \{0\}$ where $n := \nu-1$. Both, Ξ and X, are equipped with their standard metric. We then consider the class $\mathcal{A}$ of mappings $u \in H^{1,2}(\Xi, X)$ satisfying the fixed boundary condition $u(\xi) = \xi$ on Γ and the free boundary condition $u(\Sigma) \subset S$. If $\nu \geq 3$, then $u_*(\xi) := \frac{1}{|\xi|}\xi$ for $\xi \in \Delta_+^\nu$ belongs to $\mathcal{A}$ and is, in fact, energy minimizing in $\mathcal{A}$ with an isolated singularity in the free boundary. This follows immediately from the well-known fact that the extension of u_* to the full ball Δ^ν by reflection across $\mathbb{R}^{\nu-1} \times \{0\}$, i.e. $u_*(\xi) := \frac{1}{|\xi|}\xi$ for all $\xi \in \Delta^\nu$, is energy minimizing in $H^{1,2}(\Delta^\nu, S^{\nu-1})$ for the boundary condition $u(\xi) = \xi$ on $\partial\Delta^\nu$ (see [L] for a simple proof). Note that $\mathcal{A}$ does not contain mappings which are continuous on $\Xi \cup \Sigma \cup \Gamma$, because the restriction of such a map to $\overline{\Sigma}$ would define a retraction of the disc $\overline{\Sigma}$ to its boundary.

Example 2. We consider $\Xi = \Delta_+^\nu$, Σ, Γ, $S = S^{\nu-2} \times \{0\}$ with $\nu \geq 3$ as in Example 1 but target manifold $X = \mathbb{R}^\nu$ this time. The class $\mathcal{A} \subset H^{1,2}(\Sigma, \mathbb{R}^\nu)$ is again defined by the boundary conditions $\Gamma \ni \xi \mapsto \xi$, $\Sigma \to S$, and a variant of Proposition 1 gives us an energy minimizing map u in $\mathcal{A}$. This mapping is a classical harmonic vector function on Δ_+^ν, continuous on $\overline{\Delta_+^\nu} \setminus \overline{\Sigma}$, but certainly discontinuous at some point of $\overline{\Sigma}$. At its regular points $\xi \in \Sigma$ the minimizing map u must satisfy the natural boundary condition $(\partial u \mid \partial\xi_\nu)(\xi) \perp \mathrm{Tan}_{u(\xi)} S$ or, in view of $u_\nu(\xi) = \xi_\nu$, $(\partial u \mid \partial\xi_\nu)(\xi) - e_\nu \in \mathbb{R}u(\xi)$ where e_ν denotes the ν-th canonical basis vector in $\mathbb{R}^\nu$. One may check that the harmonic extension u_{**} of $u_{*|\partial\Xi}$ by Poisson's formula for the half ball satisfies this natural boundary condition, hence u_{**} is an obvious candidate for a minimizing map in $\mathcal{A}$. If one could prove that the minimizing map u above must have all the symmetries of the boundary conditions one would be able to conclude $u = u_{**}$. However, minimizers to variational problems to not inherit, as a rule, all the symmetries of the data and we are unable to decide whether u_{**} is really minimizing in $\mathcal{A}$. An analysis of the blow up tangent maps to u at 'edge points' $\xi \in S^{\nu-2} \times \{0\} = \Gamma \cap \Sigma$ reveals the triviality of these tangent maps. This suggests that u should be continuous at such points and hence must have finitely many isolated singularities in the free boundary by Theorem 2 below.

Example 3. We consider a domain Ξ in $\mathbb{R}^3$ with standard metric which is composed of a cylindrical shell $1-s \leq |(\xi_1,\xi_2)| \leq 1$, $|\xi_3| \leq h$, and two half balls $|\xi - (0,0,h)| \leq 1$, $\xi_3 \geq h$ and $|\xi - (0,0,-h)| \leq 1$, $\xi_3 \leq -h$ $(0 < s < 1$, $h > 0)$. We denote by Γ, Σ the outer and inner boundary component of Ξ respectively and define a map w of finite energy on Ξ by letting $w(\xi)$ the unit vector in the direction of the ray through ξ which emanates from the closest point to ξ on the segment $\{(0,0)\} \times [-h,h]$. Note that w has singularities at $\pm(0,0,h)$ of the same type as the map u_* in Example 1 and w is regular otherwise. The energy of w may be computed

$$\mathcal{E}(w) = 4\pi + 2\pi h \log \frac{1}{1-s}.$$

Now, w maps the inner boundary component Σ into $S := S^1 \times \{0\}$. Hence the class $\mathcal{A}$ of mappings in $H^{1,2}(\Xi, \mathbb{R}^3)$ satisfying the free boundary condition $\Sigma \to S$ and the fixed boundary condition $\Gamma \ni \xi \mapsto w(\xi)$ is non-empty and contains an energy minimizing element u. We contend that u must be discontinuous somewhere on $\overline{\Xi}$ if $\frac{2}{h} < 1 + \log(1-s)$. Indeed, if $v \in \mathcal{A}$ is continuous on $\overline{\Xi}$ then v maps the outer boundary of each annulus $A_t := \Xi \cap \mathbb{R}^2 \times \{t\}$, $-h \leq t \leq h$, onto $S^1 \times \{0\}$ with degree 1 (as w does) while on the inner boundary of this annulus v has a continuous extension to Σ with values in $S^1 \times \{0\}$. Therefore, the projection of the image $v(A_t)$ onto $\mathbb{R}^2 \times \{0\}$ must cover the disc $\Delta^2 \times \{0\}$ and we may estimate the energy of v on A_t from below by π. Consequently, $\mathcal{E}(v) \geq 2\pi h$ for every continuous $v \in \mathcal{A}$ and the assertion about u follows in view of $\mathcal{E}(u) \leq \mathcal{E}(w)$.

Since u is a classical harmonic vector function on the interior Ξ and u is smooth at the fixed boundary we conclude that u must have singularities at the free boundary Σ. This phenomenon is obviously not related to the non-smoothness of Σ, for we may smooth out the edges of Σ (in fact, we may even approximate Ξ by real analytic domains) without affecting the above reasoning. By Theorem 2 below u then has finitely many isolated singularities on Σ. Note, that all the arguments are still valid, if we replace $S^1 \times \{0\}$ by the supporting surfaces $S := S^1 \times \mathbb{R}$ of codimension 1 in $\mathbb{R}^3$.

The first example of a harmonic mapping into a Euclidean space with singularities at a free boundary was given by Gulliver & Jost [GJ] who constructed such a mapping from the half sphere S^ν_+ (with standard metric) into $\mathbb{R}^\nu$ satisfying the free boundary condition $\partial S^\nu_+ \to S^{\nu-1}$. However, this map is not minimizing since we have a purely free boundary condition here. (It may be energy minimizing locally on $S^\nu_+ \cup \partial S^\nu_+$.) Note that, in contrast to the first two examples, there is no topological obstruction to the continuity of admissible maps in Example 3: Any continuous extension of $w_{|\Gamma}$ with values in $\mathbb{R}^3$ and constant value $\in S^1 \times \{0\}$ on Σ will satisfy the boundary conditions. Thus, it is the geometry of the data which is responsible for the singularities. In our final example we now want to show that, on the other hand, singularities at free boundaries may sometimes be forced by the topology of the boundary configuration.

Example 4. We try to find Ξ, Γ, Σ, X, S and $g: \Gamma \to X$ such that the class $\mathcal{A}$ of $H^{1,2}$ mappings from Ξ to X satisfying the fixed boundary condition $\Gamma \ni \xi \mapsto g(\xi)$ and the free boundary condition $\Sigma \to S$ is non-void but contains no mapping continuous on $\Xi \cup \Gamma \cup \Sigma$. Suppose a minimizing map u exists in $\mathcal{A}$, then u cannot be regular at every point $\Xi \cup \Gamma \cup \Sigma$. If g is smooth, then u is continuous near Γ by [SU2] and if X has nonpositive sectional curvature we know complete interior regularity from [SU1]. We conclude that u must have singularities at the free boundary Σ.

We can arrange all this if we make the following choices: Σ a compact connected manifold, $\Xi = \Sigma \times]0,1[$ where we identify $\Sigma = \Sigma \times \{0\}$, $\Gamma = \Sigma \times \{1\}$, X a connected closed submanifold of $\mathbb{R}^{n+k}$ and S a closed submanifold of X such that the inclusion $S \to X$ induces an epimorphism of fundamental groups. The latter condition ensures that we can extend any smooth $g: \Gamma \to X$ to $v \in H^{1,2}(\Xi, X)$ such that $v(\Sigma) \subset S$. We first extend g to each edge $\{\xi\} \times [0,1]$ corresponding to a vertex ξ of a smooth triangulation of Σ by selecting a smooth path $v: \{\xi\} \times [0,1] \to X$ such that $v(\xi,1) = g(\xi,1)$ and $v(\xi,0) \in S$. Next, for each edge $[\xi,\eta]$ in Σ we define v on $[\xi,\eta] \times \{0\}$ as a smooth path in S joining $v(\xi,0)$ to $v(\eta,0)$ in such a way that v maps the topological circle $\partial([\xi,\eta]) \times [0,1])$ homotopically constant into X. We can therefore extend v to a Lipschitz map from $[\xi,\eta] \times [0,1]$ into X. For each 2-simplex $[\xi,\eta,\zeta]$ in Σ we now extend v to an $H^{1,2}$ mapping from $[\xi,\eta,\zeta] \times [0,1]$ into X by projecting each point into the boundary of $[\xi,\eta,\zeta] \times [0,1]$ radially away from some interior point of the face $[\xi,\eta,\zeta] \times \{0\}$ and then evaluating v as already defined. (Here, the term 'radial projection' refers to some bi-Lipschitz homeomorphism of

$[\xi,\eta,\zeta] \times [0,1]$ onto a piecewise linear triangular prism in $\mathbb{R}^3$.) With a similar procedure we can extend v to the skeletons of higher dimension and finally obtain $v \in H^{1,2}(\Xi, X)$ such that $v_{|\Gamma} = g$ and $v(\Sigma) \subset S$.

Now, the class $\mathcal{A}$ of $H^{1,2}(\Xi, X)$ mappings which coincide with g on Γ and map Σ into S is non-empty and contains an energy minimizing element u. The condition that $\mathcal{A}$ does not contain continuous functions can be achieved by choosing a boundary mapping $g: \Gamma \to X$ which is not homotopic to a continuous map factoring through the submanifold S of X. To be specific, we take flat tori $\Sigma = (S^1)^{\nu-1}$, $X = (S^1)^n$ with $n = \nu \geq 3$, g an inclusion $\Gamma = \Sigma \times \{1\} \to \Sigma \times \{z\} \subset X$ and S a connected sum of the boundaries of tubular neighborhoods to ν loops in general position which are generators of the fundamental group of X (or any submanifold S of X with dimension m, $2 \leq m \leq n-2$, carrying such generators). Since any continuous map $\Gamma \to S \to X$ induces the trivial morphism on the $(n-1)^{th}$ integer homology groups while g induces a nontrivial morphism, no continuous extension v of g to $\Xi \cup \Gamma \cup \Sigma$ can satisfy the free boundary condition $v(\Sigma) \subset S$. Therefore, the minimizing map u above must have points of discontinuity and, by what has been said before, these points can only lie in the free boundary. In case $\nu = 3$ they form a finite singular set (see Theorem 2 below). Many more examples of harmonic mappings with singularities in the free boundary can be concocted along similar lines.

We say that $\xi \in \Xi \cup \Sigma$ is a *regular point* of $u \in H^{1,2}(\Xi, X)$ if u coincides with a continuous function on some neighborhood of ξ in $\Xi \cup \Sigma$ (μ_Ξ almost everywhere). The set of regular points is denoted Reg(u) and its complement $(\Xi \cup \Sigma) \setminus \mathrm{Reg}(u)$ is called the *singular set* Sing(u). We can now formulate our main results on the regularity of harmonic mappings at a free boundary. We continue to assume the conditions on the Riemannian manifolds $\Xi, \Sigma \subset \partial\Xi$, $X \subset \mathbb{R}^{n+k}$ and $S \subset X$ described at the beginning of this article. We also assume that these manifolds are sufficiently smooth (C^3 will be enough). A map $u \in H^{1,2}(\Xi, X)$ is *locally energy minimizing* on $\Xi \cup \Sigma$ with respect to the free boundary condition $u(\Sigma) \subset S$ if every $\xi \in \Xi \cup \Sigma$ has a neighborhood in $\Xi \cup \Sigma$ with $\mathcal{E}(u) \leq \mathcal{E}(v)$ for all $v \in H^{1,2}(\Xi, X)$ such that $v(\Sigma) \subset S$ and v coincides with u outside this neighborhood of ξ.

Theorem 1. *Suppose $u \in H^{1,2}(\Xi, X)$ is bounded and locally on $\Xi \cup \Sigma$ energy minimizing with respect to the free boundary condition $u(\Sigma) \subset S$. Then the $\nu - 2$ dimensional Hausdorff measure of the singular set Sing(u) vanishes. (ν the dimension of Ξ). On its regular set, u is as smooth as the data $\Xi \cup \Sigma$, X, S allow and u satisfies the differential equation for harmonic mappings from Ξ into X there. Furthermore, at each point $\xi \in$ Reg$(u) \cap \Sigma$, u satisfies the natural boundary condition associated with the free boundary* *condition $u(\Sigma) \subset S$, i.e. the normal derivative of u at ξ is perpendicular to S at $u(\xi)$. If u is locally energy minimizing as above, but unbounded, the same conclusions hold, provided we have global bounds on the extrinsic curvature of X and the extrinsic curvature of S together with its derivative.*

By "bounded extrinsic curvature" we mean that the second fundamental form with respect to the ambient $\mathbb{R}^{n+k}$ is bounded and that the manifold has a uniform tubular neighborhood in $\mathbb{R}^{n+k}$. The "derivative" of the extrinsic curvature of S refers to the covariant derivative of the second fundamental form of S. By a process known as "dimension reduction" we can improve the estimate for the singular set by one dimension for bounded minimizers u:

Theorem 2. *Suppose $u \in H^{1,2}(\Xi, X)$ is locally on $\Xi \cup \Sigma$ energy minimizing with respect to the free boundary condition $u(\Sigma) \subset S$. If $u(\Sigma)$ is bounded and X has bounded extrinsic curvature, then the Hausdorff dimension of the singular set in the free boundary can be estimated (with $\nu = \dim \Xi$)*

$$\dim(\mathrm{Sing}(u) \cap \Sigma) \leq \nu - 3$$

and all points of $\mathrm{Sing}(u) \cap \Sigma$ are isolated in $\mathrm{Sing}(u)$ in case $\nu = 3$. If $u(\Xi)$ is bounded we also have, for the full singular set,

$$\dim(\mathrm{Sing}(u)) \leq \nu - 3$$

and $\mathrm{Sing}(u)$ is discrete in $\Xi \cup \Sigma$ in case $\nu = 3$.

Theorems 1 and 2 have been obtained in collaboration with F. Duzaar. Complete proofs appear in [DS1], [DS2]. As far as the interior singular set $\mathrm{Sing}(u) \cap \Xi$ is concerned, the results are due to Schoen & Uhlenbeck [SU1]. Our contribution is to the singular set $\mathrm{Sing}(u) \cap \Sigma$ in the free boundary. Note that the dimension estimate for $\mathrm{Sing}(u) \cap \Sigma$ is optimal (as is the corresponding estimate for $\mathrm{Sing}(u) \cap \Xi$). This is shown by the examples above from which one can easily construct, for any $\nu > 3$, examples with singular sets of dimension $\nu - 3$ in the free boundary by adding redundant independent variables (replacing e.g. Ξ of dimension 3 by $\Xi \times S^{\nu-3}$ and $u: \Xi \to X$, $u(\Sigma) \subset S$, by $\hat{u}(\xi, \eta) = u(\xi)$, $\hat{u}(\Sigma \times S^{\nu-3}) \subset S$). Hardt & Lin [HL], working independently, have also considered the regularity of harmonic maps with a free boundary and obtained results similar to Theorem 2.

We now want to explain the essential steps in the proof of Theorems 1 and 2. Clearly, we may localize the problem in the domain and henceforth assume that $\Xi = \Delta_+^\nu$ is the upper unit half ball with its bounding equatorial $\nu-1$ disc $\Sigma = \Delta^\nu \cap \mathbb{R}^{\nu-1} \times \{0\}$ and its bounding half sphere $\Gamma = \partial \Delta_+^\nu \setminus \Sigma$.

We may assume that the Riemannian metric γ on $\Xi \cup \Sigma$ is close to the standard metric, i.e. $\gamma_{\alpha\beta}(0) = \delta_{\alpha\beta}$ and $\|D\gamma\| \le \frac{1}{8}$ on $\Xi \cup \Sigma$. We may also suppose that γ is parallel to Σ, that is $\xi_\nu = \mathrm{dist}_\gamma(\xi, \Sigma)$ and $e_\nu = \mathrm{grad}_\gamma \xi_\nu$. Since interior partial regularity has been proved by Schoen & Uhlenbeck we concentrate on the free boundary Σ.

The first step is to establish the Euler equation for the minimizing map u and a version of the natural boundary condition on Σ which is associated with the free boundary condition $u(\Sigma) \subset S$. Using variations of the type $u_+(\xi) = \Phi(t\eta(\xi), u(\xi))$, where Φ is the flow of a compactly supported vector field V on X and $\eta \in H^{1,2}(\Xi, \mathbb{R}) \cap L^\infty(\Xi, \mathbb{R})$ such that V is tangential to S along S and $\eta_{|\Gamma} = 0$ or V is arbitrary and $\eta_{|\Gamma \cup \Sigma} = 0$, one derives

Lemma 1 (*Variational equation*).

$$\int_\Xi Du(\xi) \cdot D\psi(\xi) J\gamma(\xi) d\xi = 0$$

for all vector fields $\psi \in H^{1,2}(\Xi, \mathbb{R}^{n+k})$ *tangental to* X *along* u, *tangential to* S *along* $u_{|\Sigma}$ *and with compact support in* $\Xi \cup \Sigma$.

Here, $Du(\xi) \cdot D\psi(\xi) := \mathrm{trace}[Du(\xi)^\dagger D\psi(\xi)]$ and $J\gamma(\xi)d\xi$ denotes the Riemannian measure on Ξ associated with the metric γ. Specializing ψ we obtain from Lemma 1:

Lemma 2. (i) (*Euler equation*)

$$\Delta_\gamma u = \mathrm{trace}_\gamma[(A \circ u)(Du \otimes Du)] \quad \textit{weakly on } \Xi,$$

where $A(x)$ *is the second fundamental form of* $X \subset \mathbb{R}^{n+k}$ *at* $x \in X$ *and* Δ_γ *the Laplace-Beltrami operator associated with* γ.

(ii) (*natural boundary condition*)

$$\lim_{\varepsilon \searrow 0} \frac{1}{\varepsilon} \int_{\Xi \cap [\varepsilon < \xi_\nu < 2\varepsilon]} (\eta \circ u) \left[\frac{\partial(\pi \circ u)}{\partial \xi_\nu} \cdot \chi \right] J\gamma \, d\xi = 0$$

for all $\chi \in H^{1,2}(\Xi, \mathbb{R}^{n+k}) \cap L^\infty(\Xi, \mathbb{R}^{n+k})$ *with compact support in* $\Xi \cup \Sigma$ *and all* $\eta \in C^1(X, \mathbb{R})$ *with support in a tubular neighborhood of* S *and* $\sup_X(|\eta| + |D\eta|) < \infty$.

Here $\pi(x)$ denotes the nearest point in S to $x \in X$ if unique. Although $\pi \circ u$ may not be defined (almost) everywhere on Ξ the product $(\eta \circ u)\frac{\partial(\pi \circ u)}{\partial \xi_\nu}$

is defined on Ξ. (We assume that S has global tubular neighborhoods in X and in $\mathbb{R}^{n+k}$.) Additional information is gained from variations of the independent variables. This is a well-known argument which can be adapted to the free boundary case and leads to the following lemma in which we abbreviate $E^+_{\xi,\rho}(u)$ for the energy of u on the ball of radius ρ in $\mathbb{R}^{\nu-1} \times \mathbb{R}_+$ centered at $\xi \in \mathbb{R}^\nu$ and $\xi^* = \xi - 2\xi_\nu e_\nu$ for the reflection of ξ across $\mathbb{R}^{\nu-1} \times \{0\}$:

Lemma 3 (*monotonicity formulae at the free boundary*). *There exists a constant C depending on $\nu = \dim \Xi$ only such that*

(i) $e^{C\rho}\rho^{2-\nu}E^+_{0,\rho}(u) - e^{C\sigma}\sigma^{2-\nu}E^+_{0,\sigma}(u) \geq \int_{\rho\Xi\setminus\sigma\Xi} e^{C|\xi|}|\xi|^{2-\nu}|\partial_{rad}u(\xi)|^2 d\xi$ *for* $0 < \sigma < \rho < 1$,

(ii) $\sigma^{2-\nu}[E^+_{\xi,\sigma}(u) + E^+_{\xi^*,\sigma}(u)] \leq C\rho^{2-\nu}[E^+_{\xi,\rho}(u) + E^+_{\xi^*,\rho}(u)]$ *for* $0 < \sigma < \rho < \frac{1}{2}$ *and* $\xi \in \Xi \cup \Sigma$ *with* $|\xi| \leq \frac{1}{2}$.

It is known from interior regularity theory that Theorem 1 follows with an easy measure theoretic argument from a *small-energy-regularity-theorem* asserting the Hölder continuity of u on a closed (half) ball provided the energy of u on the concentric (half) ball with doubled radius is sufficiently small. As usual, the Hölder continuity will be derived from a Morrey growth condition for the energies on (half) balls of radius r as $r \downarrow 0$. To prove such a Morrey condition, one would like to compare the energy minimizing function u on a small (half) ball with the $\mathbb{R}^{n+k}$ valued harmonic function h on this (half) ball taking the same boundary values as u. h need not map into X, of course, but using a neighborhood retraction p onto X one might hope to produce a reasonable comparison mapping $p \circ h$ for u.

However, at this point one encounters a serious difficulty. Namely, one cannot localize the problem in the target manifold X as long as continuity of u is unknown. Indeed, the examples above show that arbitrarily small neighborhoods of a singular point (in the free boundary) may be mapped onto the whole manifold X. Since we are unwilling to make any special a priori assumptions about the position of the image $u(\Xi)$ in X we must face the possibility that the harmonic function h above will have values far away from X so that we cannot use the neighborhood retraction p onto X to produce a comparison mapping with values in X. Once continuity of u is known one may, of course, localize the regularity problem also in X and use standard methods from regularity theory for elliptic systems of partial differential equations to infer higher order regularity of u. This also applies to the free boundary for which we refer to Gulliver & Jost [GJ]. Henceforth, we concentrate on the proof of Hölder continuity for u which is, as usual, the most difficult part in the regularity proof.

In [SU1] Schoen & Uhlenbeck have developed a method to overcome the difficulty described above: They apply the harmonic replacement idea not to the minimizing map u directly, but to a mollified map U instead which is constructed from u by a delicate smoothing process and satisfies energy estimates as well as additional oscillation estimates in terms of the (small) energy of u. The controlled oscillation of U allows to infer that the harmonic replacement H of U on a suitable ball has values close to X so that $p \circ H$ is defined. Using the minimality of u, the Euler equation for u and the relation between u and U Schoen & Uhlenbeck finally arrive at the desired Morrey growth condition for the energy integrals of u.

In our free boundary situation, it is a natural idea to extend $u\colon \Delta_+^\nu \to X$ to the full ball Δ^ν by reflection across Σ in Δ^ν and across S in X and then to apply the Schoen & Uhlenbeck technique to the extended function $\tilde{u}$. However, here we again run into difficulties due to the fact that we have a global problem with respect to the target manifold X and the supporting manifold $S \subset X$. Namely, the extension of u by reflection $\tilde{u}(\xi^*) := u(\xi)^\sigma$, where $x \mapsto x^\sigma$ denotes the geodesic reflection across S and $\Delta_+^\nu \ni \xi \mapsto \xi^*$ the reflection across $\mathbb{R}^{\nu-1} \times \{0\}$ in $\mathbb{R}^\nu$ as before, is meaningful only if $u(\xi) \in X$ has a unique nearest point in S. Furthermore, $\tilde{u}$ will be a reasonable function only if the image $u(\Xi)$ stays away from the focal set of S in X. We do not want to make such an assumption, however, (contrary to [GJ] where full regularity up to the free boundary is proved in a special geometric situation). Therefore, we introduce a *partial reflection* $\tilde{u}(\xi^*)$ which coincides with $u(\xi)$ if $u(\xi)$ is not close to S, equals the reflection $u(\xi)^\sigma$ of $u(\xi)$ if $u(\xi)$ is very close to S and suitably interpolates inbetween. This may not seem a very promising procedure, but note that u, although possibly discontinuous at Σ, maps many points near Σ into a small neighborhood of S due to the free boundary condition $u(\Sigma) \subset S$. This turns out sufficient to secure all the properties of $\tilde{u}$ which are necessary to successfully apply the Schoen & Uhlenbeck smoothing method. For convenience we state the precise definition of $\tilde{u}$ and its essential properties only for the Euclidean case $X = \mathbb{R}^n$. As before, $\pi(x)$ is the nearest point in S to $x \in \mathbb{R}^n$ (if unique) and we abbreviate $d(x) = |\pi(x) - x|$ for the distance from x to S. $\kappa > 0$ is a bound on the norm of the second fundamental form of S in $\mathbb{R}^n$; we also assume that S admits a global tubular neighborhood in $\mathbb{R}^n$ of radius $1/\kappa$.

Definition. Fix a function $\phi \in C^2(\mathbb{R}, \mathbb{R})$ such that $\phi(t) = 1$ for $t \leq 1/4\kappa$, $\phi(t) = 0$ for $t \geq 1/2\kappa$ and $\phi' \leq 0$ on $\mathbb{R}$. The *extension $\tilde{u}$ by partial reflection* is then defined $\tilde{u}(\xi) := u(\xi)$ for $\xi \in \Delta_+^\nu \cup \Sigma$ and

$$\tilde{u}(\xi^*) := \phi \circ d \circ u(\xi)[2\pi \circ u(\xi) - u(\xi)] + (1 - \phi \circ d \circ u(\xi))u(\xi) \quad \text{for } \xi \in \Delta_+^\nu.$$

Then $\tilde{u}(\xi^*) = 2\pi \circ u(\xi) - u(\xi)$ is the reflection of $u(\xi)$ across S if $d(u(\xi)) \leq 1/4\kappa$, while $\tilde{u}(\xi^*) = u(\xi)$ if $d(u(\xi)) \geq 1/2\kappa$. In view of the free boundary condition $u(\Sigma) \subset S$ we also have $\tilde{u} \in H^{1,2}(\Delta^\nu, \mathbb{R}^n)$. One key ingredient of our regularity proof is the following lemma asserting that the local averages $\tilde{u}_{r,\xi}$ of $\tilde{u}$, taken with respect to some fixed mollifying kernel on a ball of radius r centered at ξ, are close to the supporting manifold S if ξ lies in the free boundary Σ:

Lemma 4. *There exist $\varepsilon(\nu,\kappa) > 0$ and $C(\nu,\kappa) < \infty$ such that for $\xi \in \Sigma$ with $|\xi| \leq \frac{1}{2}$ and $0 < r < 1 - |\xi|$ we have*

$$\operatorname{dist}(\tilde{u}_{r,\xi}, S) \leq C(\nu,\kappa)\mathcal{E}(u)^{2/3}$$

provided $\mathcal{E}(u) \leq \varepsilon(\nu,\kappa)$.

This lemma is needed to ensure that the function $\tilde{U}$ obtained from $\tilde{u}$ by the Schoen & Uhlenbeck smoothing has values sufficiently close to S on Σ so that we can obtain an admissible comparison function for u by pushing $\tilde{U}$ into S with a small deformation of the ambient space $\mathbb{R}^n$. The proof of Lemma 4 depends on the monotonicity formulae Lemma 3 and a precise analysis of the various contributions to the energy integrals of $\tilde{u}$ which come from the sets of $\xi \in \Delta^\nu_+$ such that $\tilde{u}(\xi^*)$ is very close to S ($d(u(\xi)) \leq 1/4\kappa$), not close to S ($d(u(\xi)) \geq 1/2\kappa$ or $\tilde{u}(\xi^*)$ is in the region where we have interpolated between $u(\xi)$ and $u(\xi)^\sigma$ in the definition of $\tilde{u}$.

Another important ingredient needed in the method of Schoen & Uhlenbeck is the Euler equation. Now, for $\tilde{u}$ itself we do not have a suitable differential equation at our disposal. The reason is that $\tilde{u}$ is only an extension of u by *partial* reflection and on the region of interpolation between the reflection across S in $\mathbb{R}^n$ and the identity mapping of $\mathbb{R}^n$ we cannot compute any reasonable equation for $\tilde{u}$. However, it is possible to derive a differential inequality for $\Delta_{\tilde{\gamma}}\tilde{u}$ in terms of $|Du|^2$ where $\tilde{\gamma}$ denotes the metric on Δ^ν obtained from extending γ by reflection across Σ:

Lemma 5. *We have*

$$\Delta_{\tilde{\gamma}}\tilde{u} = H|(Du)^*|^2 \quad \textit{weakly on } \Delta^\nu$$

with some bounded measurable $\mathbb{R}^n$ valued function H and $(Du)^*(\xi) := 0$, $(Du)^*(\xi^*) := Du(\xi)$ for $\xi \in \Delta^\nu_+$.

The proof of this lemma uses the Euler equation and the natural boundary condition from Lemma 2 in an essential way. For the boundedness of H we need a bound on the second derivative of the nearest point projection π of a tubular neighborhood of S onto S. It is at this point that we have to assume a bound on the derivative of the second fundamental form of S in $\mathbb{R}^n$.

Once Lemmas 4 and 5 are established the transfer of the reasoning in [SU1] to the free boundary presents no particular difficulties. The reader can find the complete proofs in [DS1] or provide the necessary details himself if he is familiar with the work of Schoen & Uhlenbeck on the interior partial regularity of harmonic mappings. One ends up with a Morrey energy growth condition at the free boundary which, when combined with Morrey growth conditions in the interior, finally proves the asserted small-energy-regularity-theorem. The general Riemannian case $X \subset \mathbb{R}^{n+k}$ can be treated with appropriate modifications following the same ideas. Finally, we have already explained above that one obtains higher regularity and the dimension estimate on the singular set formulated in Theorem 1 from the small-energy-regularity-theorem. Given the smoothness of u up to the free boundary, the weak natural boundary condition from Lemma 2 is equivalent to the statement that the normal derivative of u on Σ is perpendicular to S along u. With this observation the proof of Theorem 1 is complete.

The proof of the optimal dimension estimate for the singular set in Theorem 2 is based on a dimension reduction argument introduced by Federer in the context of regularity theory for area minimizing rectifiable currents. Variants of this method have been used by Schoen & Uhlenbeck [SU1] to estimate the interior singular set of harmonic mappings and by Grüter [G] for partial regularity of minimizing currents at a free boundary. It can also be used for minimizing harmonic mappings at a free boundary. We give the full proof of Theorem 2 in [DS2] and merely sketch the principal ideas in the rest of this article. Since the interior case has been treated in [SU1] we again concentrate on the free boundary.

Recall that we consider a harmonic mapping $u: \Xi \to X \subset \mathbb{R}^{n+k}$ on the unit half ball $\Xi = \Delta^\nu_+$ which is energy minimizing with respect to the free boundary condition $u(\Sigma) \subset S$ where $\Sigma = \Delta^\nu \cap \mathbb{R}^{\nu-1} \times \{0\}$. For a singular point $a \in \mathrm{Sing}(u) \cap \Sigma$ we consider the scaled maps $u_{a,\mu}(\xi) := u(a + \mu L\xi)$, $\mu > 0$, with a suitable linear automorphism L of $\mathbb{R}^\nu$ transforming $\mathbb{R}^{\nu-1} \times \mathbb{R}_+$ to itself. We can select a sequence $\mu_i \downarrow 0$ such that $u_{(i)} := u_{a,\mu_i}$ converges strongly in $H^{1,2}(\Xi, \mathbb{R}^{n+k})$ to some mapping u_1 which is singular at the origin and satisfies the free boundary condition $u_1(\Sigma) \subset S$ again. From Lemma 3, (i) one infers that $\partial_{rad} u_1 = 0$, hence the singular set of u_1 is a cone. Moreover, if the Hausdorff measure $\mathbf{H}^s(\mathrm{Sing}(u) \cap \Sigma)$ is positive for some $s \geq 0$ one can choose the point a such that also $\mathbf{H}^s(\mathrm{Sing}(u_1) \cap \Sigma) > 0$. One then repeats the process starting from u_1 and a suitable point $a_1 \in$

$\mathrm{Sing}(u_1) \cap \Sigma$, $a_1 \neq 0$, to obtain u_2 which has similar properties as u_1 and, in addition, is independent of the coordinate ξ_1 after a suitable rotation. The process terminates with u_j such that $\Sigma \cap \mathrm{Sing}(u_j) = \Sigma \cap \mathbb{R}^{j-1} \times \{0\}$ where j must be strictly smaller than $\nu - 1$ in view of Theorem 1. Since $\mathbf{H}^s(\mathrm{Sing}(u_j) \cap \Sigma)$ is positive one concludes $s \leq j - 1 \leq \nu - 3$ which is the dimension estimate for the singular set in the free boundary claimed in Theorem 2. The discreteness of $\mathrm{Sing}(u) \cap \Sigma$ in case $\nu = 3$ must hold, because otherwise we could produce a limit $u_0(\xi) = \lim_{i\to\infty} u(2|a_i|\xi)$ of scaled maps with singularities at $\frac{1}{2}a_i/|a_i| \to a_0 \neq 0$ as $i \to \infty$ and then u_0 would be homogeneous of degree zero with the segment $[0, a_0]$ contained in $\mathrm{Sing}(u_0)$ which contradicts Theorem 1.

The validity of this reasoning depends on the following: One must know that sequences of scaled harmonic mappings $u_{(i)}$ as above do not only converge weakly—this is an immediate consequence of the monotonicity formulae (passing to subsequences if necessary)—but actually in $H^{1,2}$ norm and that the strong limit is energy minimizing again or, at least, satisfies the conclusions of Theorem 1. As in the interior regularity theory these facts can be established with an improvement of the small-energy-regularity-theorem. Namely, the *small-mean-oscillation-regularity-theorem* gives a Hölder estimate for u on a neighborhood of 0 in $\Xi \cup \Sigma$, provided $u \in H^{1,2}(\Xi, X)$ is energy minimizing with respect to the free boundary condition $u(\Sigma) \subset S$ and the mean oscillation $W^+(u)$ defined below is sufficiently small (depending on $\mathcal{E}(u)$ and geometric data). We combine this regularity theorem with the differential inequality from Lemma 5 for the extensions $\tilde{u}_{(i)} \colon \Delta^\nu \to \mathbb{R}^{n+k}$ by partial reflection to prove the desired strong convergence of the $u_{(i)}$ to a limit function with singular set of vanishing $\nu - 2$ dimensional measure.

It remains to discuss the small-mean-oscillation-regularity-theorem at the free boundary. In the interior theory Schoen & Uhlenbeck used the oscillation integral

$$W(u) := \int_{\Delta^\nu} |u(\xi) - q|^2 d\xi$$

with some fixed $q \in X$. Clearly, at the free boundary we must include some term measuring the deviation of $u(\xi)$ from the supporting manifold S for the free boundary values. We therefore propose to define the *mean oscillation at the free boundary*

$$W^+(u) := \int_{\Delta^\nu_+} [|u(\xi) - q|^2 + (d \circ u(\xi))^2] d\xi$$

where $d(x)$ denotes the distance from $x \in X$ to S. (In [HL] the authors use a different definition of the mean oscillation at the free boundary which, however, can be treated similarly.) Proceeding as in [SU1] one sees readily that a small-mean-oscillation-regularity-theorem at the free boundary based on this definition of mean oscillation can be proved, provided we have an extension lemma at our disposal which in the free boundary situation replaces the interior extension lemma used in [SU1]. It turns out that the constructions of Schoen & Uhlenbeck can be supplemented with some additional effort to yield the following extension result which is sufficient to complete the proof of Theorem 2 along the lines described above. In this final lemma, Γ denotes the bounding half sphere $\partial\Delta_\nu^+ \setminus \Sigma$ of Δ_+^ν as before, κ and λ are bounds for the second fundamental forms of S and X in $\mathbb{R}^{n+k}$ respectively (where S and X are also assumed to have global tubular neighborhood of radius $1/\kappa$ and $1/\lambda$ respectively) and the notations $\mathcal{E}(v,\Xi)$, $W^+(v,\Xi)$ stand for the energy $\mathcal{E}(v)$ and the mean oscillation $W^+(v)$ of $v \in H^{1,2}(\Xi, X)$, while $\mathcal{E}(v,\Gamma)$, $W^+(v,\Gamma)$ denote the corresponding quantities defined for the trace $v_{|\Gamma}$ with respect to the standard metric on Γ.

Lemma 6 (*extension lemma at the free boundary*). *There exist positive constants $\gamma = \gamma(\nu,\kappa,\lambda)$, $\beta = \beta(\nu)$, $C = C(\nu)$ such that every map $v \in H^{1,2}(\Gamma, X)$ satisfying the free boundary condition $v(\Gamma \cap \overline{\Sigma}) \subset S$ and the smallness condition*

$$\mathcal{E}(v,\Gamma)^{1/2} W^+(v,\Gamma) \leq \gamma\varepsilon^\beta$$

for some $0 < \varepsilon \leq 1$ can be extended to $\overline{v} \in H^{1,2}(\Xi, X)$ with the free boundary condition $\overline{v}(\Sigma) \subset S$ and the estimates

$$\mathcal{E}(\overline{v},\Xi) \leq C[\varepsilon\mathcal{E}(v,\Gamma) + \varepsilon^{-2\beta} W^+(v,\Gamma)],$$
$$W^+(\overline{v},\Xi) \leq C\varepsilon^{-2\beta} W^+(v,\Gamma).$$

Acknowledgement: I want to thank the organizers of the International Workshop on "Variational Problems" for the invitation and for their hospitality during my stay in Paris. My participation was supported by the C.N.R.S. (France) and the Sonderforschungsbereich 256 at Bonn University (Germany).

REFERENCES

[BZ] F. Bethuel and X. Zheng, *On the density of smooth functions between two compact manifolds in Sobolev spaces,* C.R. Acad. Sci. Paris, **303**, Série I, no. 10, 1986.

[DS1] F. Duzaar and K. Steffen, *A partial regularity theorem for harmonic maps at a free boundary,* Asymptotic Analysis **2**, 1989, 299-343.

[DS2] F. Duzaar and K. Steffen, *An optimal estimate for the singular set of a harmonic map in the free boundary,* J. Reine Angew. Math. **401**, 1989, 157–187.

[G] M. Grüter, *Optimal regularity for codimension one currents with a free boundary,* Man. Math. **58**, 1987, 295–343.

[GJ] R. Gulliver and J. Jost, *Harmonic maps which solve a free boundary value problem,* J. Reine Angew. Math. **385**, 1987, 307–325.

[HL] R. Hardt and F.H. Lin, *Partially constrained boundary conditions with energy minimizing mappings,* Comm. Pure Appl. Math. **42**, 1989, 309–334.

[L] F.H. Lin, *A remark on the mapping* $x/|x|$; C.R. Acad. Sci. Paris, **305**, Série I, 1987, 529–532.

[SU1] R. Schoen and K. Uhlenbeck, *A regularity theorem for harmonic maps,* J. Differential Geometry **17**, 1982, 307–335.

[SU2] R. Schoen and K. Uhlenbeck, *Boundary regularity and the Dirichlet problem for harmonic maps,* J. Differential Geometry **18**, 1982, 253–266.

Klaus Steffen
Mathematisches Institut
Heinrich-Heine-Universität
D-4000 Düsseldorf
Federal Republic of Germany

Global existence and partial regularity results for the evolution of harmonic maps

Michael Struwe

Mathematik, ETH-Zentrum, CH-8092 Zürich

Consider a compact, m-dimensional Riemannian manifold M with metric $\gamma = (\gamma_{\alpha\beta})_{1\le\alpha,\,\beta\le m}$ and with $\partial M = \emptyset$, or $M = \mathbb{R}^m$, $m \ge 2$. For a compact, ℓ-dimensional manifold N, $\partial N = \emptyset$, with metric $g = (g_{ij})_{1\le i,\,j\le \ell}$ and C^1-maps $u : M \to N$ let

$$E(u) = \int_M e(u)\, dM$$

be the energy of u, with density

$$e(u) = \frac{1}{2}\gamma^{\alpha\beta} g_{ij}(u) \frac{\partial}{\partial x_\alpha} u^i \frac{\partial}{\partial x_\beta} u^j$$

and volume element

$$dM = \sqrt{|\gamma|}\, dx\,,\ |\gamma| = \det(\gamma_{\alpha\beta})\,,$$

in local coordinates.

Here $(\gamma^{\alpha\beta}) = (\gamma_{\alpha\beta})^{-1}$, and a summation convention is used. By definition, $u \in C^1(M,N)$ is harmonic iff u is a stationary point of E and satisfies the Euler-equations

$$\Delta_M u + \Gamma_N(u)\,(\nabla u, \nabla u)_M = 0, \tag{1.1}$$

where Δ_M is the Laplace-Beltrami operator on M and Γ_N is a bilinear form with coefficients involving the Christoffel symbols of the metric g on N and the metric γ on M.

By Nash's embedding theorem we may assume that $N \subset \mathbb{R}^n$ isometrically. Then E becomes the familiar Dirichlet integral

$$E(u) = \frac{1}{2}\int_M (\nabla u, \nabla u)_M\, dM$$

for maps $u : M \to N \subset \mathbb{R}^n$ with density

$$(\nabla u, \nabla u)_M = \gamma^{\alpha\beta} \frac{\partial}{\partial x_\alpha} u^i \frac{\partial}{\partial x_\beta} u^i\,.$$

Hence u is harmonic if

$$\int_M \Delta_M u \cdot \varphi\, dM = 0$$

for all smooth φ such that $\varphi(x) \in T_{u(x)}N$ for all $x \in M$, $T_pN \subset \mathbb{R}^n$ denoting the tangent space to N at $p \in N$. I.e. u is harmonic iff there exists a unit normal vector field $\nu_u(x) \perp T_{u(x)}N$ and a scalar function λ on M such that

(1.2) $$-\Delta_M u + \lambda \nu_u = 0 .$$

If $M = S^1$, harmonic maps $M \to N$ parametrize closed geodesics in N; if $N = \mathbb{R}$, harmonic maps are harmonic functions.

One would like to understand how much of the topological and differentiable structure of Riemannian manifolds N is captured in the set of harmonic maps $M \to N$. A particular problem is to find harmonic representatives in homotopy classes of maps $u_0 : M \to N$.

One approach to this problem, suggested by Eells and Sampson [ES], is to consider the evolution problem

(1.3) $$\begin{cases} \frac{\partial}{\partial t}u - \Delta_M u + \lambda \nu_u = 0 \\ u|_{t=0} = u_0 \end{cases}$$

for given $u_0: M \to N$, with $\nu_u \perp T_uN$, $|\nu_u| \equiv 1$, and a scalar function λ. (We may also use the explicit form of $\lambda \nu_u$ from (1.1)).

Theorem 1.1: (Eells-Sampson) *Suppose that the sectional curvature* K_N *of* N *is non-positive. Then for any smooth* u_0 *problem (1.3) has a unique, smooth global solution* $u : M\times[0,\infty[\to N$ *which converges to a smooth harmonic map* $u_\infty: M \to N$ *as* $t \to \infty$ *suitably.*

Since under the conditions of Theorem 1.1 the solution u_t to (1.3) furnishes a homotopy $u_0 \sim u_\infty$ of any map u_0 with a harmonic map u_∞ it follows that in non-positively curved manifolds any homotopy class contains a harmonic representative.

The main tools in proving Theorem 1.1 are the energy inequality

(1.4) $$E(u_T) + \int_M \int_0^T |\tfrac{\partial}{\partial t}u|^2 \, dMdt \le E(u_0), \ \forall\, T \ge 0,$$

and the Bochner type estimate

(1.5) $$(\partial_t - \Delta_M)\, e(u) + (\nabla^2 u, \nabla^2 u)_M + \mathrm{Ric}_M(\nabla u, \nabla u)_M \le K_N\, e(u)^2$$

where Ric_M denotes the Ricci curvature of M, and K_N is an upper bound for the sectional curvature of N. If $K_N \le 0$, (1.4), (1.5) yield a uniform bound on e(u), while (1.4) yields asymptotic convergence towards a harmonic map.

(Proof of (1.4) : Multiply (1.3) by $\frac{\partial}{\partial t}u$ and integrate. Proof of (1.5): Differentiate (1.3) with respect to x_α, multiply by $\gamma^{\alpha\beta}\frac{\partial}{\partial x_\beta}u$, sum over α, and integrate.)

Of course, we would like to know whether Theorem 1.1 is true in general, that is, without any curvature restrictions. Our answer will depend on the dimension m.

2. The case m = 2; harmonic maps of Riemann surfaces

The case $m = 2$ is special because of the uniformization theorem and conformal invariance of Dirichlet's integral. This added structure permits better results than in higher dimensions.

Complementing the Eells-Sampson result Theorem 1.1, we first have a negative result

Theorem 2.1: (Eells-Wood [EW]) *There is no harmonic map* $u : T^2 \to S^2$ *of degree* ± 1.

Hence maps $T^2 \to S^2$ of degree ± 1 cannot be represented by harmonic maps. Theorem 2.1 seems to indicate that a curvature restriction as in Theorem 1.1 may be necessary. Surprisingly, also a topological condition may suffice.

Theorem 2.2: (Lemaire [Le] , Sacks-Uhlenbeck [SaU]) *Suppose* $\pi_2(N) = 0$. *Then any map* $u_0 : M \to N$ *is homotopic to a harmonic map.*

Is there a chance to obtain generalizations of Theorem 1.1 of a similar kind? By Theorem 2.1 such generalizations must allow singularities to form. Hence it is natural to study also mappings in Sobolev spaces $H^{m,p}(M;N) = \{u \in H^{m,p}(M;\mathbb{R}^n); u(M) \subset N\}$. In particular, $H^{1,2}(M;N)$ is the completion of $C^1(M;N)$ in the norm induced by the L^2-norm and the functional E.

Theorem 2.3: (Struwe [St 1]) *For* $u_0 \in H^{1,2}(M;N)$ *the evolution problem (1.3) possesses a global weak solution* $u : M \times]0,\infty[\to N$ *satisfying (1.4) and which is regular*

with exception of finitely many points $(\bar{x}_k, \bar{t}_k)$, $1 \le k \le K$, *and unique in this class. If* u_0 *is smooth, so is* u *near* $t = 0$.

At a singularity $(\bar{x}, \bar{t})$ *a non-constant harmonic map* $\bar{u} : S^2 \to N$ *("harmonic sphere") separates in the sense that for sequences* $t_m \nearrow \bar{t}$, $R_m \searrow 0$, $x_m \to x$ *the scaled maps* $\tilde{u}_m(x) = u(\exp_{x_m}(R_m x), t_m) \to \tilde{u}$ *locally in* $H^{2,2}(\mathbb{R}^2, N)$, *where* $\tilde{u}$ *is a non-constant harmonic map* $\tilde{u} : \mathbb{R}^2 \to N$ *with finite energy which extends to a regular harmonic map* $\bar{u} : S^2 \cong \overline{\mathbb{R}}^2 \to N$.

As $t \to \infty$ *suitably,* $u_t = u(\cdot,t)$ *converges weakly in* $H^{1,2}(M; N)$ *(and strongly in* $H^{2,2}$ *away from at most finitely many points where again non-constant harmonic spheres separate) to a regular harmonic map* $u_\infty : M \to N$.

Let $[u_0]$ be the homotopy class of u_0 (which is meaningful even for $H^{1,2}$-maps, cf. [SU 1]), and let

$$\varepsilon_N = \inf \{ E(u) ; u : S^2 \to N \text{ is a non-constant (regular) harmonic map} \} .$$

Then $\varepsilon_N > 0$; hence by Theorem 2.3, if $\pi_2(N) = 0$ and if $E(u_0) < \inf \{ E(u); u \in [u_0] + \varepsilon_N$ there exists a global, regular solution to the evolution problem (1.3) which as $t \to \infty$ converges to a harmonic map $u_\infty \in [u_0]$, cf. [St 2]. In particular, we reobtain Theorem 2.2.

Sketch of the proof of Theorem 2.3:

The key estimates used in the proof of Theorem 2.3 are the energy inequality (1.4) and the (elementary) Sobolev-type estimate

$$(2.1) \qquad \int_0^T \int_\Omega |\nabla u|^4 \varphi^2 \, dxdt \le c_0 \sup_{0 \le t \le T} \Big(\int_{\operatorname{supp} \varphi} |\nabla u(\cdot,t)|^2 \, dx \Big) \cdot \Big(\int_0^T \int_\Omega |\nabla^2 u|^2 \varphi^2 \, dxdt + \|\nabla \varphi\|_{L^\infty}^2 \int_0^T \int_\Omega |\nabla u|^2 dxdt \Big),$$

on a domain $\Omega \subset \mathbb{R}^2$ which holds for all $\varphi \in C_0^\infty(\Omega)$ and all functions with u, ∇u, $\nabla^2 u$, $\frac{\partial}{\partial t} u \in L^2(\Omega \times [0, T])$. c_0 is a universal constant. (Proof of (2.1): For fixed t, $x = (x_1, x_2)$ write $|\nabla u(x, t)|^4 \varphi(x)^2 = \int_\infty^{x_1} \frac{\partial}{\partial x_1} (|\nabla u((s, x_2), t)|^2 \varphi(s, x_2)) \, ds \cdot \int_\infty^{x_2} \frac{\partial}{\partial x_2} (|\nabla u((x_1, s), t)|^2 \varphi(x_1, s)) \, ds$, and use Hölder's inequality; then integrate over $\Omega \times [0, T]$, using Fubini's theorem.) (2.1) implies regularity of u at all points (x_0, t_0) where

$$(2.2) \qquad \limsup_{t \nearrow t_0} \int_{B_R(x_0)} |\nabla u(\cdot,t)|^2 dx \le \varepsilon_0$$

for some $R > 0$, where $\varepsilon_0 > 0$ is a constant depending only on M, N and related to c_0 in (2.1). The characterization of singularities given in Theorem 2.3 is obtained by a blow-up argument, again using (2.1) to prove $H^{2,2}_{loc}$-convergence of the blown-up functions. Local solvability of (1.3) and finiteness of the singular set is obtained from a local energy inequality: Let $E(u;\Omega) = \frac{1}{2}\int_\Omega |\nabla u|^2\,dx$. Then for $R > 0$, $x_0 \in M$

$$E(u_T; B_R(x_0)) \le E(u_0; B_{2R}(x_0)) + c_1 E(u_0) \frac{T}{R^2}\,, \tag{2.3}$$

with a constant $c_1 = c_1(M, N)$. (Proof: Multiply (1.3) by $\frac{\partial}{\partial t} u\, \varphi^2$, where $\varphi \in C_0^\infty(B_{2R}(x_0))$ satisfies $0 \le \varphi \le 1$, $|\nabla\varphi| \le c/R$, $\varphi \equiv 1$ on $B_R(x_0)$, and integrate by parts to obtain

$$\int_0^T \int_\Omega |\tfrac{\partial}{\partial t} u|^2 \varphi^2\,dxdt + \int_0^T \int_\Omega \tfrac{d}{dt}\Big(\frac{|\nabla u|^2\varphi^2}{2}\Big)\,dxdt = \int_0^T \int_\Omega \tfrac{d}{dt} u\, \nabla u\, \nabla\varphi\, \varphi\,dxdt$$

$$\le \int_0^T \int_\Omega |\tfrac{\partial}{\partial t} u|^2 \varphi^2\,dxdt + c\int_0^T |\nabla u|^2 |\nabla\varphi|^2\,dxdt\,,$$

whence (2.3) is immediate.) Since for $u_0 \in H^{1,2}(M,N)$ we may choose $R > 0$ such that $E(u, B_{2R_0}(x_0)) \le \varepsilon_0/2$ uniformly, (2.3) guarantees that (2.2) holds for $t_0 \le \frac{\varepsilon_0 R^2}{2c_1 E(u_0)}$ and (2.1) gives an a-priori estimate which allows to prove local solvability of (1.3). By (2.3), if (x_0, t_0) is such that (2.2) fails to hold for all $R > 0$ then also for $t_1 < t_0$

$$\int_{B_{2R}(x_0)} |\nabla u(\cdot,t_1)|^2\,dx \ge \limsup_{t\nearrow t_0}\Big(\int_{B_R(x_0)} |\nabla u(\cdot,t)|^2\,dx\Big) - c_1 E(u_0)\frac{t_0 - t_1}{R^2}\,,$$

and taking the inferior limit with repect to $t_1 < t_0$ it follows that at such a point for all $R > 0$

$$\liminf_{t\nearrow t_0} \int_{B_R(x_0)} |\nabla u(\cdot,t)|^2\,dx > \varepsilon_0\,. \tag{2.4}$$

But now for any $t_0 > 0$, $x_1,\dots,x_K \in M$ satisfying (2.4), by weak lower semi-continuity of E and absolute continuity of the Lebesque integral

$$\begin{aligned} E(u(t_0)) &= \lim_{R\to 0} E\big(u(t_0); \Omega \setminus \bigcup_{k=1}^K B_R(x_k)\big) \\ &\le \lim_{R\to 0}\ \liminf_{t\nearrow t_0} E\big(u(t); \Omega \setminus \bigcup_{k=1}^K B_R(x_k)\big) \\ &\le \liminf_{t\nearrow t_0} E(u(t)) - \lim_{R\to 0}\Big(\sum_{k=1}^K \liminf_{t\nearrow t_0} E(u_t; B_R(x_k))\Big) \qquad (2.5) \\ &\le \liminf_{t\nearrow t_0} E(u(t)) - K\varepsilon_0\,. \end{aligned}$$

Reiterating (2.5) at singular times $t_1 \le \dots \le t_L$ with K_ℓ singular points at each time t_ℓ we obtain the estimate

$$0 \le E(u_{t_L}) \le E(u_{t_{L-1}}) - K_L \varepsilon_0 \le \ldots \le E(u_0) - \sum_{\ell=1}^{L} K_\ell \varepsilon_0$$

which permits to a-priori bound the number of singular points in terms of the initial energy $E(u_0)$. Using (1.4) and (2.1) again, finally, we can also prove asymptotic convergence of a suitable sequence $u_m = u(\cdot, t_m)$, $t_m \to \infty$, towards a harmonic map $u_\infty : M \to N$ weakly in $H^{1,2}$ and (by (2.1)) strongly in $H^{2,2}$ off a set of points x_0 where

$$\limsup_{m \to \infty} \int_{B_R(x_0)} |\nabla u_m|^2 \, dx > \varepsilon_0$$

for all $R > 0$. By (1.4) these can be at most finite in number.

Theorem 2.3 has recently been generalized to maps from Riemann surfaces with boundary into complete manifolds by K.C. Chang [Cha].

In Theorem 2.3 we have found it natural to generalize the concept of solutions and maps to include maps in Sobolev spaces. As a side-line we remark that a notion of weakly harmonic map can be defined for maps $u \in H^{1,2}(M; N)$. If such a map is E-minimizing it can be shown to be regular cf. [GG], [SU 1]; this is also the case for weakly harmonic maps of class $H^{1,2}$ which possess at most isolated singularities, cf. [SaU]. However, it is not known if *all* weakly harmonic maps $u \in H^{1,2}(M; N)$ are necessarily regular. One possible approach to this question would be through Theorem 2.3: A uniqueness result for weak solutions to (1.3) in the class of functions satisfying only (1.4) would give an affirmative answer.

Another problem of interest is the question whether (1.3) may develop singularities in finite time.

3. The case $m \ge 3$

One of the striking differences between the two-dimensional and the higher dimensional case is the fact that in higher dimensions there may be E-minimizing weakly harmonic maps with singularities. In fact for $m \ge 3$ the map $\bar{u} : B_1(0; \mathbb{R}^m) \to S^{m-1} = \partial B_1(0; \mathbb{R}^m)$ given by $x \to \bar{u}(x) = x / |x|$ has finite energy and is absolutely E-minimizing among maps $u : B_1(0; \mathbb{R}^m) \to S^{m-1}$ with the same boundary data, cp. [BCL]. Jäger, Kaul [JK], Baldes [Ba] have shown that in sufficiently high dimensions this map is even E-minimizing among maps into S^m, resp. an ellipsoid with S^{m-1} as equator.

A partial regularity theory for E-minimizing harmonic maps has been achieved by Schoen-Uhlenbeck [SU 1]. By a series of blow-up arguments they obtain estimates on the Hausdorff dimension of the singular set of E-minimizing harmonic maps $M \to N$. Moreover, they identify (minimal) harmonic spheres in N as being responsible for the occurance of singularities. The main tool in their work is a monotonicity formula for E-minimizing harmonic maps: If $u : B_R(0; \mathbb{R}^m) \to N$ is minimal, then for all $0 < \rho < r < R$ there holds

(3.1) $$\rho^{2-m} E(u; B_\rho(0)) \le r^{2-m} E(u; B_r(0)) .$$

A generalization of Theorem 1.1 to arbitrary targets hence must allow for "large" singular sets. Moreover, in order to be able to access the tools based on monotonicity estimates like (3.1) an analogue of this formula has to be obtained.

Theorem 3.1 : (Monotonicity formula, [St 3]) *Suppose* $u : \mathbb{R}^m \times [-R_0^2, 0[\to N$ *is a* C^2*- solution* (1.3) *with* $E(u_t) \le c$ *and* $|\nabla u| \le c$ *uniformly. Let*

$$G(x,t) = \frac{1}{(2\pi|t|)^{m/2}} \exp\left(-\frac{|x|^2}{4|t|}\right)$$

be the fundamental solution to the heat equation with singularity at 0 *and let*

$$\Phi(u; R) = \frac{1}{2} R^2 \int_{\mathbb{R}^m \times \{-R^2\}} |\nabla u|^2 G \, dx = R^2 \int_{\mathbb{R}^m \times \{-R^2\}} e(u) G \, dx .$$

Then for all $0 < \rho < r < R_0$ *there holds*

(3.2) $$\Phi(u; \rho) \le \Phi(u; r) .$$

This monotonicity estimate appears as a natural variation of (3.1) if we take into account that by inhomogenity of space-time the influence of the values of ∇u at different points have to be balanced with a suitable weight function. The proof of (3.2) uses invariance of (1.3) under scaling $u \to u_r(x,t) = u(rx, r^2t)$, whence e.g. at $r = 1$ we have

$$\begin{aligned} \frac{d}{dr}\Phi(u; r) &= \frac{d}{dr}\Phi(u_r, 1) = \int_{\mathbb{R}^m \times \{-1\}} \nabla u \, \nabla\left(\frac{d}{dr} u_r\right) G \, dx \\ &= \int_{\mathbb{R}^m \times \{-1\}} \left[-\Delta u \left(x \cdot \nabla u + 2t \frac{\partial}{\partial t} u\right) + \frac{x}{2|t|} \cdot \nabla u \left(x \cdot \nabla u + 2t \frac{\partial}{\partial t} u\right)\right] G \, dx \end{aligned}$$

on account of $\nabla G = -\frac{x}{2|t|} G$. But now by (1.3), since $\nabla u, \frac{\partial}{\partial t} u \perp T_u N$, $t = -1$ the latter equals

(3.3) $$= \frac{1}{2} \int_{\mathbb{R}^m \times \{-1\}} \left|x \cdot \nabla u - 2 \frac{\partial}{\partial t} u\right|^2 G \, dx \ge 0 .$$

For maps on compact manifolds M an analogous estimate holds with a (harmless) exponential factor, cf. [CS, Lemma 4.2].

Using the machinery developed by Schoen-Uhlenbeck [SU 1], in particular [Sc], now an "ε-regularity" theorem (i.e. a-priori C^1-bounds for solutions with small energy) and partial regularity results for weak solutions to (1.3) which can be weakly approximated by regular soltuions to (1.3) can be obtained, cf. [St 3, Theorem 5.1, 6.1].

More precisely, the partial regularity result is a consequence of uniform C^1-a-priori bounds for the approximating solution off a closed set Σ having Hausdorff-dimension $\leq m$ with respect to the parabolic metric $\delta((x,t),(y,s)) = |x-y| + \sqrt{|s-t|}$.

By (3.3), moreover, singularities seem to be associated with (not necessarily minimal) harmonic spheres or certain homogeneous solutions $\bar{u}(x,t) = w(x/\sqrt{|t|})$ to (1.3) on $\mathbb{R}^m \times]-\infty, 0[$, cf. [St 3, Theorem 8.1].

A complete generalization of Theorem 1.1 finally was achieved by combining the monotonicity formula (3.3) with the penalty approach to (1.3) of Chen [Che]. Chen had observed that for $N = S^{n-1} \subset \mathbb{R}^n$, global solutions to (1.3) can be obtained as weak limits of solutions $u_K : M \times [0,\infty[\to \mathbb{R}^n$ to unconstrained problems

(3.4) $$\frac{\partial}{\partial t} u_K - \Delta_M u_K + K(|u_K|^2 - 1)\, u_k = 0,$$
$$u_K|_{t=0} = u_0 : M \to S^{n-1}$$

related to an energy integral E_K with density

$$e_K(u) = \frac{1}{2}\left(|\nabla u|_M^2 + \frac{K}{2}(|u|^2-1)^2\right).$$

Indeed, (3.4) can be solved globally. By the analogue of (1.4) a sequence (u_K) of solutions to (3.4) converges weakly to a map $u : M \times [0,\infty[\to S^{n-1}$ with bounded energy. Since $u \perp T_u S^{n-1}$ for all $u \in S^{n-1}$ it is possible to pass to the limit $K \to \infty$ in (3.4) and to show that u weakly solves (1.3). Shatah [Sh], resp. Keller, Rubinstein and Sternberg [KRS] have independently found similar existence results for global weak solutions to (1.3) in special geometries along the same lines; i.e. by using the penalty method to construct a candidate for a global weak solution and then exploiting the particular geometric structure to verify (1.3).

The simple observation needed to pass from special geometries to general results is the observation that for solutions u_K to (3.4) on $M \times [-R^2, 0[$ a monotonicity formula like (3.2) holds if the density e is replaced by e_K in the definition of Φ. The analogue of (1.4) and the Bochner type inequality (1.5) also remain valid.

Thus, by the mechanism cited above, one obtains uniform C^1-a-priori bounds for the u_K off a closed set Σ of Hausdorff dimension $\leq m$ (with respect to the parabolic metric δ)

and partial regularity of the weak limit u. The uniform local a-priori estimates an u_K away from Σ allow to pass to the limit in (3.4) *without* using the special geometry of the target! For arbitrary targets N the penalty approximation now has to be suitably adapted; the rest of the argument stays the same.

This leads to:

Theorem 3.2: (Chen-Struwe [CS]) *Suppose* M,N *are compact manifolds without boundary,* $\dim M = m \geq 3$, *and let* $u_0 \in C^1(M; N)$ *be given. Then there exists a global weak solution* $u : M \times [0,\infty[\to N$ *to (1.3) satisfying (1.4) which is regular off a closed set* Σ *of Hausdorff dimension* $\leq m$ *with respect to the parabolic metric* δ.

As $t \to \infty$ *suitably,* u_t *converges weakly in* $H^{1,2}$ *to a weakly harmonic map* $u_\infty : M \to N$ *which is regular off a closed set* Σ_∞ *of Hausdorff dimension* $\leq m - 2$.

Uniqueness of the solution u is not known (even in the class of partially regular solutions.) Very recently, Coron-Ghidaglia [CG] have shown that singularities may occur in finite time. However, it is not known whether the estimates on the dimension of the singular sets Σ, Σ_∞ are optimal. Remark that for E-minimal harmonic maps the co-dimension of the singular set is ≥ 3, cf. [SU 1].

References

[Ba] Baldes, A.: Stability and uniqueness properties of the equator map from the ball into an ellipsoid, Math. Z.

[BCL] Brezis, H.- Coron, J.-M.- Lieb, E.: Harmonic maps with defects, Comm. Math. Phys. 107(1986), 649-705

[Cha] Chang, K.C.: Heat flow and boundary value problems for harmonic maps, preprint (1988)

[Che] Chen, Y.: Weak solutions to the evolution problem of harmonic maps, Math. Z. (in press)

[CS] Chen, Y.- Struwe, M.: Existence and partial regularity for the heat flow for harmonic maps, Math. Z. (in press)

[ES] Eells, J.- Sampson, J.H.: Harmonic mappings of Riemannian manifolds, Am. J. Math. 86 (1964), 109-160

[EW] Eells, J.- Wood, J.C.: Restrictions on harmonic maps of surfaces, Topology 15 (1976), 263-266

[GG] Giaquinta, M.- Giusti, E.: On the regularity of the minima of variational integrals, Acta Math. 148 (1982), 31-40

[Ha] Hamilton, R.: Harmonic maps of manifolds with boundary, Springer lecture notes 471, Berlin-Heidelberg-New York (1975)

[JK] Jäger, W.- Kaul, H.: Rotationally symmetric harmonic maps from a ball into a sphere and the regularity problem for weak solutions of elliptic systems, J. Reine Angew. Math. 343(1983), 146-161

[KRS] Keller, J.- Rubinstein, J.- Sternberg, P.: Reaction-diffusion processes and evolution to harmonic maps, preprint (1988)

[Le] Lemaire, L.: Applications harmoniques de surfaces Riemanniennes, J. Diff. Geom. 13(1978), 51-78

[SaU] Sacks, J.- Uhlenbeck, K.: The existence of minimal immersions of 2-spheres, Ann. Math. 113 (1981), 1-24

[Sc] Schoen, R.M.: Analytic aspects of the harmonic map problem, in: Seminar on nonlinear PDE, Chern (Ed.), Springer 1984

[SU 1] Schoen, R.M., Uhlenbeck, K.: A regularity theory for harmonic maps, J. Diff. Geom. 17 (1982), 307-335

[SU 2] Schoen, R.M., Uhlenbeck, K.: Boundary regularity and the Dirichlet problem for harmonic maps, J. Diff. Geom. 23 (1984)

[Sh] Shatah, J.: Weak solutions and development of singularities of the SU(2) σ-model, Comm. Pure Appl. Math. 41 (1988), 459-469

[St 1] Struwe, M.: On the evolution of harmonic mappings of Riemannian surfaces, Comm. Math. Helv. 60 (1985), 558-581

[St 2] Struwe, M.: Heat-flow methods for harmonic maps of surfaces and applications to free boundary problems, Proc. VIII Latin Amer. Conf. Math. (in press)

[St 3] Struwe, M.: On the evolution of harmonic maps in higher dimensions, J. Diff. Geom. (in press)

Part III
Hamiltonian Systems

Multiple periodic trajectories in a relativistic gravitational field

A. AMBROSETTI and U.BESSI

1. Let us consider the Newtonian potential:

$$U(x) = -\frac{1}{|x|} + \gamma W(x)$$

where $\gamma \in \mathbf{R}$ and $W: \mathbf{R}^n \to \mathbf{R}$ is smooth. If we want to take into account the relativistic correction to the motion of a particle under the potential U with given energy $h < 0$, we are led to consider a motion at the same energy governed by the potential:

$$U_\kappa(x) = (1 + \frac{4}{3}h\kappa)U(x) - \kappa U^2(x)$$

where $\kappa = \frac{3}{c^2}$, c being the speed of light (cfr. chap 2, §8 of [6]).

Theorem 1. *Suppose $W(x) = W(-x)$, $W \in C^1(\mathbf{R}^n, \mathbf{R})$ and let $\kappa > 0$ be given. Then there exist $\gamma^* > 0, h^* < 0$ such that $\forall h \in (h^*, 0)$ and $|\gamma| < \gamma^*$ the problem*

$$\begin{cases} \ddot{q} + U'_\kappa(q) = 0 \\ \frac{1}{2}|\dot{q}|^2 + U_\kappa(q) = h \end{cases} \qquad (P)$$

has at least n distinct trajectories.

Theorem 1 is a particular case of a more general result stated below (cfr Theorem 6).

Existence of periodic solutions for conservative systems with singular potentials has been stated in [1,2], whose results cover both the case of a

---

Supported by Italian Ministry of Education

relativistic gravitational field like U_κ, as well as that of classical Newtonian potentials (i.e. when $\kappa = 0$). In [1,2] there are no multiplicity results such as those by Ekeland and Lasry dealing with smooth convex Hamiltonian Systems ([3], see also [5]), and theorems 1 and 6 are a first contribution in this direction.

It is worth noticing that we are able to cover the relativistic case, only (i.e. $\kappa > 0$). The proof of the existence of multiple closed trajectories for the motion of a particle under the classical, Newtonian gravitational field would be an interesting question to pursue.

Our arguments are based on a variational principle used in [1] and recalled in section 2 below. In section 3 we discuss suitable estimates, which are prompted for a general multiplicity result (Theorem 6) proved in section 4. Such a Theorem covers a class of potentials V which, roughly, satisfy the same assumptions as in [1] and some additional geometric conditions on the set $\{V \le h\}$. Finally, in section 5 we give the proof of Theorem 1.

Notations. For $x, y \in \mathbf{R}^n$, $|x|$ and $x \cdot y$ or simply xy denote the euclidean norm and the scalar product respectively; B_r stands for $\{x \in \mathbf{R}^n | \le r\}$; $|\cdot|_\infty$ is the L_∞ norm.

2. Let $V \in C^2(\mathbf{R}^n - \{0\}, \mathbf{R})$ be a potential satisfying:

V1) $3V'(x)x + V''(x)x \cdot x > 0 \qquad \forall x \in \mathbf{R}^n - \{0\}$

V2) $V(x) = V(-x) \qquad \forall x \in \mathbf{R}^n - \{0\}$

V3) $V'(x)x > 0 \qquad \forall x \in \mathbf{R}^n - \{0\}$

V4) $\exists \alpha, \beta \in (0,2)$: $\quad -\alpha V(x) \le V'(x)x \le -\beta V(x) \quad \forall x \in \mathbf{R}^n - \{0\}$

For all $h < 0$ set $\Omega = \{x \ne 0 | V(x) \le h\}$ and $\partial\Omega = \{x = 0 | V(x) = h\}$. By V3, Ω is star-shaped with respect to $x = 0$ and hence $\forall x \ne 0$ there exists a unique $\lambda > 0$ such that $\lambda x \in \partial\Omega$. In particular, we let $P: \partial B_1 \to \partial\Omega$ be the map defined by setting $Py = \lambda y \in \partial\Omega$. Define the Minkowski functional of Ω as $j_\Omega: \mathbf{R}^n - \{0\} \to \mathbf{R}$,

$$j(x) = j_\Omega(x) = \frac{1}{\lambda} \iff \lambda x \in \partial\Omega.$$

If $B_r \subset \Omega$ then

$$j(x) \le \frac{|x|}{r} \quad \forall x \ne 0. \tag{1}$$

Furthermore, V4 implies readily

$$\frac{h}{s^\beta} \le V(sx) \le \frac{h}{s^\alpha} \quad \forall 0 < s \le 1, \forall x \in \partial\Omega \tag{2}$$

and hence, setting $\xi = sx$,

$$\frac{h}{j(\xi)^\beta} \le V(\xi) \le \frac{h}{j(\xi)^\alpha} \quad \forall \xi \in \Omega. \tag{3}$$

Let

$$V_\kappa(x) = V(x) - \frac{\kappa}{|x|^2} \quad (\kappa > 0).$$

Given any $h < 0$ we look for periodic solutions of

$$\begin{cases} \ddot{q} + V_\kappa'(q) = 0 \\ \frac{1}{2}|\dot{q}|^2 + V_\kappa(q) = h \end{cases} \tag{4}$$

Let

$$H_0 = \{u \in H^{1,2}(S^1, \mathbf{R}^n) | u(t + \frac{1}{2}) = -u(t) \quad \forall t\}$$

and

$$\Lambda_0 = \{u \in H_0 | u(t) \neq 0 \quad \forall t\}.$$

H_0 is a Hilbert space under the scalar product $(u|v) = \int_0^1 \dot{u} \cdot \dot{v} dt$ and with norm $\|u\|^2 = \int_0^1 |\dot{u}|^2 dt$. Recall also that

$$\|u\| \ge 4|u|_\infty \tag{5}$$

Define

$$M = \{u \in \Lambda_0 | \int_0^1 [V(u) + \frac{1}{2}V'(u)u] dt = h\}.$$

For future references, we note that V4 implies:

$$\frac{h}{1 - \frac{\beta}{2}} \le \int_0^1 V(u) dt \le \frac{h}{1 - \frac{\alpha}{2}} \quad \forall u \in M \tag{6}$$

as well as

$$\frac{2\alpha h}{\alpha - 2} \le \int_0^1 V'(u) u dt \le \frac{2\beta h}{\beta - 2} \quad \forall u \in M \tag{7}$$

Let

$$f_\kappa(u) = \frac{1}{2}\|u\|^2 \cdot \frac{1}{2} \int_0^1 V_\kappa'(u) u dt \qquad (u \in \Lambda_0). \tag{8}$$

Later on we will need some of the results of [1]. We collect them in the following lemma.

Lemma 2. *Let $h < 0$ be given and suppose that V1-V4 hold. Then*
i) $M \neq \emptyset$, is a C^1 manifold in H_0 and is a strong deformation retract of Λ_0;
ii) the sublevels $\{u \in M | f_\kappa(u) \leq const\}$ are complete and f_κ satisfies the Palais-Smale condition on M;
iii) $f_\kappa(u) > 0 \quad \forall u \in M$, in particular $m = \min_M f_\kappa > 0$;
iv) let $u \in M$ a critical point of $f_\kappa|_M$. Let

$$\omega_\kappa^2 = \frac{\int_0^1 V_\kappa'(u)u dt}{\|u\|^2}.$$

Then $q(t) = u(\omega_\kappa t)$ is a solution of (4).

The proof of Lemma 2 is essentially contained in [1]; however, some comments are in order:
(i) is nothing but Lemmas 2.1 and 4.3 of [1]. Note that V4 implies that $V(x) + \frac{1}{2}V'(x) \to 0$ as $|x| \to \infty$.
(ii) is Lemma 4.6 of [1].
(iii) follows from V3 and (5).
(iv) lemma 2.3 of [1] yields:

$$\omega_\kappa^2 \int_0^1 [\dot{u} \cdot \dot{v} + V_\kappa'(u)v]\mathrm{d}t = 0 \quad \forall v \in H_0. \tag{9}$$

Since V is even, (9) holds $\forall v \in H^1(S^1, \mathbf{R}^n)$ and u is a classical solution of

$$-\omega_\kappa^2 \ddot{u} + V_\kappa'(u) = 0.$$

Moreover, since $u \in M$, one has

$$\int_0^1 [V_\kappa(u) + \frac{1}{2}V_\kappa'(u)u]\mathrm{d}t = \int_0^1 [V(u) + \frac{1}{2}V'(u)u]\mathrm{d}t = h$$

and hence

$$\frac{1}{2}\omega_\kappa^2|\dot{u}|^2 + V_\kappa(u) = h$$

and (iv) follows.

3. In this section we derive some estimates which will allow us to show that different critical points give rise to geometrically different trajectories.

Let

$$A(a,b) = (1 - \frac{a}{2})^{1/b}.$$

Lemma 3. *Suppose $B_r \subset \Omega$. Then:*

$$m = \min_M f_\kappa \geq r^2 A^2(\beta, \alpha) \frac{8\alpha h}{\alpha - 2}.$$

Proof. Let $z \in M$ be such that $f_\kappa(z) = m$. From the left hand side of (6) it follows that:

$$\exists t^* \quad \text{such that} \quad V(z(t^*)) \geq \frac{h}{1 - \frac{\beta}{2}}. \tag{10}$$

Moreover, there results:

$$j(z(t^*)) \geq (1 - \frac{\beta}{2})^{1/\alpha} = A(\beta, \alpha). \tag{11}$$

In fact, otherwise, $j(z(t^*)) < (1 - \frac{\beta}{2})^{1/\alpha} < 1$ and $z(t^*) \in \Omega$. Then (3) and (10) would yield a contradiction.
Since $B_r \subset \Omega$ then (11) and (1) imply

$$|z|_\infty \geq |z(t^*)| \geq rj(x^*) \geq rA(\beta, \alpha). \tag{12}$$

From (5), (7), (8) and (12) we infer:

$$f_\kappa(z) = \frac{1}{2}\|z\|^2 \cdot \frac{1}{2}\int_0^1 V_\kappa'(z) z \mathrm{dt} \geq \frac{1}{2}\|z\|^2 \cdot \frac{1}{2}\int_0^1 V'(z) z \mathrm{dt} \geq$$

$$4|z|_\infty^2 \int_0^1 V'(z) z \mathrm{dt} \geq 4r^2 A(\beta, \alpha) \frac{2\alpha h}{\alpha - 2}.$$

Q.E.D

Consider

$$\Sigma = \{v \in \Lambda_0 | v = \xi \cos(t) + \eta \sin(t), |\xi| = |\eta| = 1, \xi \cdot \eta = 0\}$$

and let

$$\Sigma' = \{w(t) = Pv(t), \quad v \in \Sigma\}.$$

For all $w \in \Sigma'$ there exists a unique $\sigma = \sigma(w) > 0$ such that $u := \sigma w \in M$. Let:

$$\tilde{\Sigma} = \{u = \sigma(w)w, \quad w \in \Sigma'\}.$$

Using (6) and (2) (recall that $w(t) \in \partial\Omega \quad \forall t$) it follows readily that

$$0 < A(\beta,\alpha) \leq \sigma \leq A(\alpha,\beta) < 1 \tag{13}$$

Lemma 4. *Suppose $B_r \subset \Omega$ and let $\mu = \max_{|x|=1} |dP_x|$. Then the following estimate holds:*

$$\max_{u \in \tilde{\Sigma}} \leq \pi^2 \mu^2 A^2(\alpha,\beta) \left(\frac{2\beta h}{\beta - 2} + \frac{2\kappa}{r^2 A^2(\beta,\alpha)} \right).$$

Proof. For all $u = \sigma w \in \tilde{\Sigma}$, one has:

$$f_\kappa(\sigma w) = \frac{1}{2}\sigma^2 \|w\|^2 \cdot \frac{1}{2} \int_0^1 [V'(\sigma w)\sigma w + \frac{2\kappa}{\sigma^2 |w|^2}]\mathrm{dt}$$

using (13) it follows:

$$f_\kappa(\sigma w) \leq \frac{1}{4} A^2(\alpha,\beta)\|w\|^2 \int_0^1 [V'(\sigma w)\sigma w + \frac{2\kappa}{A^2(\beta,\alpha)|w|^2}]\mathrm{dt}.$$

Since $w(t) = Pu(t)$, then

$$|\dot{w}(t)| \leq |dP_{u(t)}||\dot{u}(t)| \leq 2\pi\mu,$$

and hence

$$f_\kappa(\sigma w) \leq \pi^2 \mu^2 A^2(\alpha,\beta) \int_0^1 [V'(\sigma w)\sigma w + \frac{2\kappa}{A^2(\beta,\alpha)|w|^2}]\mathrm{dt}.$$

Finally, from (7) and $B_r \in \Omega$ it follows:

$$f_\kappa(\sigma w) \leq \pi^2 \mu^2 A^2(\alpha,\beta) \left(\frac{2\beta h}{\beta - 2} + \frac{2\kappa}{r^2 A^2(\beta,\alpha)} \right)$$

proving the lemma. Q.E.D

4. To find n trajectories of (4), we use the S^1 index introduced in [4]; namely for any closed, S^1-invariant set G and for any $n \in \mathbf{N}$ we let $\tau_n(G)$

be the smallest k such that it exists a continuous $\phi: G \to \mathbf{C}^k - \{0\}$ such that

$$\phi(u(\cdot + \theta)) = e^{in\theta}\phi(u) \quad \forall u \in A, \theta \in S^1.$$

We define : $\gamma(G) = \inf_{n>0} \tau_n(G)$.

Notice that M does not contain any fixed points of the S^1 action on H_0 and recall that $\gamma(\tilde{\Sigma}) = \gamma(\Sigma) \geq n$ (cfr [4], Lemma 2.7′ and Proposition 2.2). Moreover, Lemma 2 yields that any

$$c_n = \inf_{\gamma(A)\geq n} \max_A f_\kappa$$

carries a critical point of f_κ on M which gives rise to a solution of (4).

Lemma 5. *If $u \in M$ is a critical point of f_κ such that $f_\kappa(u) < 9m$, then the minimal period of u is 1.*

Proof. Recall that any $u \in H_0$ has the form:

$$u = \sum_{k \text{ odd}} u_k e^{2\pi i k t}.$$

If the minimal period of u is smaller tan 1, then

$$u = \sum_{k \text{ odd}} u_k e^{2\pi i l k t}$$

for some $l > 3$. Set $\tilde{u} = u(\frac{t}{l}) \in H_0$. There results $\|u\|^2 = l^2\|\tilde{u}\|^2$, while $\int_0^1 V'(u)u\mathrm{dt} = \int_0^1 V'(\tilde{u})\tilde{u}\mathrm{dt}$. Hence

$$f_\kappa(\tilde{u}) = \frac{1}{2}\|\tilde{u}\|^2 \cdot \frac{1}{2}\int_0^1 V'(\tilde{u})\tilde{u}\mathrm{dt} = \frac{1}{2l^2}\|u\|^2 \cdot \frac{1}{2}\int_0^1 V'(u)u\mathrm{dt} \leq \frac{1}{9}f_\kappa(u) < m,$$

a contradiction. Q.E.D

We are now in position to prove our main general result:

Theorem 6. *Suppose V satisfies V1-V5 and $B_r \in \Omega$. If*

$$C_1 := \pi^2\mu^2 A^2(\alpha,\beta)\left(\frac{2\beta h}{\beta - 2} + \frac{2\kappa}{r^2 A^2(\beta,\alpha)}\right) <$$

$$9r^2 A^2(\beta,\alpha)\frac{8\alpha h}{\alpha - 2} := 9C_2 \tag{14}$$

then (4) has at least n distinct periodic trajectories.

Proof. Since $\gamma(\tilde{\Sigma}) \geq n$, then $c_n \leq \max_{\tilde{\Sigma}} f_\kappa(u)$. From lemmas 3 and 4 it follows

$$c_n \leq C_1 < 9C_2 \leq 9m.$$

From Critical Point Theory it follows that $f_\kappa|_M$ has at least n critical points $u_1, \dots u_n$ with $f_\kappa(u_i) \leq c_n$. From Lemma 5 they give rise to n distinct trajectories of (4). Q.E.D

5. Here we derive Theorem 1 from Theorem 6. Setting

$$\phi(\gamma, x) = (1 + \frac{4}{3}h\kappa)(-\frac{1}{|x|} + \gamma W) - \gamma^2 \kappa W^2 + 2\gamma\kappa\frac{1}{|x|}W$$

there results

$$U_\kappa = \phi(\gamma, x) - \frac{\kappa}{|x|^2}.$$

Denote by D_γ the connected component of $\{\phi(\gamma, x) - \frac{\kappa}{|x|^2} \leq h\}$ such that $0 \in \bar{D}_\gamma$, and

$$D_0 = \{-\frac{1 + \frac{4}{3}h\kappa}{|x|} - \frac{\kappa}{|x|^2} \leq h\}.$$

Then fixed R large enough, $\exists \gamma' > 0$ such that $D_\gamma \subset B_R \quad \forall |\gamma| < \gamma'$. Let $\eta: \mathbf{R}^+ \to \mathbf{R}$ be smooth and

$$\begin{cases} \eta(s) = 1 & s \in [0, R] \\ \eta(s) = 0 & s \in [2R, +\infty] \end{cases}$$

and set $\tilde{W}(x) = \eta(|x|)W(x)$. We also set:

$$V(\gamma, x) = (1 + \frac{4}{3}h\kappa)(-\frac{1}{|x|} + \gamma\tilde{W}) - \gamma^2\kappa\tilde{W}^2 + 2\gamma\kappa\frac{1}{|x|}\tilde{W}.$$

Note that V depends smoothly on γ, $V(0, x) = \phi(0, x) = -(1 + 4h\kappa)\frac{1}{|x|}$ and $V(0, x)$ satisfies V1,V3 and V4 with $\alpha = \beta = 1$. Therefore $\exists \gamma''(\leq \gamma')$ such that $\forall |\gamma| < \gamma''$, $V(\gamma, x)$ satisfies V1-3 as well as V4 with some $\alpha = \alpha(\gamma)$, $\beta = \beta(\gamma)$ such that

$$\alpha(\gamma), \beta(\gamma) \to 1 \quad \text{as} \quad \gamma \to 0. \tag{15}$$

Moreover, since W is even, then $V(\gamma, x)$ satisfies V2, too and Theorem 6 applies with constants C_1 and C_2 depending on γ through α and β. In addition, r and μ depend on γ, too; since for $\gamma = 0$ the set $\Omega_0 = \{V(0,x) \leq h\}$ is nothing but the ball $B_{h'}$ with $h' = -\frac{1+4h\kappa}{h}$, then $r \to h'$ as $\gamma \to 0$. Similarly one has that $\mu \to h'$ as $\gamma \to 0$. This and (15) imply:

$$C_1(\gamma) \to \frac{1}{4}\pi^2\mu^2 \left(-2h + \frac{8\kappa}{h'^2}\right) \quad \text{as} \quad \gamma \to 0$$

$$C_2(\gamma) \to -2hh'^2 \quad \text{as} \quad \gamma \to 0.$$

It follows that $\exists h^* < 0$ such that $\forall h \in (h^*, 0)$ one has $C_1(0) < 9C_2(0)$. Therefore $\exists 0 < \gamma^*(\leq \gamma'')$ such that

$$C_1(\gamma) < 9C_2(\gamma) \quad \forall |\gamma| \leq \gamma^*, \forall h \geq h^*.$$

This allows us to apply Theorem 6 which yields the existence of n prime closed orbits of energy h for the potential $V(\gamma, x) - \kappa|x|^{-2}$. Obviously for any such orbit q one has $V(\gamma, q) - \kappa|q|^{-2} \leq h$ and hence $q(t) \in D_\gamma \subset B_R$. Therefore $\tilde{W}(q) = W(q)$, q is a trajectory of energy h for the potential U_κ and Theorem 1 follows.

BIBLIOGRAPHY

[1] A. Ambrosetti, V. Coti Zelati, "Closed orbits of fixed energy for singular hamiltonian systems", Archive Rat. Mech. and Anal., to appear.

[2] A. Ambrosetti, V. Coti Zelati, "Closed orbits of fixed energy for a class of N-body problems", preprint S.N.S., April 1990

[3] A. Ambrosetti, G. Mancini, "On a theorem by Ekeland and Lasry concerning the number of periodic hamiltonian trajectories", J.Diff. Eq. **43**, No 2 (1982) 249-256.

[4] V.Benci "A geometrical index for the group S^1 and some applications to the study of periodic solutions of ordinary differential equations" Comm. Pure and Applied Math., **34** (1981) 393-423

[5] I. Ekeland, J.M.Lasry, "On the number of periodic trajectories for a Hamiltonian flow on a convex energy surface", Ann. of Math. **112** (1980), 283-319.

[6] T. Levi Civita, "Fondamenti di meccanica relativistica", Bologna, 1928.

Scuola Normale Superiore,
Piazza Cavalieri 7
56100 Pisa, Italy

Periodic Solutions of Some Problems of 3-Body Type

Abbas Bahri * and Paul H. Rabinowitz **

§1. Introduction

The study of time periodic solutions of the n-body problem is a classical one. See e.g [1]. The purpose of this paper is to sketch some of our recent research on the existence of time periodic solutions of Hamiltonian systems of 3-body type [2]. This work presents a new direct variational approach to the problem.

To describe it more fully, consider the Hamiltonian system of ordinary differential equations

$$(HS). \qquad m_i\ddot{q}_i + V_{q_i}(t,q) = 0, \quad 1 \leq i \leq 3$$

In (HS), $q_i \in \mathbf{R}^\ell$, $1 \leq i \leq 3$, $\ell \geq 3$, $m_i > 0$, $q = (q_1, q_2, q_3)$ and V is defined in $\mathbf{R} \times \mathbf{F_3}(\mathbf{R}^\ell)$ where $F_3(\mathbf{R}^\ell)$ is the configuration space

$$(1.1) \qquad F_3(\mathbf{R}^\ell) = \{(q_1, q_2, q_3) \in (\mathbf{R}^\ell)^3 \mid q_i \neq q_j \quad \text{if} \quad i \neq j\}.$$

Furthermore V is T-periodic in t.

We are interested in T-periodic solutions of (HS). It is assumed that V is an interaction potential:

$$(1.2) \qquad V = \sum_{\substack{i,j=1 \\ i\neq j}}^{3} V_{ij}(t, q_i - q_j)$$

* This research was supported in part by the National Science Foundation under grant NSF# DMS88# 03494 and by a grant from the Sloan Foundation.

** This research was supported in part by the National Science Foundation under grant #MCS-8110556, the US Army Research Office under Contract # DAA L03-87-K-0043, and the Office of Naval Research under Grant No. N00014-88-K-0134. Any reproduction for the purposes of the United States Government is permitted.

Each function V_{ij}, $1 \le i \ne j \le 3$, satisfies

(V_1) $V_{ij} \in C^2(\mathbf{R} \times (\mathbf{R}^\ell \backslash \{0\}, \mathbf{R})$ and is T-periodic in t,
(V_2) $V_{ij}(t,q) < 0$ for all $t \in [0,T]$, $q \in \mathbf{R}^\ell \backslash \{0\}$,
(V_3) $V_{ij}(t,q), \frac{\partial V_{ij}}{\partial q_k}(t,q) \to 0$ as $|q| \to \infty$ uniformly in t, $1 \le k \le 3$,
(V_4) $V_{ij}(t,q) \to -\infty$ as $q \to 0$, uniformly in t,
(V_5) for all $M > 0$, there is an $R > 0$ such that

$$\frac{\partial V_{ij}}{\partial q} \cdot q > M \left| \frac{\partial V_{ij}}{\partial q} \right|$$

whenever $|q| > R$,
(V_6) there is a neighborhood, W, of 0 in $\mathbf{R}^\ell$ and $U_{ij} \in C^1(W \backslash \{0\}, \mathbf{R})$ such that $U_{ij}(q) \to \infty$ as $q \to 0$ and $-V_{ij}(q) \ge |U'_{ij}(q)|^2$ for $q \in W \backslash \{0\}$.

Conditions (V_1) – (V_5) are satisfied in particular by potentials of the form

$$V(q) = -\sum_{\substack{i,j=1 \\ i \ne j}}^{3} \frac{\alpha_{ij}}{|q_i - q_j|^{\beta_{ij}}} \tag{1.3}$$

where α_{ij} and β_{ij} are positive constants. Hypothesis (V_6) is also satisfied by V in (1.3) if $\beta_{ij} \ge 2$ for each i, j. The classical 3-body problem corresponds to the case in which $\beta_{ij} = 1$, for all $i \ne j$ and $\alpha_{ij} = \alpha_{ji}$. The significance of ($V_6$) will be discussed below.

To formulate (HS) as a variational problem, let $E = W_T^{1,2}(\mathbf{R}, (\mathbf{R}^\ell)^3)$, the Hilbert space of T-periodic functions from $\mathbf{R}$ into $(\mathbf{R}^\ell)^3$ with norm:

$$\|q\| = \left(\int_0^T |\dot{q}|^2 dt + \|[q]\|^2 \right)^{1/2} \tag{1.4}$$

where

$$[q] = \frac{1}{T} \int_0^T q(s) ds. \tag{1.5}$$

The functional associated with (HS) is

$$I(q) = \int_0^T \left(\frac{1}{2} \sum_{i=1}^{3} m_i |\dot{q}_i|^2 - V(t,q) \right) dt. \tag{1.6}$$

Set

$$(1.7) \qquad \Lambda = \{q \in E \mid q(t) \in F_3(\mathbf{R}^\ell) \text{ for all } t \in [0,T]\}.$$

It is not difficult to prove that

Lemma 1.8: *If* V *satisfies* $(V_1), (V_2), (V_4)$, *and* (V_6), *then for each* $c > 0$, *there is a* $\delta(c) > 0$ *such that if* $I(q) \le c$, *then*

$$\inf_{\substack{t \in [0,T] \\ i \neq j}} |q_i(t) - q_j(t)| \ge \delta(c).$$

See e.g. [2] or [3]. An immediate consequence of Lemma 1.8 is that the variational problem can be posed on Λ rather than E. Moreover it is easy to verify that if $q \in \Lambda$ and $I'(q) = 0$, then q is a classical T-periodic solution of (HS).

Our main result is:

Theorem 1.9. *If* V *satisfies* $(V_1) - (V_6)$, *then* I *possesses an unbounded sequence of critical values.*

If condition (V_6) is dropped, it is possible that $q \in E$ with $I(q) < \infty$ but $q_i(t) = q_j(t)$ for some $i \neq j$ and $t \in [0,T]$, i.e. a *collision* occurs at time t. Thus without (V_6) it is possible that collisions can occur for periodic solutions of (HS). Since a collision orbit cannot be a classical solution of (HS), following [4], we say $q \in E$ is a *generalized* T*-periodic solution* of (HS) if

(1.10)

(i) $\mathcal{D} = \{t \in [0,T] \mid q(t) \notin F_3(\mathbf{R}^\ell)\}$ has measure 0.

(ii) $q \in C^2$ and satisfies (HS) in $[0,T] \backslash \mathcal{D}$.

(iii) $-\int_0^T V(t, q(t))dt < \infty$.

(iv) Assume that there exists a constant $C > 0$ such that $|V_t' \cdot V^{-1}| \le C$ uniformly in t, in a deleted neighborhood of the origin in $\mathbf{R}^\ell$. We then require that $\frac{1}{2}\sum_{i=1}^3 |\dot{q}_i(t)|^2 + V(t, q(t))$ can be extended as a continuous function on $t \in [0,T]$, i.e. energy is conserved through the collisions.

Remark 1.11. Conditions (1.10) (i)-(iv) are not mutually exclusive but we prefer to define generalized T-periodic solution in this way since it is these conditions that one verifies in applications.

Given Theorem 1.9, using an approximation argument from [4], it is not difficult to show:

Theorem 1.12. *If V satisfies $(V_1)-(V_5)$, then (HS) possesses a generalized T-periodic solution.*

Corollary 1.13. *If in addition, V is independent of t and $V'(q) \neq 0$ for all $q \in (\mathbf{R}^\ell)^3$, (HS) has infinitely many distinct T-periodic solutions.*

In the next two sections, we will discuss some of the preliminaries that go into the proof of Theorem 1.9. Then in §4, the proof of Theorem 1.9 itself will be sketched. Finally in §5, a few remarks will be made about the proofs of Theorem 1.12 and Corollary 1.13.

§2. The breakdown of (PS) and a Morse Lemma for neighborhoods of infinity.

A standard condition used in the study of variational problems is the Palais-Smale condition or (PS) for short. It says any sequence (q^k) satisfying

$$(2.1) \qquad I(q^k) \text{ is bounded and } I'(q^k) \to 0$$

is precompact. Unfortunately (PS) does not hold for (1.6). However the behavior of (PS) sequences can be characterized precisely.

Proposition 2.2. *Suppose V satisfies $(V_1)-(V_4)$ and (V_6). Let (q^k) satisfy (2.1). Then the following alternative holds: Either*

(i) *there exists a subsequence, still denoted by (q^k), and a sequence $(v_k) \subset \mathbf{R}^\ell$ such that $(q_i^k - v_k)$ converges in $W_T^{1,2}(\mathbf{R}, \mathbf{R}^\ell)$ for $i = 1, 2, 3$, or*
(ii) *there exists a subsequence, still denoted by (q^k), a sequence $(v_k) \subset \mathbf{R}^\ell$, and $i \in \{1, 2, 3\}$ such that*

   a. *$\|[q_i^k - v_k]\| \to \infty$ and $\|\dot{q}_i^k\|_{L^2} \to 0$ as $k \to \infty$, and*
   b. *if $j \neq r \in \{1,2,3\} \backslash \{i\}$, $(q_j^k - v_k, q_r^k - v_k)$ converges in $W_T^{1,2}(\mathbf{R}, \mathbf{R}^\ell)^2)$ to a classical solution of the two-body problem associated with the potential $V_{ji} + V_{ij}$.*

Moreover if

$$(2.3) \quad I_{jr}(q_j, q_r) \equiv \int_0^T \left(\frac{1}{2}(m_j|\dot{q}_j|^2 + m_r|\dot{q}_r|^2) - V_{jr}(t, q_j - q_r) - V_{rj}(t, q_r - q_j) \right) dt,$$

then $I_{jr}(q_j^k, q_r^k) \to c$.

Proposition 2.2 tells us that if a (PS) sequence is not precompact in the usual sense, it has a subsequence which converges to a "two-body solution at infinity". We further note that one can take $v^k = \frac{1}{2}[q_j^k + q_r^k]$.

One major new idea in our work is to use a "Morse Lemma" in a neighborhood of a sequence of type (ii) in Proposition 2.2.

Proposition 2.4. *Let V satisfy $(V_1) - (V_5)$. Then*

(1) *for all $C > 0$, there exists an $\alpha(C) > 0$ such that if $q = (q_1, q_2, q_3) \in \Lambda$ satisfies*

$$(i) \quad \sum_{i=1}^{2} \|q_i - v(q)\|_{L^\infty} \le C,$$

and

$$(ii) \quad \frac{1}{2} m_3 \|\dot{q}_3\|_{L^2}^2 + \frac{1}{1 + \|[q_3 - v(q)]\|^2} \le \alpha(C)$$

for some $v = v(q) \in \mathbf{R}^\ell$, then there is a unique $\lambda(q) > 0$, continuously differentiable in q, and satisfying

$$I(q) = I_{12}(q_1, q_2) + \frac{1}{2} \int_0^T m_3 |\dot{Q}_3|^2 dt + \frac{1}{1 + \|[Q_3 - \frac{1}{2}(q_1 + q_2)]\|^2}$$

where

$$Q_3 = \frac{1}{2}[q_1 + q_2] + \frac{1}{\lambda(q)}(q_3 - [q_3]) + \lambda(q)[q_3 - \frac{1}{2}(q_1 + q_2)].$$

(2) *Conversely for all $C > 0$, there exists $\overline{\alpha}(C) > 0$ such that if $(q_1, q_2, Q_3) \in \Lambda$ satisfies*

$$(iii) \quad \sum_{i=1}^{2} \|q_i - v(q)\|_{L^\infty} \le C,$$

and

$$(iv) \quad \frac{1}{2} m_3 \|\dot{Q}_3\|_{L^2}^2 + \frac{1}{1 + \|[Q_3 - \frac{1}{2}(q_1 + q_2)]\|^2} \le \overline{\alpha}(C)$$

for some $v = v(q) \in \mathbf{R}^\ell$, then there is a unique $\mu(q_1, q_2, Q_3) > 0$, continuously differentiable in its argument, and satisfying

$$I(q_1, q_2, q_3) = I_{12}(q_1, q_2) + \frac{1}{2} \int_0^T m_3 |\dot{Q}_3|^2 dt + \frac{1}{1 + \|[Q_3 - \frac{1}{2}(q_1 + q_2)]\|^2}$$

where

$$q_3 = \frac{1}{2}[q_1 + q_2] + \mu(q_1, q_2, Q_3)(Q_3 - [Q_3]) + \frac{1}{\mu(q_1, q_2, Q_3)}[Q_3 - \frac{1}{2}(q_1 + q_2)].$$

(3) *If* $\alpha(C) = \overline{\alpha}(C)$ *is sufficiently small, then* $\lambda(q_1, q_2, q_3)\mu(q_1, q_2, Q_3) = 1$ *and the transformations defined in 1 and 2 are inverse diffeomorphisms.*

In Proposition 2.4, we could replace $\frac{1}{2}[q_1+q_2]$ by any convex combination of $[q_1]$ and $[q_2]$. In particular we could have taken the center of mass $\frac{m_1[q_1]+m_2[q_2]}{m_1+m_2}$. If we do so, the representation provided by Proposition 2.4 has the physical interpretation that the interaction of the motion of the body q_3 with the two other bodies can be replaced by the motion of a new body Q_3 which interacts (at the level of mean values) only with the center of mass of the other bodies. Proposition 2.4 allows us to represent I in a simple fashion in a neighborhood of a sequence violating the (PS) condition, i.e. near a "critical point at infinity". In this sense we have a Morse Lemma for neighborhoods of critical points at infinity.

One final technical result will be given in this section. Let

$$I^c = \{q \in \Lambda \mid I(q) \leq c\}. \tag{2.5}$$

Proposition 2.6. *Let* V *satisfy* $(V_1) - (V_5)$. *Then there is an* $\epsilon_0 > 0$ *such that for all* $\epsilon \in (0, \epsilon_0)$, I^ϵ *is homotopy equivalent to an ENR (Euclidean neighborhood retract),* $X \subset \mathbf{R}^\ell$. *In particular the singular homology of* I^ϵ *(with rational coefficients) vanishes in all dimensions* $\geq \ell$.

§3. An abstract theorem in Morse Theory.

The proof of Theorem 1.9 involves in part the construction of a certain deformation retraction. In this section, a finite dimensional version of this result will be stated. In the next section, the extensions needed for the proof of Theorem 1.9 will be discussed. More details can be found in [2] and [5].

Let $\mathcal{M}$ be a compact n-dimensional Riemannian manifold and let $f \in C^2(\mathcal{M}, \mathbf{R})$. Assume all of the critical points of f are nondegenerate. Let Z denote a pseudogradient vector field for f, i.e. Z is a locally Lipschitz continuous vector field on $\mathcal{M}$ and

$$\langle f'(x), Z(x)\rangle \geq |f'(x)|^2 \tag{3.1}$$

and

$$|Z(x)| \leq \gamma |f'(x)| \tag{3.2}$$

for all $x \in \mathcal{M}$ where $\gamma > 0$ is a constant. Further assume the critical points of f are nondegenerate zeroes of Z. Consider the ordinary differential equation

$$\frac{dx}{ds} = -Z(x), \qquad x(0,y) = y. \tag{3.3}$$

Let $\varphi(s,y)$ denote the solution of (3.3). Suppose $Z(x_0) = 0$. Set

$$W_u(x_0) = \{x \in \mathcal{M} \mid \varphi(s,x) \to x_0 \quad \text{as} \quad s \to -\infty\}, \tag{3.4}$$

i.e. $W_u(x_0)$ is the unstable manifold for the flow (3.3) which emanates from x_0. Let $a < b$ be noncritical values of f and

$$f^c = \{x \in \mathcal{M} \mid f(x) \le c\}.$$

Let

$$C_a^b = \{x \in \mathcal{M} \mid f'(x) = 0 \quad \text{and} \quad a \le f(x) \le b\}.$$

Then we have

Theorem 3.5. *Let Z_1 be a pseudogradient vector field for f. Then in any C^1 neighborhood of Z_1, there exists a pseudogradient vector field Z for f such that f^b retracts by deformation onto*

$$f^a \cup \left(\bigcup_{x \in C_a^b} W_u(x) \right).$$

If C_a^b is a single point, then Theorem 3.5 is a classical deformation result. See e.g. [6, p.156-160]. There are many extensions of Theorem 3.5, especially in an infinite dimensional space. The theorem can also be formulated in different ways. We refer to Bahri [5] for such extensions, alternate formulations, and applications. We note that Theorem 3.5 can not be generalized as such to situations where (PS) fails. In the next section we will discuss how to extend Theorem 3.5 so as to apply to the functional (1.6).

§4. The construction of a pseudogradient vector field for I and a sketch of the proof of Theorem 1.9.

Let C_1 be a constant and $\beta \in C(\mathbf{R}^+, \mathbf{R}^+)$. Let $\mathcal{V}_3$ denote the set of $(q_1, q_2, q_3) \in \Lambda$ satisfying

$$(4.1) \qquad (i) \qquad \sum_{i=1}^{2} \|q_i - \frac{1}{2}[q_i + q_2]\|_{L^\infty} \leq C_1$$

and

$$(ii) \quad \frac{1}{2} m_3 \|\dot{q}_3\|_{L^2}^2 + \frac{1}{1 + \|[q_3 - \frac{1}{2}(q_1 + q_2)]\|^2} v \leq \beta(C_1).$$

The sets $\mathcal{V}_1, \mathcal{V}_2$ are defined in a similar way via a permutation of indices. If β is sufficiently small, Proposition 2.4 is valid on $\mathcal{V}_3$ (resp. $\mathcal{V}_1, \mathcal{V}_2$) and one can therefore use, as in the classical situation in Morse Theory [6], the new coordinates (q_1, q_2, Q_3) to define a pseudogradient vector field $\tilde{Z}$ for I on $\mathcal{V}_3$. Then $\tilde{Z}$, similarly defined on $\mathcal{V}_1, \mathcal{V}_2$ can be extended to Λ by taking convex linear combinations of I' and $\tilde{Z}$.

We will give an idea for the construction of $\tilde{Z}$ on $\mathcal{V}_3$. A detailed proof can be found in [2]. Suppose we have a pseudogradient vector field Z_{12} for I_{12}. To simplify our presentation, assume that the critical points of I_{12} and I are nondegenerate. (This, of course, is not the case due to the translational symmetry possessed by I_{ij} and I.) Let $(\overline{q}_1, \overline{q}_2)$ be a critical point of I_{12} and therefore a zero of Z_{12}. Let $W_u(\overline{q}_1, \overline{q}_2)$ be the unstable manifold associated with $(\overline{q}_1, \overline{q}_2)$ for the differential equation

$$(4.2) \qquad \frac{d}{ds}(q_1, q_2) = -Z_{12}(q_1, q_2).$$

Now we define $\tilde{Z}$ in the coordinates $(q_1, q_2, Q_3 - [Q_3], [Q_3 - \frac{1}{2}(q_1 + q_2])$ via

$$(4.3) \qquad \frac{dq}{ds} = -\tilde{Z}(q)$$

if and only if

$$(4.4) \qquad \begin{cases} \frac{d}{ds}(q_1, q_2) = -Z_{12}(q_1, q_2) \\ \frac{d}{ds}(Q_3 - [Q_3]) = -(Q_3 - [Q_3]) \\ \frac{d}{ds}[Q_3 - \frac{1}{2}(q_1 + q_2)] = 0. \end{cases}$$

Using Proposition 2.4, it is easy to verify that $\tilde{Z}$ is a pseudogradient vector field for I on $\mathcal{V}_3$. The solution $\varphi(s, q)$ of (4.3) compactifies the "critical points at infinity" in the sense of [7], i.e. the decreasing (with respect to

I as $s \to +\infty$) orbits of the gradient flow that are not compact. In doing so, we introduce new equilibrium points for $\tilde{Z}$ which are distinct from the critical points. This prevents Theorem 3.5 from being extended directly to (4.3). It is necessary to take into account the "unstable manifolds of critical points at infinity". By doing so, one can prove a version of Theorem 3.5 for the current situation. Let

$$\mathcal{C}_a^b = \{q \in \Lambda \mid I'(q) = 0 \quad \text{and} \quad a \le I(q) \le b\},$$

$$\mathcal{C}_a^b(i,j) = \{(\overline{q}_i, \overline{q}_j) \mid I_{ij}(\overline{q}_i, \overline{q}_j) = 0 \quad \text{and} \quad a \le I_{ij}(\overline{q}_i, \overline{q}_j) \le b\},$$

$$\mathcal{D}_a^b = \bigcup_{q \in \mathcal{C}_a^b} W_u(q), \quad \text{and} \quad \mathcal{D}_a^b(\infty) = \bigcup_{i \ne j} \bigcup_{(\overline{q}_i, \overline{q}_j) \in \mathcal{C}_a^b(i,j)} W_u^\infty(\overline{q}_i, \overline{q}_j).$$

Here $W_u^\infty(\overline{q}_i, \overline{q}_j)$ denotes the unstable manifolds of critical points at infinity.

Theorem 4.5. *Let $a < b$ be noncritical values of I. Then I^b retracts by deformation on*

$$I^a \cup \mathcal{D}_a^b \cup \mathcal{D}_a^b(\infty)$$

and $W_u^\infty(\overline{q}_i, \overline{q}_j)$ is a trivializable fiber over $W_u(\overline{q}_i, \overline{q}_j)$, the fiber having the homotopy type of a sphere $S^{\ell-1}$.

With the aid of Theorem 4.5, the proof of Theorem 1.9 can now be sketched. To simplify matters, assume that I has no critical points. Then, for any $b > a = \epsilon > 0$, $\mathcal{C}_\epsilon^b = \phi$ so by Theorem 4.5,

(4.6) $$I^b \simeq I^\epsilon \cup \mathcal{D}_\epsilon^b(\infty)$$

where $\simeq$ denotes retraction by deformation. The proof continues via three steps.

Step 1. Since $\Lambda = \bigcup_{b \in \mathbf{R}^+} I^b$ and (4.6) holds for all $b > \epsilon$, it can be shown that

(4.7) $$\Lambda \sim I^\epsilon \cup \mathcal{D}_\epsilon^\infty(\infty)$$

where $\sim$ denotes homotopy equivalence.

Step 2. Let

$$\mathcal{B}_{ij} = \bigcup_{(\overline{q}_i, \overline{q}_j) \in \mathcal{C}_\epsilon^\infty(i,j)} W_u(\overline{q}_i, \overline{q}_j);$$

$$\mathcal{B}_{ij}^\infty = \bigcup_{(\overline{q}_i, \overline{q}_j) \in \mathcal{C}_\epsilon^\infty(i,j)} W_u^\infty(\overline{q}_i, \overline{q}_j)$$

and

$$\mathcal{B}^\infty = \bigcup_{i \neq j} \mathcal{B}_{ij}^\infty .$$

An improved version of Theorem 4.5 [2] says that $\mathcal{B}_{ij}^\infty$ fibers over $\mathcal{B}_{ij}$, the fiber being trivializable and having the homotopy type of a sphere $S^{\ell-1}$. Set

(4.8) $\Lambda_{ij} = \{(q_i, q_j) \in W_T^{1,2}(\mathbf{R}, (\mathbf{R}^\ell)^2) \mid q_i(t) \neq q_j(t) \quad \text{for all} \quad t \in [0,T]\}.$

It is not difficult to check that I_{ij} satisfies (PS) on Λ_{ij} up to translations, i.e. if I_{ij} is bounded and $I'_{ij} \to 0$ along the sequence (q_i^m, q_j^m), then there is a sequence $(v_m) \subset \mathbf{R}^\ell$ such that $(q_i^m - v_m, q_j^m - v_m)$ possesses a convergent subsequence. Therefore an infinite dimensional version of Theorem 3.5 [4], and an argument related to (4.7), yields

$$\Lambda_{ij} \sim I_{ij}^\epsilon \cup \mathcal{B}_{ij}. \tag{4.9}$$

Step 3. By Proposition 2.6, the rational homology of I^ϵ vanishes in dimension $\geq \ell$. Applying the Mayer-Vietoris sequence to the excisive triad $(\Lambda, I^\epsilon, \mathcal{B}^\infty)$ shows that

$$H_k(\Lambda) = H_k(\mathcal{B}^\infty) \quad \text{for} \quad k \geq \ell. \tag{4.10}$$

Similarly

$$H_k(\Lambda_{ij}) = H_k(\mathcal{B}_{ij}) \quad \text{for} \quad k \geq \ell. \tag{4.11}$$

Moreover, from the fibration of $\mathcal{B}_{ij}^\infty$ over $\mathcal{B}_{ij}$, one deduces that

$$H_k(\mathcal{B}_{ij}^\infty) = H_k(\mathcal{B}_{ij}) \oplus H_{k-\ell+1}(\mathcal{B}_{ij}) \quad \text{for} \quad k \geq \ell. \tag{4.12}$$

Combining (4.10)-(4.12) yields:

$$H_k(\Lambda) = \bigoplus_{i \neq j} H_k(\Lambda_{ij}) \oplus H_{k-\ell+1}(\Lambda_{ij}) \quad \text{for} \quad k \geq \ell. \tag{4.13}$$

Let α_k be the dimension of $H_k(\Lambda)$ and β_k the dimension of $H_k(\Lambda_{ij})$. Then by (4.13)

$$\alpha_k = 3(\beta_k + \beta_{k-\ell+1}) \quad \text{for} \quad k \geq \ell. \tag{4.14}$$

However Λ_{ij} has the homotopy type of the free loop space on $S^{\ell-1}$ — see [2] — and therefore β_k is bounded independently of k [8]. On the other hand, by a Theorem of Sullivan and Vigué-Poirrier [8], the sequence (α_k) is unbounded. This contradiction shows that I has at least one positive critical value.

A more complicated variant of this argument given in [2] which takes $\mathcal{D}_a^\infty$ into account proves that I, in fact, has an unbounded sequence of critical values.

§5. The proof of Theorem 1.12 and Corollary 1.13.

We will give a brief sketch of the ideas involved in getting Theorem 1.12 from Theorem 1.9. First for all $\delta > 0$, the potentials V_{ij} are approximated by V_{ij}^δ which satisfy $(V_1) - (V_6)$, $V_{ij}^\delta(t,x) = V_{ij}(t,x)$ if $|x| \geq \delta$, and $-V_{ij}^\delta(t,x) \geq -V_{ij}(t,x)$ if $|x| < \delta$. Then Theorem 1.9 applies to the functional

$$I_\delta(q) = \int_0^T \left(\frac{1}{2} \sum_{i=1}^3 m_i |\dot{q}_i|^2 - V^\delta(t,q) \right) dt. \tag{5.1}$$

Next it is shown that there are constants M and ϵ_1 which are independent of δ such that I_δ has a critical value c_δ in $I_\delta^M \backslash I_\delta^{\epsilon_1}$. Thus

$$\epsilon_1 \leq c_\delta \leq M \tag{5.2}$$

independently of δ. Let q^δ be a critical point of I_δ corresponding to c_δ. The bounds (5.2) and the properties of V_δ lead to upper bounds depending only on ϵ_1 and M for

$$\sum_{i=1}^3 \|q_i^\delta - \frac{1}{2}[q_1^\delta + q_2^\delta]\|_{W^{1,2}} \tag{5.3}$$

and for

$$-\int_0^T V_\delta(q_\delta)dt. \tag{5.4}$$

These bounds enable us to let $\delta \to 0$ and find a subsequence of (q^δ) converging to a generalized T-periodic solution of (HS) thereby giving Theorem 1.12.

To prove Corollary 1.13, we use a standard argument. By Theorem 1.12, we have a generalized T-periodic solution q^1. By the assumption that $V'(q) \neq 0$ for $q \in (\mathbf{R}^\ell)^3$, $q^1 \not\equiv$ const.. Let T/k_1 denote its minimal period. Applying Theorem 1.12 again with T replaced by $T/(1+k_1)$, there exists a $T/(1+k_1)$ periodic solution q^2 having a minimal period $\leq \frac{T}{1+k_1}$. Clearly q^2 is geometrically distinct from q^1. Repeating this argument generates a sequence of geometrically distinct generalized T-periodic solutions of (HS).

REFERENCES

[1] Poincaré, H., *Les méthodes nouvelles de la mécanique céleste*, Lib. Albert Blanchard, Paris, 1987.

[2] Bahri, A. and P. H. Rabinowitz, *Periodic solutions of Hamiltonian systems of 3-body type*, to appear, Analyse Nonlinéaire.

[3] Greco, C., *Periodic solutions of a class of singular Hamiltonian systems*, Nonlinear Analysis, T.M.A., **12**, (1988), 259-270.

[4] Bahri, A. and P. H. Rabinowitz, *A minimax method for a class of Hamiltonian systems with singular potentials*, J. Functional Anal., **82**, (1989), 412-428.

[5] Bahri, A., work in preparation.

[6] Hirsch, M. W., *Differential Topology*, Springer-Verlag 1975.

[7] Bahri, A., *Critical points at infinity in some variational problems*, to appear, Pitman Research Notes in Mathematics.

[8] Sullivan, D. and M. Vigué-Poirrier, The homology theory of the closed geodesic problem, J. Diff. Geom. **11**, (1976), 633-644.

Abbas Bahri
Mathematics Department
Rutgers University
New Brunswick, New Jersey

Paul H. Rabinowitz
Mathematics Department and
Center for Mathematical Sciences
University of Wisconsin
Madison, WI 53706

Periodic Solutions of Dissipative Dynamical Systems

VIERI BENCI and MARCO DEGIOVANNI

1. Introduction

Let M be a compact Riemannian manifold which we suppose, for the sake of simplicity, embedded in a Euclidean space and let us consider the differential equation

$$(1.1) \qquad \begin{cases} \gamma \in C^2(\mathbf{R}; M) \\ P_\gamma(\gamma'') = F(t, \gamma, \gamma') \end{cases}$$

where $P_{\gamma(t)}$ is the orthogonal projection on the tangent space $T_{\gamma(t)}M$ and $F(t, \gamma(t), \gamma'(t)) \in T_{\gamma(t)}M$ for every t.

Equation (1.1) describes the motion of a material point constrained on the manifold M and subjected to the external force F. We suppose F to be periodic of period 1 with respect to t and we look for solutions γ of (1.1) of the same period. Let us state our main result.

1.2 Theorem. *Let us suppose that M is a compact submanifold in $\mathbf{R}^n$ of class C^∞ whose Euler-Poincaré characteristic $\chi(M; \mathbf{K})$ is different from zero for some field $\mathbf{K}$.*

Let $F_1 : S^1 \times \mathbf{R}^n \times \mathbf{R}^n \to \mathbf{R}^n$ be a map of class C^1 such that

$$(1.3) \qquad F_1(t, q, v) \in T_q M$$

whenever $t \in S^1$, $q \in M$, $v \in T_q M$. Let us assume that there exist $a \in \mathbf{R}$, $R > 0$ such that

$$(1.4) \qquad |F_1(t, q, v)|^2 + a\,(F_1(t, q, v)|v) \leq 0$$

whenever $t \in S^1$, $q \in M$, $v \in T_qM$, $|v| \geq R$.

Then there exists $\epsilon > 0$ *such that for every map* $F_2 : S^1 \times \mathbf{R}^n \times \mathbf{R}^n \to \mathbf{R}^n$ *of class* C^1 *satisfying* (1.3) *and the smallness condition*

$$\forall t \in S^1, \forall q \in M, \forall v \in T_qM, \ |F_2(t,q,v)| \leq \epsilon \cdot (1 + |v|),$$

there exists at least a solution $\gamma \in C^2(S^1; M)$ *of* (1.1) *with* $F = F_1 + F_2$.

1.5 *Remark.* If F_1 is a dissipative term of the form

$$F_1(t,q,v) = -a(t,q)v + F_0(t,q),$$

with $a(t,q) > 0$, then (1.4) is satisfied. Therefore we have an existence result for every small perturbation of F_1.

However the result holds also for $F_1 \equiv 0$. In this case we have a nonconservative perturbation of a geodesic flow. As far as we know, even in this particular case our result is new.

1.6 *Remark.* If F is independent of v and has the potential form

$$F(t,q) = -\mathrm{grad}_q V(t,q),$$

then problem (1.1) has at least a periodic solution and, under fairly general assumptions (e.g., M is simply connected), even infinitely many periodic solutions [1].

1.7 *Remark.* The condition on the Euler-Poincaré characteristic cannot be removed. In fact, if the Euler-Poincaré characteristic of M is zero, it is possible to give an example in which (1.1) has no solution of any period. For example, let us consider the two-dimensional torus

$$M = (\mathbf{R}/\mathbf{Z}) \times (\mathbf{R}/\mathbf{Z})$$

and let $F_0(q,v) = (1 - v_1,\ \pi - v_2)$. If we set $F = \delta F_0$ ($\delta \neq 0$), it is readily seen that (1.1) has no periodic solution of any period. On the other hand the assumptions of Theorem 1.2 are satisfied just setting

$$F_1 = \delta F_0 \text{ with } \delta > 0 \text{ and } F_2 = 0 \text{ (dissipative case)},$$

or

$$F_1 = 0 \text{ and } F_2 = \delta F \text{ with } \delta \text{ small (perturbative case)}.$$

1.8 *Remark.* If $M = \mathbf{R}^n$ (so that $P_\gamma = \mathrm{Id}$), problem (1.1) can be easily treated by the Leray-Schauder topological degree. Of course, as in this case M is not compact, suitable growth conditions on F must be imposed.

In our situation the topological degree does not seem to be a suitable tool. However we still use a functional analytic approach whose main tools are the parabolic flow on M generated by the equation

$$\frac{\partial u}{\partial s} = P_u\left(\frac{\partial^2 u}{\partial t^2}\right) - F\left(t, u, \frac{\partial u}{\partial t}\right)$$

and the infinite dimensional version of the Lefschetz fixed point theory (see e.g. [7]).

2. Parabolic Flows

In this section we give an existence theorem for a parabolic system on M. To this aim we recall some notions and results from [4, 5, 6, 13].

Let H be a real Hilbert space, $f : H \to \mathbf{R} \cup \{+\infty\}$ a lower semicontinuous function and $A : H \to \mathcal{P}(H)$ a map. We set

$$D(f) = \{u \in H : f(u) < +\infty\},$$
$$D(A) = \{u \in H : A(u) \neq \emptyset\}$$

and we suppose that $D(A) \subset D(f)$.

2.1 *Definition.* Let $u \in D(f)$. The function f is said to be *subdifferentiable* at u, if there exists $\alpha \in H$ such that

$$\liminf_{v \to u} \frac{f(v) - f(u) - (\alpha | v - u)}{|v - u|} \geq 0.$$

We denote by $\partial^- f(u)$ the (possibly empty) set of such α's and we set $D(\partial^- f) = \{u \in D(f) : \partial^- f(u) \neq \emptyset\}$.

Now let us suppose that f is bounded from below and that there exists a continuous function $\chi : \mathbf{R} \to \mathbf{R}^+$ such that

$$(\alpha - \beta | u - v) \geq -\{\chi(f(u))(1 + |\alpha|^2) + \chi(f(v))(1 + |\beta|^2)\} |u - v|^2$$

whenever $u, v \in D(\partial^- f)$, $\alpha \in \partial^- f(u)$, $\beta \in \partial^- f(v)$ and also whenever $u, v \in D(A)$, $\alpha \in A(u)$, $\beta \in A(v)$.

2.2 Theorem. *If (u_h) is a sequence in $D(\partial^- f)$ converging to some $u \in H$, (α_h) a sequence in H weakly converging to some $\alpha \in H$ and $\alpha_h \in \partial^- f(u_h)$, $\lim_h \sup f(u_h) < +\infty$, then we have $u \in D(\partial^- f)$, $\alpha \in \partial^- f(u)$, and $f(u) = \lim_h f(u_h)$.*

Proof. See [5, Theorem 1.18 and Remark 1.14]. QED

2.3 Theorem. *For every $u_0 \in D(f)$ and for every $g \in L^2(0,T;H)$ $(T > 0)$ there exists one and only one $u \in H^1(0,T;H)$ such that $(f \circ u) \in W^{1,1}(0,T)$ and*

$$\text{(2.4)} \qquad \begin{cases} g(s) - u'(s) \in \partial^- f(u(s)) & \text{a.e.} \\ u(0) = u_0. \end{cases}$$

Moreover

(i) $(f \circ u)'(s) = (g(s) - u'(s)|u'(s))$ a.e.;

(ii) *if $(u_0^{(h)})$ is a sequence in $D(f)$ converging to u_0 with $\lim_h f(u_0^{(h)}) = f(u_0)$, (g_h) a sequence in $L^2(0,T;H)$ converging to g and (s_h) a sequence in $[0,T]$ converging to some s, we have*

$$\lim_h u_h(s_h) = u(s), \ \lim_h f(u_h(s_h)) = f(u(s)),$$

where u_h is the solution of (2.3) corresponding to $u_0^{(h)}$ and g_h.

Proof. Since f is bounded from below, existence and uniqueness on all $[0,T]$ follow from [13, Theorem 2.25]. Property (i) is a straightforward extension of the chain rule.

If $(u_0^{(h)})$, (g_h), and (s_h) are as in (ii), we deduce by [4, Theorem 3.10] that (u_h) converges to u uniformly on $[0,T]$. Since f is lower semicontinuous and bounded from below, by the relation

$$f(u_h(s_h)) = f\left(u_0^{(h)}\right) + \int_0^{s_h} (g_h(\tau)|u_h'(\tau))\, d\tau - \int_0^{s_h} |u_h'(\tau)|^2\, d\tau$$

we conclude that $\lim_h f(u_h(s_h)) = f(u(s))$. QED

2.5 Theorem. *Let us suppose that for every $K \in \mathbf{R}$ the set*

$$\{(u,\alpha) \in H \times H : u \in D(A),\ \alpha \in A(u),\ f(u) \le K\}$$

is closed in $H \times H$ and let $u \in H^1(0,T;H)$ be a curve such that $f \circ u$ is bounded on $[0,T]$ and $-u'(s) \in A(u(s))$ a.e..

Then the following facts hold:

(i) *for every $s \in]0,T[$, the set $A(u(s))$ is not empty and has a unique element $A_0(u(s))$ of minimal norm;*

(ii) *there exists a continuous function $\zeta : \mathbf{R}^2 \to \mathbf{R}^+$ such that*

$$\forall\, s \in]0,T[,\ s\,|A_0(u(s))|^2 \le \zeta\left(\sup_{0 \le \tau \le T} f(u(\tau)),\ \int_0^T |u'(\tau)|^2\, d\tau\right).$$

Proof. Property (i) follows from [6, Theorem 2.3]. Moreover for every $s \in]0, T[$ there exists $u'_+(s)$, $u'_+(s) = -A_0(u(s))$ and

$$\forall\, \tau \in]0, s[,\ |u'_+(s)| \le |u'_+(\tau)| \exp\left\{ 2 \int_0^T \chi(f(u(\xi)))\,(1 + |u'(\xi)|^2)\, d\xi \right\}.$$

Squaring the inequality and integrating on $[0, s]$ with respect to τ, we get (ii). QED

Now let M be a compact submanifold in $\mathbf{R}^n$ of class C^∞ and let us set $H = L^2(S^1; \mathbf{R}^n)$, $\Lambda = H^1(S^1; M)$ and for every $\gamma \in \Lambda$ $T_\gamma \Lambda = \{\alpha \in H : \alpha(t) \in T_{\gamma(t)} M \text{ a.e.}\}$. We consider the energy functional $E : H \to \mathbf{R} \cup \{+\infty\}$ defined by

$$E(\gamma) = \begin{cases} \frac{1}{2} \int_{S^1} |\gamma'|^2\, dt & \text{if } \gamma \in \Lambda \\ +\infty & \text{if } \gamma \in H \setminus \Lambda. \end{cases}$$

2.6 Theorem. *The following facts hold:*

(i) *the functional E is lower semicontinuous;*

(ii) *$D(\partial^- E) = \Lambda \cap H^2(S^1; \mathbf{R}^n)$ and for every $\gamma \in D(\partial^- E)$*

$$\partial^- E(\gamma) = \{\alpha \in H : P_\gamma(\alpha) = -P_\gamma(\gamma'') \text{ a.e.}\};$$

(iii) *there exist $\sigma > 0$ and a continuous function $\chi : \mathbf{R} \to \mathbf{R}^+$ such that*

$$\| \gamma'' \| \le \chi(E(\gamma))\,(1 + \| P_\gamma(\gamma'') \|_{L^2}),$$

$$(\alpha_1 - \alpha_2 | \gamma_1 - \gamma_2)_{L^2} \ge \sigma \| \gamma_1' - \gamma_2' \|_{L^2}^2$$
$$- \left\{ \chi(E(\gamma_1)) \left(1 + \| \alpha_1 \|_{L^2}^2\right) + \chi(E(\gamma_2)) \left(1 + \| \alpha_2 \|_{L^2}^2\right) \right\} \| \gamma_1 - \gamma_2 \|_{L^2}^2$$

whenever $\gamma, \gamma_1, \gamma_2 \in D(\partial^- E)$, $\alpha_1 \in \partial^- E(\gamma_1)$, $\alpha_2 \in \partial^- E(\gamma_2)$.

Proof. It is a particular case of [3, Theorems 2.2, 2.7, 2.9 and Lemma 2.6] (see also [11]). QED

2.7 Theorem. *Let $\mathcal{F} : \Lambda \to L^2(S^1; \mathbf{R}^n)$ be a map such that $\mathcal{F}(\gamma) \in T_\gamma \Lambda$ for every $\gamma \in \Lambda$. Let us assume that there exists $c_1 \in \mathbf{R}^+$ such that*

(2.8) $$\|\mathcal{F}(\gamma)\|_{L^2} \le c_1\,(1 + \|\gamma\|_{H^1});$$

(2.9) $$\| \mathcal{F}(\gamma_1) - \mathcal{F}(\gamma_2) \|_{L^2} \le c_1\,(1 + \| \gamma_1 \|_{H^1} + \| \gamma_2 \|_{H^1})\, \| \gamma_1 - \gamma_2 \|_{H^1},$$

whenever $\gamma, \gamma_1, \gamma_2 \in \Lambda$.

Then for every $\gamma \in \Lambda$ *there exists one and only one* $u \in C([0,+\infty[;\Lambda)$ *such that* $u(s) \in H^2(S^1;\mathbf{R}^n)$ *and* $u \in H^1(0,T;L^2(S^1;\mathbf{R}^n))$ *whenever* $s, T > 0$ *and*

$$\begin{cases} \dfrac{\partial u}{\partial s} = P_{u(s)}\left(\dfrac{\partial^2 u}{\partial t^2}\right) - \mathcal{F}(u(s)) \quad \text{a.e. in} \quad]0,T[\\ u(0) = \gamma. \end{cases}$$

Moreover

(i) *the map* $\Phi : \Lambda \times [0,+\infty[\to \Lambda$ *defined by* $\Phi(\gamma, s) = u(s)$ *is continuous and satisfies* $\Phi(\gamma,0) = \gamma$, $\Phi(\Phi(\gamma,s_1),s_2) = \Phi(\gamma, s_1+s_2)$;

(ii) *the map* $E \circ u$ *belongs to* $W^{1,1}(0,T)$ *for every* $T > 0$ *and*

$$(E \circ u)'(s) = -\int_{S^1}\left(\mathcal{F}(u(s)) + \frac{\partial u}{\partial s}\,\Big|\,\frac{\partial u}{\partial s}\;dt\right) \quad \text{a.e.;}$$

(iii) *there exists a continuous function* $\zeta : \mathbf{R}^2 \to \mathbf{R}^+$ *such that*

$$\forall\, s > 0,\; s\int_{S^1}\left|\frac{\partial^2 u(s)}{\partial t^2}\right|^2 dt \le \zeta(s, E(\gamma)).$$

Proof. By (2.8) and the compactness of M, there exists $c_2 \in \mathbf{R}^+$ such that

$$\forall\, \gamma \in \Lambda,\; \|\mathcal{F}(\gamma)\|_{L^2}^2 \le c_2(1 + E(\gamma)). \tag{2.10}$$

Let $\gamma \in \Lambda$, $T > 0$ and let

$$X = \{\nu \in L^\infty\,(0,T;H^1(S^1;\mathbf{R}^n)) : \\ \nu(s) \in \Lambda,\; 1 + E(\nu(s)) \le (1 + E(\gamma))\exp(c_2 s) \quad \text{a.e.}\}.$$

Given $\nu \in X$, we have $\mathcal{F}(\nu) \in L^\infty(0,T;L^2(S^1;\mathbf{R}^n))$. Therefore by Theorems 2.3 and 2.6 there exists one and only one $u \in C([0,T];\Lambda) \cap H^1(0,T;L^2(S^1;\mathbf{R}^n))$ such that $u(s) \in H^2(S^1;\mathbf{R}^n)$ for every $s \in]0,T]$ and

$$\begin{cases} \dfrac{\partial u}{\partial s} = P_{u(s)}\left(\dfrac{\partial^2 u}{\partial t^2}\right) - \mathcal{F}(\nu(s)) \quad \text{a.e. in} \quad]0,T[\\ u(0) = \gamma. \end{cases}$$

We claim that $u \in X$. In fact $E \circ u$ is absolutely continuous and

$$\begin{aligned} (E \circ u)'(s) &= -\int_{S^1}\left(\mathcal{F}(\nu(s)) + \frac{\partial u}{\partial s}\,\Big|\,\frac{\partial u}{\partial s}\right) dt \\ &\le \tfrac{1}{2}\|\mathcal{F}(\nu(s))\|_{L^2}^2 - \tfrac{1}{2}\left\|\frac{\partial u}{\partial s}\right\|_{L^2}^2. \end{aligned} \tag{2.11}$$

Therefore

$$(E\circ u)'(s)\le c_2(1+E(\nu(s)))\le c_2(1+E(\gamma))\exp(c_3 s),$$

which implies by an integration $u\in X$.

By (2.11) we also deduce that

$$\begin{aligned}(2.12)\qquad &\int_0^T\left\|\frac{\partial u}{\partial s}+\mathcal{F}(\nu(s))\right\|_{L^2}^2 ds\\ &\le 2\int_0^T\left\|\frac{\partial u}{\partial s}\right\|_{L^2}^2 ds+2\int_0^T\|\mathcal{F}(\nu(s))\|_{L^2}^2\,ds\\ &\le 4E(\gamma)+4\int_0^T\|\mathcal{F}(\nu(s))\|_{L^2}^2\,ds\\ &\le 4E(\gamma)+4c_2\int_0^T(1+E(\nu(s))\,ds\\ &\le 4\{-1+(1+E(\gamma))\exp(c_2T)\}.\end{aligned}$$

Let us set

$$\overline{E}=-1+(1+E(\gamma))\exp(c_2T),$$
$$\overline{\chi}=\max_{[0,\overline{E}]^2}\chi,$$
$$c_3=2\overline{\chi}\{T+8(-1+(1+E(\gamma))\exp(c_2T))\},$$
$$c_4=\sup\left\{c_1\left(1+\|\gamma_1\|_{H^1}+\|\gamma_2\|_{H^1}\right):E(\gamma_i)\le\overline{E}\right\}.$$

Now let $\nu_1,\nu_2\in X$ and let $u_1,u_2\in X$ be obtained by the previous procedure. By Theorem 2.6 iii and (2.9) we have

$$\begin{aligned}\left(\|u_1-u_2\|_{L^2}^2\right)'&=2\left(\frac{\partial u_1}{\partial s}+\mathcal{F}(\nu_1)-\frac{\partial u_2}{\partial s}-\mathcal{F}(\nu_2)\,\Big|\,u_1-u_2\right)_{L^2}\\ -2\left(\mathcal{F}(\nu_1)-\mathcal{F}(\nu_2)\,|\,u_1-u_2\right)_{L^2}&\le -2\sigma\left\|\frac{\partial u_1}{\partial t}-\frac{\partial u_2}{\partial t}\right\|_{L^2}^2\\ +2\overline{\chi}\left(1+\left\|\frac{\partial u_1}{\partial s}+\mathcal{F}(\nu_1(s))\right\|_{L^2}^2+\right.&\left.\left\|\frac{\partial u_2}{\partial s}+\mathcal{F}(\nu_2(s))\right\|_{L^2}^2\right)\|u_1-u_2\|_{L^2}^2\\ &+\delta\,\|\mathcal{F}(\nu_1)-\mathcal{F}(\nu_2)\|_{L^2}^2+\frac{1}{\delta}\|u_1-u_2\|_{L^2}^2\\ &\le\left[\delta c_4^2\,\|\nu_1-\nu_2\|_{H^1}^2-2\sigma\left\|\frac{\partial u_1}{\partial t}-\frac{\partial u_2}{\partial t}\right\|_{L^2}^2\right]\\ +\left[\frac{1}{\delta}+2\overline{\chi}\left(1+\left\|\frac{\partial u_1}{\partial s}+\mathcal{F}(\nu_1(s))\right\|_{L^2}^2\right.\right.&+\left.\left.\left\|\frac{\partial u_2}{\partial s}+\mathcal{F}(\nu_2(s))\right\|_{L^2}^2\right)\right]\|u_1-u_2\|_{L^2}^2.\end{aligned}$$

Since

$$0 \leq 2\overline{\chi} \int_0^T \left(1 + \left\| \frac{\partial u_1}{\partial s} + \mathcal{F}(v_1(s)) \right\|_{L^2}^2 + \left\| \frac{\partial u_2}{\partial s} + \mathcal{F}(v_2(s)) \right\|_{L^2}^2 \right) ds \leq c_3,$$

we deduce by an integration

$$\begin{aligned} &\|u_1(s) - u_2(s)\|_{L^2}^2 \exp\left(-\frac{s}{\delta} - c_3\right) \\ &\leq \delta c_4^2 \int_0^s \|v_1(\tau) - v_2(\tau)\|_{H^1}^2 \exp\left(-\frac{\tau}{\delta}\right) d\tau \\ &-2\sigma \int_0^s \left\| \frac{\partial u_1}{\partial t}(\tau) - \frac{\partial u_2}{\partial t}(\tau) \right\|_{L^2}^2 \exp\left(-\frac{\tau}{\delta} - c_4\right) d\tau. \end{aligned}$$

By another integration we get

$$\begin{aligned} &2\sigma \int_0^T (T-s) \left\| \frac{\partial u_1}{\partial t}(s) - \frac{\partial u_2}{\partial t}(s) \right\|_{L^2}^2 \exp\left(-\frac{s}{\delta} - c_3\right) ds \\ &\quad + \int_0^T \|u_1(s) - u_2(s)\|_{L^2}^2 \exp\left(-\frac{s}{\delta} - c_3\right) ds \\ &\leq \delta c_4^2 \int_0^T (T-s) \| v_1(s) - v_2(s) \|_{H^1}^2 \exp\left(-\frac{s}{\delta}\right) ds, \end{aligned}$$

which implies

$$\begin{aligned} &2\sigma T \int_0^T (T-s) \left\| \frac{\partial u_1}{\partial t}(s) - \frac{\partial u_2}{\partial t}(s) \right\|_{L^2}^2 \exp\left(-\frac{s}{\delta}\right) ds \\ &\quad + \int_0^T (T-s) \|u_1(s) - u_2(s)\|_{L^2}^2 \exp\left(-\frac{s}{\delta}\right) ds \\ &\leq \delta T c_4^2 \exp(c_3) \int_0^T (T-s) \|v_1(s) - v_2(s)\|_{H^1}^2 \exp\left(-\frac{s}{\delta}\right) ds. \end{aligned}$$

It is readily seen that the space X endowed with the metric

$$\rho(w_1, w_2) = \left\{ \int_0^T (\|w_1(s) - w_2(s)\|_{H^1}^2) \exp\left(-\frac{s}{\delta}\right) ds \right\}^{1/2},$$

is a complete metric space and the map $\{v \mapsto u : X \to X\}$ is a contraction, provided that δ is sufficiently small. Its fixed point u is just the solution we are looking for.

Property (ii) is contained in Theorem 2.3. On the other hand, if $\gamma_1, \gamma_2 \in \Lambda$ and $E(\gamma_i) \leq E(\gamma)$, we have as in the previous argument

$$\|u_1(s) - u_2\|_{L^2}^2 \exp\left(-\frac{s}{\delta} - c_3\right) \leq \|\gamma_1 - \gamma_2\|_{L^2}^2$$
$$+\delta c_4^2 \int_0^s \|u_1(\tau) - u_2(\tau)\|_{H^1}^2 \exp\left(-\frac{\tau}{\delta}\right) d\tau$$
$$-2\sigma \int_0^s \left\| \frac{\partial u_1}{\partial t}(\tau) - \frac{\partial u_2}{\partial t}(\tau) \right\|_{L^2}^2 \exp\left(-\frac{\tau}{\delta} - c_3\right) d\tau,$$

where u_1, u_2 correspond to γ_1, γ_2 respectively.

Therefore

$$2\sigma T \int_0^T (T-s) \left\| \frac{\partial u_1}{\partial t}(s) - \frac{\partial u_2}{\partial t}(s) \right\|_{L^2}^2 \exp\left(-\frac{s}{\delta}\right) ds$$
$$+ \int_0^T (T-s) \|u_1(s) - u_2(s)\|_{L^2}^2 \exp\left(-\frac{s}{\delta}\right) ds \leq T^2 \|\gamma_1 - \gamma_2\|_{L^2}^2$$
$$+ \delta T c_4^2 \exp(c_3) \int_0^T (T-s) \|u_1(s) - u_2(s)\|_{H^1}^2 \exp\left(-\frac{s}{\delta}\right) ds.$$

Because of the arbitrarity of T, we conclude that $\forall\, T > 0$ the map $\{\gamma \mapsto u : \Lambda \to L^2(0,T; H^1(S^1; \mathbf{R}^n))\}$ is continuous. By (2.9) also the map $\{\gamma \mapsto \mathcal{F}(u) : \Lambda \to L^2(0,T; L^2(S^1; \mathbf{R}^n))\}$ is continuous and property (i) follows from Theorem 2.3.

Finally, we set $D(A) := D(\partial^- f) = \Lambda \cap H^2(S^1; \mathbf{R}^n)$ and

$$\forall\, \gamma \in D(A), \quad A(\gamma) = \partial^- E(\gamma) + \mathcal{F}(\gamma).$$

Combining Theorem 2.6 iii with (2.9), we get for every $\gamma_1, \gamma_2 \in D(A)$, $\alpha_1 \in A(\gamma_1)$, $\alpha_2 \in A(\gamma_2)$,

$$(\alpha_1 - \alpha_2 | \gamma_1 - \gamma_2)_{L^2} \geq \sigma \|\gamma_1' - \gamma_2'\|_{L^2}^2 - \{\chi(E(\gamma_1))\,(1 + \|\alpha_1 - \mathcal{F}(\gamma_1)\|_{L^2}^2)$$
$$+\chi(E(\gamma_2))\,(1 + \|\alpha_2 - \mathcal{F}(\gamma_2)\|_{L^2}^2)\}\, \|\gamma_1 - \gamma_2\|_{L^2}^2$$
$$-c_1\,(1 + \|\gamma_1\|_{H^1} + \|\gamma_2\|_{H^1})\, \|\gamma_1 - \gamma_2\|_{H^1} \|\gamma_1 - \gamma_2\|_{L^2}.$$

Therefore, by substituting χ with another continuous function and taking into account (2.10), we have also the inequality

$$(\alpha_1 - \alpha_2 \,|\, \gamma_1 - \gamma_2)_{L^2}$$
$$\geq -\{\chi(E(\gamma_1))\,(1 + \|\alpha_1\|_{L^2}^2) + \chi(E(\gamma_2))\,(1 + \|\alpha_2\|_{L^2}^2)\}\, \|\gamma_1 - \gamma_2\|_{L^2}^2$$

whenever $\gamma_1, \gamma_2 \in D(A)$, $\alpha_1 \in A(\gamma_1)$, $\alpha_2 \in A(\gamma_2)$.

Moreover, combining Theorem 2.2 with (2.8) and (2.9) we get that the set $\{(\gamma, \alpha) \in H \times H : \gamma \in D(A),\ \alpha \in A(\gamma),\ E(\gamma) \leq K\}$ is closed in $H \times H$ for every $K \in \mathbf{R}$. By Theorem 2.6 we obtain

$$\left\| \frac{\partial^2 u(s)}{\partial t^2} \right\|_{L^2} \leq \chi(E(u(s))) \left(1 + \left\| P_{u(s)}\left(\frac{\partial^2 u(s)}{\partial t^2}\right) \right\|_{L^2}\right)$$

$$\leq \chi(E(u(s))) \left(1 + \left\| P_{u(s)}\left(\frac{\partial^2 u(s)}{\partial t^2}\right) - \mathcal{F}(u(s)) \right\|_{L^2} + \|\mathcal{F}(u(s))\|_{L^2}\right)$$

and (iii) also follows, combining Theorem 2.5 with (2.10), (2.12) and the definition of X.

QED

3. Some Topological Tools

In this section we are concerned with fixed point theory on absolute neighborhood retracts. We refer the reader to [7, 10] for notions and results on *ANR*'s and to [7] for the theory of the Lefschetz index.

First of all we need a simple extension of [2, Theorem 5.3] and [7, Theorem VIII.5.4].

3.1 Theorem. *Let $\mathbf{K}$ be a field and (X, A) a pair of ANR's with A closed in X. Let us assume that $H_*(X, A; \mathbf{K})$ is of finite type and that the Euler-Poincaré characteristic $\chi(X, A; \mathbf{K})$ is different from zero.*

Let $\Phi : (X, A) \times [0, +\infty[\to (X, A)$ be a semiflow, namely a continuous map such that $\Phi(x, 0) = x$, $\Phi(\Phi(x, s_1), s_2) = \Phi(x, s_1 + s_2)$, and let us suppose that $\forall\, s > 0$ the set $\Phi(X, s)$ has compact closure in X.

Then there exists $x \in \overline{X \setminus A}$ such that $\Phi(x, s) = x$, $\forall\, s \geq 0$.

Proof. Let $K_h = \{x \in \overline{X \setminus A} : \Phi(x, 2^{-h}) = x\}$. Then $\{K_h\}$ is a decreasing sequence of compact sets. On the other hand the map $\Phi(\cdot, 2^{-h}) : (X, A) \to (X, A)$ is homotopically equivalent to the identity map of (X, A), so that its Lefschetz index is different from zero.

By [2, Theorem 4.5] or [7, Theorem VIII.3.4] we deduce that $K_h \neq \emptyset$, $\forall\, h$. Therefore $\bigcap_{h \in \mathbf{N}} K_h \neq \emptyset$ and the thesis follows by the density of $\{m 2^{-h} : m, h \geq 1\}$ in $[0, +\infty[$.

QED

Now let M be a compact submanifold in $\mathbf{R}^n$ of class C^∞ and let us consider the energy functional

$$E(\gamma) = \tfrac{1}{2} \int_{S^1} |\gamma'|^2 \, dt$$

defined on the submanifold $\Lambda = H^1(S^1; M)$ of the Hilbert space $H^1(S^1; \mathbf{R}^n)$. For every $b \in \mathbf{R}$, we set $E^b = \{\gamma \in \Lambda : E(\gamma) \leq b\}$.

3.2 Theorem. *The set of critical values of E is negligible in $\mathbf{R}$. Moreover, if $b > 0$ is not a critical value of E, then E^b is an ANR, for every field $\mathbf{K}$ the homology $H_*(E^b;\mathbf{K})$ is of finite type and $\chi(E^b;\mathbf{K}) = \chi(M;\mathbf{K})$.*

Proof. Let $t_0 < t_1$. Since M is compact, it is well known (see [9, Corollary 10.8]) that there exists $\delta > 0$ such that for every $u_0, u_1 \in M$ with $\rho(u_0, u_1) < \delta$ there exists one and only one geodesic $\gamma : [t_0, t_1] \to M$ such that

$$\gamma(t_0) = u_0, \ \gamma(t_1) = u_1, \ \int_{t_0}^{t_1} |\gamma'(t)| \, dt < \delta.$$

Moreover the map $\{(u_0, u_1) \mapsto \frac{1}{2} \int_{t_0}^{t_1} |\gamma'(t)|^2 \, dt\}$ is of class C^∞ in the open subset $\{(u_0, u_1) \in M^2 : \rho(u_0, u_1) < \delta\}$ of M^2.

Now let us identify S^1 with $[0,1]/\{0,1\}$ and let $h \in \mathbf{N}$. Let $0 = t_0 < \cdots < t_k = 1$ be such that

$$\forall\, i = 1, \ldots, k, \quad \gamma \in E^h \Rightarrow \int_{t_{i-1}}^{t_i} |\gamma'(t)| \, dt < \delta.$$

If we set

$$B = \left\{\gamma \in \Lambda : E(\gamma) < h, \gamma_{|[t_{i-1}, t_i]} \text{ is a geodesic for } 1 \le i \le k\right\}$$

then the map $\{\gamma \mapsto (\gamma(t_0), \ldots, \gamma(t_{k-1}))\}$ defines a homeomorphism between B and an open subset of M^k. Then (see [9, Lemma 16.1 and Theorem 16.2]) B has a natural structure of finite dimensional manifold of class C^∞ and $\widetilde{E} := E_{|B}$ is of class C^∞. Moreover, for every $\gamma \in \Lambda$ with $E(\gamma) < h$, γ is a critical point of E if and only if $\gamma \in B$ and γ is a critical point of $\widetilde{E}$.

By the Sard's theorem [12], the set of critical values of $\widetilde{E}$ is negligible. Therefore also the set of critical values of E in $]-\infty, h[$ is negligible, for every $h \in \mathbf{N}$.

Now let us suppose that $b \in]0, h[$ is not a critical value of E. The set E^b is a submanifold with boundary in Λ, hence an *ANR*. Moreover (see [9, Theorem 16.2]) E^b is homotopically equivalent to $\widetilde{E}^b$, which is a compact submanifold with boundary in B. Therefore $H_*(E^b;\mathbf{K})$ is of finite type for every field $\mathbf{K}$.

Now let us take (see [8, Proposition 1.4.14 and Theorem 1.4.15]) an $a \in]0, b]$ such that a is not a critical value of E and the set E^0 of point curves is a strong deformation retract of E^a. In particular $\chi(E^a;\mathbf{K}) = \chi(M;\mathbf{K})$. By Theorem 2.7 we can define a semiflow $\Phi : (E^b, E^a) \times [0, +\infty[\to (E^b, E^a)$ by means of the parabolic problem

$$\begin{cases} u \in C([0,+\infty[;\Lambda) \cap H^1(0,T;L^2(S^1;\mathbf{R}^n)) \ \forall\, T > 0 \\ u(s) \in H^2(S^1;\mathbf{R}^n) \ \forall\, s > 0 \\ \dfrac{\partial u}{\partial s} = P_{u(s)}\left(\dfrac{\partial^2 u}{\partial t^2}\right) + \dfrac{\partial u}{\partial t} \quad \text{a.e. in} \quad]0,+\infty[\\ u(0) = \gamma \end{cases}$$

where γ is given in Λ and $\Phi(\gamma, s) := u(s)$.

We have only to remark that

$$\begin{aligned} \frac{d}{ds}\{E(\Phi(\gamma,s))\} &= \int_{S^1}\left(\frac{\partial u}{\partial t} - \frac{\partial u}{\partial s}\middle|\frac{\partial u}{\partial s}\right) dt \\ &= \int_{S^1}\left(-P_{u(s)}\left(\frac{\partial^2 u}{\partial t^2}\right)\middle|P_{u(s)}\left(\frac{\partial^2 u}{\partial t^2}\right) + \frac{\partial u}{\partial t}\right) dt \\ &= -\int_{S^1}\left|P_{u(s)}\left(\frac{\partial^2 u}{\partial t^2}\right)\right|^2 dt - \frac{1}{2}\int_{S^1}\frac{d}{dt}\left(\left|\frac{\partial u}{\partial t}\right|^2\right) dt \le 0, \end{aligned}$$

so that the sets E^b and E^a are invariant under the semiflow.

Since $\forall\, \gamma \in E^b$, $\forall\, s > 0$,

$$s\int_{S^1}\left|\frac{\partial^2 u(s)}{\partial t^2}\right|^2 dt \le c(s,b),$$

we deduce that the closure of $\Phi(E^b, s)$ in E^b is compact.

Now the sets E^b and E^a are *ANR* with homology of finite type. By the exact sequence of the pair (E^b, E^a) it is readily seen that also $H_*(E^b, E^a; \mathbf{K})$ is of finite type and

$$\chi(E^b, E^a; \mathbf{K}) = \chi(E^b; \mathbf{K}) - \chi(E^a; \mathbf{K}) = \chi(E^b; \mathbf{K}) - \chi(M; \mathbf{K}).$$

On the other hand, if $\gamma \in E^b$ is such that $\Phi(\gamma, s) = \gamma \ \forall\, s \ge 0$, we have

$$\int_{S^1} |\gamma'|^2\, dt = \int_{S^1} (P_\gamma(\gamma'') + \gamma'|\gamma')\, dt = 0,$$

so that $\gamma \notin \overline{E^b \setminus E^a}$. By Theorem 3.1 it follows that $\chi(E^b; \mathbf{K}) - \chi(M; \mathbf{K}) = 0$ and the proof is complete. QED

4. Proof of the Main Result

Let M be a compact submanifold in $\mathbf{R}^n$ of class C^∞ such that $\chi(M; \mathbf{K}) \neq 0$ for some field $\mathbf{K}$ and let

$$F_1 : S^1 \times \mathbf{R}^n \times \mathbf{R}^n \to \mathbf{R}^n$$

be a map of class C^1 such that

$$F_1(t,q,v) \in T_qM, \tag{4.1}$$

for every $t \in S^1$, $q \in M$, $v \in T_qM$ and such that there exist $a \in \mathbf{R}$, $R > 0$ satisfying

$$|F_1(t,q,v)|^2 + a(F_1(t,q,v)|v) \leq 0 \tag{4.2}$$

whenever $t \in S^1$, $q \in M$, $v \in T_qM$ and $|v| \geq R$.

Let

$$F_2 : S^1 \times \mathbf{R}^n \times \mathbf{R}^n \to \mathbf{R}^n$$

be another map of class C^1 satisfying the same condition (4.1) and the estimate

$$|F_2(t,q,v)| \leq \varepsilon(1 + |v|) \tag{4.3}$$

for every $t \in S^1$, $q \in M$, $v \in T_qM$.

If we set $F = F_1 + F_2$, we have to prove the existence of some $\gamma \in C^2(S^1; M)$ such that

$$P_\gamma(\gamma'') = F(t,\gamma,\gamma'), \tag{4.4}$$

provided that $\varepsilon > 0$ is sufficiently small. The smallness of ε will depend on M, a and R.

By classical regularity results, it is sufficient to find a solution $\gamma \in H^2(S^1; M)$.

Of course F satisfies (4.1). Moreover by (4.2) and (4.3) there exists $c \geq |a|$ such that

$$|F(t,q,v)| \leq c(1 + |v|) \tag{4.5}$$

whenever $t \in S^1$, $q \in M$, $v \in T_qM$ and

$$|F(t,q,v)|^2 + a(F(t,q,v)|v) \leq \varepsilon c(1 + |v|^2) \tag{4.6}$$

whenever $t \in S^1$, $q \in M$, $v \in T_qM$ and $|v| \geq R$.

By Theorem 3.2 there exists a regular value b of the energy functional E such that

$$(2b - c)\exp(-3c) > R^2.$$

Since the energy functional verifies the Palais-Smale condition [8], there exists $\delta > 0$ such that

$$(2b - 2\delta - c)\exp(-3c) > R^2, \tag{4.7}$$

$$\forall\, \gamma \in H^2(S^1; M),\ b - \delta \le E(\gamma) \le b \Rightarrow \|P_\gamma(\gamma'')\|_{L^2} \ge \delta. \tag{4.8}$$

Let $\vartheta_1 : \mathbf{R} \to [0,1]$ be a smooth function such that $\vartheta(s) = 1$ for $s \le c + 2b\exp(3c)$, $\vartheta_1(s) = 0$ for $s \ge 1 + c + 2b\exp(3c)$ and let

$$\widetilde{F}(t,q,v) = \vartheta_1(|v|^2)F(t,q,v).$$

It is readily seen that $\widetilde{F} : S^1 \times \mathbf{R}^n \times \mathbf{R}^n \to \mathbf{R}^n$ is of class C^1 and satisfies (4.1), (4.5), (4.6) and the estimate

$$|\widetilde{F}(t,q_1,v_1) - \widetilde{F}(t,q_2,v_2)| \le \text{ const. } (|q_1 - q_2| + |v_1 - v_2|)$$

whenever $t \in S^1$, $q_i \in M$, $v_i \in T_{q_i}M$.

Now let $\vartheta_2 : \mathbf{R} \to [0,1]$ be another smooth function such that $\vartheta(s) = 1$ for $s \le 2b - 2\delta$, $\vartheta_1(s) = 0$ for $s \ge 2b$ and let us define $\mathcal{F} : \Lambda \to L^2(S^1; \mathbf{R}^n)$ by

$$\mathcal{F}(\gamma) = \vartheta_2\left(\int_{S^1} |\gamma'|^2\, dt\right) \widetilde{F}(t,\gamma,\gamma').$$

We have

$$\mathcal{F}(\gamma) \in T_\gamma\Lambda, \tag{4.9}$$

$$|\mathcal{F}(\gamma)(t)| \le c(1 + |\gamma'(t)|) \quad \text{a.e. in} \quad S^1, \tag{4.10}$$

$$\|\mathcal{F}(\gamma_1) - \mathcal{F}(\gamma_2\|_{L^2} \le \text{ const. } (1 + \|\gamma_1\|_{H^1} + \|\gamma_2\|_{H^1})\|\gamma_1 - \gamma_2\|_{H^1}, \tag{4.11}$$

whenever $\gamma, \gamma_1, \gamma_2 \in \Lambda$ and

$$\int_{S^1} |\mathcal{F}(\gamma)|^2\, dt + a\int_{S^1} (\mathcal{F}(\gamma)|\gamma')\, dt \le \varepsilon c\int_{S^1} (1 + |\gamma'|^2)\, dt \tag{4.12}$$

whenever $\gamma \in \Lambda$ and $|\gamma'(t)| \ge R$ a.e. in S^1.

4.13 Lemma. *If ε satisfies the inequality*

$$\varepsilon c(1 + 2b) < \delta^2, \tag{4.14}$$

then every solution $\gamma \in H^2(S^1; \mathbf{R}^n)$ of

$$P_\gamma(\gamma'') = \mathcal{F}(\gamma),$$

with $E(\gamma) \le b$ is in fact a solution of (4.4).

Proof. By (4.10) we have

$$2(\mathcal{F}(\gamma)(t)|\gamma'(t)) \le c(1 + 3|\gamma'(t)|^2).$$

Therefore

$$\frac{d}{dt}(|\gamma'|^2) = 2(P_\gamma(\gamma'')|\gamma') = 2(\mathcal{F}(\gamma)|\gamma') \le c(1 + 3|\gamma'|^2).$$

Since $E(\gamma) \le b$, we have $|\gamma(t_1)|^2 \le 2b$ for some $t_1 \in S^1$. Combining this fact with the Gronwall's lemma, we get

$$\forall\, t \in S^1, \quad |\gamma'(t)|^2 \le c + 2b\exp(3c),$$

so that $\widetilde{F}(t,\gamma,\gamma') = F(t,\gamma,\gamma')$. To conclude the proof, it is sufficient to show that $E(\gamma) \le b - \delta$. By contradiction, let us assume that $E(\gamma) > b - \delta$. Then $|\gamma'(t_2)|^2 > 2b - 2\delta$ for some $t_2 \in S^1$. Again by the Gronwall's lemma

$$\forall\, t \in S^1, \quad |\gamma'(t)|^2 \ge (2b - 2\delta - c)\exp(-3c).$$

Combining this fact with (4.7), we deduce that $|\gamma'(t)| \ge R$ for every t in S^1. Since

$$\int_{S^1} (\mathcal{F}(\gamma)|\gamma')\,dt = \int_{S^1} (P_\gamma(\gamma'')|\gamma')\,dt = \tfrac{1}{2}\int_{S^1} \frac{d}{dt}(|\gamma'|^2)\,dt = 0,$$

by (4.12) and (4.8) we have

$$\delta^2 \le \int_{S^1} |\mathcal{F}(\gamma)|^2\,dt \le \varepsilon c(1 + 2b),$$

which contradicts (4.14). QED

Now let us come to the proof of our main result.

By Theorem 2.7 we can define a semiflow $\Phi : E^b \times [0,+\infty[\to E^b$ by means of the parabolic problem

$$\begin{cases} u \in C([0,+\infty[;\Lambda) \cap H^1(0,T;L^2(S^1;\mathbf{R}^n)) \ \forall\, T > 0 \\ u(s) \in H^2(S^1;\mathbf{R}^n) \ \forall\, s > 0 \\ \dfrac{\partial u}{\partial s} = P_{u(s)}\left(\dfrac{\partial^2 u}{\partial t^2}\right) - \mathcal{F}(u) \quad \text{a.e. in} \quad]0,+\infty[\\ u(0) = \gamma \end{cases}$$

where γ is given in Λ and $\Phi(\gamma,s) := u(s)$.

Let us remark that $E(\gamma) \geq b \Rightarrow \mathcal{F}(\gamma) = 0$. As in the proof of Theorem 3.2, we deduce that

$$E(\Phi(\gamma, s)) \geq b \Rightarrow \frac{d}{ds}\{E(\Phi(\gamma, s))\} \leq 0,$$

so that the set E^b is invariant under the semiflow. Moreover, we have again

$$s \int_{S^1} \left| \frac{\partial^2 u(s)}{\partial t^2} \right|^2 dt \leq c(s, b)$$

which implies the relative compactness of $\Phi(E^b, s)$ in E^b.

By Theorem 3.2 E^b is an ANR with homology of finite type and $\chi(E^b; \mathbf{K}) = \chi(M; \mathbf{K}) \neq 0$. Therefore we can apply Theorem 3.1 with $X = E^b$, $A = \emptyset$, obtaining the existence of $\gamma \in E^b \cap H^2(S^1; \mathbf{R}^n)$ such that $P_\gamma(\gamma'') = \mathcal{F}(\gamma)$. By Lemma 4.13, γ is also a solution of (4.4), provided that ε satisfies (4.14). QED.

REFERENCES

1. V. Benci, *Periodic solutions of Lagrangian systems on a compact manifold*, J. Differential Equations **63** (1986), 135-161.
2. C. Bowszyc, *Fixed point theorems for the pairs of spaces*, Bull. Polish Acad. Sci. Math. **16** (1968), 845-850.
3. A. Canino, *Existence of a closed geodesic on p-convex sets*, Ann. Inst. H. Poincaré Anal. Non Linéaire **5** (1988), 501-518.
4. M. Degiovanni, *Parabolic equations with nonlinear time-dependent boundary conditions*, Ann. Mat. Pura Appl. (4) **141** (1985), 223-263.
5. M. Degiovanni, A. Marino, M. Tosques, *Evolution equations with lack of convexity*, Nonlinear Anal. **9** (1985), 1401-1443.
6. M. Degiovanni, M. Tosques, *Evolution equations for (φ, f)-monotone operators*, Boll. Un. Mat. Ital. B (6) **5** (1986), 537-568.
7. A. Granas, *Points fixes pour les applications compactes: espaces de Lefschetz et la théorie de l'indice*, Séminaire de Mathématiques Supérieures, Presses de l'Université de Montréal, Quebec, 1980.
8. W. Klingenberg, *Lectures on closed geodesics*, Grundlehren der Mathematischen Wissenschaften, **230**, Springer-Verlag, Berlin-New York, 1978.
9. J. Milnor, *Morse theory*, Annals of Mathematics Studies, **51**, Princeton University Press, Princeton, NJ, 1963.
10. R. S. Palais, *Homotopy theory of infinite dimensional manifolds*, Topology **5** (1966), 1-16.

11. D. Scolozzi, *Un teorema di esistenza di una geodetica chiusa su varietà con bordo*, Boll. Un. Mat. Ital. A (6) 4 (1985), 451-457.
12. S. Sternberg, *Lectures on differential geometry*, Englewood Cliffs, NJ, 1964.
13. M. Tosques, *Quasi-autonomous evolution equations associated with* (φ, f)*-monotone operators*, Integral Functionals in the calculus of variations (Trieste, 1985), Rend. Circ. Mat. Palermo Suppl. (2) 15 (1987), 163-180.

Vieri Benci
Istituto di Matematiche Applicate
Facoltà di Ingegneria
Università di Pisa
Italy

Marco Degiovanni
Dipartimento di Automazione Industriale
Facoltà di Ingegneria
Università di Brescia
Italy

Periodic Trajectories for the Lorentz-Metric of a Static Gravitational Field*

VIERI BENCI and DONATO FORTUNATO

0. Introduction

In General Relativity a gravitational field is described by a symmetric, second order tensor

$$g \equiv g(z)[\,\cdot\,,\,\cdot\,] \qquad z = (z_0, \ldots, z_3) \in \mathbf{R}^4$$

on the space-time manifold $\mathbf{R}^4$. The tensor g is assumed to have the signature $+,-,-,-$; namely for all $z \in \mathbf{R}^4$ the bilinear form $g(z)[\,\cdot\,,\,\cdot\,]$ possesses one positive and three negative eigenvalues. The "pseudo-metric" induced by g is called Lorentz-metric.

Let g_{ij} $(i, j = 0, \ldots, 3)$ denote the components of the metric tensor g and, as usual, we set

$$z_0 = t \qquad (z_1, z_2, z_3) = x$$

We shall assume that

(g_1) $g_{ij} \in C^1(\mathbf{R}^4, \mathbf{R})$ and there exist $\nu_1, \nu_2 > 0$ such that for all $z \in \mathbf{R}^4$ we have:

$$g_{00}(z) > \nu_1$$

and

$$\sum_{ij=1}^{3} g_{ij}(z)\,\xi_i\,\xi_j \leq -\nu_2|\xi|^2$$

for all $\xi = (\xi_1, \xi_2, \xi_3) \in \mathbf{R}^3$.

(g_2) g_{ij} $(i, j = 0, \ldots, 3)$ do not depend on t and for $i = 1, 2, 3$, $g_{i0} = 0$.
Assumption (g_1) implies that g is a Lorentz-metric.
Assumption (g_2) implies that the gravitational field is static (cf, e.g. [8]).

*Sponsored by M.P.I. (fondi 60% "Problemi differenziali nonlineari e teoria dei punti critici; fondi 40% "Equazioni differenziali e calcolo delle variazioni").

In this paper we are interested in trajectories $z = z(t)$ which are parametrized with respect to t

$$z(t) = \big(t, x(t)\big), \tag{0.2}$$

and which are time-like, i.e.,

$$g\big(z(t)\big)[\dot z(t), \dot z(t)] > 0 \qquad \text{for all } t, \tag{0.3}$$

where $\cdot$ denotes the derivative with respect to t.

We recall that the trajectories $z(t) = \big(t, x(t)\big)$ of a "test particle" in a gravitational field described by g are the time-like solutions of the Euler-Lagrange equations

$$\frac{d}{dt}\frac{\partial \mathcal{L}}{\partial \zeta}(z, \dot z) - \frac{\partial \mathcal{L}}{\partial z}(z, \dot z) = 0 \tag{0.4}$$

with Lagrangian $\mathcal{L}(z, \zeta) = \sqrt{g(z)\,[\zeta, \zeta]}\ \ z, \zeta \in \mathbf{R}^4$.

More explicitly (0.4) can be written as follows:

$$\frac{d}{dt}\sum_{j=0}^{3} \frac{g_{kj}\ \dot z_j}{\sqrt{g(z)\ [\dot z, \dot z]}} - \sum_{i,j=0}^{3} \frac{1}{2\sqrt{g(z)\ [\dot z, \dot z]}}\frac{\partial g_{ij}}{\partial z_k}\ \dot z_i \dot z_j = 0 \ \ k = 0, \ldots, 3$$

where z_0 coincides with the parameter t.

In particular we are interested in periodic trajectories. More precisely, for a prescribed $T > 0$ we shall study the following problem

$$\begin{cases} \text{find solutions } z(t) = \big(t, x(t)\big) \text{of (0.4) such that} \\ x(t) = x(t+T) \text{for all } t. \end{cases} \tag{0.5}$$

It is easy to verify that, if $\bar x \in \mathbf{R}^3$ is a critical point of g_{00} (i.e., grad $g_{00}(\bar x) = 0$), then $z = (t, \bar x)$ solves problem (0.5). Such a trajectory is called a trivial solution of (0.5). We shall prove the following theorem:

Theorem 0.1. *Suppose that g satisfies (g_1), (g_2). Suppose moreover that*

(g_3) $\operatorname{grad} g_{ij}(x) \to 0$ $(i, j = 0, \ldots, 3)$ *and* $g_{00}(x) \to C \in \mathbf{R}$ *as* $|x| \to \infty$ *with* $C = \sup g_{00}(x)$. *We can take, without loss of generality,* $C \equiv 1$.

(g_4) *The minimum C_0 of g_{00} is attained only in one point (say $x = 0$) and g_{00} has no other critical value.*
Under the above assumptions there exists $\overline{T} > 0$ such that for all $T > \overline{T}$ problem (0.5) possesses a non-trivial solution.

This paper is organized as follows. In section 1 we set up a variational formulation of problem (0.5), more precisely (0.5) is reduced to the research of the critical points of a suitable functional. Sections 2 and 3 are devoted to the proof of Theorem 0.1. This proof is carried out by using suitable minimax methods which have been developed for studying nonlinear differential equations (cf. e.g. [9] for a review of these topics). We recall at last that minimax methods have been recently used for proving the esistence of geodesics for the Lorentz-metric of a gravitational field [5].

1. The variational principle

First of all we introduce some notations. $H^1 \equiv H^1((0,1), \mathbf{R}^4)$ denotes the Sobolev space of $\mathbf{R}^4$- valued, absolutely continuous functions on (0.1) which possess square integrable derivatives.

If

$$z(s) = \big(z_o(s), \dots, z_3(s)\big) \in H^1$$

we set

$$t(s) = z_o(s), \quad x(s) = \big(z_1(s), \dots, z_3(s)\big). \tag{1.1}$$

Let $S^1 = \mathbf{R}/\mathbf{Z}$ and set

$$\begin{cases} H^1_{0,t} = \{(t(s), 0) \in H^1 \mid t(0) = t(1) = 0\} \\ H^1_x(S^1) = \{(0, x(s)) \in H^1 \text{ s.t. the 1-periodic extension of } x \\ \qquad\qquad \text{belongs to } H^1_{\text{loc}}(\mathbf{R}, \mathbf{R}^3)\}. \end{cases} \tag{1.2}$$

Let $T > 0$ and consider the manifold in H^1

$$M = \{(t, x) \in H^1 \mid t(0) = 0, \ t(1) = T \text{ and } x \in H^1_x(S^1)\}.$$

Obviously,

$$M = \bar{z} + H^1_{0,t} \oplus H^1_x(S^1) \tag{1.3}$$

where

$$\bar{z} = (sT, 0) \qquad s \in (0,1).$$

Now consider the following problem:

$$(1.4)\qquad \begin{cases} \text{find critical points with positive critical values of the} \\ \text{functional} \\ h(z) = \frac{1}{T^2}\int_0^1 g(z(s))[\dot{z}(s), \dot{z}(s)]\, ds \\ \text{on the manifold } M. \end{cases}$$

Here $\cdot$ denotes the derivative with respect to the parameter s.

We recall that $z \in M$ is a critical point of h on M if

$$\langle h'(z), \xi \rangle = 0 \qquad \text{for all } \xi \in H^1_{0,t} \oplus H^1_x(S^1)$$

where $h'(z)$ denotes the Frechét-differential of f at z and $< \cdot, \cdot >$ the standard pairing between H^1 and its dual. The solutions of (1.4) permit to construct the solutions of (0.5). More precisely, the following theorem holds:

Theorem 1.1. *Suppose that the metric tensor g satisfies the assumptions $(g_1), (g_2)$ and let $z(s) = \big(t(s), x(s)\big)$ be a solution of (1.4). Then z is C^2 and $t = t(s)$ is strictly increasing. Now set*

$$\widetilde{x}(t) = x(s(t)) \qquad t \in (0, T)$$

where $s = s(t)$ is the inverse of $t = t(s)$. Then $\widetilde{x}$ can be extended to $\mathbf{R}$ *as a T-periodic C^2 function such that*

$$\widetilde{z}(t) = (t, \widetilde{x}(t))$$

solves (0.5).

Proof. Standard calculations show that h is Frechét differentiable. Let $z(s) = (t(s), x(s)) \in M$ be a solution of (1.4), then

$$(1.5)\qquad \langle h'(z), (\tau, 0) \rangle = 0 \qquad \text{for all } \tau \in H^1_{0,t}$$

$$(1.6)\qquad \langle h'(z), (0, \xi) \rangle = 0 \qquad \text{for all } \xi \in H^1_x(S^1).$$

(1.5) can be written as follows:

$$\int_0^1 g_{00}(x(s))\, \dot{t}(s)\, \dot{\tau}(s)\, ds = 0 \qquad \text{for all } \tau \in H^1_{0,t}.$$

Then there exists a constant C_1 such that

$$(1.7)\qquad \dot{t}(s) = \frac{C_1}{g_{oo}(x(s))} \qquad \text{for all } s \in (0, 1).$$

Since $x \in H^1_x(S^1)$, from (1.7) we deduce that $\dot{t} \in H^1(S^1)$ (i.e., $\dot{t} = \dot{t}(s)$ can be extended to $\mathbf{R}$ as a 1-periodic function belonging to $H^1_{\text{loc}}(\mathbf{R})$).

By using well known regularity results from (1.6) we can deduce that x is a C^2, 1-periodic function. Then by (1.7) also $\dot{t}$ is C^2 and 1- periodic. We conclude that $z(s) = (t(s), x(s))$ is a classical solution on $\mathbf{R}$ of the Euler-Lagrange equation

$$\frac{d}{ds}\frac{\partial \mathcal{L}^*}{\partial \zeta} - \frac{\partial \mathcal{L}^*}{\partial z} = 0. \tag{1.8}$$

With respect to the Lagrangian function

$$\mathcal{L}^*(z,\zeta) = g(z)\,[\zeta,\zeta] \qquad z,\zeta \in \mathbf{R}^4. \tag{1.9}$$

Since $t(0) = 0$ and $t(1) = T$, (1.7) implies that $t = t(s)$ is increasing. Let $s = s(t)$ $(t \in (0,T))$ be its inverse and set $\widetilde{x}(t) = x(s(t))$. Since x and $\dot{t}$ are 1-periodic, we have

$$\widetilde{x}(0) = \widetilde{x}(T) \text{ and } \frac{d}{dt}\widetilde{x}(0) = \frac{d}{dt}\widetilde{x}(T).$$

We conclude that $\widetilde{x} = \widetilde{x}(t)$ is C^2 and T-periodic.

Finally, since $z(s) = (t(s), x(s))$ solves (1.8) and $h(z) > 0$, standard arguments show that $(t, \widetilde{x}(t))$ solves (0.4). QED

2. Finite dimensional approximation

Since the gravitational field is static (see assumption g_2) the functional h in (1.4) can be written as follows

$$h(z) = \frac{1}{T^2}\int_0^1 \left[\beta(x(s))\left(\dot{t}(s)\right)^2 - \left(\alpha(x(s))\dot{x}(s) \mid \dot{x}(s)\right)\right] ds$$
$$z = (t,x) \in M$$

where we have set

$$\alpha(x) = g_{ij}(x) \qquad i,j = 1,2,3 \qquad \beta(x) = g_{00}(x) \tag{2.1}$$

and $(\cdot \mid \cdot)$ denotes the standard inner product in $\mathbf{R}^3$. In order to solve problem (1.4) first we shall carry out a finite dimensional approximation.

Let $n \in \mathbf{N}$ and denote by $H_{n,x}$ and $H_{n,t}$ the "finite dimensional approximations" of $H^1_x(S^1)$ and $H^1_{0,t}$ respectively. More precisely we set

$$H_{n,x} = \text{span}\,\{\,\phi_i \sin 2\pi\ell s; \phi_i \cos 2\pi\ell s : i = 1,2,3; \ell = 0,\dots,n\}$$
$$H_{n,t} = \text{span}\,\{\,\sin \pi\ell s : \ell = 1,\dots,n\}$$

where ϕ_i, $i = 1,2,3$, is the standard basis in $\mathbf{R}^3$.
Moreover, we set

$$M_n = \bar{z} + H_{n,t} \oplus H_{n,x} \qquad \bar{z} = (sT,0) \quad s \in (0,1).$$

In this section we shall prove the following theorem:

Theorem 2.1. *Suppose that the assumptions of Theorem 0.1 hold. Then there exist positive constants $\overline{T}, \delta, \gamma$ such that for all $T > \overline{T}$ and all $n \in \mathbf{N}$ there exists a critical point $z_n \in M_n$ of h/M_n such that*

$$C_0 + \delta \leq h(z_n) \leq 1 - \gamma.$$

We would like to exploit the S^1-index theory developed in [3,7]. This is possible if $h(z)$ is invariant under a S^1-group action. However h is not S^1 invariant since the 1-periodic extension to $\mathbf{R}$ of t does not belong to $H^1_{\text{loc}}(\mathbf{R})$.

Nevertheless we can introduce a suitable substitution which permits us to overcome this difficulty.

Let $t(s) = sT + \tau$ with $\tau \in H_{n,t}$ and set

$$\eta = \dot{t} - T = \dot{\tau}.$$

Consider the functional

$$f(z) = f_n(z) = \frac{1}{T^2}\int_0^1 [\beta(x)(\eta+T)^2 - (\alpha(x)\dot{x} \mid \dot{x})]\, ds \tag{2.2}$$
$$z = (\eta, x) \in H_n = H_{n,\eta} \oplus H_{n,x}$$

where

$$H_{n,\eta} = \text{span}\,\{\cos \pi\ell s : \ell = 1,\dots,n\}.$$

Let $(\eta, x) \in H_n$ be a critical point of f_n, then

$$(t,x) \quad \text{with} \quad t(s) = \int_0^s \eta(\rho)\, d\rho + sT,$$

is a critical point of h on M_n and

$$f_n(\eta, x) = h(t,x).$$

Therefore, in order to prove Theorem 2.1 it is enough to study the critical points of f_n.

Moreover, it is easy to check that f_n is invariant under the unitary representation of the group S^1 on H_n given by the time-translations, namely, we have

$$f_n(\eta(s+\lambda), x(s+\lambda)) = f_n(\eta(s), x(s))$$

for all $\lambda \in \mathbf{R}$ and all $(\eta, x) \in H_n$ (here the functions $\eta \in H_{n,\eta}$ are extended by periodicity to all $\mathbf{R}$).

The proof of Theorem 2.1 utilizes the following abstract critical point theorem.

Theorem 2.2. *Let E be a real Hilbert space on which a unitary representation G of the group $S^1 = \mathbf{R}/\mathbf{Z}$ acts.*
Let I be a C^1 functional on E satisfying the following properties:

(I_1) *I is G-invariant (i.e., $\forall\, u \in E,\ \forall\, g \in G\ I(u) = I(g(u))$).*

(I_2) *I satisfies the Palais-Smale condition* (P.S.) *in $(0, \mu)$, $\mu > 0$ (i.e., for all $c \in (0, \mu)$ any sequence $\{u_k\} \subset E$ s.t. $I'(u_k) \to 0$ as $k -> \infty$ and $I(u_k) \to c$ contains a convergent subsequence).*

(I_3) *There exist two closed subspaces E_1 and E_2 of E with codim $E_2 < +\infty$ and a bounded S_1-invariant neighbourhood B of 0 such that :*

(a) $I(u) \geq \lambda > I(0)\ \forall\, u \in E_1 \cap \partial B$ *(∂B being the boundary of B)*

(b) $\sup I(E_2) < \mu$

(c) Fix $G \subset E_1$ *or* Fix $G \subset E_2$, *where* Fix $G = \{u \in E \mid g(u) = u$ *for all* $g \in G\}$.

Suppose moreover that

(I_4) $\forall\, u \in$ Fix G, $I'(u) = 0 : I(u) < \lambda$.

Then I posseses at least

$$\tfrac{1}{2}[\dim(E_1 \cap E_2) - \operatorname{codim}(E_1 + E_2)]$$

distinct critical points (two critical points u_1, u_2 are distinct if $u_1 \neq g(u_2)$ for all $g \in G$) whose critical values belong to the interval $[\lambda, \sup f(E_2)]$.

Variants of Theorem 2.2 have been proved in [1, Th. 2.4], [4, Th. 1.4], [3,6] and therefore we omit its proof. Now in order to verify that f_n satisfies the assumptions of Theorem 2.2 we need to prove several lemmas.

Lemma 2.3. *Suppose that g satisfies (g_1), (g_2) and (g_3). Then for any $n \in \mathbf{N}$, f_n satisfies P.S. in $\mathbf{R}\backslash\{1\}$ i.e., any sequence $\{z_k\} \subset H_n$ such that*

$$f_n'(z_k) \to 0 \qquad \text{as } k \to \infty \tag{2.3}$$

$$f_n(z_k) \to r \neq 1 \qquad \text{as } k \to \infty \tag{2.4}$$

contains a convergent subsequence.

Proof. Let $\{z_k\} \subset H_n$ be a sequence satisfying (2.3), (2.4). Since H_n is finite dimensional we need to prove only that $\{z_k\}$ is bounded.

By (2.3) there exists a sequence $\{\varepsilon_k\} \subset \mathbf{R}$ such that

$$\varepsilon_k \to 0 \qquad \text{as } k \to \infty \text{ and}$$

$$(2.5) \qquad \langle f_n'(z_k), \zeta\rangle = \varepsilon_k \|\zeta\| \qquad \text{for all} \quad \zeta \in H_n.$$

Set

$$z_k = (\eta_k, x_k)$$

and take in (2.5) $\zeta = (\eta_k, 0)$ and $\zeta = (0, \widetilde{x}_k)$ where $\widetilde{x}_k = x_k - \int_0^1 x_k \, ds$. We have

$$(2.6) \qquad \frac{4}{T^2}\int_0^1 \beta(x_k)(\eta_k + T)\eta_k = \varepsilon_k \|\eta_k\|_{L^2}$$

and

$$(2.7) \qquad \frac{2}{T^2}\int_0^1 [-(\alpha'(x_k)[\widetilde{x}_k]\dot{x}_k \mid \dot{x}_k) - (2\alpha(x_k)\dot{\widetilde{x}}_k \mid \dot{\widetilde{x}}_k) + (\beta'(x_k) \mid \widetilde{x}_k)T^2]\, ds = \varepsilon_k \|\widetilde{x}_k\|$$

where $\alpha'(x)[\widetilde{x}_k] = \{(\text{grad } g_{ij}(x) \mid \widetilde{x}_k)\}$ $i, j = 1, 2, 3$ and $\beta'(x_k) = \text{grad } g_{00}(x_k)$.

From (2.6) we have that

$$(2.8) \qquad \|\eta_k\|_{L^2}$$

is bounded. Moreover

$$(2.9) \qquad f_n(z_k) = \frac{1}{T^2}\int_0^1 [\beta(x_k)(\eta_k + T)^2 - (\alpha(x_k)\dot{x}_k \mid \dot{x}_k)]\, ds$$

is bounded. Since $\beta(x)$ is bounded and $\alpha(x)$ is positive definite (uniformly with respect to x) we deduce from (2.8), (2.9) that

$$(2.10) \qquad \|\dot{x}_k\|_{L^2}$$

is bounded. Now we shall prove that

$$\|x_k\|_{L^\infty} \tag{2.11}$$

is bounded. Arguing by contradiction we suppose that there exists a sequence (which we continue to call by $\{x_k\}$) such that

$$\|x_k\|_{L^\infty} \to +\infty \qquad \text{as } k \to \infty. \tag{2.12}$$

Now (2.10) and (2.12) imply that

$$\inf_{s\in(0,1)} |x_k(s)| \to +\infty \qquad \text{as } k \to \infty. \tag{2.13}$$

Then by (g_3) we get

$$\|\text{grad } g_{ij}(x_k)\|_{L^\infty} \to 0 \qquad \text{as } k \to \infty. \tag{2.14}$$

On the other hand (2.10) implies that $\|\dot{\widetilde{x}}_k\|_{L^2}$ is bounded. Then, since $\|\dot{\widetilde{x}}_k\|^2_{L^2} \geq \|\widetilde{x}_k\|^2_{L^2}$ (Wirtinger inequality), we deduce that

$$\|\widetilde{x}_k\|_{H^1} \tag{2.15}$$

is bounded. Then, by Sobolev embedding theorems,

$$\|\widetilde{x}_k\|_{L^\infty} \tag{2.16}$$

is bounded. By (2.7),(2.10),(2.14),(2.16) we easily deduce that

$$(\alpha\,(x_k)\,\dot{\widetilde{x}}_k \mid \dot{\widetilde{x}}_k) \to 0 \qquad \text{as } k \to \infty. \tag{2.17}$$

Then, by using (2.4), we have

$$\frac{1}{T^2}\int_0^1 \beta\,(x_k)(\eta_k^2 + 2\eta T + T^2)\,ds \to r \text{ as } k \to \infty. \tag{2.18}$$

Since $\beta(x) \to 1$ as $|x| \to \infty$, by (2.13) we deduce that

$$\|\beta(x_k) - 1\|_{L^\infty} \to 0 \qquad \text{as } k \to \infty. \tag{2.19}$$

From (2.19),(2.18), and since $\eta \in L^2_*$, we have that

(2.20) $$\frac{1}{T^2}\int_0^1 \eta_k^2\,ds + 1 \to r \qquad \text{as } k \to \infty.$$

Moreover, by (2.6),(2.19) and since $\int_0^1 \eta_k\,ds = 0$, we have

(2.21) $$\frac{4}{T^2}\|\eta_k\|_{L^2}^2 = \varepsilon_k\|\eta_k\|_{L^2} + 0(1).$$

Then

(2.22) $$\|\eta_k\|_{L^2}^2 \to 0 \qquad \text{as } k \to \infty.$$

From (2.20) and (2.22) we deduce that $r = 1$ and this contradicts (2.4).
QED

Now set

(2.23) $$\begin{cases} W_n &= \left\{x \in H_{n,x} \mid \int_0^1 x(s)\,ds = 0\right\} \\ V_n &= H_{1,x} \oplus H_{n,\eta} \end{cases}$$

Evidently

(2.24) $$\begin{cases} V_n + W_n = H_n \\ V_n \cap W_n = \text{span}\,\{\phi_i \sin 2\pi s, \phi_i \cos 2\pi s;\ i = 1,2,3\}. \end{cases}$$

Lemma 2.4. *Suppose that β satisfies (g_4), then for all $R > 1$ there exists $Q_R > 0$ such that*

(2.25) $$\mu_R \equiv \inf\left\{\int_0^1 \beta(x(s))\,ds : x \in H_{1,x},\ 1 \leq \|x\|_{H^1} \leq R\right\} = C_0 + Q_R$$

where $C_0 = \min \beta(x) = \beta(0)$.

Proof. Let $R > 1$ and set

$$C_R = \{x \in H_{1,x} \mid 1 \leq \|x\|_{H^1} \leq R\}.$$

Since $H_{1,x}$ is finite dimensional, C_R is compact then

(2.26) $$\exists \bar{x} \in C_R \quad \text{s.t.} \quad \int_0^1 \beta(\bar{x}(s))\,ds = \mu_R.$$

Since $\|\bar{x}\|_{H^1} \geq 1$, there exists $s_0 \in [0,1]$ such that $|\bar{x}(s_0)| > 0$. Then by (g_4) we have $\beta(\bar{x}(s_0)) > C_0$ and $\beta(\bar{x}(s)) \geq C_0\ \forall\ s \in [0,1]$. These inequalities and (2.26) easily imply (2.25). QED

Lemma 2.5. *Suppose that g is continuous and that it satisfies (g_4) (with $C_0 > 0$). Now let*

$$R > 1 + \sup\{\|\alpha(x)\|_{L^\infty} : \|x\|_{H^1} \le 1\} \tag{2.27}$$

where $\alpha(x) = \{g_{ij}(x)\}$ $i,j = 1,2,3$, and set for $n \in \mathbf{N}$

$$S_R = \{(\eta, x) \in V_n \mid C_0 \|\eta\|^2_{L^2} + \|x\|^2_{H^1} = R\} \tag{2.28}$$

where V_n is defined in (2.23). Then there exists $\overline{T} > 0$ such that for all $T > \overline{T}$ there exists $\delta \equiv \delta(T)$ such that

$$\inf_{z \in S_R} f_n(z) \ge C_0 + \delta \tag{2.29}$$

both $\overline{T}$ and $\delta(T)$ being independent on n.

Proof. Let $z = (\eta, x) \in S_R$ and distinguish two cases:
(I case) Suppose that

$$\|x\|_{H^1} \ge 1. \tag{2.30}$$

Then, since $\|x\| \le R$, by Lemma 2.4 we have

$$\begin{aligned} f_n(z) &= -\frac{1}{T^2}\int_0^1 (\alpha(x)\dot{x} \mid \dot{x})\, ds + \frac{1}{T^2}\int_0^1 \beta(x)\eta^2\, ds \\ &\quad + \frac{2}{T}\int_0^1 \beta(x)\eta\, ds + \int_0^1 \beta(x)\, ds \\ &\ge -\frac{1}{T^2}\int_0^1 (\alpha(x)\dot{x} \mid \dot{x})\, ds + Q_R + C_0 \\ &\quad - \frac{2}{T}\int_0^1 |\eta|\, ds. \end{aligned} \tag{2.31}$$

Now $z = (\eta, x) \in S_R$, then

$$\int_0^1 |\eta|\, ds \le \frac{R}{C_0}. \tag{2.32}$$

Moreover, since $\|x\|_{H^1} \le R$, there exists a constant $C_1 \equiv C_1(R)$ such that

$$\|\alpha(x)\|_{L^\infty} \le C_1$$

then

$$\frac{1}{T^2}\int_0^1 (\alpha(x)\dot{x} \mid \dot{x})\, ds \le \frac{C_1}{T^2}\cdot \|x\|^2_{H^1} \le \frac{C_1 R}{T^2}. \tag{2.33}$$

By (2.31), (2.32) and (2.33) we obtain

$$f_n(z) \ge -\frac{C_1 R}{T^2} - \frac{2R}{TC_0} + Q_R + C_0.$$

Then there exists $\overline{T} > 0$, depending on R and C_0, such that

$$\text{for all } T \ge \overline{T} \qquad f_n(z) \ge C_0 + \frac{Q_R}{2}. \tag{2.34}$$

(II case) Suppose now

$$\|x\|^2_{H^1} < 1. \tag{2.35}$$

Then, since $z \in S_R$, we get

$$C_0\,\|\eta\|^2_{L^2} > R - 1. \tag{2.36}$$

Now

$$\begin{aligned} f_n(z) &= \int_0^1 \beta(x)\Big(\frac{\eta+T}{T}\Big)^2 ds - \frac{1}{T^2}\int_0^1 (\alpha(x)\dot{x} \mid \dot{x})\, ds \\ &\ge C_0 \int_0^1 \Big(\frac{\eta+T}{T}\Big)^2 ds - \frac{1}{T^2}\int_0^1 (\alpha(x)\dot{x}|\dot{x})\, ds \\ &= \frac{C_0}{T^2}\,\|\eta\|^2_{L^2} + C_0 - \frac{1}{T^2}\int_0^1 (\alpha(x)\dot{x} \mid \dot{x})\, ds. \end{aligned}$$

Then, by using (2.36), we get

$$\begin{aligned} f_n(z) &> \frac{R-1}{T^2} + C_0 - \frac{1}{T^2}\cdot \sup\{\|\alpha(x)\|_{L^\infty} : \|x\|_{H^1} \le 1\} \\ &= C_0 + \frac{r}{T^2} \end{aligned} \tag{2.37}$$

where $r = R - 1 - \sup\{\|\alpha(x)\|_{L^\infty} : \|x\|_{H^1} \le 1\}$. By (2.27) r is positive. Finally, if $T \ge \overline{T}$ and $\delta = \min\left\{\frac{Q_R}{2}, \frac{r}{T^2}\right\}$, from (2.34) and (2.37) we obtain

$$f_n(z) \ge C_0 + \delta \quad \text{for all} \quad z \in S_R.$$

QED

Lemma 2.6. *Suppose that* $\sup \beta(x) = 1$ *and that* $\beta(0) = C_0 \neq 1$. *Suppose moreover that* α *is positive definite and set* $W = \left\{x \in H^1(S^1) \mid \int_0^1 x\, ds = 0\right\}$ *then*

$$\sup f(W) < 1.$$

Proof. Let $z = (0, x) \in W$. Since $\sup\left\{\beta(x) : x \in \mathbf{R}^3\right\} = 1$, we have

$$f(z) = \int_0^1 \beta(x) - \frac{1}{T^2}\int_0^1 (\alpha(x)\dot{x} \mid \dot{x}) \leq 1.$$

Arguing by contradiction suppose that $\sup f(W) = 1$. Then there exists $\{x_k\} \subset W$ such that

$$\int_0^1 (\alpha(x_k)\dot{x}_k \mid \dot{x}_k) \to 0 \quad \text{and} \quad \int_0^1 \beta(x_k) \to 1 \tag{2.38}$$

for $k \to \infty$. Since the matrix $\alpha(x)$ is positive definite (uniformly with respect to $x \in \mathbf{R}^3$)

$$\|\dot{x}_k\|^2_{L^2} \to 0 \qquad \text{as } k \to \infty.$$

Then, since $\int_0^1 x_k = 0$, we get

$$\|x_k\|_{H^1} \to 0 \qquad \text{as } k \to \infty.$$

We deduce that $\|x_k\|_{L^\infty} \to 0$ as $k \to \infty$ and

$$\int_0^1 \beta(x_k)\, ds \to \beta(0) = C_0.$$

And this contradicts (2.38). QED

We are now ready to prove Theorem 2.1.

Proof of Theorem 2.1. We have already observed that it is enough to study the critical points of f_n. We shall show that f_n satisfies assumptions (I_1), (I_2), (I_3) in Theorem 2.2. As we have already ovserved f_n satisfies (I_1). By virtue of Lemma 2.3 also (I_2) is satisfied with $\mu = 1$. By Lemmas 2.5 and 2.6 we obtain that, if $T > \overline{T} > 0$ (cf. Lemma 2.4), f_n satisfies assumption (I_3) (for all $n \in \mathbf{N}$) with

$$E_1 \equiv V_n,\ \ E_2 \equiv W_n,\ \ \lambda = C_0 + \delta,\ \ \partial B \equiv S_R$$

where V_n and W_n are defined in (2.23).

Moreover also (I_4) is satisfied. In fact $\operatorname{Fix} G = \{0\} \times \mathbf{R}^3$ and 0 is the unique critical point of β; then it is easy to check that $u \in \operatorname{Fix} G$ is a critical point of f_n if and only if $u = 0$.

Since $\operatorname{codim}(V_n + W_n) = 0$ and $\dim V_n \cap W_n = 6$ (cf. 2.24), we obtain, by using Theorem 2.2, that, if $T > \overline{T}$, for all $n \in \mathbf{N}$ f_n possesses at least three distinct critical points z_n^1, z_n^2, z_n^3 such that

$$C_0 + \delta \le f_n(z_n^i) \le \sup f(W) = 1 - \gamma$$

where $\gamma > 0$ by Lemma 2.6. QED

3. Proof of Theorem 0.1

Let $\{z_n\}$ be a sequence of critical points of h/M_n constructed as in Theorem 2.1. Then we have

$$C_0 + \delta \le h(z_n) \le 1 - \gamma. \tag{3.1}$$

By slight modifications of the arguments used in proving Lemma 2.3 we can prove that $\{z_n\}$ is bounded in H^1, then there exists a subsequence (which we continue to call $\{z_n\}$) such that

$$z_n \rightharpoonup z^* \quad \text{weakly in } H^1. \tag{3.2}$$

We shall prove that

$$z_n \to z^* \quad \text{strongly in } H^1.$$

We set

$$\begin{aligned} z_n &= \bar{z} + \zeta_n & \zeta_n &= (\tau_n, \xi_n) \\ z^* &= \bar{z} + \zeta^* & \zeta^* &= (\tau^*, \xi^*) \end{aligned}$$

and

$$(\tau_n^*, \xi_n^*) = \zeta_n^* = P_n \zeta^*$$

P_n being the projection on H_n.

Moreover we set

$$\Lambda = \begin{pmatrix} 1 & 0 & 0 & 0 \\ 0 & -1 & 0 & 0 \\ 0 & 0 & -1 & 0 \\ 0 & 0 & 0 & -1 \end{pmatrix}.$$

Since z_n are critical points of h/M_n we have

$$\begin{aligned} \frac{T^2}{2}\langle h'(z_n), \Lambda(\zeta_n - \zeta_n^*)\rangle &= \int_0^1 g(x_n)\,[\dot{z}_n, \Lambda(\dot{\zeta}_n - \dot{\zeta}_n^*)]\,ds - \\ &- \frac{1}{2}\int_0^1 \sum_{i,j=0}^{3} \sum_{\ell=1}^{3} \frac{\partial g_{ij}}{\partial x_\ell}(x_n)(\zeta_n - \zeta_n^*)_\ell \cdot (\dot{z}_n)_i\,(\dot{z}_n)_j\,ds = 0. \end{aligned} \tag{3.3}$$

H^1 is compactly embedded into L^∞, then by (3.2), $\xi_n \to \xi^*$ in L^∞ and $\{z_n\}$ is bounded in L^∞. Therefore

$$(3.4)\quad \frac{\partial g_{ij}}{\partial x_\ell}(x_n)(\xi_n - \xi_n^*)_\ell \to 0 \text{ in } L^\infty \ (i, j, = 0, \ldots, 3 \text{ and } \ell = 1, 2, 3).$$

Then from (3.3), (3.4), (3.2) we deduce that

$$(3.5)\qquad \int_0^1 g(x_n)[\dot{z}_n, \Lambda(\dot{\zeta}_n - \dot{\zeta}_n^*)]\, ds = 0(1).$$

In (3.5) and in the sequel 0(1) denotes a sequence converging to zero. Since $z_n = \bar{z} + \zeta_n$ we have

$$\int_0^1 g(x_n)[\dot{\bar{z}}, \Lambda(\dot{\zeta}_n - \dot{\zeta}_n^*)]\, ds + \int_0^1 g(x_n)[\dot{\zeta}_n, \Lambda(\dot{\zeta}_n - \dot{\zeta}_n^*)]\, ds = 0(1).$$

Then, since

$$(3.6)\qquad \Lambda(\dot{\zeta}_n - \dot{\zeta}_n^*) \rightharpoonup 0 \quad \text{weakly in } L^2$$

and $g_{ij}(x_n)(\dot{\bar{z}})_i$ converges (strongly) in L^∞, we get

$$(3.7)\qquad \int_0^1 g(x_n)[\dot{\zeta}_n, \Lambda(\dot{\zeta}_n - \dot{\zeta}_n^*)]\, ds = 0(1)$$

which can also be written as

$$\int_0^1 g(x_n)[(\dot{\zeta}_n - \dot{\zeta}_n^*), \Lambda(\dot{\zeta}_n - \dot{\zeta}_n^*)]\, ds + \int_0^1 g(x_n)[\dot{\zeta}_n^*, \Lambda(\dot{\zeta}_n - \dot{\zeta}_n^*)]\, ds = 0(1).$$

Then, by (3.6) and since $g_{ij}(x_n)(\dot{\zeta}_n^*)_i$ converges in L^2, we get

$$(3.8)\qquad \int_0^1 g(x_n)[(\dot{\zeta}_n - \dot{\zeta}_n^*), \Lambda(\dot{\zeta}_n - \dot{\zeta}_n^*)]\, ds = 0(1).$$

On the other hand it is easy to see that

$$(3.9)\qquad \int_0^1 g(x_n)[(\dot{\zeta}_n - \dot{\zeta}_n^*), \Lambda(\dot{\zeta}_n - \dot{\zeta}_n^*)]\, ds \geq \text{ const } \|\dot{\zeta}_n - \dot{\zeta}_n^*\|_{L^2}^2.$$

From (3.8) and (3.9) and since $\zeta_n^* \to \zeta^*$ in H^1 we get (at least for a subsequence)

$$z_n \longrightarrow z^* \quad \text{in } H^1. \tag{3.10}$$

Let us finally show that z^* is a critical point of h/M. By (3.10) we have

$$\forall\, \zeta \in H^1_{0,t} \times H^1(S^1) : \langle h'(z_n), \zeta \rangle \longrightarrow \langle h'(z^*), \zeta \rangle \text{ as } n \to \infty. \tag{3.11}$$

On the other hand

$$\langle h'(z_n), \zeta \rangle = \langle h'(z_n), \zeta_n \rangle + \langle h'(z_n), \zeta - \zeta_n \rangle. \tag{3.12}$$

Since z_n is a critical point of h/M_n and $\zeta - \zeta_n \to 0$ in H^1 as $n \to \infty$, from (3.12) we deduce that

$$\langle h'(z_n), \zeta \rangle = 0(1). \tag{3.13}$$

Finally from (3.11) and (3.13) we deduce that

$$\forall\, \zeta \in H^1_{0,t} \times H^1_x(S^1) : \langle h'(z^*), \zeta \rangle = 0$$

and therefore z^* is a critical point of h/M.

By (3.1) we have

$$C_0 + \delta \leq h(z^*) \leq 1 - \gamma. \tag{3.14}$$

Now set $z^* = \big(t^*(s), x^*(s)\big)$ and, arguing by contradiction, suppose that x^* is constant. Then it is easy to see that $h(z^*) = C_0$ and this contradicts (3.14). Therefore, by Theorem 1.1, we construct from z^* a nontrivial solution of problem (0.5).

QED

REFERENCES

[1] P. Bartolo, V. Benci, D. Fortunato, *Abstract critical point theorems and applications to some nonlinear problems with "strong" resonance at infinity*, Nonlinear Anal. T.M.A. **7** (1983), 981-1012.

[2] V. Benci, *A geometrical index for the group S^1 and some applications to the study of periodic solutions of ordinary differential equations*, Comm. Pure Appl. Math. **34** (1981), 393-432.

[3] V. Benci, *On the critical point theory for indefinite functionals in the presence of symmetries*, Trans. Amer. Math. Soc. **274** (1982), 533-572.

[4] V. Benci, A. Capozzi, D. Fortunato, *Periodic solutions of Hamiltonian systems with superquadratic potential*, Ann. Mat. Pura Appl. **143** (1986), 1-46.

[5] V. Benci, D. Fortunato, *Existence of geodesics for the Lorentz metric of a stationary gravitational field*, to appear in Ann. Inst. H. Poincaré Analyse Non-Linéaire.

[6] H. Berestycki, J. M. Lasry, G. Mancini, B. Ruf, *Existence of multiple periodic orbits on star-shaped Hamiltonian surface*, Comm. Pure Appl. Math. **38** (1985), 253-289.

[7] E. R. Fadell, P. H. Rabinowitz, *Generalized cohomological index theories for Lie group action with an application to bifurcation questions for Hamiltonian systems*, Inv. Math. **45** (1978), 139-174.

[8] L. Landau, E. Lifchitz, *Théorie des champs*, Editions Mir 1970.

[9] P. H. Rabinowitz, *Minimax methods in critical point theory with applications to differential equations*, Conf. Board Math. Sc. A.M.S. **65** (1986).

Vieri Benci
Istituto di Matematiche
Applicate - Università
Pisa, Italy

Donato Fortunato
Dipartimento di Matematica
Università
Bari, Italy

Morse Theory for Harmonic Maps

KUNG-CHING CHANG

In the previous paper [Ch1], we studied the Minimax Principle as well as the Ljusternik–Schnirelman category theory for harmonic maps with prescribed boundary data defined on Riemann surfaces by the heat flow method. In this paper, we shall continue our study on Morse theory. Our main results are the Morse inequalities (Theorem 1) for isolated harmonic maps, and the Morse handle body decomposition for nondegenerate harmonic maps (Theorem 2). These results are extensions of the work of K. Uhlenbeck [U1], where the harmonic maps are defined on manifolds without boundary, and are all assumed to be nondegenerate. Our method is based on the heat flow by which the deformation is constructed. In contrast to the perturbation method developed by K. Uhlenbeck [U1], our approach seems more direct than hers.

1. Preliminaries

Let (M,g) and (N,h) be two Riemannian manifolds. For a smooth map $u: M \to N$, let $e(u) = \frac{1}{2}|\nabla u|^2$ be the energy density, and let

$$E(u) = \int_M e(u)\, dV_g$$

be the energy. The critical points of the energy functional are harmonic maps, which satisfy the following Euler–Lagrange equation:

$$\Delta u^\alpha = \Delta_M u^\alpha + g^{ij}(x)\, {}^N\Gamma^\alpha_{\beta\gamma}(u) u^\alpha_{,i} u^\beta_{,j} = 0$$

where ${}^N\Gamma$ is the Christoffel symbol of N, (g^{ij}) is the inverse of the metric g, Δ_M is the Laplace–Beltrami operator w.r.t. g, and Δu is the trace of the tension tensor field.

The associate heat flow equation reads as follows

$$\partial_t f = \Delta f \tag{1.1}$$

where $f: [0,\infty) \times M \to N$.

We consider the initial-boundary value problem. Find $f \in C^{1+\frac{\gamma}{2},2+\gamma}((0,\infty)\times \bar{M}, N)$

$$f(0,x) = \phi(x), \tag{1.2}$$

$$f(t,\cdot)\big|_{\partial M} = \psi(\cdot), \tag{1.3}$$

where $\psi \in C^{2,\gamma}(\partial M, N)$, $\gamma > 0$, and $\phi \in C^{2,\gamma}_{\psi}(\bar{M}, N)$. The later is the class of $C^{2,\gamma}(\bar{M}, N)$ functions, which have the common boundary value ψ.

Let

$$m = \inf\{E(u) \mid u \in C^1_{\psi}(\bar{M}, N)\},$$
$$b = \inf\{E(v) \mid v : S^2 \to N, \text{ nonconstant harmonic}\}$$

(if there is no nonconstant harmonic map from S^2 to N, then we define $b = +\infty$), and let $\mathcal{F}$ be a component of $C^{2,\gamma}_{\psi}(\bar{M}, N)$.

Assume that $\dim M = 2$, and that

$$E(\phi) < m + b \text{ or } m_{\mathcal{F}} + b \text{ if } \pi_2(N) = 0, \text{ and } \phi \in \mathcal{F},$$

where

$$m_{\mathcal{F}} = \inf\{E(u) \mid u \in \mathcal{F}\}.$$

Then we have

(1) The heat flow, i.e. the solution of (1.1), (1.2) and (1.3), globally exists.

(2) $\exists$ a harmonic map $\tilde{u} \in C^{2,\gamma}_{\psi}(\bar{M}, N)$, and a sequence $t_j \uparrow +\infty$ such that

$$f(t_j, \cdot) \to \tilde{u}(\cdot) \text{ in } C^1(\bar{M}, N).$$

(3) If the infinitely dimensional manifold $C^{2,\gamma}_{\psi}(\bar{M}, N)$ is endowed with a weaker topology $W^2_p(M, N)$, $p > \frac{4}{1-\gamma}$, the flow

$$(t, \phi) \longmapsto f_{\phi}(t, \cdot)$$

is continuous from $[0,\infty) \times C^{2,\gamma}_{\psi}(\bar{M}, N) \to C^{2,\gamma}_{\psi}(\bar{M}, N)$, where $f_{\phi}(t,\cdot)$ denotes the flow with initial data ϕ.

(4) The set

$$K_c = \{u \in C^{2,\gamma}_{\psi}(\bar{M}, N) \mid \Delta u = 0, E(u) = c\}$$

is compact under the above topology, if $c < m + b$. (Or $c < m_{\mathcal{F}} + b$, if $\pi_2(N) = 0$, and $\phi \in \mathcal{F}$).

(5) Let $K = \bigcup_{c<m+b} K_c$ (or $\bigcup_{c<m_{\mathcal{F}}+b} K_c$, if $\pi_2(N) = 0$, and $\phi \in \mathcal{F}$). Suppose that

$$\operatorname{dist}_{W_p^2}(f_\phi(t,\cdot), K) \geq \delta > 0 \quad \forall t \in \mathbf{R}_+^1$$

then we have $\varepsilon = \varepsilon(\delta) > 0$ such that

$$\|\Delta f_\phi(t,\cdot)\|_{L^2(M,N)} \geq \varepsilon.$$

(6) For any closed neighborhood $U \subset C_\psi^{2,\gamma}(\bar{M}, N)$ of K_c, under the W_p^2-topology, where $c < m + b$ (or $m_{\mathcal{F}} + b$ if $\pi_2(N) = 0$ and $\phi \in \mathcal{F}$), $\exists \varepsilon > 0$, a closed neighborhood $V \subset U$, and a $W_p^2 - (p > \frac{4}{1-\gamma})$ strong deformation retract $\eta\colon [0,1] \times E_{c+\varepsilon} \to E_{c+\varepsilon}$ satisfying

$$\eta(1, E_c \cap V) \subset E_c \cap U, \text{ and}$$
$$\eta(1, E_{c+\varepsilon} \setminus V) \subset E_{c-\varepsilon},$$

where $E_a = \{u \in C_\psi^{2,\gamma}(\bar{M}, N) \mid E(u) \leq a\}$ is the level set, $\forall a \in \mathbf{R}_+^1$.

It follows from the proof of Th. 7.1 in [Ch1].

2. The Morse Inequalities

In this section, we establish the Morse inequalities for harmonic maps under the assumption that all harmonic maps are isolated. As shown in [Ch2], the crucial step in the proof is to prove the following deformation lemma.

Lemma 2.1. *Let $\mathcal{F}$ be a component of $C_\psi^{2+\gamma}(\bar{M}, N)$. Suppose that there is no harmonic maps with energy in the interval $(c, d]$, where $d < m_{\mathcal{F}} + b$, and that there are at most finitely many harmonic maps on the level $E^{-1}(c)$. Assume that $\pi_2(N) = 0$, then E_c is a strong deformation retract of E_d.*

In order to give the proof, first we turn out to improve the conclusion in Section 1, under the condition that the set of smooth harmonic maps is isolated. Namely

Lemma 2.2. *Let $E(\phi) < m_{\mathcal{F}} + b$, and let*

$$c = \lim_{t\to+\infty} E(f_\phi(t,\cdot)).$$

If K_c is isolated, then $f_\phi(t,\cdot) \to \tilde{u} \in K_c$ in the W_p^2-topology, $\forall p > \frac{4}{1-\gamma}$, as $t \to +\infty$.

Proof. According to the conclusion (2), in combining with a bootstrap iteration, one shows that $\exists \tilde{u} \in K_c$ and $t_j \uparrow +\infty$ such that

$$f_\phi(t_j, \cdot) \to \tilde{u}, \quad C_\psi^{2,\gamma'}(\bar{M}, N), \quad \forall \gamma' \in (0, \gamma).$$

If our conclusion was not true, then there would be a $\delta > 0$, such that the neighborhood

$$U_\delta = \left\{u \in C_\psi^{2,\gamma}(\bar{M}, N) \mid \text{dist}_{W_p^2}(u, \tilde{u}) \leq \delta\right\}$$

contains the single element $\tilde{u}$ in K_c, and a sequence $t'_j \uparrow +\infty$ such that $f_\phi(t'_j, \cdot) \notin U_\delta$. Therefore $\exists (t_i^*, t_i^{**})$ satisfying

(1) $t_i^*, t_i^{**} \to +\infty$,

(2) $f_\phi(t_i^*, \cdot) \in \partial U_{2\delta}, \quad f_\phi(t_i^{**}, \cdot) \in \partial U_\delta$,

and

(3) $f_\phi(t, \cdot) \in U_{2\delta} \setminus U_\delta \quad \forall t \in (t_i^*, t_i^{**})$.

On one hand, we had

$$\delta \leq \left\| f_\phi(t_i^*, \cdot) - f_\phi(t_i^{**}, \cdot) \right\|_{W_p^2} \leq C_\delta |t_i^* - t_i^{**}|^{\gamma/2},$$

provided by the embedding theorem. On the other hand, according to conclusion (5),

$$\begin{aligned} &E(f_\phi(t_i^{**}, \cdot)) - E(f_\phi(t_i^*, \cdot)) \\ &= \int_{t_i^*}^{t_i^{**}} \int_M |\partial_t f(t, \cdot)|^2 dV_g\, dt \\ &= \int_{t_i^*}^{t_i^{**}} \int_M |\Delta f(t, \cdot)|^2 dV_g\, dt \\ &> \varepsilon(\delta) |t_i^{**} - t_i^*|. \end{aligned}$$

Since the LHS of the inequality tends to zero as $i \to \infty$, this is a contradiction. $\square$

Now we return to the **proof of Lemma 2.1**, the basic idea is to reparametrize the heat flow $f_\phi(t, \cdot)$.

Let $\tau = \rho(t)$, where

$$\rho(t) = (E(\phi) - c)^{-1} \int_0^t \|\Delta f_\phi(s, \cdot)\|_{L^2}^2 ds,$$

if $E(\phi) > c$, and let

$$g(\tau, \cdot) = f(t, \cdot).$$

Then we have the following relations:

(1)

$$\begin{aligned}\partial_\tau g(\tau, \cdot) &= \frac{dt}{d\tau}\partial_t f(t, \cdot) \\ &= \frac{(E(\phi) - c)}{\|\Delta g(\tau, \cdot)\|_{L^2}^2}\Delta g(\tau, \cdot),\end{aligned}$$

(2)

$$\begin{aligned}\frac{d}{d\tau}E(g(\tau, \cdot)) &= -\int_M \langle \partial_\tau g(\tau, \cdot), \Delta g(\tau, \cdot)\rangle dV_g \\ &= -(E(\phi) - c).\end{aligned}$$

Therefore

$$E(g(\tau, \cdot)) = (1 - \tau)E(\phi) + \tau c, \quad \forall \tau \in [0, 1].$$

(3) The function $\rho: [0, \infty) \to \mathbf{R}^1$, is continuous and monotone increasing, which satisfies the following properties:

$$\begin{aligned}\rho(0) &= 0, \\ \rho(+\infty) &= 1 \quad \text{if} \quad f_\phi(t, \cdot) \to \tilde{u} \in K_c \text{ as } t \to +\infty, \\ \rho(+\infty) &> 1 \quad \text{if} \quad \lim_{t \to +\infty} E(f_\phi(t, \cdot)) < c.\end{aligned}$$

Let us define a function $\eta: [0, 1] \times E_d \to E_d$ as follows:

$$\eta(\tau, \phi) = \begin{cases} g_\phi(\tau, \cdot) & \text{if } (\tau, \phi) \in [0, 1] \times (E_d \setminus E_c), \\ \phi & \text{if } (\tau, \phi) \in [0, 1] \times E_c. \end{cases}$$

In order to show that E_c is a deformation retract of E_d, only the continuity at the following sets is needed:

(1) $\{1\} \times A$, where $A = \{\phi \in E_d \setminus E_c \mid f_\phi(\infty, \cdot) \in K_c\}$

(2) $[0, 1] \times E^{-1}(c)$.

Verification for Case (1). $\forall \phi_0 \in A$, $\forall \varepsilon > 0$, want to find $\delta > 0$ such that

$$\left.\begin{aligned}\operatorname{dist}_{W_p^2}(\phi, \phi_0) < \delta \\ \tau > 1 - \delta\end{aligned}\right\} \text{ implies } \operatorname{dist}_{W_p^2}(g_\phi(\tau, \cdot), \tilde{u}) < \varepsilon$$

where $\tilde{u} = f_{\phi_0}(\infty, \cdot)$.

Choose $\varepsilon_0 = \varepsilon_0(\delta_1)$ as in the conclusion (5), i.e.

$$\|\Delta f_\phi(t, \cdot)\|_{L^2} \geq \varepsilon_0 \text{ if } \operatorname{dist}_{W_p^2}(f_\phi(t, \cdot), K) \geq \delta_1 \ \forall t$$

and choose

$$0 < \delta_1 < \left(\frac{\varepsilon}{2C_\varepsilon}\right)^{2/\gamma} \frac{\varepsilon_0^2}{E(\phi) - c}$$

such that

$$\mathrm{dist}_{W_p^2}(g_{\phi_0}(1-\delta_1,\cdot),\tilde{u}) < \frac{\varepsilon}{2}.$$

Again, we choose $\delta_2 > 0$ such that $\mathrm{dist}_{W_p^2}(\phi,\phi_0) < \delta_2$ implies

$$\mathrm{dist}_{W_p^2}(g_{\phi_0}(1-\delta_1,\cdot), g_\phi(1-\delta_1,\cdot)) < \frac{\varepsilon}{2}.$$

Therefore we have

$$\mathrm{dist}_{W_p^2}(g_\phi(1-\delta_1,\cdot),\tilde{u}) < \varepsilon \quad \forall \phi \in B_{\delta_2}(\phi_0).$$

We want to prove

$$\mathrm{dist}_{W_p^2}(g_\phi(\tau,\cdot),\tilde{u}) < \varepsilon, \quad \forall(\tau,\phi) \in (1-\delta_1, 1] \times B_{\delta_2}(\phi_0).$$

If not, $\exists\, \tau'' > \tau' > 1 - \delta_1$ and $\phi_1 \in B_{\delta_2}(\phi_0)$ such that

$$g_{\phi_1}(\tau',\cdot) \in \partial B_{\varepsilon/2}(\tilde{u}),$$

and

$$g_{\phi_1}(\tau'',\cdot) \in \partial B_\varepsilon(\tilde{u}).$$

Then we have

$$\begin{aligned}\frac{\varepsilon}{2} &\le \mathrm{dist}_{W_p^2}(g_{\phi_1}(\tau',\cdot), g_{\phi_1}(\tau'',\cdot))\\ &= \mathrm{dist}_{W_p^2}(f_{\phi_1}(t',\cdot), f_{\phi_1}(t'',\cdot))\\ &\le C_\varepsilon |t'-t''|^{\gamma/2}.\end{aligned}$$

On the other hand

$$\begin{aligned}\varepsilon_0^2|t'-t''| &\le \int_{t'}^{t''} \|\Delta f_{\phi_1}(t,\cdot)\|_{L^2}^2 dt\\ &= E\big(f_{\phi_1}(t'',\cdot)\big) - E\big(f_{\phi_1}(t',\cdot)\big)\\ &= E\big(g_{\phi_1}(\tau'',\cdot)\big) - E\big(g_{\phi_1}(\tau',\cdot)\big)\\ &= (E(\phi) - c)|\tau''-\tau'|\\ &< \delta_1(E(\phi)-c),\end{aligned}$$

which implies that

$$\delta_1 > \frac{\varepsilon_0^2}{E(\phi)-c}|t'-t''| \ge \frac{\varepsilon_0^2}{(E(\phi)-c)}\left(\frac{\varepsilon}{2C_\varepsilon}\right)^{2/\gamma}.$$

This is a contradiction.

Verification for Case (2). $\forall \phi_0 \in E^{-1}(c)$, $\forall \varepsilon > 0$ we want to find $\delta > 0$ such that $\mathrm{dist}(\phi, \phi_0) < \delta$ implies $\mathrm{dist}(\eta(\tau, \phi), \phi_0) < \varepsilon$.

Similar to the above argument, let us choose

$$0 < \delta_1 < \varepsilon_0^2 \left(\frac{\varepsilon}{2C_\varepsilon} \right)^{2/\gamma}.$$

Find $0 < \delta < \varepsilon/2$ such that

$$E(\phi) - c < \delta_1 \quad \forall \phi \in B_\delta(\phi_0).$$

If our conclusion was not true, by the same procedure, we would have

(i) $\frac{\varepsilon}{2} \le C_\varepsilon |t' - t''|^{\gamma/2}$,

and

(ii) $\varepsilon_0^2 |t' - t''| \le (E(\phi) - c)|\tau'' - \tau'| < \delta_1$.

This is again a contradiction.

The continuity of η is proved, so that E_c is a strong deformation retract of E_d, $d < m_{\mathcal{F}} + b$. □

Before going to set up the Morse inequalities, we define the critical groups for an isolated harmonic map.

Let G be an Abelian group. Let $u_0 \in \mathcal{F}$ be an isolated harmonic map $c = E(u_0)$. Choose a neighborhood U of u_0 such that $K \cap U \cap \mathcal{F} = \{u_0\}$.

Definition.

$$C_q(u_0; G) = H_q(E_c \cap U, (E_c \setminus \{u_0\}) \cap U; G)$$

$q = 0, 1, 2, \ldots$ are defined to be the critical groups with coefficient group G of E at u_0, where $H_*(X, Y; G)$ stands for the singular relative homology group with coefficient group G.

The excision property of the relative homology groups assures that these critical groups are well defined, i.e. they do not depend on the special choice of U.

Suppose that $\forall d < m_{\mathcal{F}} + b$, there are only isolated harmonic maps. Since $K \cap E_d$ is compact, they are finite. There are only isolated critical values (at most with limit $m_{\mathcal{F}} + b$)

$$m_{\mathcal{F}} = c_0 < c_1 < \ldots < c_i < \ldots < m_{\mathcal{F}} + b.$$

For each c_i, there are finitely many harmonic maps:

$$K_{c_i} = \{u_{ij} \mid j = 1, 2, \ldots, m_i\}.$$

$\forall d < m_{\mathcal{F}} + b$, let

$$M_q^d = \sum_{c_i < d} \sum_{j=1}^{m_i} \text{rank } C_q(u_{ij}; G)$$

be the q-th Morse type number, $q = 0, 1, 2, \ldots$, for the manifold $E_d \cap \mathcal{F}_1$ and let

$$\beta_q^d = \text{rank } H_q(E_d \cap \mathcal{F}, G)$$

be the qth Betti number, $q = 0, 1, 2, \ldots$, for $E_d \cap \mathcal{F}$.

It follows from a direct computation (cf. [Ch2]) that there exists a formal power series with nonnegative coefficients $Q^d(t)$ such that

$$\sum_{q=0}^{\infty} M_q^d t^q = \sum_{q=0}^{\infty} \beta_q^d t^q + (1+t) Q^d(t).$$

This includes a series of Morse inequalities. Namely, we have proved:

Theorem 1. *Let $\mathcal{F}$ be a component of $C_{\psi}^{2,\gamma}(\bar{M}^2, N)$, and let $d < m_{\mathcal{F}} + b$. Assume that $\pi_2(N) = 0$, and that in the level set $E_d \cap \mathcal{F}$ there are only isolated harmonic maps. Then we have the following Morse inequalities:*

$$M_0^d \geq \beta_0^d,$$
$$M_1^d - M_0^d \geq \beta_1^d - \beta_0^d,$$
$$\cdots \quad \cdots$$
$$M_n^d - M_{n-1}^d + \cdots + (-1)^n M_0^d \geq \beta_n^d - \beta_{n-1}^d + \cdots + (-1)^n \beta_0^d,$$

And similarly.

Theorem 1′. *Let $d < m + b$. Assume that there are only isolated harmonic maps in the level set E_d, then the above inequalities hold, wherein, the Morse type numbers M_q^d count all harmonic maps in the level set E_d, and β_q^d is the Betti number of E_d, $q = 0, 1, 2, \ldots$.*

3. Morse Decomposition

In this section, we study the handle body decomposition of the level sets of the energy function, under the assumption that all harmonic maps in these level sets are nondegenerate. As a consequence, we explain the Morse type numbers M_q^d which was studied before.

Let u_0 be a harmonic map from M to N. Let $E = u_0^*TN$ be the pull back bundle over M. Let $\mathcal{O}$ be a neighborhood in $C^\infty(M,N)$ which contains the section $u_0(M)$. It is obvious that $\mathcal{O}$ is diffeomorphic to a neighborhood $\mathcal{O}_E$ of the zero section of the tangent space $T_{u_0}(E)$. The diffeomorphism is realized by the exponential map:

$$\begin{array}{ccc} \sigma \in \mathcal{O}_E & \xrightarrow{\exp_{u_0(x)}} & \mathcal{O} \\ \cap & & \cap \\ T_{u_0}(E) & & C_\psi^\infty(M,N). \end{array}$$

Since then we do not distinguish the tangent vector σ with its exponential map $\exp_{u_0(x)}\sigma(x)$. We shall restrict our studies in the neighborhood $\mathcal{O}_E$ of the vector space $T_{u_0}(E)$. The Taylor expansion of the energy functional at u_0 is as follows:

$$E(u) = E(u_0) + \frac{1}{2}d^2E(u_0)(\sigma,\sigma) + R(\sigma)$$

where $u(x) = \exp_{u_0(x)}\sigma(x)$, and the remainder $R(\sigma)$ satisfies

$$|R(\sigma)| = o\Big(\int_M |\nabla\sigma|^2\Big),$$

and

$$|dR(\sigma)| = o\Big(\Big(\int_M |\nabla\sigma|^2\Big)^{1/2}\Big).$$

As to the Hessian $d^2E(u_0)$, it is well known (see Eells–Lemaire [EL$_1$]) that, $\forall\sigma,\eta \in C^\infty(T_{u_0}(E))$,

$$d^2E(u_0)(\sigma,\eta) = \int_M \langle J_{u_0}\sigma,\eta\rangle dV_g,$$

where

$$J_{u_0}\sigma = -\Delta^{u_0}\sigma - \text{Trace } R^N(du_0,\sigma)du_0,$$

is the Jacobi operator.

Noticing that J_{u_0} is a linear self-adjoint elliptic differential operator, with domain $W_2^2 \cap \overset{\circ}{W}{}_2^1\,(T_{u_0}(E))$, J_{u_0} can be extended to be a continuous bilinear form on the Hilbert space $\overset{\circ}{W}{}_2^1\,(T_{u_0}(E))$. And since

$$d^2E(u_0)(\sigma,\sigma) \geq \|\sigma\|^2_{W_2^1} - C_1(u_0)\|\sigma\|^2$$

where $C_1(u_0)$ is a constant depending on u_0, the negative eigenspace of J_{u_0} must be finitely dimensional. The dimension of the negative eigenspace of J_{u_0} is called the Morse index of the harmonic map u_0, and is denoted by $\text{ind}(u_0)$. u_0 is called nondegenerate, if J_{u_0} is invertible.

For the self-adjoint operator J_{u_0}, it is well known that we have a spectral decomposition E_λ and two projections P_+ and P_-, which correspond to the positive and negative eigenspaces respectively. For any $\sigma \in C^{2,\gamma}_\psi(u_0^*TN)$, we have

$$\sigma_\pm := P_\pm \sigma \in C^{2,\gamma}_\psi(u_0^*TN).$$

The two square roots

$$A_\pm := (P_\pm(\pm J_{u_0})P_\pm)^{1/2}$$

are well defined, and we have that

$$\|A_\pm \sigma\|_{L^2} \text{ is equivalent to } \|\sigma_\pm\|_{\overset{\circ}{W}{}^1_2}.$$

In the following, we shall denote $\|A_\pm\sigma\|_{L^2}$ by $|\sigma_\pm|$, and let $|\sigma|^2 = |\sigma_+|^2 + |\sigma_-|^2$. Thus, the energy function is written as follows:

$$E(u) = c + \frac{1}{2}(|\sigma_+|^2 - |\sigma_-|^2) + R(\sigma).$$

For any given $0 < \gamma < 1$; we choose $\tau > 0$, satisfying

$$\frac{\tau}{1-\tau} < \sqrt{\frac{1-\gamma}{1+\gamma}},$$

and $\delta > 0$ such that for a W^2_p-ball B_δ with radius δ, centered at the zero section of $C^{2,\gamma}(u_0^*TN)$, we have,

$$|R(\sigma)| < \frac{1}{2}\gamma|\sigma|^2 \tag{3.1}$$

and

$$|dR(\sigma)| < \tau|\sigma|. \tag{3.2}$$

$\forall \sigma \in U = B_\delta$. (In the following we always denote B_δ by U). These imply that

$$\begin{aligned} &\frac{1}{2}(1-\gamma)|\sigma_+|^2 - \frac{1}{2}(1+\gamma)|\sigma_-|^2 \\ &\qquad \le E(u) - c \le \frac{1}{2}(1+\gamma)|\sigma_+|^2 - \frac{1}{2}(1-\gamma)|\sigma_-|^2. \end{aligned} \tag{3.3}$$

Now we are going to construct a series of deformations, which deform the level set $E_{c+\varepsilon}$ (for suitable $\varepsilon > 0$) to $E_{c-\varepsilon}$ attached with cells:

(1) According to Lemma 2.1 we have a strong deformation retract η_1, which deforms $E_{c+\varepsilon}$ into E_c, for $\varepsilon > 0$ small, if $E^{-1}(c, c+\varepsilon] \cap K = \emptyset$.

(2) By the conclusion (6), we have $\varepsilon > 0$ and a strong deformation retract η_2, which deforms E_c into $E_{c-\varepsilon} \cup (E_c \cap U)$, and satisfies $\eta_2(1, E_c \cap V) \subset E_c \cap U$, $\eta_2(1, E_c \setminus V) \subset E_{c-\varepsilon}$.

(3) Let us define two conical neighborhoods:

$$C_\gamma = \{\sigma \in U \mid |\sigma_+| \le \sqrt{\frac{1-\gamma}{1+\gamma}}|\sigma_-|\},$$

$$\hat{C}_\gamma = \{\sigma \in U \mid |\sigma_+| \le \sqrt{\frac{1+\gamma}{1-\gamma}}|\sigma_-|\}.$$

The inequality (3.3) implies that

$$C_\gamma \subset E_c \cap U \subset \hat{C}_\gamma.$$

Lemma 3.1. *There exists a strong deformation retract η_3, which deforms $E_{c-\varepsilon} \cup (E_c \cap U)$ into $E_{c-\varepsilon} \cup C_\gamma$.*

Proof. Noticing that $\forall \sigma \notin E_{c-\varepsilon} \cup C_\gamma$, with $\sigma \in U$, we have

$$|\sigma_-| \le \sqrt{\frac{1+\gamma}{2}}\delta.$$

Let $K = \sqrt{\frac{2}{1+\gamma}} - 1$ (> 0), and define a flow on U as follows:

$$\eta(t, \sigma) = (1-t)\sigma_+ + (1+tK)\sigma_-.$$

We have

(a) $\eta(0, \sigma) = \sigma$.

(b) $\eta(1, \sigma) = \sqrt{\frac{2}{1+\gamma}}\sigma_- \in U$ if $\sigma \notin C_\gamma$.

(c) Let $\phi(t) = E(\eta(t, \cdot))$, we have

$$\begin{aligned}
\phi'(t) &= -\frac{|\eta_+|^2}{1-t} - K|\eta_-|^2 + \langle dR(\eta(t,\cdot)), -\sigma_+ + K\sigma_-\rangle \\
&\le (1-\tau)\left[-\frac{|\eta_+|^2}{1-t} - K|\eta_-|^2 + \frac{\tau}{1-\tau}\left(\frac{1}{1-t} + K\right)|\eta_+||\eta_-|\right] \\
&= (1-\tau)\left[-\frac{|\eta_+|}{1-t}\left(|\eta_+| - \frac{\tau}{1-\tau}|\eta_-|\right) - K|\eta_-|\left(|\eta_-| - \frac{\tau}{1-\tau}|\eta_+|\right)\right]
\end{aligned}$$

where $\eta = (\eta_+, \eta_-)$. If $\eta \in (E_c \cap U) \setminus C_\gamma \subset \hat{C}_\gamma \setminus C_\gamma$, then we have

$$|\eta_+| \geq \sqrt{\frac{1-\gamma}{1+\gamma}}|\eta_-| \geq \frac{\tau}{1-\tau}|\eta_-|,$$

and

$$|\eta_-| \geq \sqrt{\frac{1-\gamma}{1+\gamma}}|\eta_+| \geq \frac{\tau}{1-\tau}|\eta_+|.$$

It follows

$$\phi'(t) < 0 \quad \forall \eta \in \hat{C}_\gamma \setminus C_\gamma. \tag{3.4}$$

Combining (a), (b) with (c), we obtain

$$\eta(t,\sigma) \in (E_c \cap U), \quad \forall (t,\sigma) \in [0,1] \times ((E_c \cap U) \setminus (C_\gamma \cup E_{c-\varepsilon}))$$

provided by the fact that $C_\gamma \subset E_c \cap U$.

From (a) and (b), we see that if $\sigma \notin E_{c-\varepsilon} \cup C_\gamma$, but $\sigma \in E_c \cap U$, then there is a unique $t^* \in (0,t)$ such that $\eta(t^*,\sigma) \in E^{-1}(c-\varepsilon) \cup \partial C_\gamma$. The uniqueness and the continuous dependence of t^* to σ are verified by the transversality: $\eta \overline{\pitchfork} E^{-1}(c-\varepsilon) \cup \partial C_\gamma$, which follows from the inequality (3.4).

Let us define

$$\eta_3(t,\sigma) = \begin{cases} \eta(t^*t,\sigma) & \text{if } \sigma \in (E_c \cap U) \setminus (C_\gamma \cup E_{c-\varepsilon}) \\ \sigma & \sigma \in E_{c-\varepsilon} \cup C_\gamma \end{cases}$$

This is the deformation we need. □

(4) Noticing that $\forall \sigma \in E_{c-\varepsilon} \cap C_\gamma$,

$$-\varepsilon \geq E(u) - c \geq \frac{1-\gamma}{2}|\sigma_+|^2 - \frac{1+\gamma}{2}|\sigma_-|^2.$$

We have

$$|\sigma_-| > \sqrt{\frac{2\varepsilon}{1+\gamma}} \tag{3.5}$$

so $E_{c-\varepsilon} \cap C_\gamma \subset S := \{\sigma \in C_\gamma \mid |\sigma_-| > \sqrt{\frac{2\varepsilon}{1+\gamma}}\}$.

On the other hand $\forall \sigma \in S$,

$$|\sigma_-| \geq k_0|\sigma_+| + \delta_0,$$

where

$$k_0 = \frac{1}{2}\sqrt{\frac{1+\gamma}{1-\gamma}} \quad \text{and} \quad \delta_0 = \frac{1}{2}\sqrt{\frac{2\varepsilon}{1+\gamma}}.$$

Let us define

$$T_{k_0,\delta_0} = \{\sigma \in C_\gamma \mid |\sigma_-| \geq k_0|\sigma_+| + \delta_0\}.$$

In the following, we turn out to prove

Lemma 3.2. *There is a strong deformation retract η_4 which deforms $E_{c-\varepsilon} \cup C_\gamma$ into $E_{c-\varepsilon} \cup T_{k_0,\delta_0} \cup \{\theta_+\} \times B^k_{\delta_0}$, where $k = \mathrm{ind}(u_0)$.*

Proof. We define

$$\eta_4(t,\sigma) = \begin{cases} \sigma & \sigma \in E_{c-\varepsilon} \cup T_{k_0,\delta_0}, \\ \sigma_- + [1 - t(1 - \frac{|\sigma_-|-\delta_0}{k_0|\sigma_+|})]\sigma_+ & \sigma \in C_\gamma, \delta_0 \leq |\sigma_-| \leq k_0|\sigma_+| + \delta_0 \\ \sigma_- + (1-t)\sigma_+ & \sigma \in C_\gamma \cap \{|\sigma_-| \leq \delta_0\} \end{cases} \quad \square$$

(5) Choose $\varepsilon > 0$ so small, that

$$\varepsilon < \frac{\delta^2(1-\gamma)}{2}. \tag{3.6}$$

Define

$$0 < \mu < \sqrt{\frac{1-(\gamma + \frac{2\varepsilon}{\delta^2})}{1+(\gamma+\frac{2\varepsilon}{\delta^2})}} \tag{3.7}$$

we consider the energy function on the conical section of the sphere ∂B_δ : $S_\mu = \{\sigma \in \partial B_\delta \mid |\sigma_+| < \mu|\sigma_-|\}$. Let $\sigma \in S_\mu$, we have

$$\begin{aligned} E(u) - c &\leq \frac{1+\gamma}{2}|\sigma_+|^2 - \frac{1-\gamma}{2}|\sigma_-|^2 \\ &\leq \left(\frac{1+\gamma}{2}\mu^2 - \frac{1-\gamma}{2}\right)|\sigma_-|^2 \\ &= -\frac{1}{2}\left(\frac{1-\mu^2}{1+\mu^2} - \gamma\right)|\sigma_-|^2(1+\mu^2). \end{aligned}$$

Since

$$\delta^2 = |\sigma_+|^2 + |\sigma_-|^2 \leq (1+\mu^2)|\sigma_-|^2,$$

and then

$$E(u) - c \leq -\frac{1}{2}\left(\frac{1-\mu^2}{1+\mu^2} - \gamma\right)\delta^2 < -\varepsilon,$$

i.e. $S_\mu \subset \overset{\circ}{E}_{c-\varepsilon}$.

Lemma 3.3. *The exit set of the flow*

$$\eta(t,\sigma) = e^{-k_1 t}\sigma_+ + e^{k_2 t}\sigma_-$$

on the ball B_δ, is the set S_μ, where $\frac{k_2}{k_1} = \mu^2$, $k_1, k_2 > 0$.

Proof. The flow η remains on the plane generated by the two vectors σ_+ and σ_-. Suppose that η meets ∂B_δ at time t_0, and let $\eta_+ = e^{-k_1 t_0}\sigma_+$, $\eta_- = e^{k_2 t_0}\sigma_-$.

Choosing suitable coordinates $(\eta_+, \eta_-) = \delta(\cos\theta, \sin\theta)$, we assume that the flow η leaves the ball B_δ. By comparing the tangents of the ball with the tangents of the flow, we see

$$-\frac{k_2}{k_1}\operatorname{tg}\theta > -\operatorname{ctg}\theta,$$

i.e.

$$|\eta_+| < \mu|\eta_-|.$$

In other words $(\eta_+, \eta_-) \in S_\mu$. □

Lemma 3.4. *There is a strong deformation retract η_5 which deforms the set $E_{c-\varepsilon} \cup T_{k_0,\delta_0} \cup (\{\theta_+\} \times B^k_{\delta_0})$ into $E_{c-\varepsilon} \cup (\{\theta_+\} \times B^k_\delta)$.*

Proof. We use the flow η defined in Lemma 3.3. Because $S_\mu \subset \overset{\circ}{E}_{c-\varepsilon}$, if $\sigma \notin E_{c-\varepsilon}$, then there must be a $t^* \in (0,\infty)$ such that $\eta(t^*,\sigma) \in E^{-1}(c-\varepsilon)$. On the other hand $\eta(t,\cdot)$ is transversal to the level set $E^{-1}(c-\varepsilon)$, provided by the fact:

$$\begin{aligned}
\frac{d}{dt}E(\eta(t,\sigma)) &= -\langle\eta_+, k_1\eta_+\rangle - \langle\eta_-, k_2\eta_-\rangle + \langle dR(\eta), -k_1\eta_+ + k_2\eta_-\rangle \\
&\le -k_1|\eta_+|^2 - k_2|\eta_-|^2 + \tau(|\eta_+| + |\eta_-|)(k_1|\eta_+| + k_2|\eta_-|) \\
&= -(1-\tau)[k_1|\eta_+|^2 + k_2|\eta_-|^2 - \frac{\tau}{1-\tau}(k_1+k_2)|\eta_+|\,|\eta_-|] \\
&= -(1-\tau)k_1[|\eta_+|^2 + \mu^2|\eta_-|^2 - \frac{\tau}{1-\tau}(1+\mu^2)|\eta_+|\,|\eta_-|] < 0,
\end{aligned}$$

if we choose

$$\frac{\tau}{1-\tau} < \frac{2\mu}{1+\mu^2}. \tag{3.8}$$

Therefore $\forall\sigma \in T_{k_0,\delta_0} \setminus E_{c-\varepsilon}$, $t^* = t^*(\sigma)$ is uniquely determined, and is continuous. We define our deformation retract as follows:

$$\eta_5(t,\sigma) = \begin{cases} \sigma & \text{if } \sigma \in E_{c-\varepsilon} \\ \eta(t^*(\sigma)t,\sigma) & \text{if } \sigma \in T_{k_0\delta_0} \setminus E_{c-\varepsilon} \\ e^{k_2 t^*(\delta_0\sigma - /|\sigma-|)}\sigma_- & \text{if } \sigma \in \{\theta_+\} \times B^k_{\delta_0}. \end{cases} \qquad \Box$$

For any two strong deformation retract

$$X_1 \xrightarrow{\phi_1} X_2 \xrightarrow{\phi_2} Y$$

we define their composition as follows

$$\phi(t,x) = \begin{cases} \phi_1(2t,x) & t \in \left[0,\frac{1}{2}\right] \\ \phi_2(2t-1,\phi_1(1,x)) & t \in \left[\frac{1}{2},1\right]. \end{cases}$$

This is again a strong deformation retract $\phi\colon X_1 \to Y$, which is denoted by $\phi = \phi_2 \circ \phi_1$.

Now we come to our main conclusion in this section.

Theorem 2. *Assume that* $\pi_2(N) = 0$, *and let* $\mathcal{F}$ *be a component of* $C^{2,\gamma}_\psi(M^2,N)$. *Suppose that on the level* $E^{-1}(c) \cap \mathcal{F}$, $c < m_{\mathcal{F}} + b$, *there are only nondegenerate harmonic maps* $u_1,\ldots,u_\ell$, *with Morse indices* $m_1,\ldots,m_\ell$ *respectively. Then the level set* $E_{c-\varepsilon} \cap \mathcal{F}$ *attached with* ℓ *handles, which dimensions correspond to these indices, is a strong deformation retract of* $E_{c+\varepsilon} \cap \mathcal{F}$, *for suitable* $\varepsilon > 0$.

Proof. We choose $\gamma = \frac{1}{5}$, $\tau = \frac{1}{3}$, and $\mu = \frac{1}{2}$. And then we have $\delta > 0$ small enough such that (3.1) and (3.2) hold. Choose $\varepsilon > 0$ small enough such that $\varepsilon < \frac{\delta^2}{10}$ and that the conclusion (6) holds. The inequalities (3.6), (3.7) and (3.8) are satisfied automatically. The strong deformation retract now is defined to be

$$\rho = \rho_5 \circ \rho_4 \circ \rho_3 \circ \rho_2 \circ \rho_1.$$

Combining Lemmas 3.1, 3.2 with 3.4, we obtain our conclusion.

Theorem 2′. *Let* $c < m + b$. *Suppose that on the level* $E^{-1}(c)$ *there are only nondegenerate harmonic maps with Morse indices. Then the same conclusions as in Theorem 2 holds.*

Corollary. *Suppose that* u_0 *is a nondegenerate harmonic map, with* $E(u_0) = c$ *and* $u_0 \in C^{2,\gamma}_\psi(\bar{M},N)$, $\gamma > 0$. *Assume that* $c < m + b$ *(or* $c < m_{\mathcal{F}} + b$, *if* $u_0 \in \mathcal{F}$ *and* $\pi_2(N) = 0$*). Then we have*

$$C_q(u_0;G) = \delta_{qk}G$$

where $k = \text{ind}(u_0)$.

Thus the Morse type number M_q^d *is the number of harmonic maps with index* q *in* E_d, $q = 0, 1, 2, \ldots$, *if there are only nondegenerate harmonic maps in the level set* E_d.

Acknowledgement. I am grateful to Professor J. Moser and Professor E. Zehnder for the invitation to the Forschungsinstitut für Mathematik, ETH Zürich.

REFERENCES

[Ch1] K.C. Chang, *Heat flow and boundary value problem for harmonic maps,* Ann. Inst. Henri Poincaré, Anal. nonlinéaire **6** (1989), 363–395.

[Ch2] K.C. Chang, *Infinite dimensional Morse theory and its applications,* Université de Montréal (1985).

[EL1] J. Eells, L. Lemaire, *Selected topics in harmonic maps,* CBMS Reg. Conf. Series no. **50**, (1983).

[U1] K. Uhlenbeck, *Morse theory by perturbation methods with applications to harmonic maps,* TAMS (1981).

Peking University
Beijing, China

Morse Theory and Existence of Periodic Solutions of Elliptic Type

B. D'ONOFRIO and I. EKELAND

1. Introduction and main results

In this paper we study the problem of finding periodic solutions of elliptic type for Hamiltonian systems of the following form:

$$(\mathcal{H}) \qquad \begin{cases} \dot{x} = J(Ax + N'(t,x)) \\ x(0) = x(T) \qquad \text{with } T > 0; \end{cases}$$

where J is the matrix

$$J \equiv \begin{pmatrix} 0 & I_n \\ -I_n & 0 \end{pmatrix}$$

and I_n is the identity of $\mathbf{R}^n$,

$$A \equiv \begin{pmatrix} A_n & 0 \\ 0 & A_n \end{pmatrix},$$

$$A_n \equiv \begin{pmatrix} \alpha_1 & & 0 \\ & \ddots & \\ 0 & & \alpha_n \end{pmatrix}$$

$\alpha_j \in \mathbf{R}$, $\alpha_j \neq 0$, $j = 1, \ldots, n$, $N \in C^2(\mathbf{R} \times \mathbf{R}^{2n}, \mathbf{R})$ and $N'(t,x)$ denotes the partial gradient with respect to the second variable.

We state the following assumptions on N:

$$(H_1) \quad \begin{array}{ll} (1) & N(t+T,x) = N(T) \qquad \forall (t,x) \in \mathbf{R} \times \mathbf{R}^{2n} \\ (2) & N \text{ is strictly convex with respect to the second variable} \end{array}$$

We shall describe the stability properties of periodic solutions of the system $(\mathcal{H})$: in particular we shall formulate sufficient conditions for the existence of periodic orbits with all Floquet multipliers on the unit circle.

We recall that the problem of existence of periodic solutions of elliptic type is an old problem which was studied by perturbation methods. The interest of this paper is to give a contribution to a global approach of the problem for strongly nonlinear systems.

Moreover we remark that the existence of solutions with Floquet multipliers on the unit circle is interesting for the application of the Birkhoff-Lewis theorem.

In the following we shall study the case in which N is subquadratic (see assumptions (H_2) in section 3). We shall associate with each periodic solution $(\bar{x}, T)$ of problem $(\mathcal{H})$ two numbers: its index i_T and the number d of linearly independent T-periodic solutions of the linearized problem around the solution $\bar{x}$.

In the generic case in which $d = 0$ we state the main result:

Theorem 1.1. *Let $N \in C^2(\mathbf{R} \times \mathbf{R}^{2n}, \mathbf{R})$ satisfy (H_1), (H_2). Suppose $E[T\alpha_j/\pi]$ is odd for $j = 1, \ldots, n$, then there exists a non constant T-periodic solution $\bar{x}$, with all the Floquet multipliers on the unit circle.*

Here we denote by $E[a]$ the integer part of a.

The proof of Theorem 1.1 makes use of the following two facts: a dual variational formulation of problem $(\mathcal{H})$ and a formulation of a Morse theory for Hamiltonian systems with non-positive definite Hamiltonians.

2. Morse theory

Suppose that $\bar{x}$ is a T-periodic solution of the problem $(\mathcal{H})$. We consider the linearized systems around $\bar{x}$, that is the following Hamiltonian system:

$$(\mathcal{H})' \qquad \begin{cases} \dot{y} = J(A + N''(t, \bar{x}(t)))y \\ y(0) = y(T) \end{cases}$$

We denote by $R(t)$ the resolvent of system $(\mathcal{H})'$ and we shall say the eigenvalues of $R(T)$ are the Floquet multipliers of system $(\mathcal{H})'$. Suppose that λ is a Floquet multiplier on the unit circle with multiplicity m and let ξ be the corresponding eigenvector. By Krein theory one can see that the hermitian form $i(J\xi, \xi)$ is non-degenerate on $\mathrm{Ker}(R(T)-\lambda I)^m$ and we shall say that λ has type (p,q) if our hermitian form has p eigenvalues which are positive and q negative.

Now to the system $(\mathcal{H})'$ we associate the quadratic form:

(2.1)

$$q_T(w,w) = \int_0^T [(\mathcal{L}_T w, w) + (N''(t)^{-1}\mathcal{L}_T w, \mathcal{L}_T w)]\,dt \quad \text{on } H^1(\mathbf{R}/T\mathbf{Z};\mathbf{R}^{2n}).$$

We have set $N(t) = N(t, \bar{x}(t))$.

Now, $\mathcal{L}_T$ is the linear operator $J\frac{d}{dt} + A$, from $W^{1,\alpha}(\mathbf{R}/T\mathbf{Z};\mathbf{R}^{2n})$ to $L^\alpha(0,T;\mathbf{R}^{2n})$, $1 < \alpha < +\infty$, defined as in $[\text{D-E}]_{1,2}$.

We shall say that the index of the quadratic form Q_T is the maximal dimension of a subspace $E \subset H^1(\mathbf{R}/T\mathbf{Z};\mathbf{R}^{2n})$ on which the restriction of $Q_T(w,w)$ is negative definite.

Definition 2.1. The index i_T of the system $(\mathcal{H})'$ on the interval $[0,T]$ is the index of q_T.

We recall that if $T \neq \frac{2\pi}{\alpha_j}\mathbf{Z}$, $j = 1,\ldots,n$, $\mathcal{L}_T$ is invertible and we can put $w = \mathcal{L}_T^{-1}v$. Then the index i_T of the system $(\mathcal{H})'$ coincides with the index i_T of the following quadratic form:

(2.2)

$$Q_T(v,v) = \int_0^T [(v, \mathcal{L}_T^{-1}v) + (N''(t)^{-1}v, v)]\,dt \qquad \text{on} \qquad L^2(0,T;\mathbf{R}^{2n})$$

The index of Q_T is finite; in fact $Q_T(v,v)$ is the sum of two terms, a compact one and a positive definite one.

Similarly the nullity of $(\mathcal{H})'$ is defined to be the dimension of $\mathrm{Ker}\, Q_T$.

Denoting by i_{kT} the index of the iterate of order k of $\bar{x}$, we shall compute i_{kT} in terms of i_T.

To do this, with every ω, which is an element of the unit circle U in $\mathbf{C}$, we associate the hermitian form:

$$Q_T^\omega(z,z) = \int_0^T [(\mathcal{L}_T z, z) + (N''(t)^{-1}\mathcal{L}_T z, \mathcal{L}_T z)]\,dt \tag{2.3}$$

on the complex Hilbert space

$$E_T^\omega = \{\, z \in H^1(0,T;\mathbf{C}^{2n}) \,/\, z(T) = \omega z(0)\,\}$$

Denote with $j_T(\omega)$ the Bott map from U to $\mathbf{N}$, that is the index of the form Q_T^ω.

One can prove the iteration formula

(2.4) $$i_{kT} = \sum_{\omega^k=1} j(\omega)$$

All the properties of the map j are summarized in the following:

Proposition 2.
(a) *j_T is a piecewise continuous function.*
(b) *$j_T(1) = i_T$.*
The discontinuity points of j_T are the points $\omega_0 \in U$ such that either ω_0 is a Floquet multiplier or ω_0 is of type $e^{i\alpha_m T}, e^{-i\alpha_m T}$, $m = 1, \ldots, n$.
(c) *if ω_0 is a Floquet multiplier and ω_0 has Krein type (p, q) then*

$$\Delta(\omega_0) = \lim_{\substack{\varepsilon \to 0 \\ \varepsilon > 0}} j_T(\omega_0 e^{i\varepsilon}) - j_T(\omega_0 e^{-i\varepsilon}) = q - p,$$

(d) *if $\omega_0 = e^{i\alpha_m T}$ then $\Delta(\omega_0) = d_m$, $\Delta(\bar{\omega}_0) = -d_m$, where d_m is the multiplicity of the frequency α_m,*
(e)

$$i_T \leq \lim_{\substack{\varepsilon \to 0 \\ \varepsilon > 0}} j_T(e^{i}) \leq i_T + d$$

when d is the number of linearly independent solutions of the linearized system $(\mathcal{H})'$.

3. Existence of periodic solutions of elliptic type

In this section we investigate the existence of periodic solutions of elliptic type of system $(\mathcal{H})$ using the Morse theory developed in the previous section.

We now introduce the Legendre transform N^* of N. It is defined by the following

(3.1) $$N^*(y) = \mathrm{Max}\{ (x, y) - N(t, x) : x \in \mathbf{R}^{2n} \}$$

with the reciprocity formula

(3.2) $$y = N'(t, x) \iff x = N^{*\prime}(t, y)$$

N^* is well-defined and belongs to $C^1(\mathbf{R}\times\mathbf{R};\mathbf{R}^{2n})$.

In addition we make use of the following assumptions on N:

$$(H_2)\quad \begin{array}{llll} \text{(a)}' & N(t,x)\geq n(\|x\|) & \text{with} & n(s)s^{-1}\to\infty \quad \text{as} \quad s\to+\infty \\ \text{(b)}' & N(t,x)\leq m(\|x\|) & \text{with} & m(s)s^{-2}\to 0 \quad \text{as} \quad s\to+\infty \end{array}$$

Note that in our case N is subquadratic at infinity, so N^* is superquadratic at infinity.

To problem $(\mathcal{H})$ we associate the following equivalent variational formulation

$$(3.3)\qquad \begin{cases} \psi'(v)=0 \\ v\in H^1(\mathbf{R}/T\mathbf{Z};\mathbf{R}^{2n}) \end{cases}$$

where

$$(3.4)\qquad \psi(v)=\int_0^T [\tfrac{1}{2}(\mathcal{L}_T^{-1}v,v)+N^*(t,v)]\,dt$$

One can prove the following.

Theorem 3.1. (Existence). *Suppose A, N verify all previous assumptions, then if $T\neq\frac{2\pi}{\alpha_j}\mathbf{Z}$, $j=1,\ldots,n$ there exists a critical point of the functional ψ which is a global minimizer.*

Let $\bar{v}$ be a critical point of ψ, that is $\psi'(\bar{v})=0$. We write the Hessian of ψ at $\bar{x}=\mathcal{L}_T\bar{v}$. $(\psi''(\bar{x})v,v)$ coincides with the quadratic form (2.2).

It is easy to prove the following.

Lemma 3.2. *The periodic solution $\bar{x}$ (Theorem 3.1) has index*

$$i_T(\bar{x})=0.$$

We now turn our attention back to the initial problem of finding periodic solutions of elliptic type. We recall that we study the generic case $d=0$, in which j is not a Floquet multiplier of the periodic solution $\bar{x}$.

Proof of Theorem 1.1. Now arguing as in the autonomous case one can calculate $j(-1)$ making use of the iteration theory

$$(3.5)\qquad 0\leq j(-1)\leq i_T+d-\Delta^++\Delta^-+x^+-x^-$$

where: Δ^+ is the number of discontinuity points of the form $e^{i\alpha_j T}$, $j = 1, \ldots, n$ on the upper half circle; Δ^- is the number of discontinuity points of the form $e^{-i\alpha_j T}$, $j = 1, \ldots, n$, on the upper half circle; x^+ is the number of Krein positive Floquet multipliers; x^- is the number of Krein negative Floquet multipliers.

Now $i_T = 0$ and generically $d = 0$, so by (3.5) one can deduce

$$x^+ - x^- \geq \Delta^+ - \Delta^- \tag{3.6}$$

Since $E[T\alpha_j/\pi]$ is odd for every j, we have $\Delta^+ = n$ and $\Delta^- = 0$, so the inequality becomes:

$$x^+ - x^- \geq n \tag{3.7}$$

On the other hand, the total number of Floquet multipliers on the upper half-circle cannot be more than n, since the dimension of the ambient space is $2n$. So $x^+ = n$ and $x^- = 0$.

REFERENCES

[B]$_1$ R. Bott, *On the iteration of closed geodesic and Sturm intersection theory*, Comm. Pure Appl. Math., 9 (1956), 176–206.

[B]$_2$ R. Bott, *Lecture on Morse theory, old and new*, Bull. Amer. Math. Soc. **7** (1982), 331–358.

[C] F. Clarke, *Periodic solutions of Hamiltonian inclusions*, J. Differential Equations, **40** (1981), 1–6.

[C-E] F. Clarke and I. Ekeland, *Hamiltonian trajectories having prescribed minimal period*, Comm. Pure Appl. Math. **33** (1980), 103–116.

[C-Z] C. Conley and E. Zehnder, *The Birkhoff-Lewis fixed point theorem and a conjecture of V. Arnold*, Invent. Math. **73** (1983), 33–49.

[D-E]$_1$ B. D'Onofrio and I. Ekeland, *La théorie de l'index pour certains systèmes hamiltoniens non définis positifs*, C.R.A.S. Paris, **305**, I, (1987), 249–251.

[D-E]$_2$ B. D'Onofrio and I. Ekeland, *Hamiltonian systems with elliptic periodic orbits*, preprint (1988).

[E]$_1$ I. Ekeland, *Une théorie de Morse pour des systèmes hamiltoniens convexes*, Ann. Inst. H. Poincaré Anal. Non Linéaire 1 (1984), 19–78.

[E]$_2$ I. Ekeland, *An index theory for periodic solutions of convex Hamiltonian systems*, Proc. of Symp. on Pure Math., **45** (1986).

[E-H]$_1$ I. Ekeland and H. Hofer, *Periodic solutions with prescribed minimal period for convex autonomous Hamiltonian systems*, Inv. Math., **81** (1985), 155–177.

[E-H]$_2$ I. Ekeland and H. Hofer, *Subharmonic solutions for convex nonautonomous Hamiltonian systems*, Comm. Pure and Appl. Math. (to appear).

[E-H]$_3$ I. Ekeland and H. Hofer, *Convex Hamiltonian energy surfaces and their periodic trajectories*, preprint (1987).

[G-M] M. Girardi and M. Matzeu, *Periodic solutions of Convex Autonomous Hamiltonian Systems and a Quadratic Growth at the Origin and Superquadratic at Infinity*, Annali di Matematica Pura e Applicata Vol. CXLVII, 1987, 21–72.

[M] J. Moser, *Proof of a generalized form of a fixed point theorem due to G.D. Birkhoff*, Lecture Notes in Math., **597**, Springer Verlag, Berlin and New York (1977), 464–494.

[R] P. Rabinowitz, *Periodic solutions of Hamiltonian systems*, Comm. Pure Appl. Math. **31** (1978), 157–184.

[T] F. Takens, *Hamiltonian systems: generic properties of closed orbits and local perturbations*, Math. Ann. **188** (1970), 304–312.

[V]$_1$ C. Viterbo, *Thèse de 3ème cycle*, Université Paris IX, 1985.

[V]$_2$ C. Viterbo, *Une théorie de Morse pour les systemès hamitoniens étoilés*, C.R.A.S. Paris (1), **301** (1985), 487–489.

[Y-S] V. Yakubovich and V. Starzhinskii, *Linear differential equations with periodic coefficients*, Halstedt Press, Wiley, 1980.

Université Paris-Dauphine
CEREMADE
Paris, France

Periodic Solutions of a Nonlinear Second Order System

L. Lassoued

Introduction

We are interested in the existence of T periodic solutions of a second order system $\ddot{x} + a(t)V'(x) = f(t)$ where $V : I\!R^N \to I\!R$ is a potential which will be assumed to be convex, and the maps $a : I\!R \to I\!R$ and $f : I\!R \to I\!R^N$ being T periodic. So we will denote in all the paper by (1) the problem

$$\begin{cases} \ddot{x} + a(t)V'(x) = f(t) & \text{a.e. on} \quad [0,T] \\ x(0) = x(T) \\ \dot{x}(0) = \dot{x}(T) \end{cases} \tag{1}$$

and (1′) the problem corresponding to $f = 0$

$$\begin{cases} \ddot{x} + a(t)V'(x) = 0 & \text{a.e. on} \quad [0,T] \\ x(0) = x(T) \\ \dot{x}(0) = \dot{x}(T). \end{cases} \tag{1′}$$

There are many results about this problem when the function $a(\cdot)$ keeps a constant sign; we can mention the works of Bahri and Beresticky, Benci, Clarke and Ekeland, and Rabinowitz which deal with the more general system $\ddot{x} + V'_x(t,x) = 0$ with the assumption $V(t,x) \geq 0$.

We assume here that the function $a(\cdot)$ changes sign: More precisely we suppose in all the following that $a(\cdot)$ belongs to $L^1([0,T], I\!R)$, is nonzero almost everywhere, and that the two sets

$$I^+ = \{t \in [0,T] / a(t) > 0\}$$

and

$$I^- = \{t \in [0,T] / a(t) < 0\}$$

are both non-negligible. So the two reals

$$\alpha^+ = \int_{I^+} a(t)\,dt \quad \text{and} \quad \alpha^- = \int_{I^-} |a(t)|\,dt$$

are strictly positive.

We shall use the function $a_0(\cdot)$ defined almost everywhere on $[0,T]$ by

$$a_0 = \begin{cases} \frac{a}{\alpha^+} & \text{on} \quad I^+ \\ \frac{a}{\alpha^-} & \text{on} \quad I^- \end{cases}.$$

This paper is divided in two parts: in the first part we consider the case of a subquadratic potential, and we give existence results of at least one solution of problem (1) supposing that f satisfies

$$\int_0^T \frac{|f|^2}{|a|}\,dt < +\infty. \tag{2}$$

The second part is concerned with the superquadratic case. Since the potential is supposed convex, the tool is in the two cases the dual variational formulation of the problem.

I. The subquadratic case

We suppose that V belong to $C^2(IR^N, IR)$ and satisfies for all x in IR^N:

$$V(x) \geq V(0) = 0 \tag{3}$$

$$\frac{K}{2}|x|^2 - C_1 \leq V(x) \leq \frac{k}{2}|x|^2 + C_2 \tag{4}$$

$$\epsilon I_N \leq V''(x) \leq A I_N \tag{5}$$

where k, K, ϵ and A are strictly positive constants, and A satisfies

$$A < \mu_1 \tag{6}$$

where μ_1 is the first value μ strictly positive for which the system $\ddot{y} + \mu|a(t)|y = 0$ has a nonzero T periodic solution. Let us denote by W the Fenchel conjugate of V, given by

$$W(y) = \max_{x \in IR^N} \{(x \cdot y) - V(x)\}$$

and define E to be the Hilbert space

$$E = \left\{ u : [0,T] \to IR^N / \int_0^T \frac{|u|^2}{|a|}\,dt < +\infty \quad \text{and} \quad \int_0^T u\,dt = 0 \right\}$$

with respect to the scalar product

$$\langle u, v \rangle = \int_0^T \frac{u \cdot v}{|a|}\, dt.$$

We consider the functional Φ defined over E by

$$\Phi(u) = -\frac{1}{2}\int_0^T |\Pi u|^2\, dt + \int_0^T aW\left(\frac{u+f}{a}\right) dt$$

where Πu denotes the primitive of u with mean value zero.

The following proposition, deduced from a general dual principle due to Clarke (see [7] and [8]) shows that problem (1) is equivalent to the search of critical points of Φ.

Proposition 1.

(i) *Φ is C^1 over E, twice Gâteaux differentiable.*

(ii) *$u \in E$ is a critical point of Φ over E if and only if there exists a unique $\xi \in IR^N$ such that $x = -\Pi^2 u + \xi$ solves problem (1).*

Now the space F splits into a direct orthogonal sum of three subspaces

$$E = E^+ \oplus E^- \oplus E^0$$

where

$$\begin{aligned} E^+ &= \{u \in E / u = 0 \quad \text{a.e. on} \quad I^-\} \\ E^- &= \{u \in E / u = 0 \quad \text{a.e. on} \quad I^+\} \\ E^0 &= \{u \in E / u = a_0\xi, \;\; \xi \in IR^N\}. \end{aligned}$$

But the second derivative of Φ is given by:

$$\begin{aligned} \forall h \in E : \Phi''(u) \cdot h \cdot h = -\int_0^T |\Pi h|^2\, dt + \int_{I+} W''\left(\frac{u+f}{a}\right)\frac{h \cdot h}{a}\, dt \\ - \int_{I-} W''\left(\frac{u+f}{a}\right)\frac{h \cdot h}{|a|}\, dt. \end{aligned}$$

Denoting, for $u^0 \in E^0$, by F_{u^0} the map:

$$\begin{aligned} &F_{u^0} : E^+ \times E^- \to IR \\ &(u^+, u^-) \to \Phi\left(u^+ + u^- + u^0\right) \end{aligned}$$

one sees that this map is strictly concave with respect to the variable $u^- \in E^-$. On the other hand by assumptions (5) and (6) one has

$$W''(\cdot) \geq \frac{1}{A} I_N$$

with

$$\frac{1}{A} > \frac{1}{\mu_1}$$

and it is easy to verify that $\lambda_1 = \frac{1}{\mu_1}$ is the greatest eigenvalue of the compact operator L of E defined by

$$(Lh, h)_E = \int_0^T |\Pi h|^2.$$

This implies that F_{u^0} is also strictly convex with respect to the variable $u^+ \in E^+$.

Using then a saddle point reduction method developed by Amann (see [1] and [2]), one obtains:

Proposition 2 (see [9]).

(i) *For each* $u^0 \in E^0$, F_{u^0} *has a unique saddle point* $(u^+(u^0),\ u^-(u^0))$.

(ii) *The functional* ϕ *defined on* E^0 *by*

$$\phi(u^0) = \Phi\left(u^0 + u^+\left(u^0\right) + u^-\left(u^0\right)\right)$$

is C^1 *over* E^0 *and its derivative is given by*

$$\phi'(u^0) = \Phi'\left(u^0 + u^+\left(u^0\right) + u^-\left(u^0\right)\right).$$

This proposition shows that critical points of Φ are in correspondence with those of ϕ. But ϕ is defined on a finite dimensional space, and we will obtain existence of a critical point of ϕ by minimization or maximization over all the space.

We need the following definition:

Definition 3.

(i) $M_1(k)$ is the strictly positive real given by

$$M_1(k) = \int_0^T \Pi a_0 \left(\Pi a_0 - y_{1,k}\right) dt$$

where $y_{1,k}$ is the unique solution of the system

$$\begin{cases} \ddot{y} + kay = ka\Pi^2 a_0 & \text{a.e. on} \quad I^+ \\ y(0) = y(T) \\ \dot{y}(0) = \dot{y}(T) \\ \ddot{y} = 0 \quad \text{a.e. on} \quad I^-. \end{cases}$$

(ii) $M_2(k)$ is the strictly positive real given by

$$M_2(k) = \int_0^T \Pi a_0 \left(\Pi a_0 - y_{2,k}\right) dt$$

where $y_{2,k}$ is the unique solution of the system

$$\begin{cases} \ddot{y} + kay = ka\Pi^2 a_0 & \text{a.e. on} \quad I^- \\ y(0) = y(T) \\ \dot{y}(0) = \dot{y}(T) \\ \ddot{y} = 0 \quad \text{a.e. on} \quad I^+ . \end{cases}$$

We can state:

Theorem 4. *If either*

$$\frac{1}{k\alpha^+} - \frac{1}{K\alpha^-} > M_1(k) \tag{7}$$

or

$$\frac{1}{k\alpha^+} - \frac{1}{K\alpha^-} < M_2(k) \tag{8}$$

system (1) *has at least a solution.*

Sketch of the proof (see [9]).

By standard properties of a saddle point of a convex-concave function, ϕ satisfies

$$\begin{aligned} \phi\left(u^0\right) &= \max_{u^- \in E^-} \min_{u^+ \in E^+} \Phi\left(u^+ + u^- + u^0\right) \\ &= \min_{u^+ \in E^+} \max_{u^- \in E^-} \Phi\left(u^+ + u^- + u^0\right) \end{aligned}$$

so

$$\phi\left(u^0\right) \geq \min_{u^+ \in E^+} \Phi\left(u^+ + u^0\right)$$

and

$$\phi\left(u^0\right) \leq \max_{u^- \in E^-} \Phi\left(u^- + u^0\right) .$$

Using (4), and after calculations, one obtains the following estimates:
For $u^0 = a_0\xi,\ \xi \in IR^N$

$$\phi\left(u^0\right) \geq \frac{1}{2}\left[\frac{1}{k\alpha^+} - \frac{1}{K\alpha^-} - M_1(k)\right] |\xi|^2 - C|\xi| - C$$

and

$$\phi\left(u^0\right) \leq \frac{1}{2}\left[\frac{1}{k\alpha^+} - \frac{1}{K\alpha^-} - M_2(k)\right]|\xi|^2 + C|\xi| + C$$

where C denotes a constant independent of ξ.

So if (7) is satisfied, ϕ achieves its minimum over E^0; if (8) is satisfied it achieves its maximum. The conclusion of the theorem follows.

Remark 5. Conditions (7) and (8) of the theorem are not explicit and quantities $M_1(k)$ and $M_2(k)$ are not always easy to compute. But one can establish that

$$M_1(k) \leq M_3 = \frac{1}{\mu_1}\left(\frac{1}{\alpha^+} + \frac{1}{\alpha^-}\right).$$

and

$$M_2(k) \geq M_4 = \sum_j \int_{I_j^+} |\Pi_j a_0|^2$$

where I_j^+ denotes the connect components of the set I^+ and Π_j the primitive of mean value zero on I_j^+.

Now in the case $f = 0$, system (1) has the constant solution $x = 0$, corresponding to the critical point zero of ϕ. So, we have to ensure that ϕ does not achieve its minimum on its maximum at the origin. In fact one can establish that ϕ is C^2 over E^0, and that

$$\phi''(0) \cdot h \cdot h = 2\phi_0(h) \tag{9}$$

where ϕ_0 is the reduced functional associated to the system $\ddot{x} + a(t)V''(0)x = 0$.

Let us denote by k_0 and K_0 respectively the greatest and the smallest eigenvalue of the operator $V''(0)$ and by B_0 the inverse of this operator

$$B_0 = [V''(0)]^{-1}.$$

Then, as in the proof of Theorem 4 and using (9), we obtain, for $h = a^0\xi,\ \xi \in IR^N$

$$\phi''(0) \cdot h \cdot h \geq \left(\frac{1}{\alpha^+} - \frac{1}{\alpha^-}\right) B_0\xi\xi - M_1\left(k_0\right)|\xi|^2$$

and

$$\phi''(0) \cdot h \cdot h \leq \left(\frac{1}{\alpha^+} - \frac{1}{\alpha^-}\right) B_0\xi\xi - M_2\left(k_0\right)|\xi|^2.$$

We can deduce:

Theorem 6. *If either*

$$\frac{1}{k\alpha^+} - \frac{1}{K\alpha^-} > M_1(k)$$

and

$$\frac{1}{k_0}\left(\frac{1}{\alpha^+} - \frac{1}{\alpha^-}\right) < M_2\left(k_0\right)$$

or

$$\frac{1}{K\alpha^+} - \frac{1}{k\alpha^-} < M_2(k)$$

and

$$\frac{1}{K_0}\left(\frac{1}{\alpha^+} - \frac{1}{\alpha^-}\right) > M_1\left(k_0\right)$$

system (1) *has for $f = 0$ at least one nonconstant solution.*

Indeed in the first case the origin is not a minimum for ϕ; in the second case it is not a maximum.

II. The superquadratic case

The result of this section is:

Theorem 7. *Suppose that $V \in C^2(IR^N, IR)$ satisfies:*

$$V(x) \geq V(0) = 0 \tag{10}$$

$V''(x)$ *is positive definite for all* $x \in IR^N \backslash \{0\}$. (11)

There exists a $\beta > 2$ *such that* (12)

$$\forall\ (x) \in IR^N \qquad V'(x) \cdot x = \beta V(x).$$

Then system (1′) *has at least one nonconstant solution. Moreover, if V is even, system* (1′) *has infinitely many nonconstant solutions.*

The proof of this theorem is as follows. We denote by γ the conjugate exponent of β:

$$\frac{1}{\gamma} + \frac{1}{\beta} = 1$$

and still by W the Fenchel conjugate of V, which satisfies, for all y in IR^N,

$$W'(y) \cdot (y) = \gamma W(y) \tag{13}$$

$$C_1|y|^\gamma \leq W(y) \leq C_2|y|^\gamma \tag{14}$$

$$|W'(y)| \leq C_3|y|^{\gamma-1} \tag{15}$$

where C_1, C_2 and C_3 are strictly positive constants.

We define E to be the Banach space

$$E = \left\{ u : [0,T] \to IR^N / \int_0^T \frac{|u|^\gamma}{|a|^{\gamma-1}}\, dt = 0 \text{ and } \int_0^T u\, dt = 0 \right\}$$

with the norm

$$\|u\| = \left[\int_0^T \frac{|u|^\gamma}{|a|^{\gamma-1}}\, dt \right]^{\frac{1}{\gamma}}.$$

The functional

$$\Phi(u) = -\frac{1}{2}\int_0^T |\Pi u|^2\, dt + \int_0^T aW\left(\frac{u}{a}\right)\, dt$$

is then C^1 over E and its nonzero critical points are in correspondence with the nonconstant solutions of (1).

The space E still splits into a direct sum of three spaces $E^+ \oplus E^- \oplus E^0$ defined as in section I, and Φ is strictly concave with respect to the variable $u^- \in E^-$. Let us set

$$X = E^+ \oplus E^0.$$

Then we have:

Proposition 8 (see [10]). (i) *For each $x \in X$, there exists a unique $\theta(x) \in E^-$ such that*

$$\Phi(x + \theta(x)) = \max_{u^- \in E^-} \Phi(x + u^-).$$

(ii) *The functional Ψ defined over X by*

$$\Psi(x) = \Phi(x + \theta(x))$$

is C^1 over X and such that

$$\Psi'(x) = \Phi'(x + \theta(x)).$$

We see from this proposition that critical points of Φ over E and Ψ over X are in correspondence. Next we prove that Ψ satisfies a weakened version of the classical Palais-Smale condition due to G. Cerami (see [5]).

Proposition 9. (i) *Each bounded sequence* (x_n) *in* X *such that* $\Psi'(x_n) \to 0$ *is precompact.*

(ii) *Each sequence* (x_n) *in* X *such that* $\Psi(x_n)$ *is bounded and* $\|\Psi'(x_n)\| \cdot \|x_n\| \to 0$ *is bounded.*

Proof. The proof of point (i) is classical by the use of Krasnoleski's theorem. Let us give that of point (ii):

Arguing by contradiction, we suppose that there exists a sequence (x_n) in X such that

$$|\Psi(x_n)| \le C \tag{16}$$

$$\|\Psi'(x_n)\| \, \|x_n\| \to 0 \tag{17}$$

and

$$\|x_n\| \to +\infty. \tag{18}$$

We shall denote by C all constants independent of n. From (17) and (18) we deduce

$$\|\Psi(x_n)\| \to 0.$$

Let us set $u_n = x_n + \theta(x_n)$. Then

$$|\Phi(u_n)| = |\Psi(x_n)| \le C \tag{19}$$

$$\Phi'(u_n) \cdot u_n = \Psi'(x_n) \cdot x_n \to 0 \tag{20}$$

$$\|\Phi'(u_n)\| = \|\Psi'(x_n)\| \to 0 \tag{21}$$

and

$$\|u_n\| \to +\infty. \tag{22}$$

But we have

$$\Phi(u_n) = -\frac{1}{2}\int_0^T |\Pi u_n|^2 \, dt + \int_0^T a\, W\left(\frac{u_n}{a}\right) dt$$

and

$$\begin{aligned}\Phi'(u_n) \cdot u_n &= -\int_0^T |\Pi u_n|^2 \, dt + \int_0^T W'\left(\frac{u_n}{a}\right) u_n \, dt \\ &= -\int_0^T |\Pi u_n|^2 \, dt + \gamma \int_0^T a\, W\left(\frac{u_n}{a}\right) dt.\end{aligned}$$

By (19) and (20) we deduce

$$\int_0^T |\Pi u_n|^2 \, dt \le C. \tag{23}$$

On the other hand $\Phi'(u_n)$ is given by

$$\Phi'(u_n) = \Pi^2 u_n + W'\left(\frac{u_n}{a}\right) - \xi_n \tag{24}$$

with

$$\xi_n = \frac{1}{\int_0^T |a|^{1/\beta} \, dt} \int_0^T |a|^{1/\beta} \left[\Pi^2 u_n + W'\left(\frac{u_n}{a}\right)\right] dt.$$

So

$$|\xi_n| \le C\left(1 + \|u_n\|^{\gamma-1}\right). \tag{25}$$

Multiplying (24) by u_n, integrating over I^+ and using (23), (25), (13) and (14) we obtain

$$\int_{I^+} \frac{|u_n|^\gamma}{|a|^{\gamma-1}} \, dt \le C\left[\|u_n\| + \left(1 + \|u_n\|^{\gamma-1}\right)\left|\int_{I^+} u_n \, dt\right|\right]. \tag{26}$$

In the same way we have

$$\int_{I^-} \frac{|u_n|^\gamma}{|a|^{\gamma-1}} \, dt \le C\left[\|u_n\| + \left(1 + \|u_n\|^{\gamma-1}\right)\left|\int_{I^-} u_n \, dt\right|\right]. \tag{27}$$

Let us now consider $w_n = \frac{u_n}{\|u_n\|}$. By (23) and (22), $w_n \to 0$ in $D'([0,T], IR^N)$. Since w_n is bounded, $w_n \to 0$ weakly in L^1. It follows that

$$\int_{I^+} w_n \, dt \to 0 \text{ and } \int_{I^-} w_n \, dt \to 0.$$

Dividing then (26) and (27) by $\|u_n\|^\gamma$ we see that

$$\int_{I^+} \frac{|w_n|^\gamma}{|a|^{\gamma-1}} \, dt \to 0 \text{ and } \int_{I^-} \frac{|w_n|^\gamma}{|a|^{\gamma-1}} \, dt \to 0$$

which contradicts $\|w_n\| = 1$.

The following lemmas explain the behavior of the functional Ψ.

Lemma 10. *For each finite dimensional subspace Y of X, one has*

$$\lim_{\substack{\|x\| \to +\infty \\ x \in Y}} \Psi(x) = -\infty.$$

Proof. Let us write I^+ as a union of intervals

$$I^+ = \bigcup_{j \in A} I_j$$

and set

$$F = \{v \in L^2([0,T], IR^N) \text{ s.t. } v = \xi_j\, IR^N \text{ on each } I_j\}.$$

For $x \in X$, and $s \in IR_+$ we have

$$\begin{aligned}
\Psi(sx) &= \max_{u^- \in E^-} \Psi(sx + u^-) \\
&\le \sup_{u^- \in E^-} -\frac{1}{2}\int_0^T |s\Pi x + \Pi u^-|^2\, dt + \int_{I+} aW\left(\frac{sx}{a}\right) dt \\
&\le -\frac{s^2}{2} \inf_{v\in F} \int_0^T |\Pi x + v|^2\, dt + C_2 s^\gamma .
\end{aligned}$$

But one sees easily that $q(x) = \inf_{v\in F} \int_0^T |\Pi x + v|^2\, dt$, which is no other than the distance of Πx to the closed subspace F of L^2, is a definite positive quadratic form on X. So if Y is a finite dimensional subspace of X, there exists an $m > 0$ such that

$$(x \in Y,\ \|x\| = 1) \Longrightarrow \Psi(sx) \le -\frac{m}{2}s^2 + C_2 s^\gamma$$

so the result, since $\gamma < 2$.

Lemma 11. *There exists a $\rho > 0$, a $\delta > 0$ such that*

$$\begin{array}{lll}
\text{(i) } \forall u^+ \in E^+ & 0 < \|u^+\| \le \rho & \Psi(u^+) > 0 \\
\text{(ii) } \forall u^+ \in E^+ & \|u^+\| = \rho & \Psi(u^+) \ge \delta .
\end{array}$$

Proof. The result follows from

$$\begin{aligned}
\Psi(u^+) \ge \Phi(u^+) &= -\frac{1}{2}\int_0^T |\Pi u^2|^2\, dt + \int_{I+} aW\left(\frac{u^+}{a}\right) dt \\
&\ge -\frac{C_4}{2}\|u^+\|^2 + C_1 \|u^+\|^\gamma
\end{aligned}$$

where $C_4 = T\|a\|_{1}^{1/\beta}$.

If the potential V is even, so is the functional Ψ and we can apply a theorem due to Ambrosetti and Rabinowitz (see [3]) which gives, since Proposition 9, Lemmas 10 and 11 are satisfied, the existence of infinitely many pairs of nonzero critical points for Ψ. So system (1′) has in this case infinitely many nonconstant solutions.

In the general case, we have to precise more the behavior of Ψ: two cases appear according to the sign of the mean value of the functions $a(\cdot)$:

Lemma 12. *Suppose*

$$\int_0^T a(t)\,dt \geq 0.$$

Then

$$E^0 \subset \{\Psi \leq 0\}.$$

Proof. For $u^- \in E$ and $u^0 = a_0\xi \in E^0$, we have

$$\Phi\left(u^0 + u^-\right) \leq \int_{I+} aW\left(\frac{u^0}{a}\right) dt - \int_{I-} |a| W\left(\frac{u^0 + u^-}{a}\right) dt.$$

By convexity of W

$$W\left(\frac{u^0 + u^-}{a}\right) \geq W\left(\frac{u^0}{a}\right) + W'\left(\frac{u^0}{a}\right)\cdot\frac{u^-}{a}.$$

So

$$\int_{I-} |a| W\left(\frac{u^0 + u^-}{a}\right) dt \geq \int_{I-} |a| W\left(\frac{u^0}{a}\right) dt = \alpha^- W\left(\frac{\xi}{\alpha^-}\right)$$

and

$$\Phi\left(u^0 + u^-\right) \leq \alpha^+ W\left(\frac{\xi}{\alpha^+}\right) - \alpha^- W\left(\frac{\xi}{\alpha^-}\right).$$

But since $\int_0^T a(t)\,dt \geq 0$, $\alpha^- \leq \alpha^+$ and once again by convexity of W, we obtain

$$\Phi\left(u^0 + u^-\right) \leq 0.$$

So

$$\Psi\left(u^0\right) = \max_{u^- \in E^-} \Phi\left(u^0 + u^-\right) \leq 0.$$

Lemma 13. *Suppose* $\int_0^T a(t)\,dt < 0$. *Then the origin is a strict local minimum of* Ψ *over* X. *More precisely, there exist a* $\rho > 0$ *and a* $\delta > 0$ *such that:*

$$\Psi(x) > 0 \ \text{ for } x \in X, \quad 0 < \|x\| \leq \rho$$

and

$$\Psi(x) \geq \delta \ \text{ for } x \in X, \quad \|x\| = \rho.$$

Proof. Let ϵ be a real satisfying

$$0 < \epsilon < \frac{1}{2} \text{ and } \left(\frac{\epsilon}{1-2\epsilon}\right)^\gamma < \frac{C_1}{C_2}.$$

For $x = u^+ + u^0$, $\|x\| = 1$ we distinguish two cases.

First Case: $\|u^0\| < \epsilon$ *so* $\|u^+\| - \|u^0\| > 1 - 2\epsilon$.

For $s \in IR_+$ we have

$$\begin{aligned}\Psi(sx) \geq \Phi(sx) &= \frac{s^2}{2}\int_0^T |\Pi x|^2\,dt + \int_{I+} aW\left(\frac{su^+ + su^0}{a}\right)dt \\ &\quad - \int_{I^-} |a| W\left(\frac{su^0}{a}\right)dt \\ &\geq -\frac{C_4}{2}s^2 + s^\gamma\left[C_1(1-2\epsilon)^\gamma - C_2\epsilon^\gamma\right] \\ &\geq -\frac{C_4}{2}s^2 + m_1 s^\gamma\end{aligned}$$

with $m_1 > 0$ with the choice of ϵ.

Second Case: $\|u^0\| \geq \epsilon$.

Writing $u^0 = a_0\xi$, we have

$$|\xi| \geq \epsilon' > 0$$

and using the convexity of W we obtain

$$\begin{aligned}\Psi(sx) \geq \Phi(sx) &\geq -\frac{C_4}{2}s^\gamma + s^\gamma\left[\alpha^+ W\left(\frac{\xi}{\alpha^+}\right) - \alpha^- W\left(\frac{\xi}{\alpha^-}\right)\right] \\ &\geq -\frac{C_4}{2}s^\gamma + m_2 s^\gamma\end{aligned}$$

with $m_2 > 0$, since $\alpha^+ > \alpha^-$ by hypothesis.

Consequently, there exists a constant $C > 0$ such that

$$\forall s \in IR_+,\ x \in X,\ \|x\| = 1,\ \Psi(sx) \geq -\frac{C_4}{2}s^2 + Cs^\gamma$$

and the required property of the lemma follows.

The proof of the theorem is now clear. In the case $\int_0^T a(t)\,dt < 0$ the mountain pass theorem (3) applies; in the other case it is a more general linking theorem (see [5]) which works: Indeed, choosing any $e \in E^+$, by Lemmas 10 and 13, one can find an $R > 0$ such that Ψ is nonpositive on the boundary of the set

$$Q = \left\{u^0 + se/u^0 \in E^0, \|u^0\| \leq R \text{ and } 0 \leq s \leq R\right\}.$$

Remark 14. The homogeneity of the potential V is essential in the proof of the Palais-Smale condition. In fact this proof still works if the condition $V'(x) \cdot x = \beta V(x)$ is satisfied only near infinity; for simplicity we supposed that it is satisfied everywhere.

BIBLIOGRAPHY

[1] Amman, H., *Saddle points and multiple solutions of differential equations*, Math. Z. **169** (1976), 127-166.

[2] Amman, H., and E. Zehnder, *Nontrivial solutions for a class of nonresonnance problems and applications to nonlinear differential equations*, Ann. Sci. Norm. Sup. Pisa 7 (1980), 539-603.

[3] Ambrosetti, A., and P. H. Rabinowitz, *Dual variational methods in critical point theory and applications*, J. Functional Analysis **14** (1973), 349-381.

[4] Bahri, A., and H. Beresticky, *Existence of forced oscillations for some nonlinear differential equations*, Comm. Pure Appl. Math. **37** (1984), 403-442.

[5] Bartolo, P., V. Benci and D. Fortunato, *Abstract critical point theorems and applications to some nonlinear problems with "strong" resonnance at infinity*, Nonlinear Analysis T.M.A **7** No. 9 (1983), 981-1012.

[6] Benci, V., *Some critical point theorems and applications*, Comm. Pure Appl. Math **33** (1980), 147-172.

[7] Clarke, F., *Periodic solutions to Hamiltonian inclusions*, J. Diff. Equat. **40** (1981), 1-6.

[8] Clarke, F., and I. Ekeland, *Nonlinear oscillations and boundary value problems for Hamiltonian systems*, Arch. Rat. Mech. Anal. (1980), 315-333.

[9] Lassoued, L., *Solutions périodiques d'un système différentiel non linéaire du second ordre avec changement de signe*, to appear in Ann. Math. Pura. Appl. Pisa.

[10] Lassoued, L., *Periodic solutions of a second order superquadratic system with a change of sign in the potential*, to appear.

[11] Rabinowitz, P. M., *Periodic solutions of Hamiltonian systems*, Comm. Pure Appl. Math. **31** (1978), 157-184.

Ceremade and Faculté des Sciences de Tunis
Département de Mathématiques
Campus Universitaire 1060
Tunisia

EXISTENCE OF MULTIPLE BRAKE ORBITS FOR A HAMILTONIAN SYSTEM

Andrzej Szulkin (*)

Department of Mathematics, University of Stockholm, Sweden

1. Introduction and statement of the main result

The purpose of this note is to describe a recent work done by the author on the problem of existence of periodic orbits for Hamiltonian systems. Arguments presented here are rather sketchy and some proofs are omitted. The details may be found in [13].

Let $H : \mathbf{R}^{2N} \to \mathbf{R}$ be a continuously differentiable function and let

$$J = \begin{pmatrix} 0 & -I \\ I & 0 \end{pmatrix}$$

be the standard symplectic matrix. Consider the Hamiltonian system of differential equations

$$(HS) \qquad \dot{x} = JH'(x).$$

Periodic solutions of (HS) will be called *periodic orbits*. If $x(t) = (p(t), q(t))$ is a T-periodic orbit of (HS) such that $p(t)$ is odd and $q(t)$ is even about 0 and $T/2$, then $x(t)$ will be called *a brake orbit*. Note that for such x we have $p(0) = p(T/2) = 0$, so $q(t)$ oscillates back and forth between two restpoints in the configuration space. Since $\frac{d}{dt}H(x(t)) = 0$ whenever $x(t)$ satisfies (HS), each solution (and in particular each periodic orbit) lies on some "energy surface" $\{x \in \mathbf{R}^{2N} : H(x(t)) = const\}$. One can therefore prescribe a hypersurface S ($S = H^{-1}(1)$ say) and look for T-periodic solutions of (HS) (of an a priori unknown period T).

The problem of existence of (at least) one periodic orbit for (HS) on a prescribed compact hypersurface $H^{-1}(1)$ has been studied by several authors, see e.g. Seifert [12], Weinstein [17], Rabinowitz [8], Viterbo [16], Hofer and Zehnder [7]. Results concerning the existence of a brake orbit (for H even with respect to the p-variable) may be found in Rabinowitz [10, 11]. In [4] Ekeland and Lasry have shown that (HS) possesses at least N periodic orbits if $H^{-1}(1)$ bounds a convex region and satisfies a certain

(*) Supported in part by the Swedish Natural Science Research Council

geometric condition. In [2] Berestycki et al. have generalized this result to the case of $H^{-1}(1)$ bounding a star-shaped region. Results of similar type, for Hamiltonians having certain symmetries, have been established by Girardi [5] and van Groesen [15].

Our main result concerns the problem of existence of N brake orbits for (HS). Suppose that $S = H^{-1}(1)$ is a compact hypersurface which bounds a star-shaped neighbourhood of the origin in $\mathbf{R}^{2N}$. Suppose also that $x \cdot H'(x) \neq 0$ on S (the dot denotes the inner product in $\mathbf{R}^{2N}$). Let r and R be the largest and the smallest number respectively such that

$$r \leq |x| \leq R \quad \forall x \in S, \tag{1}$$

and let ρ be the largest number for which

$$T_y(S) \cap \{x \in \mathbf{R}^{2N} : |x| < \rho\} = \emptyset \quad \forall y \in S. \tag{2}$$

Here $T_y(S)$ is the tangent hyperplane to S at y and $|x| = (x \cdot x)^{1/2}$.

Theorem 1. *Let $H \in C^2(\mathbf{R}^{2N}, \mathbf{R})$ be such that:*

(i) $H(-p,q) = H(p,q)$ $\forall (p,q) \in \mathbf{R}^{2N}$,

(ii) The set $\mathcal{A} = \{x \in \mathbf{R}^{2N} : H(x) \leq 1\}$ is nonempty, compact, star-shaped with respect to the origin and $S = H^{-1}(1)$ is the boundary of $\mathcal{A}$,

(iii) $x \cdot H'(x) \neq 0$ $\forall x \in S$.

If $R^2 < 2\rho^2$, where R and ρ are as in (1)-(2), then (HS) has at least N distinct brake orbits on S.

The proofs of existence of multiple periodic orbits in [2, 4, 5, 15] were effected by reducing the problem to the one of finding critical points of a functional which is S^1 or $\mathbf{Z}_2$-symmetric. When looking for brake orbits of (HS) it is natural to work in the space of periodic functions $(p(t), (q(t))$ such that p is odd and q is even in t. The functional associated with (HS) is no longer symmetric in this space. However, it has a partial $\mathbf{Z}_2$-symmetry which we will exploit.

The main tool in the proof of Theorem 1 is an index theory which we describe in the next section. It is a variant of the relative index introduced by Berestycki et al. [2] and further developed by Tarantello [14] (see also Benci [1] for a related concept of pseudoindex). A special feature of our index is a strong dimension property (see Proposition 7 below).

2. An index theory

Let E be a real Hilbert space and T a unitary representation of $\mathbf{Z}_2$ in E. That is, $T_0 = I_E$ (the identity mapping on E) and T_1 is a linear isometry such that $T_1 = T_1^{-1}$. A subset $A \subset E$ is said to be $T-$*invariant* (or simply *invariant*) if $T_1 A \subset A$. Let

$$E^G = \{x \in E : T_1 x = x\}$$

be the fixed point set of T. To T there corresponds an orthogonal decomposition $E = E^G \oplus F$ such that F is invariant and

$$T_1(x+y) = x - y \quad \forall x \in E^G,\ y \in F.$$

Let

$$\Sigma = \{A \subset E : A \text{ is closed and invariant}\}.$$

For $A \in \Sigma$ we define *the index of* A, denoted $\gamma(A)$, to be the smallest integer k such that there exists a continuous mapping $f : A \to \mathbf{R}^k - \{0\}$ satisfying $f(T_1 x) = -f(x)$. If there is no such k, then $\gamma(A) = \infty$. For the empty set $\emptyset$ we define $\gamma(\emptyset) = 0$. Observe that if $A \cap E^G \neq \emptyset$, then $\gamma(A) = \infty$ (because $f(x) = f(T_1 x) = -f(x)\ \forall x \in A \cap E^G$). It is easy to verify that γ satisfies the usual properties of index (which may be found e.g. in [9]). In particular, denoting $N_\delta(A) = \{x \in E : d(x, A) \le \delta\}$, where $d(x, A)$ is the distance from x to A, we have

Proposition 1. *(Continuity property) If $A \in \Sigma$ is compact, then $\gamma(A) = \gamma(N_\delta(A))$ for some $\delta > 0$.*

Let Y be a closed subspace of E. By P_Y we will denote the orthogonal projection from E to Y and Y^G will be the set $Y \cap E^G$. Suppose $A \in \Sigma$. A function $\xi : A \to \mathbf{R}$ is said to be $T-$*invariant* (or *invariant*) if $\xi(T_1 x) = \xi(x)\ \forall x \in A$. A mapping $f : A \to E$ is $T-$*equivariant* (or *equivariant*) if $f(T_1(x)) = T_1 f(x)\ \forall x \in A$, and f is *compact* if the image of each bounded subset of A is contained in a compact set.

Let $E = Y \oplus X$, where X, Y are orthogonal to each other and invariant. For $A \in \Sigma$, let $\mathcal{F}_k(A)$ be the set of all continuous mappings $f = (f_1, f_2) : A \to Y \times \mathbf{R}^k - \{(0,0)\}$ satisfying the following conditions:

(i) f is equivariant in the sense that $f_1(T_1 x) = T_1 f_1(x)$ and $f_2(T_1 x) = -f_2(x)\ \forall x \in A$,

(ii) $f_1 = P_Y - K$, where K is compact and $K(A)$ is bounded in Y,

(iii) $f_1(x) = x\ \forall x \in A \cap Y^G$ (and $f_2(x) = 0$ by equivariance).

Let $A \in \Sigma$. We define *the index of A relative to X*, denoted $\gamma_r(A, X)$, or shortly $\gamma_r(A)$ when no ambiguity can arise, to be the smallest integer k such that $\mathcal{F}_k(A) \neq \emptyset$. If there is no such k, we set $\gamma_r(A) = \infty$, and we define $\gamma_r(\emptyset) = 0$.

Below we collect some basic properties of γ_r.

Proposition 2. *(Mapping property) Let $A, B \in \Sigma$ and let $g : A \to B$ be a continuous mapping such that $g(x) = e^{-\xi(x)L} x - K(x)$, where*

(i) $L : E \to E$ is linear, equivariant, selfadjoint and $LY \subset Y$,

(ii) $\xi : A \to \mathbf{R}$ is invariant and $\xi(A)$ is bounded,

(iii) $K : A \to E$ is equivariant, compact and $K(A)$ is bounded.

If $\xi|_{A \cap Y^G} = 0$ and $K|_{A \cap Y^G} = 0$ (i.e., if $g|_{A \cap Y^G} = I_{A \cap Y^G}$), then $\gamma_r(A) \le \gamma_r(B)$.

Proposition 3. *(Monotonicity) If $A, B \in \Sigma$ and $A \subset B$, then $\gamma_r(A) \leq \gamma_r(B)$.*

Proposition 4. *(Subadditivity) If $A, B \in \Sigma$, then $\gamma_r(A \cup B) \leq \gamma_r(A) + \gamma(B)$.*

Proposition 5. *If $A, B \in \Sigma$ and $\gamma(B) < \infty$, then $\gamma_r(\overline{A - B}) \geq \gamma_r(A) - \gamma(B)$.*

Proposition 6. *(Intersection property) Let $A \in \Sigma$. Suppose that $X = X_0 \oplus X_1$, where X_0, X_1 are orthogonal to each other, invariant and $X_0 \cap E^G = \{0\}$. If $\dim X_0 = k < \infty$ and $\gamma_r(A) > k$, then $A \cap X_1 \neq \emptyset$.*

In order to state the next property of γ_r we will need the following geometric condition:

Definition. A set A is said to satisfy *Condition* $(\mathcal{G})$ if for each finite dimensional subspace E_0 of E and each $r > 0$ there exists an $R > 0$ such that if $\|x\| \leq r$, then $A \cap (x + E_0) \subset B_R$.

Here $x + E_0 = \{x + y \in E : y \in E_0\}$ and $B_R = \{x \in E : \|x\| < R\}$.

Proposition 7. *(Dimension property) Let $X_0 \subset X$ be an invariant subspace with* $\dim X_0 = k$ *and $X_0 \cap E^G = \{0\}$. Let U be an open invariant neighbourhood of the origin in $Y \oplus X_0$. If $\overline{U}$ satisfies Condition $(\mathcal{G})$, then $\gamma_r(\partial U) = k$, where ∂U is the boundary of U in $Y \oplus X_0$.*

In the proof of Proposition 7 we will make use of the following

Lemma. *(Generalized Borsuk-Ulam Theorem) Let W be an open bounded neighbourhood of $0 \in \mathbf{R}^m \times \mathbf{R}^n$ such that if $(x, y) \in W$, then $(x, -y) \in W$. Let $f = (g, h) : \partial W \to \mathbf{R}^m \times \mathbf{R}^{n-1}$ be a continuous mapping with $g(x, -y) = g(x, y)$, $h(x, -y) = -h(x, y)$ $\forall (x, y) \in \partial W$ and $f(x, 0) = (x, 0)$ $\forall (x, 0) \in \partial W$. Then $f(x, y) = 0$ for some $(x, y) \in \partial W$.*

The proof of the lemma is an immediate consequence of the Generalized Borsuk Theorem (whose appropriate version may be found in [13]).

Proof of Proposition 7. For simplicity we assume that E is separable. The proof of the general case is given in [13].

By Proposition 6, $\gamma_r(\partial U) \leq k$. Suppose $\gamma_r(\partial U) < k$ and let $f = (P_Y - K, f_2) : \partial U \to Y \times \mathbf{R}^{k-1}$, where $f \in \mathcal{F}_{k-1}(\partial U)$. Since X_0 is invariant and $X_0 \cap E^G = \{0\}$, we may identify X_0 with $\mathbf{R}^k$ and assume that $f = P_Y - K + f_2 : \partial U \to Y \oplus X_0'$, where X_0' is a $k-1$ dimensional subspace of X_0. Let $Y = Y^G \oplus Z$, where $Z \subset F$. Since E is separable, there exist two sequences, (Y_n^G) and (Z_n), of subspaces of Y^G and Z respectively, such that $\dim Y_n^G = \dim Z_n = n$ and

$$Y^G = \overline{\bigcup_{n=1}^{\infty} Y_n^G}, \qquad Z = \overline{\bigcup_{n=1}^{\infty} Z_n}.$$

Let P_n be the orthogonal projection from $Y \oplus X_0$ to $Y_n^G \oplus Z_n \oplus X_0$. Then

$$P_n f = P_Y - P_n K + f_2 : \partial U \cap (Y_n^G \oplus Z_n \oplus X_0) \to Y_n^G \oplus Z_n \oplus X_0'.$$

By the Generalized Borsuk-Ulam Theorem, there exists an $x_n \in \partial U \cap (Y_n^G \oplus Z_n \oplus X_0)$ such that $f(x_n) = 0$, i.e.,

$$P_Y x_n - P_n K(x_n) = 0 \quad \textit{and} \quad f_2(x_n) = 0. \tag{3}$$

Let $n \to \infty$. Since $K(\partial U)$ is bounded, so are $P_n K(x_n)$ and $P_Y x_n$. Therefore $\|P_Y x_n\| \leq r$, where the constant r is independent of n. By Condition $(\mathcal{G})$, $\partial U \cap (P_Y x_n + X_0) \subset B_R$ for some R. Since $x_n = P_Y x_n + P_{X_0} x_n \subset P_Y x_n + X_0$ and $x_n \in \partial U$, $\|x_n\| \leq R$. After passing to a subsequence, $P_{X_0} x_n \to w$ and $P_n K(x_n) \to y$ for some $w \in X_0$ and $y \in Y$. It follows from (3) that $P_Y x_n \to y$, so $x_n \to x = y + w \in \partial U$ and $f(x) = P_Y x - K(x) + f_2(x) = 0$. Therefore $f \notin \mathcal{F}_{k-1}(\partial U)$, a contradiction.

Remarks. (i) The main difference between our index and the one in [2, 14] is that we require the set $K(A)$ in (ii) of the definition of $\mathcal{F}_k$ to be bounded.

(ii) For the index in [2, 14] the conclusion of Proposition 7 remains valid if U is bounded but fails in general (see Bögle [3]). In the proof of Theorem 1 we will employ the above proposition to a set which satisfies Condition $(\mathcal{G})$ and is unbounded.

(iii) A similar index theory (at least for separable E) may be defined for other symmetry groups like e.g. S^1 and $\mathbf{Z}_p$ with p a prime integer.

3. Proof of Theorem 1

In the proof of Theorem 1 we will use a minimax principle which we state below.

Let E be a real Hilbert space with an inner product $\langle\ ,\ \rangle$ and T a unitary representation of $\mathbf{Z}_2$ in E. Let $L : E \to E$ be an equivariant and selfadjoint bounded linear operator. Define $\Phi(x) = \frac{1}{2}\langle Lx, x\rangle$. Then Φ is an invariant functional and $\Phi'(x) = L(x)$ (we assume via the Riesz representation theorem that $\Phi'(x) \in E$). Let ψ be an invariant functional on E such that $\psi^{-1}(1) \neq \emptyset$. Set $M = \psi^{-1}(1)$. Suppose further that $\psi \in C^{1,1}(E, \mathbf{R})$ and ψ' is a compact mapping which is bounded away from 0 on bounded subsets of M. Then M is an invariant $C^{1,1}$-manifold. For $x \in M$ denote

$$\lambda(x) = \frac{\langle Lx, \psi'(x)\rangle}{\|\psi'(x)\|^2}$$

and observe that $\langle Lx - \lambda(x)\psi'(x), \psi'(x)\rangle = 0$. So $Lx - \lambda(x)\psi'(x)$ is an element of the tangent space $T_x(M)$. In particular, $x \in M$ is a critical point of $\Phi|_M$ if and only if $Lx = \lambda(x)\psi'(x)$. We will need the following compactness hypothesis which is stronger than the usual Palais-Smale condition:

(C^*) If $(x_n) \subset M$ is a sequence such that $\Phi(x_n) \to c \in \mathbf{R}$ and $\frac{Lx_n - \lambda(x_n)\psi'(x_n)}{(\|x_n\|+1)^{1/2}} \to 0$, then (x_n) has a convergent subsequence.

Let

$$\Phi_c = \{x \in M : \Phi(x) \le c\} \text{ and } K_c = \{x \in M : \Phi(x) = c,\ Lx = \lambda(x)\psi'(x)\}.$$

Theorem 2. *(Minimax principle) Suppose that Φ, ψ and M satisfy the assumptions above and $\Phi|_M$ satisfies (C^*). Let $E = Y \oplus X$, where $Y = X^\perp$, X and Y are invariant and $LY \subset Y$. Define*

$$\Gamma_j = \{A \in \Sigma : A \subset M,\ \gamma_r(A, X) \ge j\}$$

and

$$c_j = \inf_{A \in \Gamma_j} \sup_{x \in A} \Phi(x), \qquad j = 1, ..., N.$$

Suppose also that there exist finite numbers a, b such that $c_j \in (a, b)$ for $1 \le j \le N$ and $Y^G \cap \Phi^{-1}([a,b]) = \emptyset$. Then all c_j are critical values of $\Phi|_M$. Furthermore, if $c_j = ... = c_{j+p}$ for some j and some $p \ge 0$, then $\gamma(K_{c_j}) \ge p+1$.

The proof is standard except for one point: it is not possible to employ the usual deformation lemma. The reason is that the argument which may be found e.g. in [9, Appendix A] or [10] would give a deformation η such that $\eta(1,x) = e^{-\xi(x)L}x - K(x)$, where K is compact but $K(M)$ need not be bounded. Therefore Proposition 2 cannot be applied. In order to overcome this difficulty we modify the construction of η. Recall that in [10] η is obtained by solving the initial value problem

$$\frac{d\eta}{dt} = -\chi(\eta)(L\eta - \lambda(\eta)\psi'(\eta)), \qquad \eta(0,x) = x,$$

where χ is a suitable cutoff function. Let us consider instead a slightly different problem:

$$\frac{d\eta}{dt} = -\frac{\chi(\eta)}{\|\eta\|+1}(L\eta - \lambda(\eta)\psi'(\eta)), \qquad \eta(0,x) = x.$$

Then $\eta(1,x) = e^{-\xi(x)L}x - K(x)$ as in [9, 10], but now $K(M)$ is bounded because

$$\frac{\|\lambda(x)\psi'(x)\|}{\|x\|+1} \le \frac{\|Lx\|}{\|x\|+1}.$$

Moreover, it follows from (C^*) that there exists a constant $b > 0$ such that

$$\begin{aligned} \frac{d}{dt}\Phi(\eta(t,x)) &= \langle L\eta, \frac{d\eta}{dt}\rangle = -\frac{\chi(\eta)}{\|\eta\|+1}\langle L\eta, L\eta - \lambda(\eta)\psi'(\eta)\rangle \\ &= -\frac{\chi(\eta)}{\|\eta\|+1}\|L\eta - \lambda(\eta)\psi'(\eta)\|^2 \le -b^2 \end{aligned} \tag{4}$$

whenever $\chi(\eta) = 1$. Therefore Φ decreases rapidly enough as t increases and the usual argument can be used. Observe that the inequality in (4) need not be fulfilled if Φ satisfies the usual Palais-Smale condition instead of (C^*).

In order to find a variational framework for our problem we first introduce a suitable function space. Let $H^{1/2}(S^1, \mathbf{R}^{2N})$ be the Sobolev space of 2π-periodic $\mathbf{R}^{2N}$-valued functions

$$x = \sum_{k \in \mathbf{Z}} c_k e^{ikt}, \quad \textit{where } c_k \in \mathbf{C}^{2N} \textit{ and } c_{-k} = \bar{c}_k,$$

such that

$$\sum_{k \in \mathbf{Z}} (1 + |k|)|c_k|^2 < \infty.$$

Let

$$E = \{x = (p, q) \in H^{1/2}(S^1, \mathbf{R}^{2N}) : p(-t) = -p(t),\ q(-t) = q(t)\ \forall t\}.$$

For $x = (p, q) \in E$, set $z = p + iq$. Then

$$z = \sum_{k \in \mathbf{Z}} a_k e^{ikt}, \quad \textit{where } a_k \in \mathbf{C}^N. \tag{5}$$

Since p is odd and q is even, it is easy to see that $Re\, a_k = 0\ \forall k$. A convenient norm in E is given by

$$\|x\|^2 = 2\pi(|a_0|^2 + \sum_{k \neq 0} |k|\,|a_k|^2).$$

Let $E = E^- \oplus E^0 \oplus E^+$ be the orthogonal decomposition of E into the parts corresponding to $k < 0$, $k = 0$ and $k > 0$ in (5). If $e_1, ..., e_N$ is the standard basis in $\mathbf{R}^N$, then E^0 is spanned by $(p, q) = (0, e_j)$, $1 \le j \le N$, and $E^{\pm}$ by

$$(p, q) = (e_j \sin kt, \mp e_j \cos kt), \qquad 1 \le j \le N, \ \ 1 \le k < \infty.$$

By changing H outside the hypersurface S we may assume that H is homogeneous of degree two and $H'(x)/|x|$ is bounded [8, p. 160]. It follows from (1) and the homogeneity of H that

$$\frac{|x|^2}{R^2} \le H(x) \le \frac{|x|^2}{r^2} \ \ \forall x \in \mathbf{R}^{2N}. \tag{6}$$

By [10, Lemma 2.3], there is a one-to-one correspondence between brake orbits for (HS) on S (of unknown period T) and 2π-periodic brake orbits for

$$\dot{x} = \lambda J H'(x), \quad \lambda > 0, \tag{7}$$

on S. The number λ in (7) is unknown and related to T.

Let
$$\Phi(x) = \frac{1}{2}\int_0^{2\pi} (-J\dot{x}\cdot x)\,dt,$$
$$\psi(x) = \frac{1}{2\pi}\int_0^{2\pi} H(x)\,dt \qquad \textit{and} \qquad M = \psi^{-1}(1).$$

It is easy to see that $\Psi(x) = \frac{1}{2}\|x^+\|^2 - \frac{1}{2}\|x^-\|^2$, where $x = x^- + x^0 + x^+ \in E^- \oplus E^0 \oplus E^+$. It follows by a modification of a standard argument (see [10]) that each critical point of $\Phi|_M$ corresponds to a 2π-periodic brake orbit for (7) on S, and therefore to a brake orbit for (HS) on S. Moreover, $\psi \in C^{1,1}(E, \mathbf{R})$, ψ' is compact and, by homogeneity, $\langle \psi'(x), x\rangle = 2\psi(x) = 2$ on M. So M is a $C^{1,1}$-manifold and ψ' is bounded away from zero on bounded subsets of M.

Now we apply Theorem 2 to $\Phi|_M$. A modification of the argument of [2, Lemma 3.7] shows that $\Phi|_M$ satisfies (C^*). For $x \in E$, let $T_0 x = x$ and $T_1 x(t) = x(t+\pi)$. Then $T = \{T_0, T_1\}$ is a unitary representation of $\mathbf{Z}_2$ in E. The fixed point set E^G of T consists of those $x \in E$ which are π-periodic. Let $Y = E^- \oplus E^0$ and $X = E^+$. Since $Lx = \Phi'(x) = x^+ - x^-$, $LY \subset Y$. Suppose that $A \in \Gamma_j$. Then $\gamma_r(A) \geq 1$, so $A \cap E^+ \neq \emptyset$ by Proposition 6. A computation using (6) gives $\Phi(x) \geq \pi r^2$ whenever $x \in A \cap E^+$. So $c_j \geq \pi r^2$. Let E_1 be the N-dimensional subspace of E^+ corresponding to $k = 1$ in (5). Let $A = M \cap (E^- \oplus E^0 \oplus E_1)$. Then A is the boundary of the set $B = \{x \in E^- \oplus E^0 \oplus E_1 : \psi(x) \leq 1\}$, and it can be shown that B satisfies Condition $(\mathcal{G})$. Therefore $\gamma_r(A) = N$ according to Proposition 7. Using (6) one verifies that $\Phi|_A \leq \pi R^2$. It follows that $\pi r^2 \leq c_j \leq \pi R^2$ for $1 \leq j \leq N$. Since $\Phi|_{Y^G} \leq 0 < \pi r^2$, Φ, ψ and M satisfy all assumptions of Theorem 2.

Using the hypothesis $R^2 < 2\rho^2$ and an argument similar to that in [6, Lemma 6] one shows that all $x \in K_{c_j}$, $1 \leq j \leq N$, have minimal period 2π. Therefore, for such j, $E^G \cap K_{c_j} = \emptyset$ and two critical points x_1, x_2 give rise to the same orbit for (7) if and only if $x_2 = T_1 x_1$. So if all c_j, $1 \leq j \leq N$, are distinct, $\Phi|_M$ has at least $2N$ critical points which correspond to N distinct orbits. If some c_j coincide, then there exists a j with $\gamma(K_{c_j}) \geq 2$, and K_{c_j} is an infinite set (because $E^G \cap K_{c_j} = \emptyset$). This completes the proof of Theorem 1.

References

1. V. Benci, *On critical point theory for indefinite functionals in the presence of symmetries*, Trans. Amer. Math. Soc. 274 (1982), 533-572.
2. H. Berestycki, J.M. Lasry, G. Mancini and B. Ruf, *Existence of multiple periodic orbits on star-shaped Hamiltonian surfaces*, Comm. Pure Appl. Math. 38 (1985), 253-289.

3. J.G. Bögle, *Indextheorien zur Existenz von periodischen Lösungen Hamiltonscher Systeme*, Thesis, Ludwig-Maximilians-Universität München, 1987.
4. I. Ekeland and J.M. Lasry, *On the number of periodic trajectories for a Hamiltonian flow on a convex energy surface*, Ann. Math. 112 (1980), 283-319.
5. M. Girardi, *Multiple orbits for Hamiltonian systems on starshaped surfaces with symmetries*, Ann. Inst. H. Poincaré, Analyse non linéaire 1 (1984), 285-294.
6. M. Girardi and M. Matzeu, *Solutions of minimal period for a class of nonconvex Hamiltonian systems and applications to the fixed energy problem*, Nonl. Analysis 10 (1986), 371-382.
7. H. Hofer and E. Zehnder, *Periodic solutions on hypersurfaces and a result by C. Viterbo*, Inv. Math. 90 (1987), 1-9.
8. P.H. Rabinowitz, *Periodic solutions of Hamiltonian systems*, Comm. Pure Appl. Math. 31 (1978), 157-184.
9. ———, *Minimax Methods in Critical Point Theory with Applications to Differential Equations*, CBMS 65, Amer. Math. Soc., Providence, R.I., 1986.
10. ———, *On the existence of periodic solutions for a class of symmetric Hamiltonian systems*, Nonl. Analysis 11 (1987), 599-611.
11. ———, *On a theorem of Hofer and Zehnder*, to appear in Periodic Solutions of Hamiltonian Systems and Related Topics.
12. H. Seifert, *Periodische Bewegungen mechanischer Systeme*, Math. Z. 51 (1948), 197-216.
13. A. Szulkin, *An index theory and existence of multiple brake orbits for star-shaped Hamiltonian systems*, to appear in Math Ann. 283 (1989).
14. G. Tarantello, *Subharmonic solutions for Hamiltonian systems via a* $\mathbf{Z}_p$ *pseudoindex theory*, Preprint.
15. E.W.C. van Groesen, *Existence of multiple normal mode trajectories on convex energy surfaces of even, classical Hamiltonian systems*, J. Diff. Eq. 57 (1985), 70-89.
16. C. Viterbo, *A proof of Weinstein's conjecture in* $\mathbf{R}^{2n}$, Ann. Inst. H. Poincaré, Analyse non linéaire 4 (1987), 337-356.
17. A. Weinstein, *Periodic orbits for convex hamiltonian systems*, Ann. Math. 108 (1978), 507-518.

Progress in Nonlinear Differential Equations and Their Applications

Editor

Haim Brezis
Département de Mathématiques
Université P. et M. Curie
4, Place Jussieu
75252 Paris Cedex 05
France
and
Department of Mathematics
Rutgers University
New Brunswick, NJ 08903
U.S.A.

Progress in Nonlinear Differential Equations and Their Applications is a book series that lies at the interface of pure and applied mathematics. Many differential equations are motivated by problems arising in diversified fields such as Mechanics, Physics, Differential Geometry, Engineering, Control Theory, Biology, and Economics. This series is open to both the theoretical and applied aspects, hopefully stimulating a fruitful interaction between the two sides. It will publish monographs, polished notes arising from lectures and seminars, graduate level texts, and proceedings of focused and refereed conferences.

We encourage preparation of manuscripts in some form of TEX for delivery in camera ready copy, which leads to rapid publication, or in electronic form for interfacing with laser printers or typesetters.

Proposals should be sent directly to the editor or to: Birkhäuser Boston, 675 Massachusetts Avenue, Suite 601, Cambridge, MA 02139.

PNLDE1 Partial Differential Equations and the Calculus of Variations, Volume I
Essays in Honor of Ennio De Giorgi
F. Colombini, A. Marino, L. Modica, and S. Spagnolo, editors

PNLDE2 Partial Differential Equations and the Calculus of Variations, Volume II
Essays in Honor of Ennio De Giorgi
F. Colombini, A. Marino, L. Modica, and S. Spagnolo, editors

PNLDE3 Propagation and Interaction of Singularities in Nonlinear Hyperbolic Problems
Michael Beals

PNLDE4 Variational Methods
Henri Berestycki, Jean-Michel Coron, and Ivar Ekeland, editors

PNLDE5 Composite Media and Homogenization Theory
Gianni Dal Maso and Gian Fausto Dell'Antonio, editors